Cytokine Frontiers

Takayuki Yoshimoto • Tomohiro Yoshimoto
Editors

Cytokine Frontiers

Regulation of Immune Responses in Health and Disease

Springer

Editors
Takayuki Yoshimoto
Department of Immunoregulation
Institute of Medical Science
Tokyo Medical University
Tokyo, Japan

Tomohiro Yoshimoto
Laboratory of Allergic Diseases
Institute for Advanced Medical Sciences
Hyogo College of Medicine
Hyogo, Japan

ISBN 978-4-431-54441-8 ISBN 978-4-431-54442-5 (eBook)
DOI 10.1007/978-4-431-54442-5
Springer Tokyo Heidelberg New York Dordrecht London

Library of Congress Control Number: 2013947467

Printed on acid-free paper

Springer is part of Springer Science+Business Media (www.springer.com)

Preface

The immune system is a complex network whose primary functions are to protect the host from external threats such as bacteria and viruses as well as internal threats, namely, allergic responses and malignant transformation. Immune cells produce various biological products that allow them to communicate with one another and orchestrate immune responses. Cytokines, which are secreted by immune cells in response to microbes and other antigens, are soluble protein mediators that are important in such intercellular communication. Different cytokines stimulate diverse responses in various phases of inflammation and immunity, including the innate immune response, the generation of cytotoxic T cells, and the development of antibodies by the humoral immune system. Cytokines have pleiotropic effects and functional redundancy, and the combination of cytokines produced in response to various stimuli determines the kind of immune response that is activated. In addition, cytokines operate in a complex network, in which one cytokine can affect the production of many other cytokines as well as modulate the responses to them.

Historically, cytokine identification has undergone several stages of development. Initially, cytokines were purified by several biochemical processes and identified simply according to their biological activities. Subsequently, cytokines were identified through expression cloning following their biological activities or by using specific antibodies. Cytokine identification then entered the most recent stage, which was primarily driven by the Human Genome Project in the late 1990s to the early 2000s. Identification of the complete sequence of human genes made it possible to assign a function to each member of the huge database of previously unrecognized proteins through sequence comparison with previously named genes. Thus, numerous candidate cytokines were identified based on their homology to known cytokines. The most important means of assessing the biological function of these newly identified cytokines has been to evaluate the phenotype of mice genetically engineered to either knock out a particular gene or overexpress it. Thus far, more than 100 cytokines and their cell-surface receptors have been analyzed, including the most recently assigned interleukin (IL), IL-38.

It is now also clear that the pathophysiology of many infectious, autoimmune, allergic, and malignant diseases can be largely explained by which cytokines are induced and subsequently regulate the cellular responses. In clinical medicine, cytokines are involved in a wide spectrum of diseases; cytokines are therefore considered important as therapeutic targets for specific agonists or antagonists in numerous immune and inflammatory diseases. For instance, ample basic and clinical evidence indicates that excess tumor necrosis factor (TNF)-α underlies the pathogenesis of chronic inflammatory diseases, such as rheumatoid arthritis. The therapeutic effectiveness of TNF-α inhibitors has subsequently been demonstrated in the treatment of this disease. The success of anti-TNF-α therapy has led to recognition of the importance of cytokines as therapeutic targets in disease.

Although this field has undergone enormous expansion, insufficient understanding of cytokine biology and its complicated network continues to be the rate-limiting step in developing cytokine therapeutics. The present book thus aims to provide the reader not only with information about the original properties of cytokines, but also with up-to-date findings on their roles under physiological and pathological conditions—with the ultimate goal of helping to create strategies for therapeutic treatment.

This book is not a survey of individual cytokines; rather, it guides the reader through the latest research on the cytokine network, covering signaling pathways, control of the immune response, and potential therapeutics. Among various important cytokines, we selected 15 key examples whose profiles we believed would stimulate the reader's interest. In this book, these cytokines are divided into three groups based on their physiological roles in the immune system. The first group includes cytokines associated with inflammatory disorders, proinflammatory cytokines, and the recently identified new helper T (Th) subset, Th17 cells. The second group is associated with allergic disorders, including Th2 responses and recently identified types of innate cells. The third group of cytokines is associated with immunological tolerance and anti-inflammation, including regulatory T (Treg) cells, IL-10-producing Treg (Tr1) cells, and inducible IL-35-producing Treg (iTr35) cells.

We wish to sincerely thank all the authors for their invaluable contributions to this book.

Tokyo, Japan Takayuki Yoshimoto
Hyogo, Japan Tomohiro Yoshimoto

Foreword

Immunology is the quintessential medical science. It stands at the interface between the most modern of the advances in basic biological science and an enormous range of disease states. It has achieved what no other scientific discipline has done, the eradication of disease. Both smallpox and the cattle disease rinderpest have been eradicated, and if societal issues could be dealt with, polio would be a disease of the past! Furthermore, the control of hepatitis B infection by the highly effective vaccine has probably prevented more cancers than any medical intervention other than cessation of smoking.

Although the preoccupation of the immune system with the microbial world represents the great driving force in its evolution, immunology has critical roles to play in both the control and progression of tumors, in the inflammatory aspects of atherosclerotic heart disease and type 2 diabetes, and, of course, in autoimmune and allergic disorders. Perhaps most remarkably, my personal favorite of the cytokines, IL-4, has been shown to be important in learning and memory. A recent study indicated that IL-4-deficient mice are "stupid."

Determining how the immune system mediates this vast array of functions has driven much of modern immunological research. What is increasingly clear is that cells of the immune system coordinate a multitude of biological functions through their production of a set of proteins designated the cytokines. The cytokines are members of several distinct families including the type 1 cytokines, the interferons, the immunoglobulin supergene family members, the members of the TNF family, the chemokines, and the IL-17 set of cytokines. They coordinate multiple biological functions, control cellular homeostasis, regulate inflammation, and participate in the response of the immune system to both pathogens and self-antigens.

In this volume, *Cytokine Frontiers: Regulation of Immune Responses in Health and Disease*, Takayuki Yoshimoto and Tomohiro Yoshimoto have emphasized a particularly important set of the cytokines, those that control inflammation, both the pro- and anti-inflammatory cytokines, and the closely related molecules which regulate allergic responses. This presentation is particularly timely as we have relatively recently appreciated the centrality of the orchestration of both "conventional" and allergic inflammation in the overall homeostasis of the body and the critical role played in disease progression and in the development of tumors.

Making a response appropriate to the type and magnitude of a threat, be it a particular type of pathogen, an allergen, or a tumor cell, is essential to effective control of the threatening agent. One can easily imagine the disastrous effects of dysregulation or "inappropriateness" in immunity. For example, mounting a robust Th17 response, although of great protective value in infections with extracellular bacteria, could easily be disastrous if it were to occur in response to a viral lung infection, as it would fill the lung with granulocytes when what was needed was a robust interferon-gamma response. Indeed, the problem of dual infections and the tragic example of the 1918 influenza pandemic where, almost certainly, responses inappropriate to the threat were responsible for a terrible toll of deaths, illustrate the importance of precise regulation of the nature of the immune/inflammatory response.

As the distinguished panel of authors assembled in this volume show, immunologists are obtaining a deep knowledge of the functions of the different cytokines and of how their responses are regulated. Furthermore, targeting cytokines has already proven to be highly effective in the treatment of rheumatoid arthritis, inflammatory bowel disease, psoriasis, and other disorders. The manipulation of the pattern of cytokine production and of the functional targets of the cytokines will surely be a fertile area for the development of new generations of drugs that can control and appropriately regulate inflammatory responses.

I am confident that the growing sophistication in our understanding of the cytokine universe, so clearly portrayed in this book, will have implications for improvements in medicine as great as the earlier and increasingly successful efforts to use antibodies as therapeutic agents. This volume will provide the underpinning for those who wish to have a deeper understanding of this important field.

Bethesda, MD, USA William E. Paul

Introduction

In the context of host defense and pathology, cytokines produced very early in response to multiple stresses are important in both innate and acquired immunity. These cytokines—including interleukin (IL)-1, IL-6, and tumor necrosis factor (TNF)-α—are termed pro-inflammatory cytokines and are characterized as being promptly inducible. In the presence of various pro-inflammatory cytokines and antigen-presenting cell-derived cytokines, antigen-stimulated naive $CD4^+$ T cells differentiate into distinct helper T (Th) cell subsets, which have specific cytokine profiles and distinct effector functions.

Until recently, effector T cells were classified into Th1 or Th2 subtypes depending on the cytokines they produced. However, this paradigm had to be revised with the discovery of a new subset of effector T cells, designated Th17 cells. IL-17-producing Th17 cells play an important role in clearing extracellular pathogens and in tissue inflammation. Transforming growth factor (TGF)-β and IL-6 are the factors that induce differentiation of naive T cells into Th17 cells. The differentiation and expansion of Th17 cells from naive T cells also require IL-6, IL-1, IL-21, and IL-23. Among them, IL-23 is critically essential for the maturation and stabilization of differentiated pathogenic Th17 cells. Th17 cells produce IL-17, IL-17F, IL-21, and IL-22 and thereby mediate distinct effector functions compared with Th1 and Th2 cells. Although IL-17, IL-17F, and IL-22 induce a tissue reaction, IL-21 produced by Th17 cells is essential for amplification of Th17 cells and B-cell function.

In the past decade, genes encoding a novel cytokine cluster with structural and functional similarities to IL-1 have been discovered; they were initially termed IL-1 family members 5–10 (IL1F5–IL1F10). IL-18, initially described as interferon (IFN)-γ-inducing factor, is a pro-inflammatory cytokine and also belongs to the IL-1 family. These cytokines are synthesized as precursor molecules and cleaved by the enzyme caspase-1. The NACHT, LRR, and PYD domain-containing protein 3, NALP3, inflammasome is of crucial importance in generating active caspase-1. These cytokines have recently been reclassified according to an updated cytokine nomenclature scheme. This novel IL-1-like cytokine cluster contains IL-36α (IL-1F6), IL-36β (IL-1F8), IL-36γ (IL-1F9), IL-37 (IL-1F7), and IL-38 (IL-1F10).

Similar to the classical IL-1 cytokines, IL-36 cytokines are also involved in the initiation or regulation of immune responses.

Th2-type cytokine responses play a major role in the development of allergic diseases, such as asthma, and in immunity to parasitic helminthic infections. The hallmark of Th2-mediated immune responses is the activation of mast cells, eosinophils, basophils, and goblet cells, which is mediated by the cytokines IL-4, IL-5, IL-9, and IL-13; these are necessary for immunoglobulin E production, goblet cell hyperplasia, and airway eosinophilia. Until recently, Th2 cells were thought to be the most abundant and important source of these Th2-type cytokines. However, it has become clear that there is a group of innate cells that lacks the conventional markers typical of the main lymphoid cell types and is able to produce large amounts of Th2-type cytokines. Such responses were shown to be dependent on the epithelial cell-derived cytokine IL-25, IL-33, and thymic stromal lymphopoietin (TSLP). These innate cell populations were called innate lymphoid cells (ILCs) and they were further divided into the following three subsets: group 1 ILCs (comprising ILC1s and NK cells); group 2 ILCs (comprising ILC2s); and group 3 ILCs (comprising ILC3s and LTi cells). These cells are characterized by a lymphoid morphology and an absence of T-cell receptors, but they can secrete a variety of effector cytokines that are also produced by Th cell subsets. These ILCs play important roles in tissue remodeling and innate immunity at barrier surfaces, and they provide a first line of defense against foreign pathogens by triggering a prompt innate immune response and bridging innate and adaptive immunity. Among them, ILC2s play a key role in type 2 immune responses by prompt, abundant production of type 2 cytokines—especially IL-5 and IL-13—in response to IL-25, IL-33, and TSLP.

Regulatory T (Treg) cells play an important role in the maintenance of self-tolerance. Thus, negative feedback pathways provided by Treg cells and suppressive cytokines are an intrinsic part of the immune system along with effector functions. Regulatory cytokines induce tolerance to self-components as well as eradication of pathogens with minimal collateral damage to the host. There are several types of Treg cells. The main types of Treg cells express the transcription factor Foxp3, which can be induced naturally in the thymus (natural Treg, nTreg) and in the periphery (inducible Treg, iTreg). IL-10-producing Treg cells called Tr1 cells also have regulatory activity without expressing Foxp3, and they secrete IL-10 and TGF-β in response to antigenic stimulation. Several mechanisms have been identified that contribute to suppressive functions of Treg cells, including both cell contact-dependent and cell factor-dependent mechanisms, such as the production of IL-10, TGF-β, and IL-35. Although IL-10 was initially reported to be produced by Th2 cells to inhibit the function of Th1 cells, almost all types of immune cells are now recognized as producing IL-10. For example, IL-10 deficiency leads to colitis after colonization by particular microorganisms, which suggests an important role for IL-10 in intestinal homeostasis. Although IL-10 and TGF-β are well characterized, IL-27 and IL-35 are newcomers among the anti-inflammatory cytokines. IL-27 promotes IL-10 production and differentiation of naive $CD4^+$ T cells into Tr1 cells. IL-35 is produced in Treg cells and exerts suppressive functions, and it also differentiates naive $CD4^+$ T cells into IL-35-producing induced Treg cells called iTr35.

Contents

Part II Cytokines in Allergic Disorders

Part III Cytokines in Immunological Tolerance and Anti-inflammation

Part I
Cytokines in Inflammatory Disorders

Chapter 1
The Interleukin-1 Family

Role in Inflammation and Disease

Charles A. Dinarello and Mihai G. Netea

Abstract The biological properties of the interleukin-1 (IL-1) family ligands and receptors are characteristically pro-inflammatory and act as adjuvants for specific immune responses to antigen. Thus, the IL-1 family of ligands and receptors is fundamental to innate immunity. Of the 11 members of the IL-1 family, IL-1β has emerged as a therapeutic target for an expanding number of systemic and local inflammatory conditions termed "auto-inflammatory" diseases. These diseases are distinct from autoimmune diseases and include several hereditary conditions. Howver, auto-inflammatory diseases are also common diseases such as heart failure, gouty arthritis, and type 2 diabetes. For these, neutralization of IL-1β results in a rapid and sustained reduction in disease severity. Another member of the IL-1 family, IL-1α, is also a mediator of inflammation but is classified as an "alarmin" because the cytokine is present in most cells and readily released upon cell death. Although treatment for autoimmune diseases often includes immunosuppressive drugs, blocking the IL-1 receptor is effective as an anti-inflammatory therapy for either IL-1α or IL-1β.

Keywords Auto-immune • Auto-inflammation • Cytokine • Inflammation

---

C.A. Dinarello (✉)
Department of Medicine, University of Colorado Denver, 12700 East 19th Ave., B168, Aurora, CO 80045, USA

Department of Medicine, University Medical Center Nijmegen, Nijmegen, The Netherlands
e-mail: cdinare333@aol.com

M.G. Netea
Department of Medicine, University Medical Center Nijmegen, Nijmegen, The Netherlands

T. Yoshimoto and T. Yoshimoto (eds.), *Cytokine Frontiers: Regulation of Immune Responses in Health and Disease*, DOI 10.1007/978-4-431-54442-5_1, © Springer Japan 2014

1.1 Introduction

As shown in Table 1.1, more than any other cytokine family, the interleukin-1 (IL-1) family plays a fundamental role in innate inflammation as well as reducing inflammation (Dinarello et al. 2012). It is this innate inflammation that facilitates specific immunological responses such as antibodies and cytotoxic T lymphocytes. In many ways, another interpretation of the innate response by IL-1 is its action as an adjuvant. Initially termed the nonspecific response to infection, a new name now used is "the innate immune response." The cytosolic segment of each member of the IL-1-receptor family contains the Toll-IL-1-receptor (TIR) domain. This domain is also present in each Toll-like receptor (TLR), receptors that respond to microbial products, viruses, and nucleic acids. TIR is the functional domain for both the TLR and IL-1 receptor families, as mutations in this domain result in loss of response to IL-1 and TLR agonists.

With one exception, all members of the IL-1 family are initially translated as precursors lacking a signal peptide for secretion via the Golgi apparatus. The precursors are found in the cytosol and exit the cell following its death by necrosis, not apoptosis. For example, once released, IL-1α, IL-33, and IL-36 can be processed extracellularly by neutrophil proteases into active cytokines. Although IL-1β is primarily processed intracellularly by the cysteine protease caspase-1, the IL-1β precursor can also be cleaved extracellularly into an active cytokine by similar serine proteases of neutrophils (Joosten et al. 2009). The one member of the IL-1 family that is readily secreted is the IL-1-receptor antagonist (IL-1Ra). IL-1Ra is translated with a signal peptide (Fig. 1.1), although an intracellular form also exists (Arend 2002). IL-1Ra is produced in health and is found circulating in mice and humans where the antagonist serves as a brake on inflammation driven by endogenous IL-1α or IL-1β. IL-1Ra binds to IL-1RI and blocks the receptor from binding to either

Table 1.1 Interleukin (IL)-1 family

Family name	Written name	Abbreviation	Property
IL-1 F1	Interleukin-1α	IL-1α	Pro-inflammatory
IL-1 F2	Interleukin-1β	IL-1β	Pro-inflammatory
IL-1 F3	IL-1-receptor antagonist	IL-1Ra	Anti-inflammatory
IL-1 F4	Interleukin-18	IL-18	Pro-inflammatory
IL-1 F5	IL-36-receptor antagonist	IL-36Ra	Anti-inflammatory
IL-1 F6	Interleukin-36α	IL-36α	Pro-inflammatory
IL-1 F7	Interleukin-37	IL-37	Anti-inflammatory
IL-1 F8	Interleukin-36β	IL-36β	Pro-inflammatory
IL-1 F9	Interleukin-36γ	IL-36γ	Pro-inflammatory
IL-1 F10	Interleukin-38	IL-38	Anti-inflammatory
IL-1 F11	Interleukin-33	IL-33	Pro-inflammatory

[a]In some studies, IL-18 exhibits protective properties

[b]The biological properties of IL-38 remain unclear; IL-38 binds to the IL-36 receptor and can act as an anti-inflammatory cytokine

[c]IL-33 is studied for its role in the Th2 paradigm because IL-33 binds to ST2, a member of the IL-1 family of receptors linked to allergic diseases

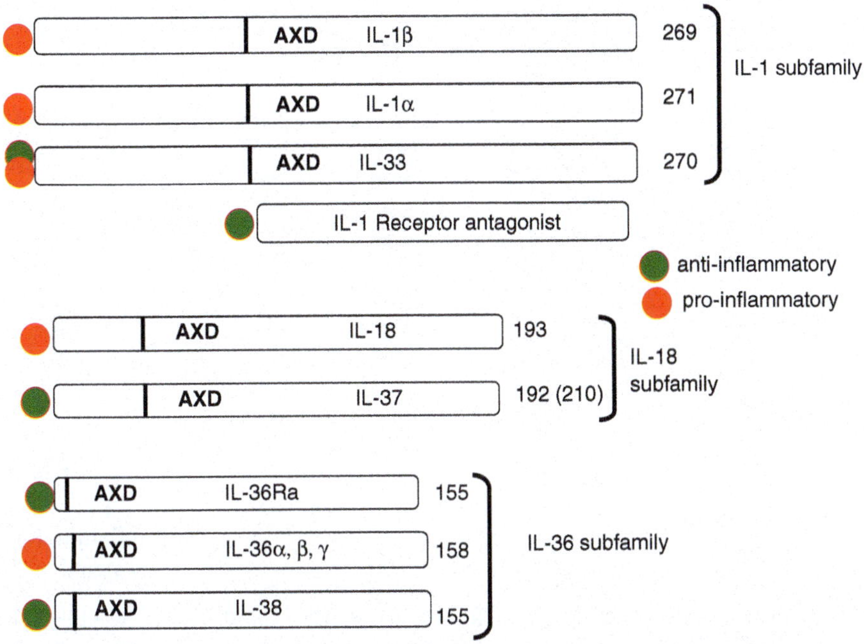

Fig. 1.1 Organization of the interleukin (IL)-1 family into three subfamilies. The number of amino acids of the full length of each member is shown at the C-terminal end. The consensus sequence (*AXD*) is common to all IL-1 family members and serves to locate the N-terminus nine amino acids forward from this site (*dark vertical bars*). The N-terminus results in propieces of various lengths. The IL-1Ra has a bona fide signal peptide and is shown by comparison

IL-1α or IL-1β (see Fig. 1.2c). Mice as well as humans born with a deficiency in functional IL-1Ra exhibit increased systemic and local inflammation; in humans a deficiency in IL-1Ra is lethal. The IL-36 receptor antagonist (IL-36Ra), another member of the IL-1 family, inhibits the activity of endogenous IL-36α, -β, and -γ. Although IL-36Ra is not readily secreted, individuals with a mutation in IL-36Ra develop a severe form of psoriasis. One may conclude that most members of the IL-1 family primarily promote inflammation and enhance specific acquired immune responses. However, there are also members that provide a brake on inflammation. The primary characteristics of each member of the IL-1 family are depicted in Table 1.1.

1.2 Interleukin-1 Family and Innate Responses

Independent of the type of organism or its products, the innate response is one of inflammation in which the host musters its defenses to increase the production and infiltration of phagocytic cells to the area of the invading microbe in an attempt limit

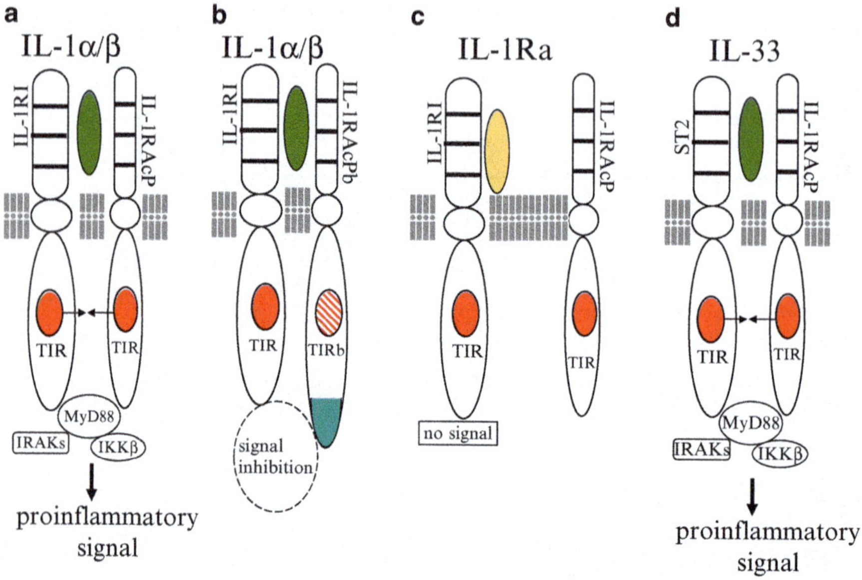

Fig. 1.2 Interleukin (IL)-1 subfamily. (**a**) IL-1α or IL-1β binds to the IL-1RI and recruits the co-receptor IL-1RAcP. The heterodimeric IL-1 receptor complex results in a close approximation of the Toll-like receptor (TIR) domains on each receptor chain (*arrows*), resulting in the binding of intracellular MyD88 to the complex followed by phosphorylation of MyD88. Subsequent phosphorylations of IRAKs and IKKβ increase NFκB and IL-1R AcP-1 translocation to the nucleus, followed by expression of pro-inflammatory genes. (**b**) In the central nervous system, IL-1α or IL-1β binds IL-1RI, recruiting IL-1RAcP, but can also recruit the co-receptor IL-1RAcPb. IL-1RAcPb contains an altered TIR domain, which results in a reduced signal. (**c**) IL-1Ra binds to IL-1RI: there is no recruitment of the co-receptor IL-1RAcP, no approximation of the TIR domains, and there is no signal. (**d**) IL-33 binds to its specific receptor, ST2, recruits the co-receptor IL-1RAcP, and the TIR domains approximate: signal transduction is initiated, resulting in the induction of the pro-inflammatory gene profile

infection and kill off the invader. Systemically, the liver increases the synthesis of acute-phase proteins, include anti-proteases. Even in humans, in most cases this process protects the subject without the use of antibiotics. For example, a break in the skin allows bacteria to gain access to the dermis and subsequent inflammation provides activation of complement, the release of preformed cytokines from keratinocytes, an increase in vascular wall adhesion molecules, and the extravasation of neutrophils. This response has functioned to battle against invaders for millions of years and can be traced back to fruit flies.

The skin, lung, and intestinal tract each provide a first line of defense against microbial invasion, and the lining cells, whether keratinocytes of the skin, the alveolar epithelial cells of the pulmonary tree, or the epithelial cells of the entire gastrointestinal tract, each contain preformed IL-1α, IL-18, and IL-33 as well as the members of the IL-36 subfamily. Because these members of the IL-1 family are each preformed in these cells, their release is a consequence of injury and is immediate. Therefore, they are termed "alarmins" as they alert the host to initiate the response. There are other alarmins from the lining cells that participate in defense, for example, defensins, which are directly antimicrobial. Each of the constitutively present IL-1 family members in lining cells is present as a precursor. In the case of IL-1α, the precursor is fully active; in the case of the other members, the precursors are weakly active at first but are converted to more active cytokines upon the infiltration of neutrophils and processing by extracellular neutrophil proteases. In the end, the infection is contained, the invading microorganism is eliminated, and the skin begins its process of repair.

Following the cloning of the mouse IL-1 receptor (Sims et al. 1988), the cytosolic domain of the IL-1 receptor was found to be homologous to Toll of the fruit fly (Gay and Keith 1991). Moreover, at the same time, the TIR domain for IL-1 signaling (see Fig. 1.2) was shown by Heguy to be required for IL-1 signaling (Heguy et al. 1992). Toll had been initially studied since its discovery in 1985 because of its central role in establishing dorsal ventral polarity in *Drosophila*. Only since 1996 was Toll linked to survival in fruit flies infected with fungi (Lemaitre et al. 1996). However, it had already been reported, back in 1988, that a member of the IL-1/TLR family, human IL-1β, protected mice from lethal *Pseudomonas* infection (van der Meer et al. 1988). As already noted, the TIR domain is essential for both IL-1-receptor family and TLR family signaling: a mutation in the TIR domain severely impairs responses to IL-1 family ligands as well to a large number of microbial products (O'Neill 2008).

The TIR domain binds MyD88 (Fig. 1.2), itself a TIR domain-containing protein, through TIR–TIR interactions triggering a cascade of kinases that propagate the IL-1 signal and result in transcription of a large number of genes, the majority of which code for other cytokines, chemokines, and a host of inflammatory mediators. Of these is IL-1 itself and other members of the IL-1 family such as IL-36 and IL-18.

The "innate immune response" regulates to the "acquired immune response." The late Charles Janeway proposed that the innate response assists the host in mounting an acquired immune response. This relationship between a nonspecific cytokine providing help for a specific response to a microbial antigen is simply the adjuvant property of some cytokines. The adjuvant property of some cytokines functions by upregulating lymphocyte growth factors such as IL-2, IL-4, and IL-6 or lymphocyte receptors, resulting in expansion of lymphocyte clones, which will either rid the host of the invading microorganism with neutralizing antibodies or in generation of cytotoxic T cells to eliminate viral infections. In 1979, purified human IL-1β, a nonspecific macrophage product, was shown to augment the T-cell response to a specific antigen (Rosenwasser et al. 1979). It was nearly 20 years later that TLR were identified as inducing IL-1β from monocytes.

1.3 Organization of the IL-1 Family of Ligands and the Consensus Sequence

As depicted in Fig. 1.1, the IL-1 family can be divided into subfamilies according to the length of the precursor and the length of the propiece for each precursor. The IL-1 subfamily is composed of IL-1α, IL-1β, and IL-33. This subfamily has the longest proteins with the longest propieces. In the case of IL-1β, the propiece is cleaved intracellularly by caspase-1 and then the mature cytokine is secreted. In the case of IL-1α, cleavage appears to occur by the membrane protease calpain, but extracellular neutrophil proteases can also cleave the IL-1α precursor. Extracellular neutrophil proteases account for the cleavage of the propiece of IL-33. The exception in the IL-1 subfamily is the IL-1Ra, which contains a signal peptide.

The IL-18 subfamily is composed of IL-18 and IL-37. By comparison, this subfamily has a smaller propiece. IL-18 requires the cleavage of its propiece by caspase-1 to be active. IL-37 is part of the IL-18 subfamily because the cytokine binds to the IL-18Rα chain. It is unclear how the propiece of IL-37 is removed. The IL-36 subfamily is composed of IL-36α, -β, and -γ as well as IL-36 Ra. In addition, IL-38 likely belongs to this family because of its binding to the IL-36R. The IL-36 subfamily has the shortest propiece.

A consensus sequence in all members of the IL-1 family is A-X-D, where A is an aliphatic amino acid such as isoleucine, methionine, or leucine, X is any amino acid, and D is aspartic acid. The aspartic acid of the consensus sequence is not the aspartic acid of the caspase-1 cleavage recognition site. The A-X-D motif is conserved in the IL-1 family where it plays a role in the three-dimensional structure of the active cytokine. The actual N-terminus is often located nine amino acids before the A-X-D site. By eliminating the amino acids before the N-terminus, the first beta-sheet structure common to all members of the IL-1 family can form. For example, with the tenth amino acid before the A-X-D consensus site as the N-terminus, the specific activity of the IL-36 subfamily (IL-36α, IL-36β, IL-36γ, and IL-36Ra) is low. However, with the ninth amino acid as the N-terminus there was a marked increased in the activity (Towne et al. 2011). In the case of IL-1β, the ninth amino acid before the A-X-D site coincides exactly with the N-terminal alanine generated by the caspase-1 site.

1.4 Interleukin-1α

From an evolutionary point of view, IL-1α is the oldest member of the IL-1 family, and its primary amino acid sequence is closely related to that of the fibroblast growth factor (FGF) family. Similar to FGF, IL-1α does not have a signal peptide, binds to nuclear DNA, exits the cell upon death, and binds to its receptor as an unprocessed precursor. As shown in Fig. 1.2, IL-1α binds to the IL-1RI and recruits the IL-1R accessory protein (IL-1RAcP) to form a heterodimeric complex, which signals to induce inflammation. In health, primary cells contain constitutive levels of the IL-1α precursor but not IL-1β (Hacham et al. 2002). The IL-1α precursor is present in keratinocytes, thymic epithelium, hepatocytes, endothelial cells, the epithelial cells of mucous membranes, including the entire gastrointestinal tract, and fibroblasts regardless of their location. The propiece of IL-1α precursor can be cleaved extracellularly by neutrophil proteases, a step that increases its biological activity. However, IL-1α can also be active as a membrane-associated cytokine. Most cell lines including tumor cell lines contain constitutive levels of IL-1α (Hurgin et al.

2007; Lonnemann et al. 1995; Werman et al. 2004). Using an epithelial cell line, what were considered to be intrinsic interferon (IFN)-γ activities depended largely on constitutively expressed IL-1α. IFN-γ activities were inhibited by antibodies to IL-1α but not to IL-1β (Hurgin et al. 2007). The concept that IL-1α acts as an autocrine growth factor assumes that the intracellular IL-1α precursor regulates normal cellular differentiation, particularly in epithelial and ectodermal cells. In support of the concept, an antisense oligonucleotide to IL-1α reduces senescence in endothelial cells (Maier et al. 1990). In fibroblasts, the constitutive IL-1α precursor binds to HAX-1, a non-receptor substrate for tyrosine kinases in hematopoietic cells. In fibroblasts, the IL-1α HAX-1 complex translocates to the nucleus (Kawaguchi et al. 2006). Although the concept is that IL-1α acts as an autocrine growth factor in fibroblasts or endothelial cells in vitro, the data should be interpreted carefully because mice deficient in IL-1α show no demonstrable defects in growth and development, including skin, fur, epithelium, and gastrointestinal function (Horai et al. 1998). However, mice deficient in IL-1α still retain the N-terminal propiece, which functions as a nuclear factor (Werman et al. 2004). In fact, in another study, the N-terminal propiece of IL-1α was shown to bind HAX-1 (Yin et al. 2001).

Is there is a role for the intracellular precursor IL-1α in normal cell function? The IL-1α precursor is present in cells that also contain large amounts of the intracellular form of the IL-1Ra (icIL-1Ra), as reviewed by Arend (2002). This form of IL-1Ra also binds to the IL-1 receptor and prevents signal transduction. In fact, icIL-1Ra is thought to compete with the intracellular pool of precursor IL-1α for nuclear-binding sites.

1.4.1 Membrane-Associated IL-1α

Precursor IL-1α can be found on the surface of several cells, particularly on monocytes and B lymphocytes, where it is referred to as membrane IL-1α (Kurt-Jones et al. 1985). Membrane IL-1α is biologically active (Kaplanski et al. 1994); its biological activities are neutralized by antibodies to IL-1α but not those to IL-1β. Endothelial cells undergoing stress-induced apoptosis release membrane apoptotic body-like particles containing nuclear fragments and histones as well as the full-length IL-1α precursor and the processed mature form (Berda-Haddad et al. 2011). When injected into mice, apoptotic body-like particles containing the IL-1α precursor induce neutrophilic infiltration that was prevented by neutralization of IL-1α but not IL-1β (Berda-Haddad et al. 2011).

1.4.2 Processing and Secretion of IL-1α

Although the IL-1α precursor is biologically active, the processed form is more active. Furthermore, the binding of IL-1α to IL-1RI has been modeled using recombinant IL-1α with an N-terminus at 113. The processing of the IL-1α precursor is accomplished by calpain II, a membrane-associated, calcium-dependent cysteine protease (Miller et al. 1994). In macrophages treated with hydroquinone, calpain II levels fall and are associated with inhibition of IL-1α precursor processing (Miller et al. 1994). Not surprisingly, calcium influx induced IL-1α secretion of the processed form (Gross et al. 2012). The secretion of IL-1α requires the presence of IL-1β, because IL-1β-deficient mice do not secrete IL-1α (Fettelschoss et al. 2011). IL-1α binding to IL-1β has been reported in which IL-1β acts as a chaperone for the secretion mechanism via caspase-1 (Fettelschoss et al. 2011). In another study, IL-1β was shown to bind to, and enhance the activity of, HMGB1 (Sha et al. 2008). It is thus possible that both IL-1α exits the cell bound to IL-1β and HMGB1.

1.4.3 Biological Functions of Constitutive IL-1α: IL-1α and Sterile Inflammation

Large numbers of reports mention the use of bacterial and fungal products to induce cytokines as models of inflammatory disease; however, most inflammatory diseases are sterile. For example, the inflammation associated with atherosclerosis, myocardial infarction, stroke, cancer, renal, and liver failure is sterile. The hypoxic insult that takes place in ischemia results in local necrosis and release of cellular contents, including nucleic acids. Members of the IL-1 family contribute to sterile inflammation, and IL-1α plays a significant role in this regard. Upon cell death by necrosis, the IL-1α precursor is released (Carmi et al. 2009; Cohen et al. 2010) and binds to the IL-1 receptor on nearby tissue macrophages and epithelial cells, triggering a response (Luheshi et al. 2011; Rider et al. 2011). For example, infiltration of neutrophils occurs first and is followed by influx of monocytes (Rider et al. 2011). Extracts of tumor cells induce neutrophilic inflammation, which does not occur in mice deficient in IL-1RI and is prevented by neutralization of IL-1α, but not neutralization of IL-1β (Chen et al. 2007). Sterile inflammation is independent on TLR2 and TLR4 (Chen et al. 2007).

Thus, IL-1α, either the unprocessed precursor or the calpain cleavage form, is classified as an alarmin because the cytokine is preformed and triggers an inflammatory response rapidly. Endothelial cells subjected to nutritional stress release inflammatory apoptotic bodies, which contain both the precursor and processed forms of IL-1α (Berda-Haddad et al. 2011). Inflammatory apoptotic bodies induce chemokine and neutrophilic infiltration into the peritoneal cavity, both of which are IL-1α dependent (Berda-Haddad et al. 2011). Platelets also contain IL-1α as well as IL-1β (Hawrylowicz et al. 1989). Platelet-derived IL-1 induces chemokines such as IL-8 from endothelial cells (Kaplanski et al. 1993) and MCP-1 from monocytes (Hawrylowicz et al. 1991). Platelet-derived IL-1α is important in brain injury in stroke models (Thornton et al. 2010) and in atherosclerosis (Gawaz et al. 2000).

1.4.4 Studies in IL-1α-Deficient Mice

Mice deficient in IL-1α are born healthy and develop normally. In some models of local inflammatory responses, wild-type and IL-1α-deficient mice develop fever and acute-phase proteins, whereas IL-1β-deficient mice do not (Horai et al. 1998). In addition, although the inflammation-associated induction of glucocorticoids was suppressed in IL-1β-deficient mice, this suppression was not observed in IL-1α-deficient mice. However, expression of IL-1β mRNA in the brain decreased 1.5 fold in IL-1α-deficient mice whereas expression of IL-1α mRNA decreased more than 30 fold in IL-1β-deficient mice. These data suggest that IL-1β exerts greater control over production of IL-1α than does IL-1α over the production of IL-1β. In caspase-1-deficient mice, IL-1α production is also reduced (Kuida et al. 1995), further suggesting that production of IL-1α is under the control of IL-1β. It is important that caspase-1-deficient mice are also deficient in caspase-11 (Kenneth et al. 2012).

In mice fed a high-fat diet, serum amyloid A protein, a marker of inflammation in atherogenesis, was markedly lower in IL-1α-deficient mice compared to wild-type or IL-1β-deficient mice (Kamari et al. 2007). IL-1α-deficient mice had significantly higher levels of non-high density lipoprotein cholesterol. The beneficial effect of IL-1α deficiency was the result of hematopoietic cells transferred from the bone marrow of IL-1α-deficient mice, resulting in a reduction in aortic lesion size twice that observed in mice transplanted with IL-1β-deficient bone marrow cells. Therefore, IL-1α appears to play a greater role in the pathogenesis of lipid-mediated atherogenesis than IL-1β, and this may be the result of an effect of membrane IL-1α.

1.5 Interleukin-1β

1.5.1 IL-1β, the Master Cytokine in the IL-1 Family

More than any other member of the IL-1 family, IL-1β has been the focus of most studies. IL-1β is a highly inflammatory cytokine, particularly in humans, as reviewed by (Dinarello 2011a). As shown in Fig. 1.2a, IL-1β and IL-1α bind to the same IL-1RI and trigger a proinflammatory signal. The interest in IL-1β also results, in part, because it is a secreted cytokine from macrophages and from the importance of the macrophage in antigen presentation before the era of dendritic cells. The inactive IL-1β precursor is converted into an active cytokine by the intracellular cysteine protease caspase-1. In particular, persons with activating mutations in one of the key genes that control the activation of caspase-1 can develop life-threatening systemic inflammation, which is reversed by either blocking the IL-1 receptor or through the use of a neutralizing antibody to IL-1β. Other chronic inflammatory diseases are mediated by IL-1β, as neutralizing antibodies have been used to treat a broad spectrum of diseases.

The IL-1β-mediated illnesses fall into the category of "auto-inflammatory" diseases, which are to be distinguished from the classic "autoimmune" diseases. Although inflammation is common to both auto-inflammatory and autoimmune diseases, in the case of IL-1-mediated disease, there is no evidence for role of adaptive immunity in its induction.

1.5.2 IL-1β is an Inducible Cytokine

In contrast to IL-1α, the IL-1β precursor is not present in health. Also differing from IL-1α, IL-1β is primarily a product of monocytes, macrophages, and dendritic cells (DC) as well as B lymphocytes and natural killer (NK) cells. In health, circulating human blood monocytes or bone marrow cells do not constitutively express mRNA for IL-1β. Endothelial cells, skin keratinocytes, fibroblasts, and epithelial cells contain constitutive IL-1α and constitutive IL-33 as precursors as well as mRNA, but these cells do not express IL-1β mRNA, even upon stimulation with TLR ligands. Melanoma cells do express IL-1β as a precursor, and the more aggressive and metastatic the melanoma, the greater the likelihood of active caspase-1 and IL-1β secretion (Okamoto et al. 2009). In the bone marrow neutrophil precursors, IL-1β gene expression is inducible but mature neutrophils in the circulation no longer produce IL-1β. Neutrophil IL-1β plays a pathological role in the severe inflammation of mice with a mutant form of the phosphatase SHP1 (Croker et al. 2011). Several malignant tumors do express IL-1β as part of their neoplastic nature, particularly acute myelogenous leukemia, melanoma, multiple myeloma, and juvenile myelogenous leukemia, each of which exhibit constitutive expression of IL-1β. In contrast to most cytokine promoters, IL-1β regulatory regions are distributed more than several thousand base pairs upstream from the transcriptional start site. In addition to a cAMP response element, there are NF-κB-like and activating protein-1 (AP-1) sites. IL-1β gene regulation has been reviewed in detail (Unlu et al. 2007). Although steady-state mRNA levels for IL-1β may be present, there is distinct dissociation between transcription and translation of the IL-1β precursor. Non-TLR ligands such as the complement component C5a, hypoxia, adherence to surfaces, or clotting of blood induce the synthesis of large amounts of IL-1β mRNA in monocytic cells without significant translation into the IL-1β protein. In these cells, the IL-1β mRNA assembles into large polyribosomes, but there is no significant elongation of the peptide (Kaspar and Gehrke 1994). This failure to complete the translation into IL-1β protein may be caused by the instability element present in the coding region. This instability region is also found in IL-18 and IL-37 and appears to limit the mRNA of these cytokines (Bufler et al. 2004). However, completion of translation of the mRNA into the respective cytokines can be accomplished by adding low concentrations of TLR ligands or IL-1 itself to the "primed" monocytes (Schindler et al. 1990).

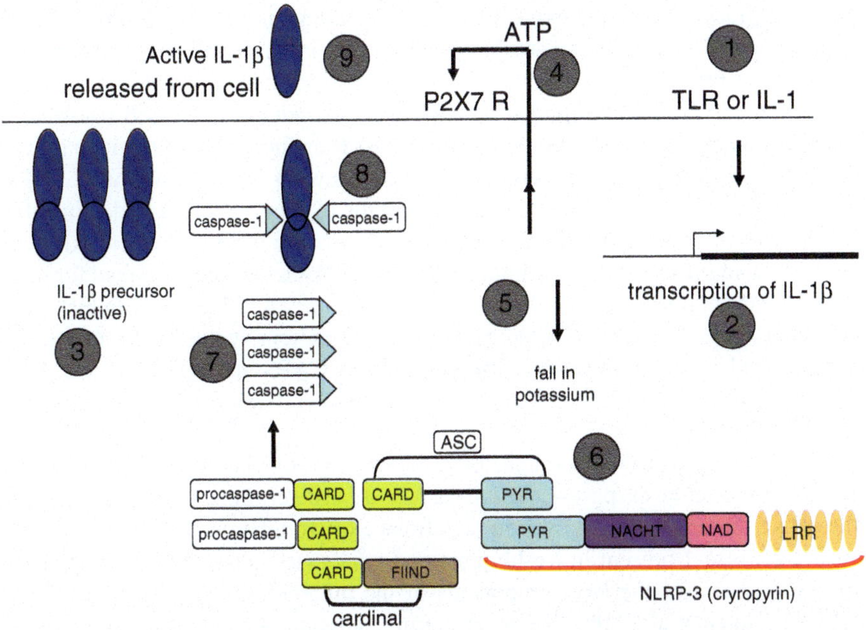

Fig. 1.3 Activation of caspase-1 by the NLRP3 inflammasome. *1* Activation of the cells following receptor binding for TLR or IL-1 receptors. *2* Transcription of the IL-1β gene. *3* Synthesis of the inactive IL-1β precursor. *4* Extracellular ATP binds P2X7 receptor. *5* Efflux of potassium. *6* Oligomerization of inflammasome components. *7* Activation of caspase-1. *8* Cleavage of the IL-1β precursor by caspase-1. *9* Maure IL-1β is released from the cell

1.5.3 Processing and Secretion of IL-1β via Caspase-1

Nearly all microbial products induce IL-1β via TLR activation; in addition, IL-1 (either IL-1α or IL-1β) induces itself both in vivo and in monocytes in vitro (Dinarello et al. 1987). Other studies supporting this concept of IL-1-induced IL-1 have been reported (Boni-Schnetzler et al. 2008; Gattorno et al. 2007; Goldbach-Mansky et al. 2006; Greten et al. 2007). Regardless of the stimulus, processing and secretion of IL-1β require conversion of procaspase-1 to active caspase-1, although in some studies processing of the IL-1β precursor is caspase-1 independent (Wewers et al. 1997). The activation to active caspase-1 is dependent on a complex of intracellular proteins termed the "inflammasome" by the late Juerg Tschopp (Agostini et al. 2004; Martinon et al. 2009). The critical component of the inflammasome is NLRP3 (see Fig. 1.3). NLRP3 is also termed cryopyrin because the gene was initially discovered in patients with familial cold auto-inflammatory syndrome, a genetic disease characterized by constitutional symptoms, fevers, and elevated acute-phase proteins following exposure to cold (Hoffman et al. 2001).

As monocytes exit the bone marrow, they circulate in the bloodstream for approximately 3 days. In the absence of disease, it is likely that these cells do not enter tissues but are destroyed in the spleen or undergo apoptosis. There is no dearth of reports that circulating human blood monocytes release processed IL-1β upon stimulation starting 4 h after stimulation with TLR agonists and continue to release the cytokine during the following 20–40 h. Following lipopolysaccharide (LPS) stimulus, IL-1β mRNA levels rise rapidly within 15 min but begin to decline after 4 h because of the short half-life of their mRNA or the action of micro RNA. In contrast, using IL-1 itself as a stimulant, IL-1β mRNA levels are sustained for more than 24 h (Schindler et al. 1990). Raising intracellular cAMP levels with histamine enhances IL-1-induced IL-1 gene expression and protein synthesis. Monocytes of patients with auto-inflammatory diseases such as CAPS and HIDS release IL-1β even without TLR stimulation during a 24-h incubation (Drenth et al. 1996; Hoffman and Wanderer 2011).

When obtained from the venous blood of healthy subjects, human blood monocytes contain active caspase-1. Active caspase-1, as determined by its cleavage into the active dimer, is present even in the absence of stimulation (Netea et al. 2009). Active caspase-1 present in freshly obtained monocytes is nevertheless dependent on the presence of the key components of the inflammasome, namely, ASC and NLRP3 (Netea et al. 2009). However, during subsequent incubation, extracellular levels of ATP increase in the supernatant as IL-1β also increases and inhibition of ATP by oxidized ATP reduces the secretion of IL-1β (Netea et al. 2009). The inhibition of IL-1β secretion by oxidized ATP is consistent with the role of the P2X7 receptor, which binds ATP and opens the potassium channel for release of intracellular potassium. The presence of active caspase-1 in circulating blood monocytes suggests that the rate-limiting step in the processing and release of IL-1β is at the level of gene expression.

However, upon differentiation of the same blood monocytes into macrophages in vitro, TLR-induced IL-1β release requires activation of caspase-1 by exogenous ATP (Netea et al. 2009). The assembly of the inflammasome components with inactive pro-caspase-1 takes place following a fall in intracellular potassium triggered by ATP binding to the P2X7 receptor. ATP activation of the P2X7 receptor opens the potassium channel, and simultaneously, as potassium levels fall, caspase-1 is activated by the inflammasome (Andrei et al. 1999; Andrei et al. 2004; Elssner et al. 2004; Gardella et al. 2000; Perregaux et al. 2000). Without exogenous ATP, there is little or no processing of the IL-1β precursor in differentiated monocyte-derived macrophages. Alveolar macrophages obtained from the lungs of healthy human also do not release IL-1β with LPS stimulation unless exogenous ATP is added (Netea et al. 2009). In addition to ATP activation of P2X7, activation of IL-1β processing can also take place with a cathelicidin-derived peptide termed LL37, which is released from neutrophils (Elssner et al. 2004).

The cleavage of the IL-1β precursor by active caspase-1 can take place in the specialized secretory lysosomes or in the cytoplasm. However, more than one pathway seems available for processed IL-1β to exit the cell: these include by exocytosis of the secretory lysosomes (Andrei et al. 1999, 2004), shedding of plasma membrane microvesicles, direct release via transporters, or multivesicular bodies containing exosomes (Qu et al. 2007). In general, the release of processed IL-1β takes place before there is significant release of lactate dehydrogenase (Brough and Rothwell 2007), although in vitro cell death eventually takes place. Pyroptosis is a caspase-1-dependent form of cell death and is induced by certain bacteria using Ipaf, a member of the NLR family of intracellular receptors (Suzuki et al. 2007). An increase in intracellular calcium is also required for the mature IL-1β to exit the cell, and this is phopholipase C dependent (Andrei et al. 2004).

1.5.4 Gain-of-Function Mutation in Cryopyrin

Diseases associated with single amino acid-activating mutations in cryopyrin are termed cryopyrin-associated periodic syndromes (CAPS). In monocytes from patients with CAPS, activation of caspase-1 occurs without a requirement for a rapid fall in the level of intracellular potassium (Gattorno et al. 2007). Therefore, mutated cryopyrin allows for the assembly of the complex of interacting proteins in the presence of normal intracellular levels of potassium. Although LPS-induced synthesis of the IL-1β precursor is often studied (Kahlenberg et al. 2005), it is unlikely that LPS plays a role in auto-inflammatory diseases. On the other hand, spontaneous secretion of IL-1β from monocytes of patients is the result of endogenous IL-1β stimulation. In patients with CAPS, there is a decrease in steady-state levels of pro-caspase-1 mRNA with IL-1Ra treatment (Goldbach-Mansky et al. 2006), suggesting that IL-1β stimulates its own production and processing. Thus, in any disease process that includes an increase in the steady-state levels of pro-caspase-1 mRNA, components of the inflammasome or the IL-1β precursor explain the "auto-inflammatory" nature of the disease. Type 2 diabetes appears to be an example of an auto-inflammatory disease in which glucose induces IL-1β production from the insulin-producing beta cell and IL-1β induces the beta cell to produce its own IL-1β (Maedler et al. 2002).

1.5.5 Polymorphisms in P2X7 and the Activation of the Inflammasome

Patients with classic auto-inflammatory diseases such as FMF or CAPS have nearly identical clinical parameters, secrete more IL-1β, and respond dramatically to IL-1 receptor blockade, yet have no mutation in NALP3. It is therefore possible that mutations in P2X7 itself or regulation of the other genes controlling potassium channels (Pascual et al. 2005) may account for dysfunctional secretion of IL-1β. For example, monocytes from patients with rheumatoid arthritis are more sensitive to release of IL-1β following ATP activation of the P2X7 receptor compared to monocytes from healthy controls (Al-Shukaili et al. 2008). However, monocytes from subjects with a P2X7 Glu496Ala loss-of-function polymorphism secrete significantly less IL-1β (Sluyter et al. 2004b). Monocytes from subjects homozygous for this polymorphism also released significantly less IL-18 (Sluyter et al. 2004a). Another P2X7 receptor polymorphism is associated with increased mortality in patients undergoing allogeneic stem cell transplantation (Lee et al. 2007). Bacteremia was documented in 68 % of patients with this polymorphism compared to 18 % in wild-type control patients (Lee et al. 2007).

In mice deficient in P2X7 receptors, inflammation, pain, and IL-1β-mediated IL-6 production are markedly reduced (Chessell et al. 2005). In addition to a fall in intracellular potassium, ATP triggers formation of peroxynitrite, which is required for caspase-1 activation because peroxynitrite scavengers prevent IL-1β secretion (Hewinson et al. 2008). Pannexin-1, a mammalian protein that functions as a hemi-channel for the uptake of dyes, is required for caspase-1 processing and release of IL-1β via the P2X7 receptor (Pelegrin and Surprenant 2006). Pannexin-1 can also function for LPS-induced IL-1β synthesis in the absence of TLR4 (Kanneganti et al. 2007). P2X7 receptor activity is also regulated by a "regeneration and tolerance factor" (Derks and Beaman 2004).

1.5.6 Polymorphism in NLRP1 and the Release of IL-1β

NLRP1 is genetically associated with risk of several autoimmune diseases including generalized vitiligo, Addison disease, type 1 diabetes, and rheumatoid arthritis. Predicted functional variations in NLRP1 reside in several common high-risk haplotypes, and the haplotypes that are high risk for disease share two substitutions, L155H and M1184V. Peripheral blood monocytes from healthy subjects homozygous for the predominant high-risk haplotype 2A released significantly greater amounts ($P<0.001$) of the IL-1β precursor to mature bioactive IL-1β under basal (resting) conditions as well as in response to TLR2 and TLR4 agonists compared with monocytes from subjects homozygous for the reference haplotype 1 (Levandowski et al. 2013). The increase in basal release was 1.8 fold greater in haplotype 2A monocytes, and these differences between the two haplotypes were consistently observed three times during a 3-month period; no differences were observed for IL-1α or tumor necrosis factor (TNF)-α. NLRP1 RNA and protein levels were not altered by the predominant high-risk haplotype, indicating that altered function of the corresponding multivariant NLRP1 polypeptide predisposes to autoimmune diseases by activation of the NLRP1 inflammasome.

1.5.7 Non-caspase-1 Processing of IL-1β

Non-caspase-1 mechanisms also exist to generate active forms of IL-1β. For example, sterile inflammation induces fever, elevated IL-6, and increased production of hepatic acute-phase proteins. These responses are absent in mice deficient in IL-1β but present in mice deficient in caspase-1 (Fantuzzi et al. 1997a; Joosten et al. 2009). Sterile inflammation is often associated with neutrophilic infiltration, and the neutrophils produce IL-1β. Because neutrophils are short-lived cells, dying within hours upon emigration, release of the IL-1β precursor from intracellular stores is not unexpected. Processing of the IL-1β precursor extracellularly into an active cytokine has been reported for the common neutrophil protease, proteinase-3 (Coeshott et al. 1999; Joosten et al. 2009). Proteinase-3 also contributes to the processing of IL-18 (Sugawara et al. 2001). Other proteases such as elastase, matrix metalloprotease 9, and granzyme A process the IL-1β precursor extracellularly. In addition, a mast cell chymase generates active IL-1β.

Mice with a targeted IKK-β deletion in myeloid cells are more susceptible to LPS-induced shock than control mice (Greten et al. 2007) and markedly elevated levels of IL-1β are found in the circulation associated with a prominent neutrophilia (Greten et al. 2007). The elevated levels of IL-1β are lethal because blockade with IL-1Ra protects these mice from death. The source of the IL-1β in these mice is the neutrophil. When incubated with proteinase-3, cleavage of the IL-1β precursor is observed, yielding molecular weights of 25,000 and 15,000 Da (Greten et al. 2007). Because the cleavage of the IL-1β precursor by proteinase-3, elastase, and cathepsin G are within three amino acids of the caspase-1 cleavage site, the products of the non-caspase-1 cleavage are biologically active (Coeshott et al. 1999; Joosten et al. 2009). Therefore, in inflammatory conditions such as urate crystal arthritis, which is characterized by a prominent neutrophilic infiltration, proteinase-3 cleavage of extracellular IL-1β precursor likely takes place (Joosten et al. 2010). Mice deficient in caspase-1 are not protected against urate-induced inflammation. Although IL-1Ra is effective in treating gout, IL-1Ra would be equally effective in any disease with extracellular processing of the precursor (Schlesinger et al. 2011; So et al. 2007, 2010). The importance of extracellular processing of the IL-1β precursor by serine proteases may explain, in part, the anti-inflammatory properties of alpha-1-antitrypsin (Numanami et al. 2003).

1.5.8 Effects in Mice Deficient in IL-1β

After 10 years of continuous breeding, mice deficient in IL-1β exhibit no spontaneous disease. However, upon challenge, IL-1β-deficient mice exhibit specific differences from their wild-type controls. The most dramatic is the response to local inflammation induced by subcutaneous injection of an irritant. Within the first 24 h, IL-1β-deficient mice do not manifest an acute-phase response, do not develop anorexia, have no circulating IL-6, and no fever (Fantuzzi et al. 1997a; Zheng et al. 1995). These findings are consistent with those reported in the same model using anti-IL-1R type I antibodies in wild-type mice (Fantuzzi et al. 1997a; Zheng et al. 1995). IL-1β-deficient mice also have reduced inflammation because of zymosan-induced peritonitis (Fantuzzi et al. 1997a, b). In contrast, IL-1β-deficient mice have elevated febrile responses to LPS, IL-1β, or IL-1α compared to wild-type mice

(Fantuzzi et al. 1996). Nevertheless, IL-1β-deficient mice injected with LPS have little or no expression of leptin mRNA or protein (Faggioni et al. 1998).

Mice deficient in IL-1β were compared to mice deficient in IL-1α after exposure to chemical carcinogens (Krelin et al. 2007). In IL-1β-deficient mice, tumors developed more slowly or did not develop in some mice. A deficiency in IL-1α, on the other hand, did not impair tumor development compared to wild-type mice injected with the same carcinogen. In IL-1Ra-deficient mice, tumor development was the most rapid. A leukocyte infiltrate was found at the site of carcinogen injection. The neutrophilic infiltrate was almost absent in IL-1β-deficient mice, whereas in IL-1Ra-deficient mice, a dense neutrophilic infiltrate was observed. In wild-type mice, the leukocytic infiltrate was sparse and the infiltrate that was observed in IL-1α-deficient mice was similar to that of control mice. These findings may reflect the fact that IL-1β is secreted into the microenvironment, resulting in the emigration of monocytes and neutrophils, whereas IL-1α, remaining cell associated, is less likely to affect the microenvironment.

1.5.9 IL-1β and Autophagy

Autophagy is an ancient process of recycling cellular components, such as cytosolic organelles and protein aggregates, through degradation mediated by lysosomes. Autophagy is activated in conditions of cell stress, hypoxia, starvation, or growth factor deprivation; it promotes cell survival by generating free metabolites and energy through degradation of the endogenous cellular components (Klionsky 2007). However, in addition to its role in the pathophysiology of cancer, neurodegenerative diseases, or aging, autophagy is also a modulator of inflammation (Schmid and Munz 2007). A role for autophagy in production of proinflammatory cytokines, particularly of IL-1β, has emerged with deletion of ATG16-L1. For example, macrophages from *ATG16L1*-deficient mice produce higher levels of IL-1β and IL-18 after stimulation with TLR4 ligands (Saitoh et al. 2008). The data suggest that higher activation of caspase-1 in the *ATG16L1*-deficient mice accounts for the higher production level (Saitoh et al. 2008). This observation was related to the specific degradation of the IL-1β precursor in autophagosomes in mouse macrophages (Harris et al. 2011). Additional studies in the *ATG16L1*-deficient mice point toward a regulatory effect of autophagy on caspase-1 activation through modulation of the NLRP3 inflammasome (Nakahira et al. 2011; Tschopp and Schroder 2010; Zhou et al. 2010).

This role of autophagy in the secretion of IL-1β was also observed in human primary monocytes, in which specific inhibition of autophagy leads to increased production of IL-1β (Crisan et al. 2011). However, in the same cells TNF-α production was decreased by autophagy inhibition. These data suggest divergent effects of autophagy on the production of these two important proinflammatory cytokines. In mice, the increase in IL-1β production is ascribed to increased activation of the inflammasome, but in human cells, it is IL-1β mRNA transcription that is elevated when autophagy was inhibited, whereas no effects were observed on caspase-1 activation (Crisan et al. 2011; Harris et al. 2011; Saitoh et al. 2008). Despite these differences between mouse and human cells, the inhibition of autophagy increases the production of IL-1β but not TNF-α.

The modulation of inflammation by autophagy in humans has been studied in Crohn's disease. Genome-wide association studies in large cohorts of Crohn's disease patients have revealed that genetic variants in two autophagy genes, *ATG16L1* and *IRGM*, result in increased susceptibility to the disease. A nonsynonymous polymorphism in *ATG16L1* on chromosome 2q37.1 and two polymorphisms in *IRGM* on chromosome 5q33.1 were significantly associated with Crohn's disease risk (Hampe et al. 2007; Rioux et al. 2007). Another study revealed a significant association of Crohn's disease susceptibility with an intronic polymorphism in the autophagy gene *ULK1* (Henckaerts et al. 2011). Moreover, autophagy defects have been reported in individuals bearing *NOD2* mutations and are consistent with the concept that impaired bacterial clearance and increased bacterial persistence are part of the pathogenesis of Crohn's disease (Lapaquette et al. 2010).

The mechanism through which polymorphisms in autophagy genes influence susceptibility to Crohn's disease appear to involve IL-1β production. The *ATG16L1* 300Ala risk allele was associated with elevated production of IL-1β and IL-6; however, this finding was only observed in cells stimulated with the NOD2 ligand muramyl dipeptide (MDP). In contrast, the expected levels of IL-1β and IL-6 were produced upon stimulation with TLR2 and TLR4 ligands (Plantinga et al. 2011). The increased production of IL-1β was associated with an increase in the steady-state levels of IL-1β mRNA rather than increased activation of the inflammasome (Plantinga et al. 2011). Studying the same polymorphism (*ATG16L1* Thr300Ala) in human dendritic cells, Cooney et al. reported defective NOD2-induced, but not TLR-induced, autophagy and antigen presentation (Cooney et al. 2010). Furthermore, effects of this polymorphism on antibacterial autophagy in epithelial cells have been observed (Homer et al. 2010). The specific effect of the *ATG16L1* polymorphism on the NOD2 pathway, and not on TLR-induced stimulation, is likely related to the fact that NOD2 and ATG16L1 form a protein complex that is essential for NOD2-induced autophagosome formation (Travassos et al. 2010). Because the *ATG16L1* Thr300Ala polymorphism affects protein stability (Kuballa et al. 2008), defective induction of autophagy and therefore enhanced IL-1β mRNA transcription upon triggering of NOD2 may be caused by the presence of defective complex.

1.6 Interleukin-33

1.6.1 IL-33 as a Member of the IL-1 Subfamily

Formerly termed IL-1F11, IL-33 belongs to the IL-1 subfamily and has been studied for its role in the Th2 paradigm of immune responses. IL-1β is also linked to the Th2 response. The existence of IL-33 was predicted in 1994 following the discovery

Table 1.2 IL-1-receptor family

Name	Designation	Ligands	Co-receptor
IL-1RI	IL-1R1	IL-1α, IL-1β, IL-1Ra	IL-1RAcP (IL-1R3)
IL-1RII	IL-1R2	IL-1β, IL-1β precursor	IL-1RAcP (IL-1R3)
IL-1RAcP	IL-1R3	IL-1α, IL-1β, IL-33, IL-36	Not applicable
ST2/IL-33Rα	IL-1 R4	IL-33	IL-1RAcP (IL-1R3)
IL-18Rα	IL-1R5	IL-18, IL-37	IL-18Rβ (IL-1R7)
IL-1Rrp-2	IL-1R6	IL-36α, -β, -γ	IL-1RAcP (IL-1R3)
IL-18Rβ	IL-1R7	IL-18	Not applicable
TIGIRR-2/IL-1RAPL	IL-1R8	Unknown	Unknown
TIGIRR-1	IL-1R9	Unknown	Unknown
SIGIRR	TIR8	Unknown	Unknown

of a novel member of the IL-1 receptor family termed ST2 (Bergers et al. 1994). ST2 is the ligand-binding chain for IL-33 (see Table 1.2) and is structurally similar to the ligand-binding chain of IL-1α and IL-1β. In addition, the co-receptor for IL-33 is the IL-1RAcP, which is also the co-receptor for IL-1α and IL-1β. It was not until 2005 that IL-33 was reported as the ligand for ST2 (Schmitz et al. 2005). ST2 is regulated by the estrogen-inducible transcription factor Fos (Bergers et al. 1994), and this property of estrogens may be related to the large number of studies on the effect of estrogens to regulate IL-1 and inflammation.

Similar to most members of the IL-1 receptor family, ST2 is composed of three extracellular Ig domains and an intracellular TIR domain. Although the name ST2 is still used, the correct term is the IL-33 receptor α-chain (IL-33Rα). As shown in Fig. 1.2d, the IL-33Rα chain is similar to the IL-1R1 in that it is the ligand-binding chain for IL-33 but requires IL-1RAcP to signal (Ali et al. 2007; Chackerian et al. 2007).

Before the discovery of IL-33, several studies suggested that the putative ligand (IL-33) for the ST2 orphan receptor was playing a role in allergic-type diseases. It became clear that activation of ST2 was uniquely driving Th2 responses. Structurally, IL-33 is closer to IL-18 than IL-1β. Biologically, IL-33 is closest to IL-1α, as the precursors for IL-1α and IL-33 are constitutively present in all endothelial cells. As discussed below, similar to IL-1α, IL-33 functions as a DNA-binding molecule. The dominant property of IL-33 is the induction of IL-4, IL-5, and IL-13 as well as other properties anticipated for a Th2-type cytokine. Diseases thought to be caused by increased immunoglobulin production may also be related to IL-33. IL-33 induces the production of IL-6, IL-1β, and PGE2 from mast cells.

1.6.2 IL-33 and Th2 Responses

The properties of recombinant IL-33 recapitulate much of the existing data that ST2 promotes Th2-type responses. For example, before its discovery, a role for IL-33 in the Th2 response was observed using soluble extracellular forms of ST2 [reviewed in (Schmitz et al. 2005)]. However, IL-33 has properties that go beyond its role in the Th2 paradigm, because similar to IL-1α, IL-1β, and IL-36, IL-33 forms a heterodimeric complex with IL-1RAcP for signal transduction (Ali et al. 2007; Chackerian et al. 2007). Although the IL-1RAcP is expressed on most nucleated cells, ST2 is somewhat restricted to low expression on most cells with the notable exception of mast cells.

There are several mechanisms by which IL-33 favors the Th2 response. Similar to IL-1β, IL-33 induces IL-6, an adjuvant for antibody production. IL-33 induction of IL-6 is prevented by a blocking antibody to IL-1RAcP (Ali et al. 2007). IL-33 initiates signal transduction via activation of NF-κB, which is typical of IL-1α, IL-1β, and IL-18 (Schmitz et al. 2005), but other studies have shown that antibody cross-linking of ST2 does not result in activation of NF-κB but rather AP-1. IL-33 treatment also increased serum IgA and IgE, an expected response for a switch from Th1 to Th2.

1.6.3 Processing of the IL-33 Precursor

Initially, IL-33 was considered closely related to IL-1β and IL-18 because the IL-33 precursor contains a caspase-1 site, which upon activation would cleave the IL-33 precursor and release the active cytokine (Schmitz et al. 2005), similar to that for IL-1β and IL-18. Indeed, the first recombinant forms of IL-33 were produced with an N-terminus at the caspase-1 site (Schmitz et al. 2005). Although recombinant IL-33 was active, the concentrations required for activity were considerably higher than those of other members of the IL-1 family. Indeed, subsequent studies revealed that caspase-1 actually results in loss of IL-33 activity and that the full length IL-33 precursor binds to ST2 and is active (Cayrol and Girard 2009), similar to the ability of the IL-1α precursor to bind to IL-1RI. In addition, it was reported that the caspase-1 cleavage site at 178 is similar to the consensus sequence for caspase-3 and that intracellular IL-33 precursor is a substrate for caspase-3 (Cayrol and Girard 2009).

Using immobilized IL-33 precursor, neutrophil proteinase 3 (PR3) was isolated from human urinary proteins (Bae et al. 2012). Neutrophil PR3 is known to process the IL-1β precursor into an active cytokine (Joosten et al. 2009). PR3 converted human and mouse precursor IL-33 proteins to biological active forms; however, increasing the incubation time of PR3 abrogated IL-33 activities (Bae et al. 2012). Using the consensus amino acid sequence sites for PR3, six human and mouse recombinant IL-33 proteins were produced and assessed for biological activities; varying levels of activity were reported (Bae et al. 2012). Another study also demonstrated cleavage of the IL-33 precursor by neutrophil proteases such as PR3, neutrophil elastase, and cathepsin G (Lefrancais et al. 2012), resulting in the generation of IL-33 with different N-termini and varying levels of activity. These studies support the concept that extracellular IL-33 is released as a precursor, is rapidly processed by neutrophil enzymes, and generates active forms with varying levels of activity. The implications for generation of active IL-33 by neutrophil enzymes for Th2 polarization remain unclear. It may be more relevant to study the effect of proteases from eosinophils in the processing of the IL-33 precursor. Nevertheless, the IL-33 precursor binds to ST2 and recruits the accessory chain for signal transduction, but compared to IL-33 generated by neutrophil proteases, the activity of IL-33 precursor is weak (Bae et al. 2012; Lefrancais et al. 2012).

There is no dearth of studies on ST2 tissue-specific localization, regulation of its expression, and effects in transgenic mice overexpressing ST2 as well as deletion, neutralization, and antibody cross-linking of ST2. Elevated levels of the soluble form of ST2 were present in the circulation of patients with various inflammatory diseases, and exogenous administration of pharmacological doses of soluble ST2 neutralized endogenous levels of the then putative ligand IL-33 and reduced inflammation (Leung et al. 2004). IL-33 activates Th2 lymphocytes, mast cells, basophils, and eosinophils as well as NK T cells and blood monocytes. One of the best studied properties of IL-33 is the induction of IL-5 and IL-13 and their respective roles in lung inflammation, such as allergic-type asthma. For example, instillation of IL-33 into the airways triggers an immediate allergic response in the lung of naïve mice and worsens the response in mice sensitized to antigen peripherally but challenged by exposure of antigen in the lung (Louten et al. 2011).

Mice deficient in ST2 do not develop a Th2 response to *Schistosoma* egg antigen. Indeed, several studies have focused on the role of IL-33 in the pathogenesis of helminth worm infections. The Th2 response by the host contributes to the elimination of these worm infestations, which are worldwide and afflict hundreds of millions. The role of IL-33 in the induction of IL-4, IL-5, and IL-13 is of paramount importance in terms of pulmonary and intestinal complications that reduce lifespan. Using mice deficient in IL-33, a crucial role was demonstrated in mice to rid them of infection with *Strongyloides venezuelensis* (Yasuda et al. 2012). The infection induces a unique class of cells called natural helper cells or nuocytes, which upon activation by IL-33 produce IL-5 and IL-13, resulting in eosinophilic infiltration into the lungs. In this model, pulmonary inflammation causes damage via eosinophilic infiltration, which is IL-33 and IL-5 dependent (Yasuda et al. 2012).

Mice injected with human IL-33 exhibit impressive pathological changes in the arterial walls, lungs, and intestinal tissues (Schmitz et al. 2005). Of particular relevance to the concept that IL-33 drives a Th2 response, esosinophilic infiltration was a prominent finding in the lung and in allergic rhinitis as well as allergic conjunctivitis (Matsuba-Kitamura et al. 2010). These initial observations have been confirmed by other reports (Kim et al. 2012). Although the interpretation of in vivo effects following the administration of an exogenous cytokine should be conservative, the findings are clearly consistent with IL-33 being a pro-inflammatory ligand of the IL-1 receptor family. Even before the ability to test IL-33-mediated activation, others had reported that neutralization of the putative ST2 ligand using soluble ST2 markedly reduced joint inflammation, synovial hyperplasia, and joint erosion when given in the therapeutic phase of collagen-induced arthritis in mice (Leung et al. 2004).

1.6.4 IL-33 as an Anti-inflammatory Cytokine

Members of the IL-1 family of ligands bind to their specific cell-surface receptors and recruit an accessory chain. The IL-1RIAcP is used by IL-1α and IL-1β but also IL-36 and IL-33. The accessory chain for IL-18 is related to the IL-1RIAcP but is encoded by a distinct gene. We now recognize that other members of the IL-1 receptor family will bind more than one cytokine. The best example is IL-1α and IL-1β. Both bind with similar affinities to IL-1RI, but the three-dimensional structures of IL-1α and IL-1β are hardly identical (Wang et al. 2010). The IL-1β precursor binds to IL-1RII as well as a processed form with the first 112 amino acids cleaved from the precursor. IL-37 binds to the IL-18-receptor α-chain (Kumar et al. 2002), and both IL-36 and IL-38 bind to the IL-36 receptor (van de Veerdonk et al. 2012).

IL-33 forms a complex with ST2 IL-1RIAcP but also with SIGIRR (Bulek et al. 2009). This complex plays a role in the Th2 response by reducing IL-33 signaling (Bulek et al. 2009) and, consistent with these observations, Th2 responses are increased in mice deficient in SIGIRR. Furthermore, there is high expression of SIGIRR in Th2 polarized cells, and in models of Th2 antigen sensitization, SIGIRR-deficient mice exhibit a greater Th2 response (Bulek et al. 2009). The complex with SIGIRR and IL-33 may explain the anti-inflammatory properties of IL-33. ST2 can sequester TLR adaptor molecules such as MyD88 and Mal (Gadina and Jefferies 2007).

In mice deficient in ST2, there is myocardial hypertrophy, ventricle dilation, and fibrosis upon pressure overload, suggesting that IL-33 plays a protective role in the heart (Sanada et al. 2007). Furthermore, elevated levels of the extracellular domain of ST2 predict outcomes in patients with systolic heart failure or following a myocardial infarction (Sanada et al. 2007). In a model of cardiomyocyte hypertrophy induced by chronic administration of phenylephrine, administration of recombinant IL-33 inhibited the phosphorylation of IκB and reduced the hypertrophy and fibrosis (Sanada et al. 2007). One of the more challenging aspects of the properties of IL-33 to act as a Th2 cytokine is its role as an antagonist in the ApoE-deficient mouse model of artherosclerosis. In this model, arterial wall plaques of mice on a high-fat diet contain IL-33 and ST2. In mice treated with IL-33, the atherosclerotic plaques were markedly reduced (Miller et al. 2008). In mice treated with soluble ST2 to neutralize IL-33, the disease worsened (Miller et al. 2008).

1.6.5 IL-33 as a Transcription Factor

Similar to IL-1α, there is another side to IL-33. Although IL-33 binds to its specific surface receptor, IL-33 is identical to a nuclear factor dominantly expressed in high endothelial venules (HEV) (Carriere et al. 2007). This nuclear factor is termed NF-HEV. In addition to endothelial cells, constitutive nuclear localization of IL-33 has been reported in several cell types such as type II lung epithelial cells (Yasuda et al. 2012), epithelial cells (Moussion et al. 2008), and pancreatic stellate cells (Masamune et al. 2010). In fact, IL-33 binding to DNA and acting as a nuclear factor is similar to IL-1α binding to chromatin and functioning as a nuclear factor (Cohen et al. 2010; Stevenson et al. 1997; Werman et al. 2004). A short IL-33 peptide similar to a sequence in Kaposi sarcoma virus binds chromatin (Roussel et al. 2008). The full-length IL-33 precursor, but not mature IL-33, binds to the N-terminal Rel homology domain of NF-κB p65 (Ali et al. 2011). In cells overexpressing the IL-33 precursor, there was a reduction in IL-1β-induced TNF-α (Ali et al. 2011). These data are consistent with other data that IL-33 possesses anti-inflammatory properties (see foregoing), and the mechanism for this property of IL-33 appears to be nuclear sequestration similar to that of IL-1α (Cohen et al. 2010).

1.7 IL-18 and IL-37 Subfamily

1.7.1 IL-18

1.7.1.1 Background

IL-18 was first described in 1989 as "IFN-γ-inducing factor" isolated in the serum of mice following an injection of endotoxin. The mice had been pretreated with *Proprionibacterium acnes*, which stimulates the reticuloendothelial system, particularly the Kupffer cells of the liver. Many investigators concluded that the serum factor was IL-12. With molecular cloning of "IFN-γ-inducing factor" in 1995 (Okamura et al. 1995), the name was changed to IL-18. Surprisingly, the new cytokine was related to IL-1 and particularly to IL-1β. Similar to IL-1β, IL-18 lacks a signal peptide, is first synthesized as an inactive precursor, and remains as an intracellular cytokine. The tertiary structure of the mature form of IL-18 closely resembles that of IL-1β (Okamura et al. 1995), although the IL-18 precursor is closely related to the IL-37 precursor. Since 1995, many studies have used neutralization of endogenous IL-18- or IL-18-deficient mice to demonstrate the role for this cytokine in promoting inflammation and immune responses [reviewed by (Dinarello 2007)]. However, the biology of IL-18 is hardly the recapitulation of IL-1β. There are several unique and specific differences between IL-18 and IL-1β. For example, in healthy human subjects and also in healthy mice, gene expression for IL-1β in blood mononuclear cells and hematopoietic cells is absent and there is no evidence that the IL-1β precursor is constitutively present in epithelial cells (Puren et al. 1999). In contrast, in the same blood cells large amounts of the IL-18 precursor are present. Peritoneal macrophages and mouse spleen contain the IL-18 precursor in the absence of disease (Puren et al. 1999). The IL-18 precursor is also present in keratinocytes and nearly all epithelial cells. In this regard, IL-18 is similar to IL-1α and IL-33.

1.7.1.2 Processing of the IL-18 Precursor

The IL-18 precursor has a molecular weight of 24,000 and is processed by caspase-1 cleavage into a mature molecule of 18,000. Compared to wild-type mice, following an injection of endotoxin into caspase-1-deficient mice, circulating IFN-γ is absent. IL-12-induced IFN-γ is also absent in caspase-1-deficient mice (Fantuzzi et al. 1999). Importantly, any phenotypic characteristic of capsase-1-deficient mice must be studied as whether the deficiency is caused by reduced IL-1β or IL-18 activity. For example, the caspase-1-deficient mouse is resistant to colitis (Siegmund et al. 2001b) but the IL-1β-deficient mouse is susceptible in the same disease. Because neutralizing antibodies to IL-18 are protective in the colitis model, caspase-1 deficiency appears to prevent processing of IL-18 (Siegmund et al. 2001a, b). On the other hand, there are examples where caspase-1 processing of IL-18 is not required. For example, Fas ligand stimulation results in release of biologically active IL-18 in caspase-1-deficient murine macrophages (Tsutsui et al. 2000). Similar to IL-1β processing, proteinase-3 appears to activate processing to mature IL-18 (Sugawara et al. 2001).

Similar to IL-1α and IL-33, the IL-18 precursor is constitutively expressed in endothelial cells, keratinocytes, and intestinal epithelial cells throughout the gastrointestinal tract. Macrophages and dendritic cells are the primary sources for the release of active IL-18, whereas the inactive precursor remains in the intracellular compartment of mesenchymal cells. Also, similar to IL-1α and IL-33, the IL-18 precursor is released from dying cells and processed extracellularly, most likely by neutrophil proteases such as proteinase-3.

1.7.1.3 Signal Transduction by IL-18

As shown in Fig. 1.4a, IL-18 forms a signaling complex by binding to the IL-18 α-chain (IL-18Rα), which is the ligand-binding chain for mature IL-18; however, this binding is of low affinity. In cells that express the co-receptor, termed IL-18-receptor β-chain (IL-18Rβ), a high-affinity complex is formed, which then signals. The complex of IL-18 with the IL-18Rα and IL-18Rβ chains is similar to that formed by other members of the IL-1 family with the co-receptor, the IL-1R accessory chain IL-1RAcP. Following the formation of the heterodimer, the TIR domains approximate, and it appears that the cascade of sequential recruitment of MyD88, the four IRAKs, and TRAF-6 followed by the degradation of IκB and release of NFκB are nearly identical as that for IL-1 (Weber et al. 2010). There are differences between IL-1 and IL-18 signaling that remain unexplained. With few exceptions, IL-1α or IL-1βis active on cells in the low nanogram/ml range and often in the picogram/ml range. In contrast, IL-18 activation of cells expressing the two IL-18-receptor chains requires 10–20 ng/ml and sometime higher levels (Lee et al. 2004; Morel et al. 2001).

Although nearly all cells express IL-1RI, not all cells express IL-1RAcP. Similarly, most cells express IL-18Rα but not all cell express IL-18Rβ. IL-8Rβ is

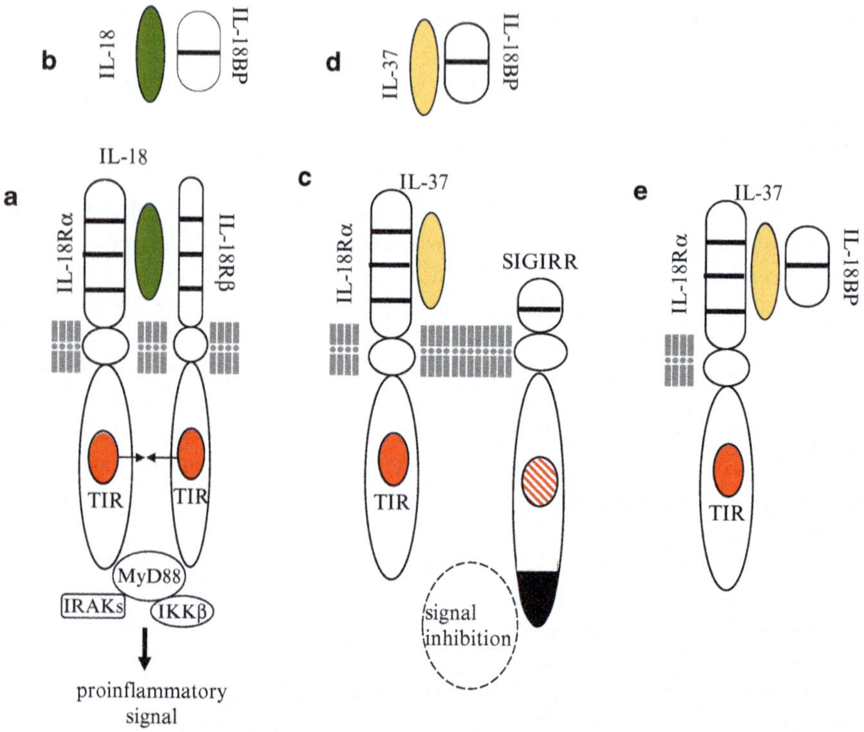

Fig. 1.4 IL-18 subfamily. (**a**) IL-18 binds to the IL-18Rα chain and recruits the co-receptor IL-18Rβ. The signaling cascade of the IL-18 receptor complex is nearly the same as that of IL-1α and IL-1β, resulting in the expression of pro-inflammatory genes. (**b**) The naturally occurring IL-18BP binds IL-18, thus neutralizing the activity of the cytokine. (**c**) IL-37 also binds to the IL-18Rα but with an affinity lower than that of IL-18 binding to the same receptor. Furthermore, the binding of IL-37 to IL-18Rα does not recruit the co-receptor, IL-18Rβ, and therefore there is no pro-inflammatory signal. The anti-inflammatory properties of IL-37 require *SIGIRR*, which may act as a "decoy" for MyD88. (**d**) IL-18BP also binds to IL-37, thus preventing binding of IL-37 to IL-18Rα. (**e**) IL-37 binds to IL-18BP, forming a complex, which then binds to IL-18Rα, enhancing the anti-inflammatory property of IL-18BP

expressed on T cells and dendritic cells but is not commonly expressed in mesenchymal cells. The best example is the A549 cell. This cell line, derived from a lung carcinoma epithelial cell, does not express IL-18Rβ (Kim et al. 2005), and there is no signal unless IL-12 is added to induce IL-18Rβ (Nakanishi et al. 2001b). In the absence of IL-18Rβ, IL-18 binds to IL-18Rα without a pro-inflammatory signal. In A549 cells transfected with IL-18Rβ, IL-18 induces IL-8 and a large number of genes. One of these genes is the former IL-2-induced gene termed NK4 (Dahl et al. 1992), now termed IL-32 (Kim et al. 2005). IL-32 is not a member of the IL-1 family but plays an important role in the regulation of cytokines such as IL-1β and TNF-α.

1.7.1.4 IL-18 as an Immunoregulatory Cytokine

Together with IL-12, IL-18 participates in the Th1 paradigm. This property of IL-18 is the result of its ability to induce IFN-γ with either IL-12 or IL-15. Without IL-12 or IL-15, IL-18 does not induce IFN-γ. IL-12 or IL-15 increases IL-18Rβ, which is essential for IL-18 signal transduction. Without IL-12 or IL-15, IL-18 plays a role in Th2 diseases (Nakanishi et al. 2001a). The importance of IL-18 as an immunoregulatory cytokine is derived from its prominent biological property of inducing IFN-γ from NK cells. Macrophage colony-stimulating factor (M-CSF) induces human blood monocytes to develop into a subset of macrophages; these cells express a membrane-bound form of IL-18 (Bellora et al. 2012). Membrane IL-18 is expressed in 30–40 % of M-CSF-primed macrophages. In contrast, monocytes, dendritic cells, and monocytes differentiated into M1 macrophages did not express membrane IL-18. Although the expression of membrane IL-18 is caspase-1 dependent (Bellora et al. 2012), LPS treatment was necessary for the release of membrane IL-18 (Bellora et al. 2012). A major immunoregulating role for IL-18 is on the NK cell. Upon shedding of membrane IL-18 into a soluble form, NK cells expressed CCR7 and produced high levels of IFN-γ. As expected, IFN-γ production was prevented by neutralization of IL-18. This mechanism may account for the role of IL-18 as a major IFN-γ-inducing factor from NK cells and the role of NK cells in the pathogenesis of autoimmune diseases.

The induction of IFN-γ by IL-18 has been studied with co-inducer IL-12. For example, mice injected with the combination of IL-18 plus IL-12 develop high levels of IFN-γ and die of hypoglycemia, intestinal inflammation, and inanition (Nakamura et al. 2000). In leptin-deficient mice, IL-18 plus IL-12 induce acute pancreatitis (Sennello et al. 2008). Several human autoimmune diseases are associated with elevated production of IFN-γ and IL-18. Diseases such as systemic lupus erythematosus, rheumatoid arthritis, type 1 diabetes, Crohn's disease, psoriasis, and graft-versus-host disease are thought to be mediated, in part, by IL-18.

1.7.1.5 Pro-inflammatory Properties of IL-18

IL-18 exhibits characteristics of other pro-inflammatory cytokines, such as increases in cell adhesion molecules, nitric oxide synthesis, and chemokine production. Blocking IL-18 activity reduces metastasis in a mouse model of melanoma, caused by a reduction in IL-18-induced expression of vascular cell adhesion molecule-1 (Vidal-Vanaclocha et al. 2000). A unique property of IL-18 is the induction of Fas ligand (FasL), which may account for the hepatic damage that takes place in macrophage activation syndrome (Mazodier et al. 2005; Tsutsui et al. 2000). The induction of fever, a well-studied property of IL-1α and IL-1β as well as IL-6, is not a property of IL-18. Injection of IL-18 into mice, rabbits, or humans does not produce fever (Gatti et al. 2002; Li et al. 2003). In contrast to IL-1 and TNF-α, IL-18 does not induce cyclooxygenase-2 and hence there is no production of prostaglandin E_2 (Lee et al. 2004; Reznikov et al. 2000). IL-18 has been administered to humans for the treatment of cancer to increase the activity and expansion of cytotoxic T cells. Not unexpectedly, and similar to several cytokines, the therapeutic focus on IL-18 has shifted from its use as an immune stimulant to inhibition of its activity (Dinarello 2007; Tak et al. 2006).

Because IL-18 can increase IFN-γ production, blocking IL-18 activity in autoimmune diseases is an attractive therapeutic target as anti-IL-12/23 reduces the severity of Crohn's disease as well as that of psoriasis. As discussed next, there appears to be a role for blocking IL-18 in Crohn's disease. However, there are several activities of IL-18 that are independent of IFN-γ. For example, IL-18 inhibits proteoglycan synthesis in chondrocytes (Joosten et al. 2000), and proteoglycan synthesis is essential for maintaining healthy cartilage. IL-18 also increases VCAM-1 expression in endothelial cells independently of IFN-γ. VCAM-1 plays a major role in multiple sclerosis and other autoimmune diseases as well as in the metastatic process (Carrascal et al. 2003).

1.7.1.6 IL-18, Hyperphagia, and the Metabolic Syndrome

Although there is no constitutive gene expression for IL-1β in freshly obtained human peripheral blood mononuclear cells (PBMC), the same cells express constitutive mRNA for IL-18 (Puren et al. 1999). In Western blot analysis from the same cells, the IL-18 precursor was present but not the IL-1β precursor. Similar observations were also made in mice (Puren et al. 1999). These findings suggest that IL-18 may act as regulator of homeostasis. Starting at 16 weeks of age, IL-18-deficient mice start to overeat, become obese, and exhibit lipid abnormalities; there is increased atherosclerosis, insulin resistance, and diabetes mellitus, reminiscent of the metabolic syndrome (Netea et al. 2006). IL-18Rα-deficient mice also develop a similar phenotype. The higher body weight is attributed to enhanced food intake, in which the IL-18-deficient mice begin to diverge from wild-type animals at a relatively early age, and to reach values 30–40 % higher than those of wild-type mice. Others have observed similar findings (Zorrilla et al. 2007). A striking finding was an increase of more than 100 % in the percent of adipose tissue in the IL-18-deficient animals, which was accompanied by fat deposition in the arterial walls. The insulin resistance in these mice is corrected by exogenous recombinant IL-18. Mice deficient in IL-18 respond normally to a challenge with exogenous leptin, suggesting that expression of the leptin receptor is unaffected. The unexpected and unique mechanism responsible for the higher food intake in the IL-18-deficient animals appears to be caused by a central nervous system loss of appetite control. IL-18 deficient-mice eat throughout the day whereas wild-type mice eat once, nocturnally.

1.7.1.7 IL-18 as a Protected Cytokine

As already stated, mice deficient in caspase-1 experience increased disease severity when subjected to dextran sulfate sodium (DSS)-induced colitis and that administration of exogenous IL-18 restored mucosal healing in these mice (Dupaul-Chicoine et al. 2010). In addition, mice deficient in NLRP3 were more susceptible to DSS colitis, which is thought to be caused by decreased IL-18 (Hirota et al. 2011). Mice deficient in NLRP6 are also more vulnerable to DSS (Chen et al. 2007; Elinav et al. 2011), and the susceptibility appears to be lack of sufficient IL-18. Thus, a growing number of studies support a protective role for IL-18. The fact that mice deficient in IL-18 develop a metabolic syndrome-like phenotype is consistent with a role for IL-18 in homeostasis. A study in age-related macular degeneration is also consistent with a protective role for IL-18. In that study, drusen, which is a mixture of complement-derived apolipoproteins and lipids, was shown to activate NLRP3 and induce the production of mature IL-1β and IL-18 (Doyle et al. 2012). In a mouse model of "wet" age-related macular degeneration, the disease was worse in mice deficient in NLRP3 but not in IL-1RI-deficient mice (Doyle et al. 2012). Therefore, IL-18 rather than IL-1α or IL-1β was protective and, upon administration of IL-IL-18, the disease severity improved. Taken together, there is a case for IL-18 being a protective rather than inflammatory cytokine.

1.7.1.8 IL-18-Binding Protein

The discovery of IL-18BP took place during the search for the soluble receptors for IL-18 (Novick et al. 1999). IL-18BP is a constitutively secreted protein with an exceptionally high affinity for IL-18 (400 pM) (Fig. 1.4b). Present in the serum of healthy humans at a 20-fold molar excess compared to IL-18 (Novick et al. 2001), IL-18BP may contribute to a default mechanism by which a Th1 response to foreign organisms is blunted to reduce triggering an autoimmune response to a routine infection. Although IL-18BP is readily secreted, it falls into the functional category of being a shed soluble receptor. As shown in Fig. 1.4b, IL-18BP contains only one IgG domain whereas the type II IL-1 receptor contains three domains. In this regard, the single IgG domain of IL-18BP is similar to SIGIRR, which also has a single IgG domain and also functions as a decoy receptor. The salient property of IL-18BP in immune responses is in downregulating Th1 responses by binding to IL-18 and thus reducing the induction of IFN-γ (Nakanishi et al. 2001a). Because IL-18 also affects Th2 responses, IL-18BP also has properties controlling a Th2 cytokine response (Nakanishi et al. 2001a). IL-18BP has a classic signal peptide and therefore is readily secreted. Serum levels in healthy subjects are in the range of 2,000–3,000 pg/ml compared to the levels of IL-18 in the same sera of 80–120 pg/ml. Moreover, IL-18BP binds IL-18 with an affinity of 3–5 nM (Novick et al. 1999), an affinity significantly higher than that of IL-18Rα. Because a single IL-18BP molecule binds a single IL-18 molecule, one can calculate bound versus free IL-18 in a mixture of both molecules (Novick et al. 2001).

If one examines immunologically mediated diseases where IFN-γ plays a pathological role such as Wegener's granulomatosis and systemic lupus erythematosus, one must consider the level of free IL-18 compared to IL-18 bound to IL-18BP. In fact, in these diseases both IL-18BP and IL-18 are high (Novick et al. 2009, 2011) but the level of IL-18BP is not sufficiently high enough to neutralize IL-18 and, therefore, the level of free IL-18 is higher than in healthy subjects. In macrophage activation syndrome where IFN-γ plays a pathological role, both IL-18BP and IL-18 are also high, but the clinical and hematological abnormalities correlate with elevated free IL-18 (Mazodier et al. 2005).

A unique property of IL-18BP is that the molecule also binds IL-37 (Bufler et al. 2002) and, in so doing, enhances the ability of IL-18BP to inhibit the induction of IFN-γ by IL-18. IL-37 binds to the IL-18Rα with a very low affinity, but in mice expressing human IL-37, a profound anti-inflammatory effect is observed (Nold et al. 2010), particularly of LPS-induced cytokines and dendritic cell maturation (Nold et al. 2010). Human IL-37-expressing mice are also resistant to colitis (McNamee et al. 2011). Thus, the anti-inflammatory property of IL-37 can be affected by the concentration of IL-18BP. As the concentration of IL-18BP increases and binds IL-37, there is the possibility that IL-37 becomes less available as an anti-inflammatory cytokine. Indeed, this has been observed in mice injected with IL-18BP. At low dosing of IL-18BP, there is reduced inflammation in a model of rheumatoid arthritis, but as the doing of IL-18BP increases, the anti-inflammatory properties of IL-18BP are lost (Banda et al. 2003).

IL-18BP is highly regulated at the level of gene expression and, unexpectedly, IFN-γ increases gene expression and synthesis of IL-18BP (Hurgin et al. 2002; Muhl et al. 2000). Therefore, IFN-γ driving an increase in the natural and potent inhibitor of IL-18 falls into the category of a negative feedback loop. The concept is supported by clinical data showing that patients being treated with IFN-α for hepatitis have elevated levels of IL-18BP (Kaser et al. 2002; Ludwiczek et al. 2002). IL-27, similar to IFN-γ, functions as both a pro- and an anti-inflammatory cytokine, and both may accomplish their roles as anti-inflammatory cytokines at the level of increased production of IL-18BP. In the skin, IL-27 also acts through a negative feedback loop for inflammation. IL-27 is acting, as is IFN-γ, by induction of IL-18BP gene expression and synthesis (Wittmann et al. 2012).

1.7.1.9 Viral IL-18BP

Natural neutralization of human IL-18 by IL-18BP takes place during a common viral infection. In molluscum contagiosum infection, characterized by raised but bland eruptions, there are large numbers of viral particles in the epithelial cells of the skin but histologically there are few inflammatory or immunologically active cells in or near the lesions. Clearly, the virus fails to elicit an inflammatory or immunological response. Amino acid similarity exists between human IL-18BP and a gene found in various members of the poxviruses; the greatest degree of homology is found to be expressed by the molluscum contagiosum gene (Xiang and Moss 2001). The ability of viral IL-18BP to reduce the activity of mammalian IL-18 likely explains the lack of inflammatory and immune cells in the infected tissues and provides further evidence for the natural ability of IL-18BP to interfere with IL-18 activity.

1.7.2 IL-37

1.7.2.1 IL-37

IL-37 was formerly termed IL-1F7. IL-37 lacks a signal peptide, has a caspase-1 site, but the secretion of IL-37 has not been documented with any certainty. It is likely, however, that similar to IL-1α and IL-33, with loss of membrane integrity upon cell death, the IL-37 precursor exits from the cell. The recombinant form of the IL-37 precursor suppresses LPS-induced IL-1β, IL-6, and TNF-α. However, this is observed primarily in macrophages that have been differentiated into the M1 phenotype by 5 days in the presence of GM-CSF. There are two consensus sequences (A-X-D) in the N-terminal domain of IL-37, IHD, and LED. A recombinant form of IL-37 with an N-terminus nine amino acids from the IHD site is active in suppressing LPS-induced TNF-α and IL-6. Whether this short form of recombinant IL-37 exists in nature is unclear.

1.7.2.2 IL-37 Reduces IL-1β- and LPS-Induced Inflammation In Vivo

A mouse homologue for human IL-37 has not been identified. Therefore, to define the in vivo functional role of IL-37, a strain of transgenic mice was generated (Nold et al. 2010). The full-length IL-37 cDNA was inserted into a vector using the standard CMV promoter for constitutive expression of the transgene in all cells. Both heterozygous and homozygous IL-37 transgenic mice (IL-37 tg) mice breed normally and exhibit no obvious phenotype. Despite the presence of the CMV promoter, the IL-37 transcript is not constitutively expressed in the tissues of the IL-37 transgenic mice. The failure to express IL-37 is likely caused by a functional instability sequence found in IL-37, which limits the half-life of IL-37 mRNA (Bufler et al. 2004). Nevertheless, upon stimulation with LPS or IL-1β, levels of IL-37 increase after 4–24 h. Once the transcript is present, the IL-37 precursor can be found in peripheral blood cells taken from the transgenic mice (Nold et al. 2003).

IL-37 transgenic mice are protected against LPS challenge compared to similarly challenged wild-type mice. IL-37 transgenic mice exhibit significantly less hypothermia, acidosis, hyperkalemia, hepatitis, and dehydration (Nold et al. 2010). In addition, circulating cytokines are significantly reduced as well as cytokines induced in whole blood cultures and in lung and spleen cell homogenates. In addition to LPS-induced cytokines, whole blood cultures from IL-37 transgenic mice produce significantly less IL-6 and TNF-α when stimulated by IL-1β or the combination of IL-12 plus IL-18. The anti-inflammatory activity of IL-37 was not limited to a reduction of the cytokines and chemokines of innate immunity. Dendritic cells isolated from the spleen of IL-37 transgenic mice upon LPS stimulation revealed a marked reduction (75 % and 89 %) in expression of CD86 and MHC II, respectively (Nold et al. 2010). The total numbers of dendritic cells, macrophages, natural killer cells, and $CD4^+$ T cells were similar in all strains and experimental conditions.

1.7.2.3 A Role for IL-37 During Experimental Colitis

IL-37 transgenic mice have been subjected to dextran sulfate sodium (DSS)-induced colitis. Despite the presence of a CMV promoter to drive expression of IL-37, mRNA transcripts were not present in colons in the resting state (McNamee et al. 2011). Expression was observed only upon disruption of the epithelial barrier, with a 6- to 7-fold increase on days 3 and 5 after continuous exposure to DSS. During the development of colitis, clinical disease scores were reduced by 50 % and histological indices of colitis were one-third less in IL-37 transgenic mice compared with wild-type counterparts. Reduced inflammation was associated with decreased leukocyte recruitment into the colonic lamina propria. In addition, release of IL-1β and TNF-α from ex vivo colonic explant tissue was decreased 5- and 13 fold, respectively, compared with wild-type mice, whereas IL-10 was increased 6 fold. However, IL-10 was not required for the anti-inflammatory effects of IL-7 because IL-10 receptor antibody blockade did not reverse IL-37-mediated protection. Mechanistically, IL-37 originating from hematopoietic cells was sufficient to exert anti-inflammatory effects because wild-type mice reconstituted with bone marrow from IL-37 transgenic mice were protected from colitis.

1.7.2.4 A Nuclear Role for IL-37

In stable transfectants of human IL-37 in RAW macrophages stimulated with LPS, levels of TNF-α, IL-1α, IL-6, and the chemokine MIP-2 were substantially reduced (72–98 %) compared with LPS-stimulated cells transfected with the empty plasmid (Sharma et al. 2008). Similar to IL-1α and IL-33, IL-37 translocates to the nucleus following stimulation (Sharma et al. 2008). In mouse RAW macrophages stably expressing IL-37, the mature carboxyl terminal was detected in the nucleus. Furthermore, a specific caspase-1 inhibitor markedly reduced nuclear entry of IL-37 (Sharma et al. 2008). The data demonstrate that IL-37 translocates to the nucleus after caspase-1 processing and may act as a transcriptional modulator reducing the production of LPS-stimulated pro-inflammatory cytokines, consistent with IL-37 being an anti-inflammatory member of the IL-1 family.

IL-37 was identified in a proteomics-based search for proteins that interacted with Smad3 (Grimsby et al. 2004). To test for a functional interaction of Smad3 with IL-37, IL-37 was transfected into A549 cells. IL-37 colocalized with phospho-Smad3 was found in perinuclear and cytosolic regions and a IL-37–Smad3 complex was observed (Nold et al. 2010). A specific inhibitor of Smad3 reversed the inhibition of IL-6 and IL-1β expression in RAW cells stably transfected with IL-37. In stable human macrophage lines expressing IL-37, depletion of Smad3 by lentiviral introduction of short hairpin (sh)RNA that inhibits Smad3 expression prevented the ability of IL-37 to reduce IL-1β- or LPS-induced production of IL-8, IL-6, and TNF. These in vitro findings were confirmed in vivo. IL-37 transgenic mice were pretreated intranasally with a Smad3-specific small interfering (si)RNA and then challenged with intranasal LPS. The reduction of lung cytokines in IL-37 transgenic mice was reversed in transgenic mice with a lung knockdown of Smad3 (Nold et al. 2010).

1.7.2.5 Role of IL-18Rα for IL-37

From the first reports on IL-37, it was observed that the recombinant forms bound to the IL-18Rα (Kumar et al. 2002; Pan et al. 2001). The binding of IL-37 to IL-18Rα has also been observed in cells from IL-37 transgenic mice using immunofluoresence, immunoprecipitation, and FRET analysis (Nold et al. 2011). IL-37 specifically binds to the third domain of the IL-18Rα (Bufler et al. 2002). Despite these studies showing binding of IL-37 to the IL-18Rα-chain, IL-37 does not act as a classical receptor antagonist for IL-18 in that the ability of recombinant IL-18 to induce IFN-γ is not inhibited by high concentrations of IL-37. However, in the presence of low concentrations of IL-18BP, recombinant IL-37 modestly reduces IL-18-induced IFN-γ (Bufler et al. 2002). The concept that IL-37 binds to the IL-18Rα and reduces cytokine production is supported, in part, with the finding that embryonic fibroblasts from mice deficient in IL-18Rα produce tenfold more IL-6 in response to IL1β than do wild-type embryonic fibroblasts (Nold-Petry et al. 2009). In addition, silencing of IL-18Rα in primary human blood monocytes results in a fourfold increase in the secretion of LPS-induced IL-1β, IL-6, IFN-γ, and CD40 ligand (Nold-Petry et al. 2009). Thus, the seemingly paradoxical hyperresponsive state in cells deficient in IL-18Rα supports the concept that IL-18Rα participates in both pro- and anti-inflammatory responses and that the endogenous ligand IL-37 engages the IL-18Rα to deliver an inhibitory signal.

1.7.2.6 Role of SIGIRR in the Anti-inflammatory Property of IL-37

The mechanism by which an IL-1β or an LPS signal is suppressed by IL-37 requires an understanding of SIGIRR. The IL-1RAcP serves as the co-receptor for IL-1α, IL-1β, IL-36α, IL-36β, IL-36γ, and IL-33, each a pro-inflammatory cytokine. However, in the IL-1 family of receptors, three co-receptors contain unusually long intracellular domains: these are SIGIRR, a variant of the IL-1RAcP termed IL-1RAcPb and receptors termed "three Ig IL-1 related receptor" (TIGIRR). There are two TIGIRRs, TIGIRR-1 and TIGIRR-2. IL-1RAcPb is expressed only in the brain and TIGIRR has limited expression. However, SIGIRR is expressed in most cells. The TIR domain of the three co-receptors is also different from that of other members of the IL-1 co-receptor family in that the TIR domain contains an amino acid sequence different from that of wild-type TIR (Smith et al. 2009). IL-1RAcPb forms the expected complex with IL-1 and IL-1R1 but does not recruit MyD88 or phosphorylate IRAK4 (Smith et al. 2009). Therefore, most IL-1 signaling is arrested. However, as some genes are increased in response to the formation of the IL-1RI/IL–1RAcPb complex, partial IL-1 signaling must take place. Nevertheless, IL-1RAcPb functions as an inhibitory receptor chain but only in the brain. Mice deficient in IL-1RAcPb exhibit a normal inflammatory response in the periphery but greater neurodegeneration in the brain. As such, IL-1RAcPb could play a role in chronic inflammatory responses in the brain by "buffering" IL-1-mediated neurodegeneration.

Similar to IL-1RAcPb, SIGIRR contains the same amino acid differences in its TIR domain, termed TIRb. Compared to wild-type TIR, TIRb likely has reduced binding of MyD88 (Smith et al. 2009). In addition to an altered TIR domain, SIGIRR has a carboxyl extension of 140 amino acids. Carboxyl extensions are also present in IL-1RAcPb as well as the two TIGIRRs. TIGIRR-2, which is associated with an X-linked cognitive deficiency, is apparently independent of IL-1 function. Little is known whether these C-terminal segments contribute to the inhibitory properties of these receptors. Nevertheless, it seems likely that the alternative sequence in the TIRb domain of SIGIRR may act as a partial decoy for MyD88. MyD88 is phosphorylated upon TLR4 as well as IL-1β and IL-18 signaling and results in downstream phosphorylation of IRAK-4. In cells expressing SIGIRR and activated by IL-37 binding to the IL-18Rα, the signal from either IL-1 or LPS initiates phosphorylation of MyD88. However, the decoy effect by the mutated TIRb of SIGIRR reduces the degree of phosphorylation of MyD88 and thus the phosphorylation of IRAK-4. The reduction, however, is partial. Indeed, the suppression by IL-37 added to blood macrophages is in the range of 20 % to 50 % and is unlike the total loss of activation by a deficiency in MyD88.

Upon binding to the IL-18Rα, the IL-37 precursor may activate SIGIRR and provide a negative signal. An inhibitory signal from IL-37 is enhanced by a low concentration of IL-18BP (Bufler et al. 2002). As shown in Fig. 1.4c,d, IL-18BP binds IL-37 (Bufler et al. 2002) and likely presents the complex of the cytokine with the binding protein to the IL-18Rα. Because A549 cells express SIGIRR, it is likely that the inhibitory signal from IL-37 activates SIGIRR or alternatively IL-37 recruits SIGIRR as the accessory chain. The inhibitory signal of SIGIRR is established in several mouse models of inflammation in which SIGIRR deficient mice exhibit more inflammation compared to wild-type control mice (Garlanda et al. 2009). In differentiated human blood M1 macrophages, recombinant IL-37 suppresses LPS-induced TNF-α and IL-6 production by 50–70 %. A source of IL-18BP in this culture may be fetal calf serum. During the differentiation of macrophages into the M1 subset by GM-CSF, it is possible that SIGIRR expression increases whereas the level of IL-18Rβ decreases (see Fig. 1.4a). In the absence of IL-18Rβ, a proinflammatory complex is not formed with IL-18Rα and thus IL-37 binding to IL-18Rα may recruit or activate SIGIRR. Thus, expression of SIGIRR and the absence of IL-18Rβ would best explain the inhibitory properties of recombinant IL-37 reducing the response to LPS induction of IL-6 and TNFα.

1.8 IL-36 Subfamily

The IL-1 family members IL-1F5, IL-1F6, IL-1F8, and IL-1F9 are now termed IL-36Ra, IL-36α, IL-36β, and IL-36γ, respectively (Dinarello et al. 2010). Each member of the IL-36 subfamily binds to the IL-1Rpr2, now termed IL-36R (Towne et al. 2004). As IL-38 also binds to the IL-36R (van de Veerdonk et al. 2012), IL-38 is included in the IL-36R subfamily. The IL-36 subfamily is closely related to the IL-1 subfamily because, similar to IL-1α and IL-1β and IL-33, IL-36R forms a signaling complex with IL-1RAcP (Ali et al. 2007; Towne et al. 2004). Thus, of the 11 members of the entire IL-1 family, 6 members use IL-1RAcP as the co-receptor for signal transduction.

1.8.1 IL-36

IL-36R is the ligand-binding chain and therefore is comparable to the ligand-binding IL-1R1 and IL-18Rα. However, two members of the IL-36 subfamily bind to IL-36R but do not signal: the IL-36 receptor antagonist (IL-36Ra) and IL-38. As such, these function as receptor antagonists (Towne et al. 2011; van de Veerdonk et al. 2012). An unusual property of IL-38 is that low concentrations (1–10 ng/ml) are able to reduce the activity of endogenous IL-36 (van de Veerdonk et al. 2012), whereas in the case of IL-1Ra, higher concentrations are required to prevent the activation of endogenous IL-1α or IL-1β.

None of the members of the IL-36 subfamily has a signal peptide indicating the generation of the N-terminus and secretion via the Golgi. In addition, each member of the IL-36 subfamily has an unusually short propiece compared to those of IL-1α, IL-1β, and IL-33 (see Fig. 1.1). Similar to IL-1β and IL-18, there is no true caspase-1 cleavage site for generation of an N-terminus in the IL-36 subfamily. It remains unknown which specific proteases generate the various N-termini of the IL-36 subfamily; nevertheless, each member has a unique N-terminus with a different levels of biological activity (Towne et al. 2011). What determines the N-terminus with optimal biological activity in the IL-36 subfamily? Each member of the IL-36 subfamily contains the IL-1 family consensus sequence of A-X-D. The aspartic acid is not the recognition amino acid for caspase-1 or caspase-3, but rather participates in the stabilization of the first beta-sheet of the three-dimensional structure that characterizes the entire IL-1 family.

The "A" of the consensus sequence is for any aliphatic amino acid, for example, leucine or isoleucine. Nine amino acids before the "A" of the consensus sequence is the N-terminal site, which results in the cytokine with the greatest activity (Towne et al. 2011). For example, the biological activity of IL-36-γ increases by a factor of 1,000 when the N-terminus is at the site nine amino acids before the consensus sequence and, in the case of the IL-36β, there is a 10,000 fold increase (Towne et al. 2011). The site nine amino acids forward from the consensus sequence is not only the N-terminus for the agonist members of the IL-36 family but also the IL-36 receptor antagonist (IL-36Ra) (Towne et al. 2011), which increases from a low level of blocking of IL-36 family ligands to a high degree of blockade. It is unclear what specific protease cleaves at this site as the amino acid is different for each member of the IL-36 subfamily. Moreover, the site for the N-terminus of the IL-36Ra (valine) is but one amino acid from the N-terminal precursor methionine, and yet IL-36Ra with an N-terminus at the valine site is 10,000 fold more potent than the IL-36 precursor. It is also an unusual situation that proteases that usually are inflammatory in processing members of the IL-1 family in that case of IL-36Ra generate an anti-inflammatory molecule.

1.8.1.1 IL-36α, -β, and -γ, Proinflammatory Members of the IL-36 Subfamily

IL-36α was highly expressed in the murine model of glomerulonephritis (Ichii et al. 2010), where the cytokine was localized to the kidney epithelium, and also in CD3 T cells surrounding the tubules. IL-36α is also found in the kidneys of the MRL/lupus, nephritic syndrome, and streptozotocin-induced diabetic models (Ichii et al. 2010). IL-36γ increases IL-8, CXCL3, and the Th17 chemokine CCL20 in human lung fibroblasts (Chustz et al. 2011) and thus may account for acute neutrophilic lung inflammation. In addition to $CD4^+$ T cells, human articular chondrocytes and synovial fibroblasts express the IL-36R (Magne et al. 2006). In chondrocytes, there is also constitutive gene expression of IL-36β. Following stimulation with IL-1β or TNF-α, levels of the IL-36β precursor rise intracellularly but the cytokine is not secreted. Although IL-36β levels were detected in the joint fluids of patients with rheumatoid arthritis as well as in serum samples, there was no correlation with disease severity (Magne et al. 2006). It is likely that IL-36 ligands are functional only when released from dying cells and can be processed extracellularly by enzymes present in inflammatory conditions such as the joint of patients with rheumatoid arthritis. It unclear to what extent IL-36β plays a role in joint disease, although constitutive expression in primary chondrocytes may indicate a role for the cytokine in osteoarthritis (Magne et al. 2006).

High levels of this cytokine are found in mouse embryonic tissues rich in epithelial cells (Debets et al. 2001). In humans, IL-36α is observed in keratinocytes, not fibroblasts, and is thought to contribute to the inflammation of psoriasis. Upon forming the heterodimer with IL-36R and IL-1RAcP, IL-36α activates NF-κB similar to that of IL-1β (Towne et al. 2004). In addition to NF-κB activation, IL-36α also activates MAPK, JNK, and ERK1/2 (Towne et al. 2004). In the mouse, bone marrow-derived dendritic cells and $CD4^+$ T cells express IL-36 receptors in health. In a comparison with IL-1 as a stimulant, the three IL-36 ligands are more active in inducing IL-1β, IL-6, IL-12, TNF-α, and IL-23 (Vigne et al. 2011). In addition, IL-36 ligands induced the production of IFN-γ, IL-4, and IL-17 from $CD4^+$ T cells. Not unexpectedly, cytokines induced by IL-36 ligands were prevented by 100- to 1,000-fold excess IL-36Ra (Vigne et al. 2011).

1.8.1.2 IL-36 in Psoriasis

Several studies implicate IL-36 ligands in the pathogenesis of psoriasis (Blumberg et al. 2007; Johnston et al. 2011; Muhr et al. 2011). IL-36γ is highly expressed in keratinocytes from healthy human skin and increases upon stimulation with TLR polyI:C (Lian et al. 2012). Furthermore, polyI:C induced caspase-3, which resulted in cell death and the release of IL-36γ. Unexpectedly, stimulation of IL-36γ gene expression was dependent on caspase-1 (Lian et al. 2012). The caspase-1 dependency may be caused by IL-18 as this cytokine is constitutively present in keratinocytes as is IL-1α. With the release of IL-36γ by polyI:C and the subsequent death of the cell, IL-36γ falls into the category of being an alarmin in the skin, particularly because of infection (Lian et al. 2012). There is also a role for IL-36 in the production of IL-17: studies suggest that each of the IL-36 ligands is expressed in the skin and dependent on IL-22 (Carrier et al. 2011). Furthermore, the expression of IL-36 ligands in the psoriatic skin correlated with IL-17 (Carrier et al. 2011). Similar to other models in the IL-1 family, auto- and co-induction accounts for a role in a pathological process.

Overexpression of IL-36 in mice results in inflammatory skin lesions that resemble psoriasis in humans, as reviewed by (Towne and Sims 2012).Similarly, mice deficient in endogenous IL-36Ra exhibit a severe lesion similar to that of humans with pustular psoriasis. The role of IL-36 in pustular psoriasis may include IL-1α, as humans with pustular psoriasis respond to an antibody that neutralizes IL-1α. Both IL-36 and IL-1α are found in the keratinocytes in healthy skin.

1.8.1.3 A Role for IL-36R in Metabolic Regulation

Obesity is characterized by chronic low-grade inflammation originating from expanding adipose tissue. Human adipogenic tissue levels of IL-36α are primarily present in adipose tissue-resident macrophages and are induced by inflammation; however, IL-36β is absent (van Asseldonk et al. 2010). IL-36α, but not IL-36β, reduces adipocyte differentiation, as shown by a significant decrease in PPAR-γ gene expression. Both IL-36α and IL-36β induce inflammatory gene expression in mature adipocytes (van Asseldonk et al. 2010). Therefore, IL-36α and IL-36β are present in adipose tissue and are involved in the regulation of adipose tissue gene expression. Importantly, IL-36α inhibits PPAR-γ expression, which may lead to reduced adipocyte differentiation, suggesting metabolic effects of this cytokine.

Although IL-36Ra is known to occupy the IL-36R and act as a receptor antagonist, earlier studies revealed that IL-36Ra inhibited the induction of IL-1β by LPS. The role of IL-36Ra was also examined in the brain. IL-36Ra injected into the rat brain induced IL-4 and also in glial cells in vitro (Costelloe et al. 2008). Moreover, the reduction in LPS-induced IL-1β was not observed in cells deficient in IL-4 and also not observed in cells deficient in SIGIRR (Costelloe et al. 2008). However, these unique properties of IL-36Ra were not observed in peripheral monocytes or dendritic cells but only in the brain.

1.8.1.4 Role of IL-36 in Human Disease

The importance of any cytokine in human biology can be found in mutations that result in a profound clinical picture. In IL-1, a mutation in the natural IL-1Ra results in severe systemic inflammation with erosive bone lesions, sterile meningitis, and death; the syndrome is called deficiency of IL-1Ra (Aksentijevich et al. 2009; Reddy et al. 2009). In case of the IL-36 family, persons with a mutation in the naturally occurring IL-36Ra suffer with a severe form of pustular psoriasis (Marrakchi et al. 2011; Onoufriadis et al. 2011). These human studies are consistent with the data from transgenic mice overexpressing IL-36α in the skin and the ability of IL-36Ra to suppress the severity of the inflammation (Blumberg et al. 2007). In mice overexpressing IL-36α in the skin, crossing the mice to generate a strain of mice heterozygous for natural IL-36Ra knockout results in worsening of the skin lesions(Blumberg et al. 2007).

1.9 IL-38

IL-8 is the name for the IL-1 family member 10. During the nomenclature revision of the IL-1 family in 2010 (Dinarello et al. 2010), the term IL-38 was assigned to IL-1F10 without any known biological function. Since then, IL-38 has been shown to bind to the IL-36 receptor (formerly IL-1Rrp2) (van de Veerdonk et al. 2012). To find the receptor for IL-38, each member of the IL-1 receptor family was immobilized, recombinant IL-38 precursor containing 152 amino acids was added, and binding was assessed using an antibody to the ligand. IL-38 bound only to the IL-36 receptor, as did IL-36Ra (van de Veerdonk et al. 2012). To assess the biological function of IL-38, heat-killed *Candida albicans* was used to stimulate memory T-lymphocyte cytokine production in freshly obtained human peripheral blood mononuclear cells from healthy subjects. The addition of recombinant IL-38 inhibited the production of T-cell cytokines IL-22 and IL-17. The dose–response suppression of IL-38 as well as that of IL-36Ra of *Candida*-induced IL-22 and IL-17 was not that of the classic IL-1 receptor antagonist, because low concentrations were optimal for inhibiting IL-22 production (van de Veerdonk et al. 2012). These data provide evidence that IL-38 binds to the IL-36R, as does IL-36Ra, and that IL-38 and IL-36Ra have similar biological effects on immune cells by engaging the IL-36 receptor.

1.10 The Influence of the IL-1 Family on Th17 Responses

The IL-1 family plays a significant role in IFN-γ production, which is essential for the defense against intracellular pathogens. On the other hand, Th2 cells are characterized by the production of IL-4 and are important in the host defense against parasitic infections. For more than one decade, the dichotomy between Th1 and Th2 has been the focus of studies on differentiation of CD4+ T lymphocytes. More recently, Th17 helper cells have been described and are characterized by their production of IL-17. IL-17 plays a major role in neutrophil recruitment and host defense against extracellular bacteria and fungi. Th17 cells produce a distinct cytokine profile, namely, IL-17A, IL-17F, IL-21, and IL-22. The cytokines produced by Th17 cells, in addition to activating neutrophils, are also crucial for nonimmune cells, for example, induction of defensins by IL-22 in epithelial cells and keratinocytes, which are part of mucosal and skin defenses. It has become apparent that Th17 responses are associated with chronic inflammation and autoimmune diseases such as multiple sclerosis, type 1 diabetes, Crohn's disease, and psoriasis. Furthermore, Th17 responses are fundamental for host defense against many microorganisms, although they also contribute to the inflammation during infection.

Although IL-4 and IL-12 were the first cytokines described as influencing Th cell differentiation, cytokines of the IL-1 family also influence cytokine differentiation. IL-18 was initially described as an IFN-γ-inducing factor because of its strong stimulatory effect on Th1/IFN-γ responses (Nakanishi et al. 2001b). It is now known that IL-18 is, in fact, a crucial cytokine directing the development of Th1 cells, and one role of IL-12 is the induction of the expression of IL-18 receptors. In contrast, binding of IL-33, another member of the IL-1 family, to its ST2 receptor plays a role in inducing Th2 responses (Schmitz et al. 2005), and it thus appeared as if distinct members of the IL-1 family of cytokines directed Th1 versus Th2 differentiation. Considering these effects of IL-18 and IL-33, it came as no surprise that IL-1, the most well known member of the family, participates in the function of Th cells.

It has been known for more than 30 years that IL-1 enhances T-cell activation and recognition of antigen: one of the early names of IL-1 was lymphocyte activation factor (LAF). The specificity of this response was, however, not known. Although initially only IL-23, IL-6, IL-21, and TGF-β were suggested to play a role in the development of Th17 responses in mice, there is no dearth of data that a more complex picture exists. Thus, IL-1β, IL-6, and TGF-β have been reported to induce the development of Th17 cells, whereas IL-23 has been reported to be important for the maintenance of Th17 cells. The combination of IL-23 and IL-1β induces the development of human Th17 cells expressing IL-17A, IL-17F, IL-22, IL-26, the chemokine CCL20, and transcription factor RORγt (Wilson et al. 2007). Interestingly, these cells also released IFN-γ, displaying a phenotype common to both Th17 and Th1 cells (Wilson et al. 2007). The strong capacity of IL-1 to induce Th17 differentiation has been also linked to its well-known capacity to induce the release of prostaglandins, as reviewed by Dinarello (2011b). PGE_2 induced by COX-2 is a stimulator of Th17 induction and inhibitors of cyclooxyugenase decrease IL-17 production (Chizzolini et al. 2008). On the other hand, engagement of the aryl-hydrocarbon receptor, a pathway demonstrated to be crucial for the generation of Th17 cells, has been shown to strongly induce IL-1β (Henley et al. 2004). In addition to inducing IL-17 production from the Th17 subset of lymphocytes, IL-1β is required for the production of IL-17 by NKT cells (Moreira et al. 2011) and of IL-22 from NK cells (Hughes et al. 2010).

Thus, cytokines of the IL-1 family have an important role in the differentiation of the Th subsets, with IL-1β strongly inducing Th17 responses, IL-18 being crucial for the generation of Th1 cells, and IL-33 being important in Th2 responses. Interestingly, reciprocal regulation has been demonstrated between the various Th subsets, with cytokines released by Th2 cells inhibiting Th1 responses, whereas IFN-γ release from Th1 cells impairs both Th2 and Th17 responses.

References

Agostini L, Martinon F, Burns K, McDermott MF, Hawkins PN, Tschopp J (2004) NALP3 forms an IL-1beta processing inflammasome with increased activity in Muckle–Wells autoinflammatory disorder. Immunity 20:319–325

Aksentijevich I, Masters SL, Ferguson PJ, Dancey P, Frenkel J, van Royen-Kerkhoff A, Laxer R, Tedgard U, Cowen EW, Pham TH, Booty M, Estes JD, Sandler NG, Plass N, Stone DL, Turner ML, Hill S, Butman JA, Schneider R, Babyn P, El-Shanti HI, Pope E, Barron K, Bing X, Laurence A, Lee CC, Chapelle D, Clarke GI, Ohson K, Nicholson M, Gadina M, Yang B, Korman BD, Gregersen PK, van Hagen PM, Hak AE, Huizing M, Rahman P, Douek DC, Remmers EF, Kastner DL, Goldbach-Mansky R (2009) An autoinflammatory disease with deficiency of the interleukin-1-receptor antagonist. N Engl J Med 360:2426–2437

Ali S, Huber M, Kollewe C, Bischoff SC, Falk W, Martin MU (2007) IL-1 receptor accessory protein is essential for IL-33-induced activation of T lymphocytes and mast cells. Proc Natl Acad Sci USA 104:18660–18665

Ali S, Mohs A, Thomas M, Klare J, Ross R, Schmitz ML, Martin MU (2011) The dual function cytokine IL-33 interacts with the transcription factor NF-kappaB to dampen NF-kappaB-stimulated gene transcription. J Immunol 187:1609–1616

Al-Shukaili A, Al-Kaabi J, Hassan B (2008) A comparative study of interleukin-1beta production and p2x7 expression after ATP stimulation by peripheral blood mononuclear cells isolated from rheumatoid arthritis patients and normal healthy controls. Inflammation 31:84–90

Andrei C, Dazzi C, Lotti L, Torrisi MR, Chimini G, Rubartelli A (1999) The secretory route of the leaderless protein interleukin 1beta involves exocytosis of endolysosome-related vesicles. Mol Biol Cell 10:1463–1475

Andrei C, Margiocco P, Poggi A, Lotti LV, Torrisi MR, Rubartelli A (2004) Phospholipases C and A2 control lysosome-mediated IL-1 beta secretion: implications for inflammatory processes. Proc Natl Acad Sci USA 101:9745–9750

Arend WP (2002) The balance between IL-1 and IL-1Ra in disease. Cytokine Growth Factor Rev 13:323–340

Bae S, Kang T, Hong J, Lee S, Choi J, Jhun H, Kwak A, Hong K, Kim E, Jo S, Kim S (2012) Contradictory functions (activation/termination) of neutrophil proteinase 3 enzyme (PR3) in interleukin-33 biological activity. J Biol Chem 287:8205–8213

Banda NK, Vondracek A, Kraus D, Dinarello CA, Kim SH, Bendele A, Senaldi G, Arend WP (2003) Mechanisms of inhibition of collagen-induced arthritis by murine IL-18 binding protein. J Immunol 170:2100–2105

Bellora F, Castriconi R, Doni A, Cantoni C, Moretta L, Mantovani A, Moretta A, Bottino C (2012) M-CSF induces the expression of a membrane-bound form of IL-18 in a subset of human monocytes differentiating in vitro toward macrophages. Eur J Immunol 42:1618–1626

Berda-Haddad Y, Robert S, Salers P, Zekraoui L, Farnarier C, Dinarello CA, Dignat-George F, Kaplanski G (2011) Sterile inflammation of endothelial cell-derived apoptotic bodies is mediated by interleukin-1alpha. Proc Natl Acad Sci USA 108:20684–20689

Bergers G, Reikerstorfer A, Braselmann S, Graninger P, Busslinger M (1994) Alternative promoter usage of the Fos-responsive gene *Fit-1* generates mRNA isoforms coding for either secreted or membrane-bound proteins related to the IL-1 receptor. EMBO J 13:1176–1188

Blumberg H, Dinh H, Trueblood ES, Pretorius J, Kugler D, Weng N, Kanaly ST, Towne JE, Willis CR, Kuechle MK, Sims JE, Peschon JJ (2007) Opposing activities of two novel members of the IL-1 ligand family regulate skin inflammation. J Exp Med 204:2603–2614

Boni-Schnetzler M, Thorne J, Parnaud G, Marselli L, Ehses JA, Kerr-Conte J, Pattou F, Halban PA, Weir GC, Donath MY (2008) Increased interleukin (IL)-1beta messenger ribonucleic acid expression in beta-cells of individuals with type 2 diabetes and regulation of IL-1beta in human islets by glucose and autostimulation. J Clin Endocrinol Metab 93:4065–4074

Brough D, Rothwell NJ (2007) Caspase-1-dependent processing of pro-interleukin-1beta is cytosolic and precedes cell death. J Cell Sci 120:772–781

Bufler P, Azam T, Gamboni-Robertson F, Reznikov LL, Kumar S, Dinarello CA, Kim SH (2002) A complex of the IL-1 homologue IL-1F7b and IL-18-binding protein reduces IL-18 activity. Proc Natl Acad Sci USA 99:13723–13728

Bufler P, Gamboni-Robertson F, Azam T, Kim SH, Dinarello CA (2004) Interleukin-1 homologues IL-1F7b and IL-18 contain functional mRNA instability elements within the coding region responsive to lipopolysaccharide. Biochem J 381:503–510

Bulek K, Swaidani S, Qin J, Lu Y, Gulen MF, Herjan T, Min B, Kastelein RA, Aronica M, Kosz-Vnenchak M, Li X (2009) The essential role of single Ig IL-1 receptor-related molecule/Toll IL-1R8 in regulation of Th2 immune response. J Immunol 182:2601–2609

Carmi Y, Voronov E, Dotan S, Lahat N, Rahat MA, Fogel M, Huszar M, White MR, Dinarello CA, Apte RN (2009) The role of macrophage-derived IL-1 in induction and maintenance of angiogenesis. J Immunol 183:4705–4714

Carrascal MT, Mendoza L, Valcarcel M, Salado C, Egilegor E, Telleria N, Vidal-Vanaclocha F, Dinarello CA (2003) Interleukin-18 binding protein reduces b16 melanoma hepatic metastasis by neutralizing adhesiveness and growth factors of sinusoidal endothelium. Cancer Res 63: 491–497

Carrier Y, Ma HL, Ramon HE, Napierata L, Small C, O'Toole M, Young DA, Fouser LA, Nickerson-Nutter C, Collins M, Dunussi-Joannopoulos K, Medley QG (2011) Inter-regulation of Th17 cytokines and the IL-36 cytokines in vitro and in vivo: implications in psoriasis pathogenesis. J Invest Dermatol 131:2428–2437

Carriere V, Roussel L, Ortega N, Lacorre DA, Americh L, Aguilar L, Bouche G, Girard JP (2007) IL-33, the IL-1-like cytokine ligand for ST2 receptor, is a chromatin-associated nuclear factor in vivo. Proc Natl Acad Sci USA 104:282–287

Cayrol C, Girard JP (2009) The IL-1-like cytokine IL-33 is inactivated after maturation by caspase-1. Proc Natl Acad Sci USA 106:9021–9026

Chackerian AA, Oldham ER, Murphy EE, Schmitz J, Pflanz S, Kastelein RA (2007) IL-1 receptor accessory protein and ST2 comprise the IL-33 receptor complex. J Immunol 179:2551–2555

Chen CJ, Kono H, Golenbock D, Reed G, Akira S, Rock KL (2007) Identification of a key pathway required for the sterile inflammatory response triggered by dying cells. Nat Med 13:851–856

Chessell IP, Hatcher JP, Bountra C, Michel AD, Hughes JP, Green P, Egerton J, Murfin M, Richardson J, Peck WL, Grahames CB, Casula MA, Yiangou Y, Birch R, Anand P, Buell GN (2005) Disruption of the P2X7 purinoceptor gene abolishes chronic inflammatory and neuropathic pain. Pain 114:386–396

Chizzolini C, Chicheportiche R, Alvarez M, de Rham C, Roux-Lombard P, Ferrari-Lacraz S, Dayer JM (2008) Prostaglandin E2 synergistically with interleukin-23 favors human Th17 expansion. Blood 112:3696–3703

Chustz RT, Nagarkar DR, Poposki JA, Favoreto S Jr, Avila PC, Schleimer RP, Kato A (2011) Regulation and function of the IL-1 family cytokine IL-1F9 in human bronchial epithelial cells. Am J Respir Cell Mol Biol 45:145–153

Coeshott C, Ohnemus C, Pilyavskaya A, Ross S, Wieczorek M, Kroona H, Leimer AH, Cheronis J (1999) Converting enzyme-independent release of TNFα and IL-1β from a stimulated human monocytic cell line in the presence of activated neutrophils or purified proteinase-3. Proc Natl Acad Sci USA 96:6261–6266

Cohen I, Rider P, Carmi Y, Braiman A, Dotan S, White MR, Voronov E, Martin MU, Dinarello CA, Apte RN (2010) Differential release of chromatin-bound IL-1alpha discriminates between necrotic and apoptotic cell death by the ability to induce sterile inflammation. Proc Natl Acad Sci USA 107:2574–2579

Cooney R, Baker J, Brain O, Danis B, Pichulik T, Allan P, Ferguson DJ, Campbell BJ, Jewell D, Simmons A (2010) NOD2 stimulation induces autophagy in dendritic cells influencing bacterial handling and antigen presentation. Nat Med 16:90–97

Costelloe C, Watson M, Murphy A, McQuillan K, Loscher C, Armstrong ME, Garlanda C, Mantovani A, O'Neill LA, Mills KH, Lynch MA (2008) IL-1F5 mediates anti-inflammatory activity in the brain through induction of IL-4 following interaction with SIGIRR/TIR8. J Neurochem 105:1960–1969

Crisan TO, Plantinga TS, van de Veerdonk FL, Farcas MF, Stoffels M, Kullberg BJ, van der Meer JW, Joosten LA, Netea MG (2011) Inflammasome-independent modulation of cytokine response by autophagy in human cells. PLoS One 6:e18666

Croker BA, Lewis RS, Babon JJ, Mintern JD, Jenne DE, Metcalf D, Zhang JG, Cengia LH, O'Donnell JA, Roberts AW (2011) Neutrophils require SHP1 to regulate IL-1beta production and prevent inflammatory skin disease. J Immunol 186:1131–1139

Dahl CA, Schall RP, He HL, Cairns JS (1992) Identification of a novel gene expressed in activated natural killer cells and T cells. J Immunol 148:597–603

Debets R, Timans JC, Homey B, Zurawski S, Sana TR, Lo S, Wagner J, Edwards G, Clifford T, Menon S, Bazan JF, Kastelein RA (2001) Two novel IL-1 family members, IL-1 delta and IL-1 epsilon, function as an antagonist and agonist of NF-kappa B activation through the orphan IL-1 receptor-related protein 2. J Immunol 167:1440–1446

Derks R, Beaman K (2004) Regeneration and tolerance factor modulates the effect of adenosine triphosphate-induced interleukin 1 beta secretion in human macrophages. Hum Immunol 65:676–682

Dinarello CA (2007) Interleukin-18 and the pathogenesis of inflammatory diseases. Semin Nephrol 27:98–114

Dinarello CA (2011a) A clinical perspective of IL-1beta as the gatekeeper of inflammation. Eur J Immunol 41:1203–1217

Dinarello CA (2011b) Interleukin-1 in the pathogenesis and treatment of inflammatory diseases. Blood 117:3720–3732

Dinarello CA, Ikejima T, Warner SJ, Orencole SF, Lonnemann G, Cannon JG, Libby P (1987) Interleukin 1 induces interleukin 1. I. Induction of circulating interleukin 1 in rabbits *in vivo* and in human mononuclear cells *in vitro*. J Immunol 139:1902–1910

Dinarello C, Arend W, Sims J, Smith D, Blumberg H, O'Neill L, Goldbach-Mansky R, Pizarro T, Hoffman H, Bufler P, Nold M, Ghezzi P, Mantovani A, Garlanda C, Boraschi D, Rubartelli A, Netea M, van der Meer J, Joosten L, Mandrup-Poulsen T, Donath M, Lewis E, Pfeilschifter J, Martin M, Kracht M, Muehl H, Novick D, Lukic M, Conti B, Solinger A, Peyman K, van de Veerdonk F, Gabel C (2010) IL-1 family nomenclature. Nat Immunol 11:973

Dinarello CA, Simon A, van der Meer JW (2012) Treating inflammation by blocking interleukin-1 in a broad spectrum of diseases. Nat Rev Drug Discov 11:633–652

Doyle SL, Campbell M, Ozaki E, Salomon RG, Mori A, Kenna PF, Farrar GJ, Kiang AS, Humphries MM, Lavelle EC, O'Neill LA, Hollyfield JG, Humphries P (2012) NLRP3 has a protective role in age-related macular degeneration through the induction of IL-18 by drusen components. Nat Med 18:791–798

Drenth JP, van der Meer JW, Kushner I (1996) Unstimulated peripheral blood mononuclear cells from patients with the hyper-IgD syndrome produce cytokines capable of potent induction of C-reactive protein and serum amyloid A in Hep3B cells. J Immunol 157:400–404

Dupaul-Chicoine J, Yeretssian G, Doiron K, Bergstrom KS, McIntire CR, LeBlanc PM, Meunier C, Turbide C, Gros P, Beauchemin N, Vallance BA, Saleh M (2010) Control of intestinal homeostasis, colitis, and colitis-associated colorectal cancer by the inflammatory caspases. Immunity 32:367–378

Elinav E, Strowig T, Kau AL, Henao-Mejia J, Thaiss CA, Booth CJ, Peaper DR, Bertin J, Eisenbarth SC, Gordon JI, Flavell RA (2011) NLRP6 inflammasome regulates colonic microbial ecology and risk for colitis. Cell 145:745–757

Elssner A, Duncan M, Gavrilin M, Wewers MD (2004) A novel P2X7 receptor activator, the human cathelicidin-derived peptide LL37, induces IL-1 beta processing and release. J Immunol 172:4987–4994

Faggioni R, Fantuzzi G, Fuller J, Dinarello CA, Feingold KR, Grunfeld C (1998) IL-1b mediates leptin induction during inflammation. Am J Physiol 274:R204–R208

Fantuzzi G, Zheng H, Faggioni R, Benigni F, Ghezzi P, Sipe JD, Shaw AR, Dinarello CA (1996) Effect of endotoxin in IL-1b-deficient mice. J Immunol 157:291–296

Fantuzzi G, Ku G, Harding MW, Livingston DL, Sipe JD, Kuida K, Flavell RA, Dinarello CA (1997a) Response to local inflammation of IL-1b converting enzyme-deficient mice. J Immunol 158:1818–1824

Fantuzzi G, Sacco S, Ghezzi P, Dinarello CA (1997b) Physiological and cytokine responses in interleukin-1b-deficient mice after zymosan-induced inflammation. Am J Physiol 273: R400–R406

Fantuzzi G, Reed DA, Dinarello CA (1999) IL-12-induced IFNγ is dependent on caspase-1 processing of the IL-18 precursor. J Clin Invest 104:761–767

Fettelschoss A, Kistowska M, LeibundGut-Landmann S, Beer HD, Johansen P, Senti G, Contassot E, Bachmann MF, French LE, Oxenius A, Kundig TM (2011) Inflammasome activation and IL-1beta target IL-1alpha for secretion as opposed to surface expression. Proc Natl Acad Sci USA 108:18055–18060

Gadina M, Jefferies CA (2007) IL-33: a sheep in wolf's clothing? Sci STKE 2007:pe31

Gardella S, Andrei C, Costigliolo S, Olcese L, Zocchi MR, Rubartelli A (2000) Secretion of bioactive interleukin-1beta by dendritic cells is modulated by interaction with antigen specific T cells. Blood 95:3809–3815

Garlanda C, Anders HJ, Mantovani A (2009) TIR8/SIGIRR: an IL-1R/TLR family member with regulatory functions in inflammation and T cell polarization. Trends Immunol 30:439–446

Gatti S, Beck J, Fantuzzi G, Bartfai T, Dinarello CA (2002) Effect of interleukin-18 on mouse core body temperature. Am J Physiol Regul Integr Comp Physiol 282:R702–R709

Gattorno M, Tassi S, Carta S, Delfino L, Ferlito F, Pelagatti MA, D'Osualdo A, Buoncompagni A, Alpigiani MG, Alessio M, Martini A, Rubartelli A (2007) Pattern of interleukin-1beta secretion in response to lipopolysaccharide and ATP before and after interleukin-1 blockade in patients with CIAS1 mutations. Arthritis Rheum 56:3138–3148

Gawaz M, Brand K, Dickfeld T, Pogatsa-Murray G, Page S, Bogner C, Koch W, Schomig A, Neumann F (2000) Platelets induce alterations of chemotactic and adhesive properties of endothelial cells mediated through an interleukin-1-dependent mechanism. Implications for atherogenesis. Atherosclerosis 148:75–85

Gay NJ, Keith FJ (1991) *Drosophila Toll* and IL-1 receptor. Nature (Lond) 351:355–356

Goldbach-Mansky R, Dailey NJ, Canna SW, Gelabert A, Jones J, Rubin BI, Kim HJ, Brewer C, Zalewski C, Wiggs E, Hill S, Turner ML, Karp BI, Aksentijevich I, Pucino F, Penzak SR, Haverkamp MH, Stein L, Adams BS, Moore TL, Fuhlbrigge RC, Shaham B, Jarvis JN, O'Neil K, Vehe RK, Beitz LO, Gardner G, Hannan WP, Warren RW, Horn W, Cole JL, Paul SM, Hawkins PN, Pham TH, Snyder C, Wesley RA, Hoffmann SC, Holland SM, Butman JA, Kastner DL (2006) Neonatal-onset multisystem inflammatory disease responsive to interleukin-1beta inhibition. N Engl J Med 355:581–592

Greten FR, Arkan MC, Bollrath J, Hsu LC, Goode J, Miething C, Goktuna SI, Neuenhahn M, Fierer J, Paxian S, Van Rooijen N, Xu Y, O'Cain T, Jaffee BB, Busch DH, Duyster J, Schmid RM, Eckmann L, Karin M (2007) NF-kappaB is a negative regulator of IL-1beta secretion as revealed by genetic and pharmacological inhibition of IKKbeta. Cell 130:918–931

Grimsby S, Jaensson H, Dubrovska A, Lomnytska M, Hellman U, Souchelnytskyi S (2004) Proteomics-based identification of proteins interacting with Smad3: SREBP-2 forms a complex with Smad3 and inhibits its transcriptional activity. FEBS Lett 577:93–100

Gross O, Yazdi AS, Thomas CJ, Masin M, Heinz LX, Guarda G, Quadroni M, Drexler SK, Tschopp J (2012) Inflammasome activators induce interleukin-1alpha secretion via distinct pathways with differential requirement for the protease function of caspase-1. Immunity 36: 388–400

Hacham M, Argov S, White RM, Segal S, Apte RN (2002) Different patterns of interleukin-1alpha and interleukin-1beta expression in organs of normal young and old mice. Eur Cytokine Netw 13:55–65

Hampe J, Franke A, Rosenstiel P, Till A, Teuber M, Huse K, Albrecht M, Mayr G, De La Vega FM, Briggs J, Gunther S, Prescott NJ, Onnie CM, Hasler R, Sipos B, Folsch UR, Lengauer T, Platzer M, Mathew CG, Krawczak M, Schreiber S (2007) A genome-wide association scan of nonsynonymous SNPs identifies a susceptibility variant for Crohn disease in ATG16L1. Nat Genet 39:207–211

Harris J, Hartman M, Roche C, Zeng SG, O'Shea A, Sharp FA, Lambe EM, Creagh EM, Golenbock DT, Tschopp J, Kornfeld H, Fitzgerald KA, Lavelle EC (2011) Autophagy controls IL-1beta secretion by targeting pro-IL-1beta for degradation. J Biol Chem 286:9587–9597

Hawrylowicz CM, Santoro SA, Platt FM, Unanue ER (1989) Activated platelets express IL-1 activity. J Immunol 143:4015–4018
Hawrylowicz CM, Howells GL, Feldmann M (1991) Platelet-derived interleukin 1 induces human endothelial adhesion molecule expression and cytokine production. J Exp Med 174:785–790
Heguy A, Baldari CT, Macchia G, Telford JL, Melli M (1992) Amino acids conserved in interleukin-1 receptors (IL-1Rs) and the Drosophila Toll protein are essential for IL-1R signal transduction. J Biol Chem 267:2605–2609
Henckaerts L, Cleynen I, Brinar M, John JM, Van Steen K, Rutgeerts P, Vermeire S (2011) Genetic variation in the autophagy gene ULK1 and risk of Crohn's disease. Inflamm Bowel Dis 17: 1392–1397
Henley DV, Bellone CJ, Williams DA, Ruh TS, Ruh MF (2004) Aryl hydrocarbon receptor-mediated posttranscriptional regulation of IL-1beta. Arch Biochem Biophys 422:42–51
Hewinson J, Moore SF, Glover C, Watts AG, MacKenzie AB (2008) A key role for redox signaling in rapid P2X7 receptor-induced IL-1 beta processing in human monocytes. J Immunol 180:8410–8420
Hirota SA, Ng J, Lueng A, Khajah M, Parhar K, Li Y, Lam V, Potentier MS, Ng K, Bawa M, McCafferty DM, Rioux KP, Ghosh S, Xavier RJ, Colgan SP, Tschopp J, Muruve D, MacDonald JA, Beck PL (2011) NLRP3 inflammasome plays a key role in the regulation of intestinal homeostasis. Inflamm Bowel Dis 17:1359–1372
Hoffman HM, Wanderer AA (2011) Inflammasome and IL-1beta-mediated disorders. Curr Allergy Asthma Rep 10:229–235
Hoffman HM, Mueller JL, Broide DH, Wanderer AA, Kolodner RD (2001) Mutation of a new gene encoding a putative pyrin-like protein causes familial cold autoinflammatory syndrome and Muckle–Wells syndrome. Nat Genet 29:301–305
Homer CR, Richmond AL, Rebert NA, Achkar JP, McDonald C (2010) ATG16L1 and NOD2 interact in an autophagy-dependent antibacterial pathway implicated in Crohn's disease pathogenesis. Gastroenterology 139:1630–1641, 1641 e1631–e1632
Horai R, Asano M, Sudo K, Kanuka H, Suzuki M, Nishihara M, Takahashi M, Iwakura Y (1998) Production of mice deficient in genes for interleukin (IL)-1a, IL- 1b, IL-1a/b, and IL-1 receptor antagonist shows that IL-1b is crucial in turpentine-induced fever development and glucocorticoid secretion. J Exp Med 187:1463–1475
Hughes T, Becknell B, Freud AG, McClory S, Briercheck E, Yu J, Mao C, Giovenzana C, Nuovo G, Wei L, Zhang X, Gavrilin MA, Wewers MD, Caligiuri MA (2010) Interleukin-1beta selectively expands and sustains interleukin-22+ immature human natural killer cells in secondary lymphoid tissue. Immunity 32:803–814
Hurgin V, Novick D, Rubinstein M (2002) The promoter of IL-18 binding protein: activation by an IFN-gamma-induced complex of IFN regulatory factor 1 and CCAAT/enhancer binding protein beta. Proc Natl Acad Sci USA 99:16957–16962
Hurgin V, Novick D, Werman A, Dinarello CA, Rubinstein M (2007) Antiviral and immunoregulatory activities of IFN-gamma depend on constitutively expressed IL-1alpha. Proc Natl Acad Sci USA 104:5044–5049
Ichii O, Otsuka S, Sasaki N, Yabuki A, Ohta H, Takiguchi M, Hashimoto Y, Endoh D, Kon Y (2010) Local overexpression of interleukin-1 family, member 6 relates to the development of tubulointerstitial lesions. Lab Invest 90:459–475
Johnston A, Xing X, Guzman AM, Riblett M, Loyd CM, Ward NL, Wohn C, Prens EP, Wang F, Maier LE, Kang S, Voorhees JJ, Elder JT, Gudjonsson JE (2011) IL-1F5, -F6, -F8, and -F9: a novel IL-1 family signaling system that is active in psoriasis and promotes keratinocyte antimicrobial peptide expression. J Immunol 186:2613–2622
Joosten LA, van De Loo FA, Lubberts E, Helsen MM, Netea MG, van Der Meer JW, Dinarello CA, van Den Berg WB (2000) An IFN-gamma-independent proinflammatory role of IL-18 in murine streptococcal cell wall arthritis. J Immunol 165:6553–6558
Joosten LA, Netea MG, Fantuzzi G, Koenders MI, Helsen MM, Sparrer H, Pham CT, van der Meer JW, Dinarello CA, van den Berg WB (2009) Inflammatory arthritis in caspase 1 gene-deficient

mice: contribution of proteinase 3 to caspase 1-independent production of bioactive interleukin-1beta. Arthritis Rheum 60:3651–3662

Joosten LA, Netea MG, Mylona E, Koenders MI, Malireddi RK, Oosting M, Stienstra R, van de Veerdonk FL, Stalenhoef AF, Giamarellos-Bourboulis EJ, Kanneganti TD, van der Meer JW (2010) Engagement of fatty acids with Toll-like receptor 2 drives interleukin-1beta production via the ASC/caspase 1 pathway in monosodium urate monohydrate crystal-induced gouty arthritis. Arthritis Rheum 62:3237–3248

Kahlenberg JM, Lundberg KC, Kertesy SB, Qu Y, Dubyak GR (2005) Potentiation of caspase-1 activation by the P2X7 receptor is dependent on TLR signals and requires NF-kappaB-driven protein synthesis. J Immunol 175:7611–7622

Kamari Y, Werman-Venkert R, Shaish A, Werman A, Harari A, Gonen A, Voronov E, Grosskopf I, Sharabi Y, Grossman E, Iwakura Y, Dinarello CA, Apte RN, Harats D (2007) Differential role and tissue specificity of interleukin-1alpha gene expression in atherogenesis and lipid metabolism. Atherosclerosis 195:31–38

Kanneganti TD, Lamkanfi M, Kim YG, Chen G, Park JH, Franchi L, Vandenabeele P, Nunez G (2007) Pannexin-1-mediated recognition of bacterial molecules activates the cryopyrin inflammasome independent of Toll-like receptor signaling. Immunity 26:433–443

Kaplanski G, Porat R, Aiura K, Erban JK, Gelfand JA, Dinarello CA (1993) Activated platelets induce endothelial secretion of interleukin-8 *in vitro* via an interleukin-1-mediated event. Blood 81:2492–2495

Kaplanski G, Farnarier C, Kaplanski S, Porat R, Shapiro L, Bongrand P, Dinarello CA (1994) Interleukin-1 induces interleukin-8 from endothelial cells by a juxacrine mechanism. Blood 84:4242–4248

Kaser A, Novick D, Rubinstein M, Siegmund B, Enrich B, Koch RO, Vogel W, Kim SH, Dinarello CA, Tilg H (2002) Interferon-alpha induces interleukin-18 binding protein in chronic hepatitis C patients. Clin Exp Immunol 129:332–338

Kaspar RL, Gehrke L (1994) Peripheral blood mononuclear cells stimulated with C5a or lipopolysaccharide to synthesize equivalent levels of IL-1b mRNA show unequal IL-1b protein accumulation but similar polyribosome profiles. J Immunol 153:277–286

Kawaguchi Y, Nishimagi E, Tochimoto A, Kawamoto M, Katsumata Y, Soejima M, Kanno T, Kamatani N, Hara M (2006) Intracellular IL-1alpha-binding proteins contribute to biological functions of endogenous IL-1alpha in systemic sclerosis fibroblasts. Proc Natl Acad Sci USA 103:14501–14506

Kenneth NS, Younger JM, Hughes ED, Marcotte D, Barker PA, Saunders TL, Duckett CS (2012) An inactivating caspase 11 passenger mutation originating from the 129 murine strain in mice targeted for c-IAP1. Biochem J 443:355–359

Kim SH, Han SY, Azam T, Yoon DY, Dinarello CA (2005) Interleukin-32: a cytokine and inducer of TNF-alpha. Immunity 22:131–142

Kim YH, Yang TY, Park CS, Ahn SH, Son BK, Kim JH, Lim DH, Jang TY (2012) Anti-IL-33 antibody has a therapeutic effect in a murine model of allergic rhinitis. Allergy 67:183–190

Klionsky DJ (2007) Autophagy: from phenomenology to molecular understanding in less than a decade. Nat Rev Mol Cell Biol 8:931–937

Krelin Y, Voronov E, Dotan S, Elkabets M, Reich E, Fogel M, Huszar M, Iwakura Y, Segal S, Dinarello CA, Apte RN (2007) Interleukin-1beta-driven inflammation promotes the development and invasiveness of chemical carcinogen-induced tumors. Cancer Res 67:1062–1071

Kuballa P, Huett A, Rioux JD, Daly MJ, Xavier RJ (2008) Impaired autophagy of an intracellular pathogen induced by a Crohn's disease associated ATG16L1 variant. PLoS One 3:e3391

Kuida K, Lippke JA, Ku G, Harding MW, Livingston DJ, Su MS-S, Flavell RA (1995) Altered cytokine export and apoptosis in mice deficient in interleukin-1b converting enzyme. Science 267:2000–2003

Kumar S, Hanning CR, Brigham-Burke MR, Rieman DJ, Lehr R, Khandekar S, Kirkpatrick RB, Scott GF, Lee JC, Lynch FJ, Gao W, Gambotto A, Lotze MT (2002) Interleukin-1F7B (IL-1H4/IL-1F7) is processed by caspase-1 and mature IL-1F7B binds to the IL-18 receptor but does not induce IFN-gamma production. Cytokine 18:61–71

Kurt-Jones EA, Beller DI, Mizel SB, Unanue ER (1985) Identification of a membrane-associated interleukin-1 in macrophages. Proc Natl Acad Sci USA 82:1204–1208

Lapaquette P, Glasser AL, Huett A, Xavier RJ, Darfeuille-Michaud A (2010) Crohn's disease-associated adherent-invasive *E. coli* are selectively favoured by impaired autophagy to replicate intracellularly. Cell Microbiol 12:99–113

Lee JK, Kim SH, Lewis EC, Azam T, Reznikov LL, Dinarello CA (2004) Differences in signaling pathways by IL-1beta and IL-18. Proc Natl Acad Sci USA 101:8815–8820

Lee KH, Park SS, Kim I, Kim JH, Ra EK, Yoon SS, Hong YC, Park S, Kim BK (2007) P2X7 receptor polymorphism and clinical outcomes in HLA-matched sibling allogeneic hematopoietic stem cell transplantation. Haematologica 92:651–657

Lefrancais E, Roga S, Gautier V, Gonzalez-de-Peredo A, Monsarrat B, Girard JP, Cayrol C (2012) IL-33 is processed into mature bioactive forms by neutrophil elastase and cathepsin G. Proc Natl Acad Sci USA 109:1673–1678

Lemaitre B, Nicolas E, Michaut L, Reichhart JM, Hoffmann JA (1996) The dorsoventral regulatory gene cassette spatzle/Toll/cactus controls the potent antifungal response in *Drosophila* adults. Cell 86:973–983

Leung BP, Xu D, Culshaw S, McInnes IB, Liew FY (2004) A novel therapy of murine collagen-induced arthritis with soluble T1/ST2. J Immunol 173:145–150

Levandowski CB, Mailloux CM, Ferrara TM, Gowan K, Ben S, Jin Y, McFann KK, Holland PJ, Fain PR, Dinarello CA, Spritz RA (2013) NLRP1 haplotypes associated with vitiligo and autoimmunity increase interleukin-1beta processing via the NLRP1 inflammasome. Proc Natl Acad Sci USA 110(8):2952–2956

Li S, Goorha S, Ballou LR, Blatteis CM (2003) Intracerebroventricular interleukin-6, macrophage inflammatory protein-1 beta and IL-18: pyrogenic and PGE(2)-mediated? Brain Res 992: 76–84

Lian LH, Milora KA, Manupipatpong KK, Jensen LE (2012) The double-stranded RNA analogue polyinosinic-polycytidylic acid induces keratinocyte pyroptosis and release of IL-36gamma. J Invest Dermatol 132(5):1346–1353

Lonnemann G, Engler-Blum G, Müller GA, Koch KM, Dinarello CA (1995) Cytokines in human renal interstitial fibrosis. II. Intrinsic Interleukin (IL)-1 synthesis and IL-1-dependent production of IL-6 and IL-8 by cultured kidney fibroblasts. Kidney Int 47:845–854

Louten J, Rankin AL, Li Y, Murphy EE, Beaumont M, Moon C, Bourne P, McClanahan TK, Pflanz S, de Waal MR (2011) Endogenous IL-33 enhances Th2 cytokine production and T-cell responses during allergic airway inflammation. Int Immunol 23:307–315

Ludwiczek O, Kaser A, Novick D, Dinarello CA, Rubinstein M, Vogel W, Tilg H (2002) Plasma levels of interleukin-18 and interleukin-18 binding protein are elevated in patients with chronic liver disease. J Clin Immunol 22:331–337

Luheshi NM, Kovacs KJ, Lopez-Castejon G, Brough D, Denes A (2011) Interleukin-1alpha expression precedes IL-1beta after ischemic brain injury and is localised to areas of focal neuronal loss and penumbral tissues. J Neuroinflammation 8:186

Maedler K, Sergeev P, Ris F, Oberholzer J, Joller-Jemelka HI, Spinas GA, Kaiser N, Halban PA, Donath MY (2002) Glucose-induced beta cell production of IL-1beta contributes to glucotoxicity in human pancreatic islets. J Clin Invest 110:851–860

Magne D, Palmer G, Barton JL, Mezin F, Talabot-Ayer D, Bas S, Duffy T, Noger M, Guerne PA, Nicklin MJ, Gabay C (2006) The new IL-1 family member IL-1F8 stimulates production of inflammatory mediators by synovial fibroblasts and articular chondrocytes. Arthritis Res Ther 8:R80

Maier JAM, Voulalas P, Roeder D, Maciag T (1990) Extension of the life span of human endothelial cells by an interleukin-1α antisense oligomer. Science 249:1570–1574

Marrakchi S, Guigue P, Renshaw BR, Puel A, Pei XY, Fraitag S, Zribi J, Bal E, Cluzeau C, Chrabieh M, Towne JE, Douangpanya J, Pons C, Mansour S, Serre V, Makni H, Mahfoudh N, Fakhfakh F, Bodemer C, Feingold J, Hadj-Rabia S, Favre M, Genin E, Sahbatou M, Munnich A, Casanova JL, Sims JE, Turki H, Bachelez H, Smahi A (2011) Interleukin-36-receptor antagonist deficiency and generalized pustular psoriasis. N Engl J Med 365:620–628

Martinon F, Mayor A, Tschopp J (2009) The inflammasomes: guardians of the body. Annu Rev Immunol 27:229–265
Masamune A, Watanabe T, Kikuta K, Satoh K, Kanno A, Shimosegawa T (2010) Nuclear expression of interleukin-33 in pancreatic stellate cells. Am J Physiol Gastrointest Liver Physiol 299:G821–G832
Matsuba-Kitamura S, Yoshimoto T, Yasuda K, Futatsugi-Yumikura S, Taki Y, Muto T, Ikeda T, Mimura O, Nakanishi K (2010) Contribution of IL-33 to induction and augmentation of experimental allergic conjunctivitis. Int Immunol 22:479–489
Mazodier K, Marin V, Novick D, Farnarier C, Robitail S, Schleinitz N, Veit V, Paul P, Rubinstein M, Dinarello CA, Harle JR, Kaplanski G (2005) Severe imbalance of IL-18/IL-18BP in patients with secondary hemophagocytic syndrome. Blood 106:3483–3489
McNamee EN, Masterson JC, Jedlicka P, McManus M, Grenz A, Collins CB, Nold MF, Nold-Petry C, Bufler P, Dinarello CA, Rivera-Nieves J (2011) Interleukin 37 expression protects mice from colitis. Proc Natl Acad Sci USA 108:16711–16716
Miller AC, Schattenberg DG, Malkinson AM, Ross D (1994) Decreased content of the IL1 alpha processing enzyme calpain in murine bone marrow-derived macrophages after treatment with the benzene metabolite hydroquinone. Toxicol Lett 74:177–184
Miller AM, Xu D, Asquith DL, Denby L, Li Y, Sattar N, Baker AH, McInnes IB, Liew FY (2008) IL-33 reduces the development of atherosclerosis. J Exp Med 205:339–346
Moreira AP, Cavassani KA, Ismailoglu UB, Hullinger R, Dunleavy MP, Knight DA, Kunkel SL, Uematsu S, Akira S, Hogaboam CM (2011) The protective role of TLR6 in a mouse model of asthma is mediated by IL-23 and IL-17A. J Clin Invest 121:4420–4432
Morel JC, Park CC, Woods JM, Koch AE (2001) A novel role for interleukin-18 in adhesion molecule induction through NFkappa B and phosphatidylinositol (PI) 3-kinase-dependent signal transduction pathways. J Biol Chem 276:37069–37075
Moussion C, Ortega N, Girard JP (2008) The IL-1-like cytokine IL-33 is constitutively expressed in the nucleus of endothelial cells and epithelial cells in vivo: a novel "alarmin"? PLoS One 3:e3331
Muhl H, Kampfer H, Bosmann M, Frank S, Radeke H, Pfeilschifter J (2000) Interferon-gamma mediates gene expression of IL-18 binding protein in nonleukocytic cells. Biochem Biophys Res Commun 267:960–963
Muhr P, Zeitvogel J, Heitland I, Werfel T, Wittmann M (2011) Expression of interleukin (IL)-1 family members upon stimulation with IL-17 differs in keratinocytes derived from patients with psoriasis and healthy donors. Br J Dermatol 165:189–193
Nakahira K, Haspel JA, Rathinam VA, Lee SJ, Dolinay T, Lam HC, Englert JA, Rabinovitch M, Cernadas M, Kim HP, Fitzgerald KA, Ryter SW, Choi AM (2011) Autophagy proteins regulate innate immune responses by inhibiting the release of mitochondrial DNA mediated by the NALP3 inflammasome. Nat Immunol 12:222–230
Nakamura S, Otani T, Ijiri Y, Motoda R, Kurimoto M, Orita K (2000) IFN-gamma-dependent and -independent mechanisms in adverse effects caused by concomitant administration of IL-18 and IL-12. J Immunol 164:3330–3336
Nakanishi K, Yoshimoto T, Tsutsui H, Okamura H (2001a) Interleukin-18 is a unique cytokine that stimulates both Th1 and Th2 responses depending on its cytokine milieu. Cytokine Growth Factor Rev 12:53–72
Nakanishi K, Yoshimoto T, Tsutsui H, Okamura H (2001b) Interleukin-18 regulates both Th1 and Th2 responses. Annu Rev Immunol 19:423–474
Netea MG, Joosten LA, Lewis E, Jensen DR, Voshol PJ, Kullberg BJ, Tack CJ, van Krieken H, Kim SH, Stalenhoef AF, van de Loo FA, Verschueren I, Pulawa L, Akira S, Eckel RH, Dinarello CA, van den Berg W, van der Meer JW (2006) Deficiency of interleukin-18 in mice leads to hyperphagia, obesity and insulin resistance. Nat Med 12:650–656
Netea MG, Nold-Petry CA, Nold MF, Joosten LA, Opitz B, van der Meer JH, van de Veerdonk FL, Ferwerda G, Heinhuis B, Devesa I, Funk CJ, Mason RJ, Kullberg BJ, Rubartelli A, van der Meer JW, Dinarello CA (2009) Differential requirement for the activation of the inflammasome for processing and release of IL-1beta in monocytes and macrophages. Blood 113:2324–2335

Nold M, Hauser IA, Hofler S, Goede A, Eberhardt W, Ditting T, Geiger H, Pfeilschifter J, Muhl H (2003) IL-18BPa:Fc cooperates with immunosuppressive drugs in human whole blood. Biochem Pharmacol 66:505–510

Nold MF, Nold-Petry CA, Zepp JA, Palmer BE, Bufler P, Dinarello CA (2010) IL-37 is a fundamental inhibitor of innate immunity. Nat Immunol 11:1014–1022

Nold MF, Nold-Petry CA, Zepp JA, Lo C, Garlanda C, Mantovani A, Bufler P, Dinarello CA (2011) Interleukin 37 exerts its anti-inflammatory functions by associating with IL-18R alpha and SIGIRR. Cytokine 56:12 (Abstract)

Nold-Petry CA, Nold MF, Nielsen JW, Bustamante A, Zepp JA, Storm KA, Hong JW, Kim SH, Dinarello CA (2009) Increased cytokine production in interleukin-18 receptor alpha-deficient cells is associated with dysregulation of suppressors of cytokine signaling. J Biol Chem 284:25900–25911

Novick D, Kim S-H, Fantuzzi G, Reznikov L, Dinarello CA, Rubinstein M (1999) Interleukin-18 binding protein: a novel modulator of the Th1 cytokine response. Immunity 10:127–136

Novick D, Schwartsburd B, Pinkus R, Suissa D, Belzer I, Sthoeger Z, Keane WF, Chvatchko Y, Kim SH, Fantuzzi G, Dinarello CA, Rubinstein M (2001) A novel IL-18BP ELISA shows elevated serum IL-18BP in sepsis and extensive decrease of free IL-18. Cytokine 14:334–342

Novick D, Elbirt D, Dinarello CA, Rubinstein M, Sthoeger ZM (2009) Interleukin-18 binding protein in the sera of patients with Wegener's granulomatosis. J Clin Immunol 29:38–45

Novick D, Elbirt D, Miller G, Dinarello CA, Rubinstein M, Sthoeger ZM (2011) High circulating levels of free interleukin-18 in patients with active SLE in the presence of elevated levels of interleukin-18 binding protein. J Autoimmun 34:121–126

Numanami H, Koyama S, Sato E, Haniuda M, Nelson DK, Hoyt JC, Freels JL, Habib MP, Robbins RA (2003) Serine protease inhibitors modulate chemotactic cytokine production by human lung fibroblasts in vitro. Am J Physiol Lung Cell Mol Physiol 284:L882–L890

Okamoto M, Liu W, Luo Y, Tanaka A, Cai X, Norris DA, Dinarello CA, Fujita M (2009) Constitutively active inflammasome in human melanoma cells mediating autoinflammation via caspase-1 processing and secretion of interleukin-1beta. J Biol Chem 285:6477–6488

Okamura H, Nagata K, Komatsu T, Tanimoto T, Nukata Y, Tanabe F, Akita K, Torigoe K, Okura T, Fukuda S, Kurimoto M (1995) A novel costimulatory factor for gamma interferon induction found in the livers of mice causes endotoxic shock. Infect Immun 63:3966–3972

O'Neill LA (2008) The interleukin-1 receptor/toll-like receptor superfamily: 10 years of progress. Immunol Rev 226:10–18

Onoufriadis A, Simpson MA, Pink AE, Di Meglio P, Smith CH, Pullabhatla V, Knight J, Spain SL, Nestle FO, Burden AD, Capon F, Trembath RC, Barker JN (2011) Mutations in IL36RN/IL1F5 are associated with the severe episodic inflammatory skin disease known as generalized pustular psoriasis. Am J Hum Genet 89:432–437

Pan G, Risser P, Mao W, Baldwin DT, Zhong AW, Filvaroff E, Yansura D, Lewis L, Eigenbrot C, Henzel WJ, Vandlen R (2001) IL-1H, an interleukin 1-related protein that binds IL-18 receptor/IL-1Rrp. Cytokine 13:1–7

Pascual V, Allantaz F, Arce E, Punaro M, Banchereau J (2005) Role of interleukin-1 (IL-1) in the pathogenesis of systemic onset juvenile idiopathic arthritis and clinical response to IL-1 blockade. J Exp Med 201:1479–1486

Pelegrin P, Surprenant A (2006) Pannexin-1 mediates large pore formation and interleukin-1beta release by the ATP-gated P2X7 receptor. EMBO J 25:5071–5082

Perregaux DG, McNiff P, Laliberte R, Conklyn M, Gabel CA (2000) ATP acts as an agonist to promote stimulus-induced secretion of IL-1 beta and IL-18 in human blood. J Immunol 165: 4615–4623

Plantinga TS, Crisan TO, Oosting M, van de Veerdonk FL, de Jong DJ, Philpott DJ, van der Meer JW, Girardin SE, Joosten LA, Netea MG (2011) Crohn's disease-associated ATG16L1 polymorphism modulates pro-inflammatory cytokine responses selectively upon activation of NOD2. Gut 60:1229–1235

Puren AJ, Fantuzzi G, Dinarello CA (1999) Gene expression, synthesis and secretion of IL-1b and IL-18 are differentially regulated in human blood mononuclear cells and mouse spleen cells. Proc Natl Acad Sci USA 96:2256–2261

Qu Y, Franchi L, Nunez G, Dubyak GR (2007) Nonclassical IL-1 beta secretion stimulated by P2X7 receptors is dependent on inflammasome activation and correlated with exosome release in murine macrophages. J Immunol 179:1913–1925

Reddy S, Jia S, Geoffrey R, Lorier R, Suchi M, Broeckel U, Hessner MJ, Verbsky J (2009) An autoinflammatory disease due to homozygous deletion of the IL1RN locus. N Engl J Med 360:2438–2444

Reznikov LL, Kim SH, Westcott JY, Frishman J, Fantuzzi G, Novick D, Rubinstein M, Dinarello CA (2000) IL-18 binding protein increases spontaneous and IL-1-induced prostaglandin production via inhibition of IFN-gamma. Proc Natl Acad Sci U S A 97:2174–2179

Rider P, Carmi Y, Guttman O, Braiman A, Cohen I, Voronov E, White MR, Dinarello CA, Apte RN (2011) IL-1α and IL-1β recruit different myeloid cells and promote different stages of sterile inflammation. J Immunol 187:4835–4843

Rioux JD, Xavier RJ, Taylor KD, Silverberg MS, Goyette P, Huett A, Green T, Kuballa P, Barmada MM, Datta LW, Shugart YY, Griffiths AM, Targan SR, Ippoliti AF, Bernard EJ, Mei L, Nicolae DL, Regueiro M, Schumm LP, Steinhart AH, Rotter JI, Duerr RH, Cho JH, Daly MJ, Brant SR (2007) Genome-wide association study identifies new susceptibility loci for Crohn disease and implicates autophagy in disease pathogenesis. Nat Genet 39:596–604

Rosenwasser LJ, Dinarello CA, Rosenthal AS (1979) Adherent cell function in murine T-lymphocyte antigen recognition IV. Enhancement of murine T-cell antigen recognition by human leukocytic pyrogen. J Exp Med 150:709–714

Roussel L, Erard M, Cayrol C, Girard JP (2008) Molecular mimicry between IL-33 and KSHV for attachment to chromatin through the H2A-H2B acidic pocket. EMBO Rep 9:1006–1012

Saitoh T, Fujita N, Jang MH, Uematsu S, Yang BG, Satoh T, Omori H, Noda T, Yamamoto N, Komatsu M, Tanaka K, Kawai T, Tsujimura T, Takeuchi O, Yoshimori T, Akira S (2008) Loss of the autophagy protein Atg16L1 enhances endotoxin-induced IL-1beta production. Nature (Lond) 456:264–268

Sanada S, Hakuno D, Higgins LJ, Schreiter ER, McKenzie AN, Lee RT (2007) IL-33 and ST2 comprise a critical biomechanically induced and cardioprotective signaling system. J Clin Invest 117:1538–1549

Schindler R, Clark BD, Dinarello CA (1990) Dissociation between interleukin-1b mRNA and protein synthesis in human peripheral blood mononuclear cells. J Biol Chem 265:10232–10237

Schlesinger N, Mysler E, Lin HY, De Meulemeester M, Rovensky J, Arulmani U, Balfour A, Krammer G, Sallstig P, So A (2011) Canakinumab reduces the risk of acute gouty arthritis flares during initiation of allopurinol treatment: results of a double-blind, randomised study. Ann Rheum Dis 70:1264–1271

Schmid D, Munz C (2007) Innate and adaptive immunity through autophagy. Immunity 27: 11–21

Schmitz J, Owyang A, Oldham E, Song Y, Murphy E, McClanahan TK, Zurawski G, Moshrefi M, Qin J, Li X, Gorman DM, Bazan JF, Kastelein RA (2005) IL-33, an interleukin-1-like cytokine that signals via the IL-1 receptor-related protein ST2 and induces T helper type 2-associated cytokines. Immunity 23:479–490

Sennello JA, Fayad R, Pini M, Gove ME, Ponemone V, Cabay RJ, Siegmund B, Dinarello CA, Fantuzzi G (2008) Interleukin-18, together with interleukin-12, induces severe acute pancreatitis in obese but not in nonobese leptin-deficient mice. Proc Natl Acad Sci USA 105: 8085–8090

Sha Y, Zmijewski J, Xu Z, Abraham E (2008) HMGB1 develops enhanced proinflammatory activity by binding to cytokines. J Immunol 180:2531–2537

Sharma S, Kulk N, Nold MF, Graf R, Kim SH, Reinhardt D, Dinarello CA, Bufler P (2008) The IL-1 family member 7b translocates to the nucleus and down-regulates proinflammatory cytokines. J Immunol 180:5477–5482

Siegmund B, Fantuzzi G, Rieder F, Gamboni-Robertson F, Lehr HA, Hartmann G, Dinarello CA, Endres S, Eigler A (2001a) Neutralization of interleukin-18 reduces severity in murine colitis and intestinal IFN-γ and TNF-α production. Am J Physiol Regul Integr Comp Physiol 281:R1264–R1273

Siegmund B, Lehr HA, Fantuzzi G, Dinarello CA (2001b) IL-1beta -converting enzyme (caspase-1) in intestinal inflammation. Proc Natl Acad Sci U S A 98:13249–13254
Sims JE, March CJ, Cosman D, Widmer MB, MacDonald HR, McMahan CJ, Grubin CE, Wignall JM, Jackson JL, Call SM (1988) cDNA expression cloning of the IL-1 receptor, a member of the immunoglobulin superfamily. Science 241:585–589
Sluyter R, Dalitz JG, Wiley JS (2004a) P2X7 receptor polymorphism impairs extracellular adenosine 5′-triphosphate-induced interleukin-18 release from human monocytes. Genes Immun 5: 588–591
Sluyter R, Shemon AN, Wiley JS (2004b) Glu496 to Ala polymorphism in the P2X7 receptor impairs ATP-induced IL-1 beta release from human monocytes. J Immunol 172:3399–3405
Smith DE, Lipsky BP, Russell C, Ketchem RR, Kirchner J, Hensley K, Huang Y, Friedman WJ, Boissonneault V, Plante MM, Rivest S, Sims JE (2009) A central nervous system-restricted isoform of the interleukin-1 receptor accessory protein modulates neuronal responses to interleukin-1. Immunity 30:817–831
So A, De Smedt T, Revaz S, Tschopp J (2007) A pilot study of IL-1 inhibition by anakinra in acute gout. Arthritis Res Ther 9:R28
So A, De Meulemeester M, Pikhlak A, Yucel AE, Richard D, Murphy V, Arulmani U, Sallstig P, Schlesinger N (2010) Canakinumab for the treatment of acute flares in difficult-to-treat gouty arthritis: results of a multicenter, phase II, dose-ranging study. Arthritis Rheum 62:3064–3076
Stevenson FT, Turck J, Locksley RM, Lovett DH (1997) The N-terminal propiece of interleukin 1 alpha is a transforming nuclear oncoprotein. Proc Natl Acad Sci USA 94:508–513
Sugawara S, Uehara A, Nochi T, Yamaguchi T, Ueda H, Sugiyama A, Hanzawa K, Kumagai K, Okamura H, Takada H (2001) Neutrophil proteinase 3-mediated induction of bioactive IL-18 secretion by human oral epithelial cells. J Immunol 167:6568–6575
Suzuki T, Franchi L, Toma C, Ashida H, Ogawa M, Yoshikawa Y, Mimuro H, Inohara N, Sasakawa C, Nunez G (2007) Differential regulation of caspase-1 activation, pyroptosis, and autophagy via Ipaf and ASC in *Shigella*-infected macrophages. PLoS Pathog 3:e111
Tak PP, Bacchi M, Bertolino M (2006) Pharmacokinetics of IL-18 binding protein in healthy volunteers and subjects with rheumatoid arthritis or plaque psoriasis. Eur J Drug Metab Pharmacokinet 31:109–116
Thornton P, McColl BW, Greenhalgh A, Denes A, Allan SM, Rothwell NJ (2010) Platelet interleukin-1alpha drives cerebrovascular inflammation. Blood 115:3632–3639
Towne J, Sims J (2012) IL-36 in psoriasis. Curr Opin Pharmacol 12(4):486–490
Towne JE, Garka KE, Renshaw BR, Virca GD, Sims JE (2004) Interleukin (IL)-1F6, IL-1F8, and IL-1F9 signal through IL-1Rrp2 and IL-1RAcP to activate the pathway leading to NF-kappaB and MAPKs. J Biol Chem 279:13677–13688
Towne JE, Renshaw BR, Douangpanya J, Lipsky BP, Shen M, Gabel CA, Sims JE (2011) Interleukin-36 (IL-36) ligands require processing for full agonist (IL-36alpha, IL-36beta, and IL-36gamma) or antagonist (IL-36Ra) activity. J Biol Chem 286:42594–42602
Travassos LH, Carneiro LA, Ramjeet M, Hussey S, Kim YG, Magalhaes JG, Yuan L, Soares F, Chea E, Le Bourhis L, Boneca IG, Allaoui A, Jones NL, Nunez G, Girardin SE, Philpott DJ (2010) Nod1 and Nod2 direct autophagy by recruiting ATG16L1 to the plasma membrane at the site of bacterial entry. Nat Immunol 11:55–62
Tschopp J, Schroder K (2010) NLRP3 inflammasome activation: the convergence of multiple signalling pathways on ROS production? Nat Rev Immunol 10:210–215
Tsutsui H, Matsui K, Okamura H, Nakanishi K (2000) Pathophysiological roles of interleukin-18 in inflammatory liver diseases. Immunol Rev 174:192–209
Unlu S, Kumar A, Waterman WR, Tsukada J, Wang KZ, Galson DL, Auron PE (2007) Phosphorylation of IRF8 in a pre-associated complex with Spi-1/PU.1 and non-phosphorylated Stat1 is critical for LPS induction of the IL1B gene. Mol Immunol 44:3364–3379
van Asseldonk EJ, Stienstra R, Koenen TB, van Tits LJ, Joosten LA, Tack CJ, Netea MG (2010) The effect of the interleukin-1 cytokine family members IL-1F6 and IL-1F8 on adipocyte differentiation. Obesity 18:2234–2236

van de Veerdonk FL, Stoeckman AK, Wu G, Boeckermann AN, Azam T, Netea MG, Joosten LA, van der Meer JW, Hao R, Kalabokis V, Dinarello CA (2012) IL-38 binds to the IL-36 receptor and has biological effects on immune cells similar to IL-36 receptor antagonist. Proc Natl Acad Sci USA 109:3001–3005

van der Meer JWM, Barza M, Wolff SM, Dinarello CA (1988) A low dose of recombinant interleukin 1 protects granulocytopenic mice from lethal gram-negative infection. Proc Natl Acad Sci USA 85:1620–1623

Vidal-Vanaclocha F, Fantuzzi G, Mendoza L, Fuentes AM, Anasagasti MJ, Martin J, Carrascal T, Walsh P, Reznikov LL, Kim SH, Novick D, Rubinstein M, Dinarello CA (2000) IL-18 regulates IL-1beta-dependent hepatic melanoma metastasis via vascular cell adhesion molecule-1. Proc Natl Acad Sci USA 97:734–739

Vigne S, Palmer G, Lamacchia C, Martin P, Talabot-Ayer D, Rodriguez E, Ronchi F, Sallusto F, Dinh H, Sims JE, Gabay C (2011) IL-36R ligands are potent regulators of dendritic and T cells. Blood 118:5813–5823

Wang D, Zhang S, Li L, Liu X, Mei K, Wang X (2010) Structural insights into the assembly and activation of IL-1beta with its receptors. Nat Immunol 11:905–911

Weber A, Wasiliew P, Kracht M (2010) Interleukin-1 (IL-1) pathway. Sci Signal 3:cm1

Werman A, Werman-Venkert R, White R, Lee JK, Werman B, Krelin Y, Voronov E, Dinarello CA, Apte RN (2004) The precursor form of IL-1alpha is an intracrine proinflammatory activator of transcription. Proc Natl Acad Sci USA 101:2434–2439

Wewers MD, Dare HA, Winnard AV, Parker JM, Miller DK (1997) IL-1b-converting enzyme (ICE) is present and functional in human alveolar macrophages: macrophage IL-1b release limitation is ICE independent. J Immunol 159:5964–5972

Wilson NJ, Boniface K, Chan JR, McKenzie BS, Blumenschein WM, Mattson JD, Basham B, Smith K, Chen T, Morel F, Lecron JC, Kastelein RA, Cua DJ, McClanahan TK, Bowman EP, de Waal MR (2007) Development, cytokine profile and function of human interleukin 17-producing helper T cells. Nat Immunol 8:950–957

Wittmann M, Bachmann M, Doble R, Pfeilschifter J, Werfel T, Mühl H (2012) IL-27 regulates IL-18 binding protein in skin resident cells. PLoS One 7(6):e38751

Xiang Y, Moss B (2001) Correspondence of the functional epitopes of poxvirus and human interleukin-18-binding proteins. J Virol 75:9947–9954

Yasuda K, Muto T, Kawagoe T, Matsumoto M, Sasaki Y, Matsushita K, Taki Y, Futatsugi-Yumikura S, Tsutsui H, Ishii KJ, Yoshimoto T, Akira S, Nakanishi K (2012) Contribution of IL-33-activated type II innate lymphoid cells to pulmonary eosinophilia in intestinal nematode-infected mice. Proc Natl Acad Sci USA 109:3451–3456

Yin H, Morioka H, Towle CA, Vidal M, Watanabe T, Weissbach L (2001) Evidence that HAX-1 is an interleukin-1 alpha N-terminal binding protein. Cytokine 15:122–137

Zheng H, Fletcher D, Kozak W, Jiang M, Hofmann K, Conn CC, Siszynski D, Grabiec C, Trumbauer MA, Shaw AR, Kostura MJ, Stevens K, Rosen H, North RJ, Chen HY, Tocci MJ, Kluger MJ, Van der Ploeg LHT (1995) Resistance to fever induction and impaired acute-phase response in interleukin-1b deficient mice. Immunity 3:9–19

Zhou R, Tardivel A, Thorens B, Choi I, Tschopp J (2010) Thioredoxin-interacting protein links oxidative stress to inflammasome activation. Nat Immunol 11:136–140

Zorrilla EP, Sanchez-Alavez M, Sugama S, Brennan M, Fernandez R, Bartfai T, Conti B (2007) Interleukin-18 controls energy homeostasis by suppressing appetite and feed efficiency. Proc Natl Acad Sci USA 104:11097–11102

Chapter 2
IL-6 and Inflammatory Diseases

Daisuke Kamimura, Yasunobu Arima, Toshio Hirano, Hideki Ogura, and Masaaki Murakami

Abstract Interleukin-6 (IL-6) is a multifunctional cytokine that plays key roles not only in the immune system but also in a variety of biological processes. It is a primary regulator of both acute and chronic inflammations. Moreover, it has proven an excellent target for clinical treatment, as the anti-IL-6 receptor antibody has been successfully used against autoimmune disorders such as rheumatoid arthritis, juvenile idiopathic arthritis, and Castleman's disease. In fact, it could be argued that IL-6 is the best example of basic cytokine research extending into clinical application. Here, we summarize IL-6 and its biological functions, with particular emphasis on inflammation and chronic inflammatory diseases, and a recently discovered inflammation control mechanism, the inflammation amplifier (formerly known as the IL-6 amplifier). We also describe a recent finding that indicates neural stimulations can modulate the activation of the inflammation amplifier at local blood vessels, creating a gate for the influx of immune cells into the central nervous system, which suggests the entry of immune cells into target organs can be artificially manipulated by local neural stimulation.

Keywords Chronic inflammation • Interleukin-6 • Neuroimmune interaction • NF-κB • STAT3 • The inflammation amplifier

D. Kamimura • Y. Arima • H. Ogura • M. Murakami (✉)
Laboratory of Developmental Immunology, JST-CREST, Graduate School of Frontier Biosciences, Graduate School of Medicine, and WPI Immunology Frontier Research Center, Suita Osaka University, 2-2 Yamada-oka, Osaka 565-0871, Japan
e-mail: murakami@molonc.med.osaka-u.ac.jp

T. Hirano
JST-CREST, Osaka University, 2-2 Yamada-oka, Suita Osaka, 565-0871 Osaka, Japan

T. Yoshimoto and T. Yoshimoto (eds.), *Cytokine Frontiers: Regulation of Immune Responses in Health and Disease*, DOI 10.1007/978-4-431-54442-5_2, © Springer Japan 2014

2.1 General Aspects of IL-6 and Its Signal Transduction

2.1.1 *Discovery of IL-6, Its Receptor Subunits, and Related Signal Transduction Molecules*

We now know that IL-6 has a wide variety of biological roles in numerous systems including the immune, nervous, and endocrine systems (Kamimura et al. 2003; Taga and Kishimoto 1997). As a result, many research groups from different fields had independently sought the molecular cloning of IL-6, which led to a number of different names for the same molecule. We give special consideration to Kishimoto and Hirano, as they were the first to publish on the matter (Hirano et al. 1986). Hirano who is one of authors originally found IL-6 to be a soluble factor present in the culture supernatant of cells in pleural effusion and isolated from patients with pulmonary tuberculosis. IL-6 was seen to induce B-cell growth and antibody production, which is why it was originally named B-cell stimulatory factor 2, or BSF2. Other groups subsequently molecularly identified factors such as interferon (IFN)β2, which was later recognized to be BSF2 based on its 26-kDa mass and comparisons of cDNA sequences. Eventually, the research community settled on the name IL-6.

The IL-6 receptor consists of a ligand-binding IL-6-receptor chain (IL-6R, gp80, or CD126) and signal-transducing subunit gp130 (CD130). After the molecular cloning of IL-6R by Kishimoto's group, it was revealed that IL-6R has a cytoplasmic tail that was considered too short to transmit intracellular signaling (Yamasaki et al. 1988). They also later found a protein of 130 kDa co-precipitated with IL-6–IL-6R complexes, gp130 (Taga et al. 1989). Through cDNA cloning, they showed gp130 has a long cytoplasmic domain and is an essential signal-transducing component of the IL-6 receptor (Hibi et al. 1990). The extracellular regions of gp130 and IL-6R contain a four-cysteine motif and a WSXWS motif, which constitute cytokine-binding domains, whereas the intracellular domain of gp130 has the box regions responsible for Jak kinase binding (Murakami et al. 1991) and multiple tyrosine residues that are phosphorylated upon ligand binding (Fig. 2.1a). In addition to the receptor components, it was also successfully identified a key transcription factor for IL-6 signaling, STAT3 (Akira et al. 1994), a transcription factor that leads to IL-6 expression, C/EBPβ (Akira et al. 1990), and factors that negatively regulate the IL-6 signal including SOCS family molecules (Yoshimura et al. 2007). They also revealed that gp130 forms homodimers upon IL-6 binding, which activates tyrosine kinase activity (Murakami et al. 1993), and had a central role in the development of the anti-IL-6R antibody, which is currently a popular treatment against autoimmune diseases. Hirano's group also did a comprehensive study of the intracellular signaling, finding Gab1 and Gab2 are adaptor molecules for gp130-mediated ERK signaling (Nishida et al. 1999; Takahashi-Tezuka et al. 1998), and established mutant versions of gp130 knock-in mice to dissect gp130 signaling in vivo (Ohtani et al. 2000). One knock-in strain, F759 mice, spontaneously developed rheumatoid arthritis-like joint disease (Atsumi et al. 2002) to provide an excellent

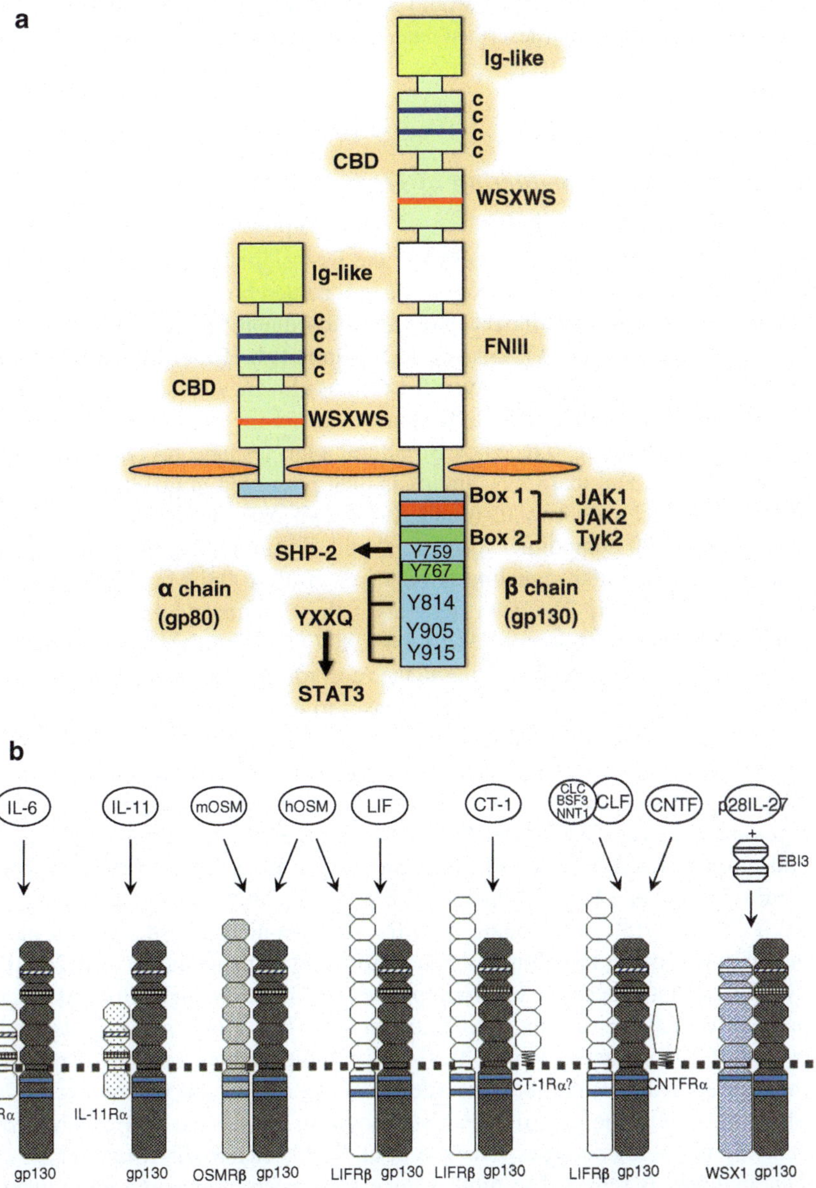

Fig. 2.1 Components of the interleukin (IL)-6 receptor. (**a**) The IL-6 receptor α-chain (gp80, *left*) and β-chain (gp130, *right*), and their conserved motifs. IL-6 binding to cytokine-binding domain (*CBD*) transduces intracellular signaling through gp130, which has a relatively long cytoplasmic tail with multiple tyrosine residues that can be phosphorylated by JAK family kinases. (**b**) gp130 is shared by many cytokines besides IL-6, including IL-11, LIF, CNTF, OSM, CT-1, CLC/NNT1/BSF3, neuropoietin/CT-2, IL-27, and IL-35. These IL-6 family members all use gp130 to transmit signal transduction

model for inflammation and would eventually lead to the discovery of the inflammation amplifier (described later), a novel inflammation-inducing mechanism (Murakami et al. 2011; Ogura et al. 2008; Sawa et al. 2006).

2.1.2 *Signal Transduction of IL-6*

Experiments using mutated versions of the IL-6 signal transducer gp130 revealed important regions within its cytoplasmic tail, named box 1 and box 2, that are conserved among other cytokine receptors and subsequently led to the identification of the binding regions for JAK kinases (Murakami et al. 1993; Murakami et al. 1991) (Fig. 2.1a). gp130 contains six tyrosine residues in its cytoplasmic domain. All but the first are phosphorylated by JAK kinases and are involved in signal transduction emanating from the IL-6 receptor. Structural analyses have revealed that IL-6 first binds to IL-6R, and that this complex is then presented in gp130 to form a high-affinity hexamer with two IL-6 and IL-6R molecules each. The ligand binding makes the gp130 homodimer formation bend, bringing them into close proximity to enable intracellular signaling (Boulanger et al. 2003; Murakami et al. 1993; Skiniotis et al. 2005). The main signaling pathway induced by the IL-6 receptor is mediated by the activation of STAT3 via YXXQ motifs located in the gp130 cytoplasmic region. Upon IL-6 binding to the receptor, the four distal tyrosine residues of gp130 are phosphorylated by JAK kinases to form binding sites for STAT3. The gp130-bound STAT3 is then tyrosine phosphorylated by JAK kinases and subsequently forms a dimer that translocates into the nucleus, where it induces the transcription of its target genes (Fig. 2.2a). One such gene is a negative regulator of cytokine signaling, SOCS3. SOCS3 binds to the JAK-kinase-phosphorylated second tyrosine residue of gp130 (Y759 in human gp130) and works as a negative feedback molecule for IL-6 signaling. SOCS3 actually binds to JAK kinases and the Y759 of gp130 simultaneously to form a high-affinity complex that mediates the inhibition of the signaling in a noncompetitive manner (Babon et al. 2012). This SOCS3-mediated negative feedback prevents immune dysregulation. As discussed next, mice devoid of this second tyrosine (F759 mice) show prolonged and enhanced STAT3 activation by IL-6 and exhibit spontaneous development of a rheumatoid arthritis-like joint disorder with age (Fig. 2.2b) (Atsumi et al. 2002). The second tyrosine of gp130 is also known to induce the SHP2/Gab/Ras/ERK pathway in some cell types (Nishida et al. 1999; Takahashi-Tezuka et al. 1998). This pathway is reported to protect mice from gastric adenoma by inducing the tissue-protecting trefoil factor, pS2/TFF1 (Tebbutt et al. 2002).

In addition to these conventional IL-6 signal transduction pathways, we recently identified a novel role of IL-6 signaling in assisting NF-κB signaling to synergistically induce the transcription of proinflammatory genes. We named this signaling the "inflammation amplifier" (see Sect. 2.5). Although gp130 is expressed in most cell types, IL-6R expression is relatively limited. However, IL-6R can be solubilized from the cell membrane by both alternative splicing and shedding via the action of ADAM family proteases (Chalaris et al. 2011). Theoretically, the presence

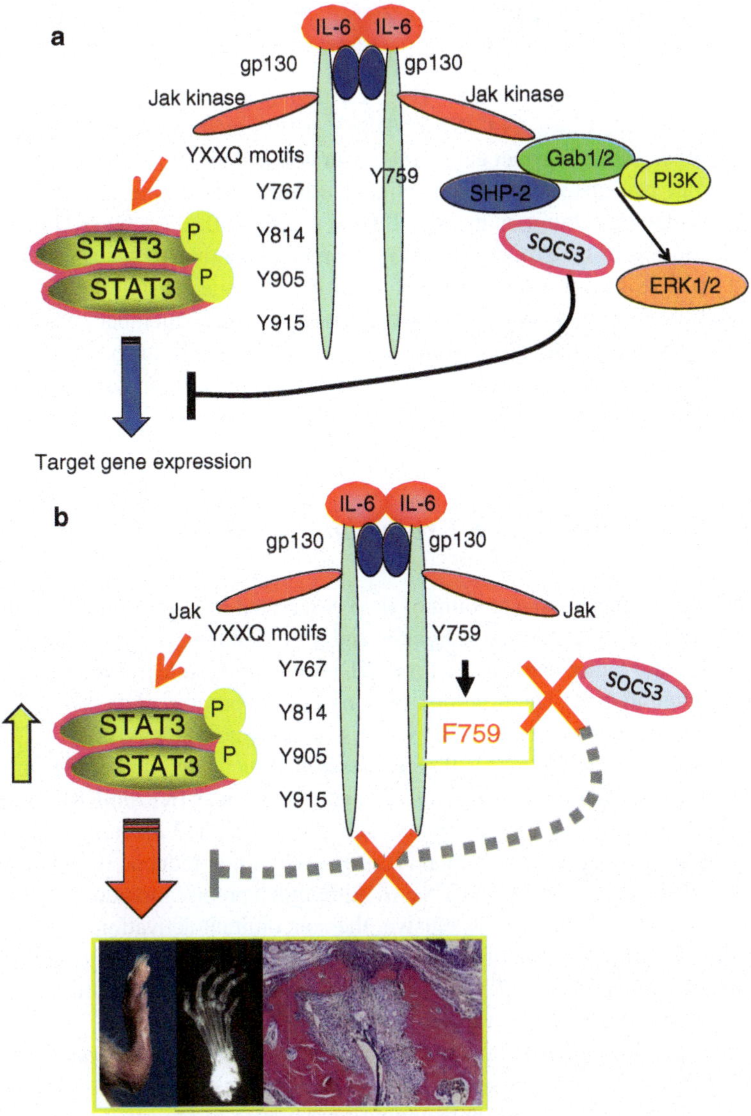

Fig. 2.2 Intracellular signaling pathways of gp130. (**a**) Signal transduction from wild-type gp130. Tyrosine 759 is involved in the gp130-mediated ERK pathway and is essential for the SOCS3-mediated negative feedback loop. The distal four tyrosine residues form the YXXQ motif required for STAT3 binding. (**b**) Signal transduction from a gp130 F759 mutant. Because of the absence of the SOCS3-binding site (i.e., Y759), STAT3 activation is prolonged, and F759 mice suffer from autoimmune arthritis that clinically resembles rheumatoid arthritis (*bottom pictures*)

of soluble IL-6R enables IL-6 to signal into cells that express gp130 in vivo, which is the basis for multiple functions of IL-6. The soluble form of gp130 is also present naturally and acts as an antagonist for signaling mediated by IL-6/soluble IL-6R complexes, but not for IL-6/membrane IL-6R complexes (Jostock et al. 2001).

2.1.3 The IL-6 Family

gp130 is shared by many cytokines (Fig. 2.1b) in addition to IL-6, including IL-11, LIF, CNTF, OSM, CT-1, CLC/NNT1/BSF3, neuropoietin/CT-2, IL-27 and IL-35 (Collison et al. 2012; Derouet et al. 2004; Murakami et al. 2004). These IL-6 family members all use gp130 to transmit signal transduction. Some require a second signal-transducing subunit, namely, LIFRβ or OSMRβ. IL-27, a heterodimeric protein that consists of IL-27p28 and EBI3, was recently shown to act as an antagonist against gp130 signaling (Stumhofer et al. 2010). Despite sharing the signal transducer gp130, the biological functions of these IL-6 family members only partially overlap (Kamimura et al. 2003; Murakami et al. 2004). As described in detail next, IL-6, for example, has a nonredundant role in promoting inflammation, particularly chronic inflammation.

2.1.4 Regulation of IL-6 Expression

It is well known that a large amount of IL-6 is secreted in response to inflammatory stimuli such as Toll-like receptor ligands and proinflammatory cytokines including IL-1, IL-17, and tumor necrosis factor (TNF)-α to combat infections and, finally, to promote inflammation. The promoter region of IL-6 contains a NF-κB-binding site in addition to response elements for various transcription factors such as C/EBPβ. NF-κB activation is essential for IL-6 expression, which is greatly compromised in cells lacking vital components for NF-κB signaling such as IKKγ/NEKO (Lee et al. 2012; Ogura et al. 2008). In addition, the activation of STAT3 is found to synergistically enhance IL-6 production when NF-κB activation is induced by other inflammatory stimuli such as IL-17 and TNF-α. Although a precise molecular mechanism of this synergistic effect remains elusive, the concomitant activation of NF-κB and STAT3 for the amplification of IL-6 expression in non-immune cells (inflammation amplifier, discussed in Sect. 2.5) is considered central for the induction and maintenance of inflammatory disease conditions.

IL-6 production can be stimulated by routine activities such as physical exercise, which results in a larger amount of IL-6 production in skeletal muscle than other cytokines (Pedersen et al. 2001). Running upregulates plasma IL-6 levels in mice to stimulate the secretion of glucagon-like peptide-1 (GLP-1), an insulin-regulating hormone, in intestinal L cells to improve insulin secretion and glycemia, suggesting IL-6 is a key regulator for glucose homeostasis, which when disrupted can cause metabolic syndromes (Ellingsgaard et al. 2011). IL-6 levels are also modulated by social interactions and stress in humans. Competitive social events, for example, increase baseline IL-6 levels (Chiang et al. 2012). Experimentally imposed stress paradigms in humans have been shown to cause potential increases in IL-6 circulation (Steptoe et al. 2007). Anxiety also promotes IL-6 production in humans (O'Donovan et al. 2010). In addition, it is known that serum IL-6 levels are controlled by a circadian cycle in humans, with a biphasic pattern that peaks at about 0800 and 2100 and bottoms out at about 1900 and 0500 (Vgontzas et al. 2005). How these mental and daily events influence IL-6 production is not well defined at the molecular level, however. Because mental conditions can influence the immune system and potentially trigger relapse of autoimmune disorders (Srivastava and Boyer 2010), the elucidation of the molecular mechanisms underlying the neuroimmune interactions that induce pro-inflammatory cytokines including IL-6 may open a new avenue for the treatment of many chronic inflammatory diseases. One example can be seen in our discussion of the inflammation amplifier and its neural-mediated activation in endothelial cells (see Sect. 2.5), as inflammation may be induced via accumulation of immune cells in affected tissues of the central nervous system.

Posttranscriptional regulation of IL-6 is also important for controlling IL-6 levels and thus maintaining immune homeostasis in both the steady state and disease conditions. For example, the RNase zc3h12a, also known as regnase-1, is a Toll-like receptor (TLR)-inducible gene that controls IL-6 mRNA decay. Accordingly, zc3h12a-deficient mice show highly increased production of IL-6 and IL-12p40, leading to the development of autoimmunity (Matsushita et al. 2009). The stability of regnase-1 is controlled by IKK complex-mediated phosphorylation, which causes ubiquitination and degradation, and regnase-1 mRNA is targeted by regnase-1 itself. Thus, IL-6 mRNA levels can be finely tuned by RNases. Another example of the posttranscriptional control of IL-6 is its dependency on micro RNA (miR). Let-7a directly inhibits IL-6 expression whereas IL-6 is shown to activate NF-κB, which in turn represses let-7a levels and promotes IL-6 production. This positive feedback loop is considered important for maintaining the transformed state in certain cancer cells (Iliopoulos et al. 2009). IL-6 mRNA is also targeted by miR-26. A zinc-finger protein with an RNA interacting motif, Zcchc-11, induces the addition of uridines to the miR-26 3′-end, which abrogates IL-6 repression by miR-26 (Jones et al. 2009). In addition, IL-6R mRNA levels are suppressed by miR-124 in hepatocellular cancer cell lines, and it has been shown that systemic delivery of miR-124 prevents hepatocellular carcinogenesis without any side effects (Hatziapostolou et al. 2011). The regulation of IL-6 production can be achieved at a translational level as well. An RNA-binding protein, KSRP, which is known to have the ability to degrade mRNA with AU-rich elements, also participates in translational silencing. In KSRP-depleted cells, IL-6 mRNA is redistributed to polysomes, a phenomenon associated with increased production of IL-6. This translational silencing effect is dependent on the 3′-untranslated region of IL-6 mRNA (Dhamija et al. 2011).

2.2 Biological Functions of IL-6 in Inflammation and Disease

As already mentioned, gp130 plays an important role in the signal transduction by IL-6 that regulates a variety of biological functions. Because space is limited here, we discuss the function of IL-6 by focusing only on recent findings about its role in inflammation and disease. Its other biological functions are summarized elsewhere (Kamimura et al. 2003; Taga and Kishimoto 1997).

2.2.1 IL-6 as a Pro-inflammatory Mediator

It is widely accepted that CD4$^+$ T cells produce a large amount of cytokines that promote both acute and chronic inflammations. When activated, effector CD4$^+$ T cells can be divided into different subsets based on their cytokine profiles. A relatively new subset, type 17 helper CD4$^+$ T cells (Th17), mainly produce IL-17 and are responsible for the development of various autoimmune disorders, at least in mice (Nishihara et al. 2007; Veldhoen et al. 2006). Upon antigen stimulation of undifferentiated naïve CD4$^+$ T cells, IL-6 and another cytokine, transforming growth factor (TGF)-β, direct these cells to express the transcription factor Rorγt, which is necessary for Th17 differentiation (Ivanov et al. 2006). TGF-β is dispensable for Th17 generation in certain culture conditions (Ghoreschi et al. 2010). STAT3 activation emanating from the gp130 YXXQ motifs, but not the gp130 Y759-mediated SHP2/ERK pathway, is important for IL-6 activation (Fig. 2.2a). Accordingly, mice lacking gp130-STAT3 signaling in T cells have a substantially decreased number of Th17 cells (Nishihara et al. 2007). IL-6 was also reported to contribute to human Th17 differentiation (Zielinski et al. 2012). Interestingly, in the absence of TGF-β, IL-6, together with IL-21, promotes naïve helper T cells to differentiate toward follicular helper T cells (Tfh), which requires the transcription factor Bcl-6 (Nurieva et al. 2009). Tfh plays an important role in T-cell–B-cell cooperation, resulting in enhanced formation of germinal centers and high-affinity antigen-specific immunoglobulin secretion. Because many cell types are known to produce IL-6, the cellular source of the IL-6 that promotes Th17 or Tfh differentiation and subsequent pathogenic inflammation in vivo remains a matter of debate. In a murine model of multiple sclerosis, IL-6 from a subset of B cells is shown to contribute to persistent inflammation in the central nervous system (Barr et al. 2012). It was also recently shown that deficiency of IL-6 and IL-21 in mice fails to induce Tfh cell-dependent immune responses against viral infection, and that IL-6 from follicular B cells is important for Tfh development (Karnowski et al. 2012). Although many studies have noted the significance of IL-6 production and its effect in immune cells, our recent findings (see Sect. 2.5) suggest nonimmune cells, including endothelial cells and fibroblasts, also make an important contribution to IL-6 production, particularly during inflammation. To formally demonstrate the functional cellular source(s) of IL-6 during inflammation, a conditional knockout of the IL-6 gene in mice such as the one established recently should be examined (Quintana et al. 2013).

2.2.2 IL-6 as an Anti-inflammatory Mediator

Under certain conditions, IL-6 exhibits an anti-inflammatory role in myeloid cells, such as dendritic cells and macrophages. Exposure to IL-6 before stimulation with microbial products [e.g., lipopolysaccharide (LPS)] inhibits major histocompatibility complex (MHC) class II expression and pro-inflammatory mediators in bone marrow-derived dendritic cells in vitro (Park et al. 2004). In fact, because IL-6 is present at low levels under normal conditions, IL-6-deficient mice show a higher MHC class II expression on dendritic cells than that in wild-type mice, whereas mice with enhanced IL-6 signaling caused by the loss of the SOCS3-binding site in gp130 (F759 mice) show a lower level (Park et al. 2004). Mechanistically, IL-6 reduces the level of cystatin C, an endogenous inhibitor of cathepsins, thereby increasing cathepsin S activity and subsequent degradation of MHC class II components in IL-6-treated dendritic cells (Kitamura et al. 2005). Similarly, prolonged action of IL-6 has been shown to mimic the anti-inflammatory effects of IL-10, which also activates STAT3, in macrophages (Yasukawa et al. 2003). Moreover, the anti-inflammatory effects of IL-6 are manifested in a murine model of allergic asthma. In this model, IL-6-deficient mice show exaggerated lung inflammation whereas lung-specific overexpression of IL-6 reduced the disease symptoms (Wang et al. 2000). Importantly, IL-6 stimulation is also known to suppress T-cell-receptor-mediated signaling via SOCS3 molecules (Atsumi et al. 2009). Thus, direct IL-6 stimulation in certain immune cell populations can induce an anti-inflammatory signal.

2.2.3 *IL-6 and Cancer*

As already mentioned, gp130, the signal-transducing receptor subunit of IL-6, is expressed in almost all cell types in the body, whereas the expression of the IL-6-binding subunit, IL-6R, is more restricted. IL-6R expression is abundant in immune cells but not in nonimmune cells such as fibroblasts. However, shedding of IL-6R is detectable under normal conditions. The resulting soluble forms of IL-6R enable IL-6 to transmit intracellular signaling in cell types expressing gp130 alone, a phenomenon sometimes called IL-6 trans-signaling (Waetzig and Rose-John 2012). Because IL-6 is a well-known growth factor for cancer cells, it is likely that cells that do not express IL-6R exploit IL-6 trans-signaling for survival and expansion. Upon genotoxic stimuli such as those caused by drugs used for chemotherapy, IL-6 is produced from thymic endothelial cells in response to DNA damage response such that the resulting microenvironment provides a chemo-resistant niche for cancer cells. Therefore conventional chemotherapy can at the same time lead to tumor suppression and potentially create a tumor-promoting environment in an IL-6-dependent manner (Gilbert and Hemann 2010). It has been reported that excess body weight is correlated with an increase of a risk for cancer-associated death in humans, particularly hepatocellular carcinoma (Calle et al. 2003). One possible mechanism for this outcome has come from the observation that the risk of hepatocellular carcinoma was higher in high-fat diet mice. Interestingly, tumor-bearing mice fed with the high-fat diet showed elevated levels of IL-6 and TNF-α in their serum and around the tumor area. Similarly, it was seen in IL-6-deficient animals that obesity induces tumor progression via production of IL-6 (Park et al. 2010). Cancer progression is generally ascribed to primary tumor growth and secondary metastasis. One mechanism, termed tumor-self seeding, describes the ability of circulating tumor cells to infiltrate an established tumor and enrich it, potentially causing tumor growth and the breeding of metastatic cells. IL-6 functions as an autocrine factor in tumor cells and also acts as an attractant for the circulating tumor cells. In fact, knockdown of IL-6R or gp130 significantly inhibits the seeding activity of cancer cells (Kim et al. 2009). IL-6 also promotes breast cancer by suppressing miR-200c, which leads to constitutive activation of NF-κB and JNK2. Activated JNK2 in turn phosphorylates HSF1, which induces demethylation of the IL-6 promoter that facilitates transcription. This signaling circuit is present in human cancer cells and a mouse model of ErbB2-mediated tumorigenesis (Rokavec et al. 2012).

2.2.4 IL-6 and Infectious Diseases

Chronic viral infection including human immunodeficiency virus (HIV)-1 infection often impairs the function of T cells. In such cases, T cells produce fewer effector cytokines, leading to the persistence of the virus. Recently, it was shown, by studying the lymphocytic choriomeningitis virus in mice, that IL-7 prevents this effect by downregulating SOCS3. Remarkably, this effect was found to depend on IL-6, as the beneficial effect of recombinant IL-7 administration on T-cell expansion and viral clearance was diminished in IL-6-deficient mice (Pellegrini et al. 2011). Although the mechanism of action by IL-6 remains elusive, it is possible that IL-7 is produced from nonimmune cells including fibroblasts by IL-6 stimulation, just as in an arthritis model (Sawa et al. 2006), or from hepatocytes in a manner dependent on type I interferon (IFN) molecules (Sawa et al. 2009). Another report indicates an important role of IL-6 in murine chronic viral infection. Upon chronic infection with the lymphocytic choriomeningitis virus, IL-6 is produced. This production enhances the Tfh responses known to promote the germinal center reaction and subsequent antibody production from B cells, and therefore is particularly important for viral control. Experiments using bone marrow chimera mice have suggested that the main source of IL-6 is irradiation-resistant stroma cells, rather than irradiation-sensitive immune cells (Harker et al. 2011). Human herpesvirus 8 (HHV8), also known as Kaposi sarcoma-associated herpesvirus, encodes a molecule similar to IL-6 called viral IL-6 (vIL-6). vIL-6 is reported to directly bind to gp130 and transmit signals in the absence of IL-6R, and it is widely accepted that it plays a significant role in the pathology of HHV8-associated diseases such as multicentric Castleman's disease (MCD). vIL-6 transgenic mice have been seen to spontaneously develop plasma cell-type MCD symptoms. Interestingly, the diseased phenotype was abrogated when the mice were crossed with IL-6-deficient mice, indicating that endogenous mouse IL-6 is required for vIL-6-mediated MCD pathology. Therefore, it may be that a combination of classic signaling mediated by endogenous IL-6 and vIL-6-mediated direct stimulation of gp130 is required for MCD (Suthaus et al. 2012). The effect of IL-6 on the control of pathogen infection is, of course, not limited to viruses. The obligate intracellular parasite *Toxoplasma gondii* can infect and persist in neurons, subsequently leading to chronic encephalitis.

Using a conditional knockout of gp130 in neurons, it was demonstrated that IL-6 signaling in neurons protects against apoptosis during the infection, which would explain why synapsin-Cre gp130fl/fl mice are unusually susceptible to *Toxoplasma* encephalitis (Handel et al. 2012). gp130-mediated signaling is also important in astrocytes during *Toxoplasma* encephalitis because astrocyte-specific deletion of gp130 in mice using GFAP-Cre failed to control the parasites and the mice died of encephalitis (Drogemuller et al. 2008).

2.3 IL-6 Signaling and Human Diseases: Genetic Evidence

From the studies described here, it should come as no surprise to learn that IL-6 is involved in autoimmune diseases and infections associated with inflammation. In addition, accumulating evidence has revealed that a certain level of inflammation is evident in many diseases that had not been hitherto considered associated with immune cells or inflammatory mediators: these include metabolic syndromes, such as obesity and atherosclerosis, and neurodegenerative diseases, such as Alzheimer's disease and Parkinson's disease.

Genome-wide association studies have found a genetic link between the gene polymorphisms of IL-6 and its signaling molecules and human disorders. For example, a polymorphism located in the IL-6 promoter at −174 has been reported to associate with many diseases, including the −174G/C polymorphism and Alzheimer's disease (Dai et al. 2012). An association of the −174G/C polymorphism with coronary heart disease has been revealed by meta-analysis, which found this polymorphism is associated with a higher risk of the disease in Asian populations (Yin et al. 2012). In addition, the STAT3 locus and IL6ST (gp130) locus are found to associate with ulcerative colitis and rheumatoid arthritis, respectively (Franke et al. 2008) (Stahl et al. 2010). Moreover, the IL-6R gene has been identified as a risk locus for asthma (Ferreira et al. 2011). Recent meta-analyses including data of more than a hundred thousand volunteers suggested that the minor allele of IL-6R in humans, namely Asp358Ala, whose frequency is reported to be 39 %, is associated with an increase of circulating sIL-6R levels and reduced risk of coronary heart disease (Hingorani and Casas 2012; Sarwar et al. 2012). Constitutive activation of gp130/STAT3 signaling by somatic mutations has been demonstrated in benign liver tumors such as inflammatory hepatocellular adenomas (IHCS). The Zucman-Rossi lab reported that 60 % of IHCS harbor somatic mutations in gp130, which renders ligand-independent activation of STAT3 (Rebouissou et al. 2009). The group subsequently showed that a subset of IHCS lacking these gp130 mutations has gain-of-function mutations in STAT3, most of which are located in the SH2 domain, which induces STAT3 dimerization (Pilati et al. 2011). Conversely, a dominant-negative version of STAT3 gene mutations has been reported in hyper IgE syndrome, which is a de novo mutation and characterized by high serum IgE

(Minegishi et al. 2007). Combined, that IL-6 and its signal transduction pathway are essential in animal models of multiple autoimmune diseases, and evidence from the genome-wide association studies (GWAS), demonstrate IL-6 signaling is an attractive therapeutic target for many inflammatory diseases.

2.4 IL-6 Signaling as a Therapeutic Target

Clinical efficacy of blocking IL-6 signaling has already been demonstrated in humans. The human version of the anti-IL-6R monoclonal antibody (tocilizumab, or Actemra) has been approved in some countries for treating patients with moderate to severe rheumatoid arthritis. Notably, tocilizumab often shows substantial and better efficacy in patients than other disease-modifying anti-rheumatic drugs (DMARDs). In Japan, tocilizumab has also been used for the treatment of juvenile idiopathic arthritis and Castleman's disease.

As effective as tocilizumab is, there is also an effort to develop antibodies or gp130:Fc fusion proteins that can block IL-6 signaling (Jones et al. 2011). Although tocilizumab may be a "silver bullet" against some chronic inflammatory diseases, including rheumatoid arthritis, a nonbiological small compound that inhibits IL-6 signaling or production is also awaited because of its potential orally active properties and cost-effectiveness, just as other small compound drugs. A novel JAK kinase inhibitor, tofacitinib (or CP690,550), from Pfizer is expected to be launched soon for the treatment of moderate to severe active rheumatoid arthritis in some countries including the United States, Japan, and Europe. Clinical trials of this compound are also underway for psoriasis and inflammatory bowel diseases. STAT3 is another attractive target to interfere with IL-6 signaling. Although many efforts have been made to develop STAT3 inhibitors, to date none has reached the clinical drug stage. Recently, a promising orally available STAT3 inhibitor, BP-1-102, was reported (Zhang et al. 2012). The classical JAK/STAT pathway, which is a major signaling cascade of IL-6, is a relatively simple mechanism because only a couple of factors, namely, JAK and STAT, are involved. Although the pathway therefore makes for a simple model in the understanding of the cellular functions of IL-6, it also means that the number of target molecules is limited. We have recently discovered a novel mechanism for IL-6 and chemokine production under inflammatory situations. This mechanism, termed the inflammation amplifier, involves simultaneous activation of two key transcription factors, STAT3 and NF-κB (Ogura et al. 2008). We have already performed functional genome-wide screening to identify the genes that control its activation (Murakami et al. 2013). Remarkably, the number of genes regulating the inflammation amplifier far exceed the number we had expected, and may offer a vast number of targets for the inhibition of IL-6 production. We discuss the inflammation amplifier and its regulation in the next section.

2.5 The Inflammation Amplifier

2.5.1 Establishment of an Animal Model to Study the Pathogenesis of Rheumatoid Arthritis

Although the anti-IL-6R antibody has successfully treated more than 50 % of patients in a clinical setting (Nishimoto et al. 2000), the molecular mechanism responsible remains elusive. One reason is the lack of adequate experimental animal models. Because IL-6 signaling undoubtedly contributes to the pathogenesis of rheumatoid arthritis, we hypothesized that hyperactivation of IL-6 signaling in mice may cause the disease. To prove this hypothesis, we created a knock-in mouse line that has mutated human gp130 via the amino acid substitution Y759F (F759 mice). The tyrosine is an important docking site for the negative feedback regulation by SOCS3 in IL-6 signaling, meaning the mutation is expected to augment the signal transduction (Fig. 2.2a). Indeed, F759 showed hyperactivated STAT3 mediated by IL-6/gp130 signaling in vivo (Ohtani et al. 2000). Furthermore, these F759 mice spontaneously develop an age-dependent autoimmune joint disease in an IL-6-dependent manner (Fig. 2.2b) (Sawa et al. 2006). Also, F759-arthritis is accompanied by an accumulation of memory/activated T cells, and its clinical course is chronic and progressive. The disease symptoms begin as mild swelling and redness in the paws, but eventually lead to decreased joint mobility. Larger joints were affected symmetrically, and all these joints eventually became ankylotic. Radiologic analysis of the affected joints revealed characteristics that resemble advanced human rheumatoid arthritis, as also did histological examination, which showed leukocytes infiltrating the joint space, hyperplasia of the synovium with pannus formation, destruction of the cartilage and bone, and bony ankylosis (Atsumi et al. 2002).

2.5.2 The Discovery of the Inflammation Amplifier

The F759 arthritis model gives us a chance to investigate the role of IL-6 signaling in autoimmune disease development. We found that in F759 mice the number of memory/activated $CD4^+$ T cells dramatically increased with age. Additionally, when F759 mice were crossed with CD4-deficient or MHC class II-deficient mice, a significant suppression of disease development was observed in the offspring (Sawa et al. 2006). These observations suggest that excessive IL-6 signaling could have a role in inducing autoimmune arthritis via $CD4^+$ T-cell activation. Because IL-6 is a multifunctional cytokine and regulates the immune system, we considered whether IL-6 directly activates immune cells including $CD4^+$ T cells or dendritic cells, but surprisingly found that it in fact inactivates these cells (Atsumi et al. 2009; Kitamura et al. 2005; Park et al. 2004). Subsequent bone marrow chimera experiments demonstrated that wild-type mice transfused with F759 bone marrow did not develop

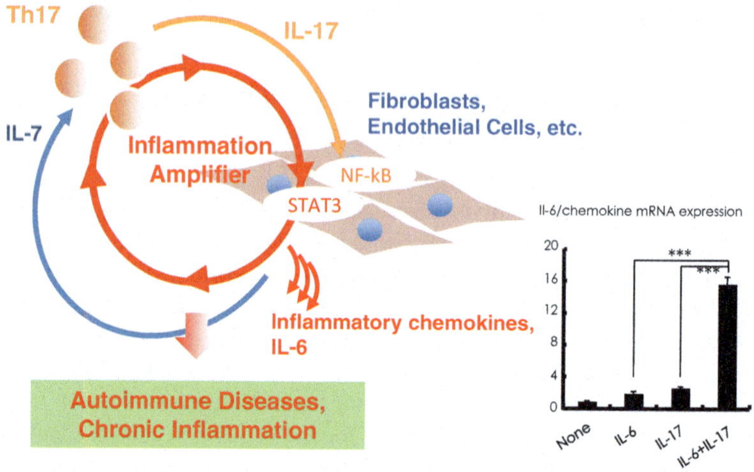

Fig. 2.3 The inflammation amplifier. The inflammation amplifier is defined as hyperinduction of IL-6 and chemokines in nonimmune cells that arise from simultaneous activation of STAT3 and NF-κB. IL-7 from nonimmune cells also contributes to enhance the inflammation amplifier by generating Th17 or sustaining Th17 survival. The amplifier is known to be essential for the pathogenesis of F759 arthritis, autoimmune encephalomyelitis (EAE), and chronic graft rejection

arthritis, whereas F759 mice that received wild-type bone marrow did. In spite of the necessity for $CD4^+$ T cells, it was clearly demonstrated that the gp130 F759 mutation, which causes hyperactivation of IL-6 signaling, is required only in nonimmune cells for F759 arthritis development to occur. Further studies have since shown that excessive IL-6 signaling in nonimmune cells promotes the production of IL-7, which is known to be important for T-cell proliferation and survival, and results in excess homeostatic proliferation of $CD4^+$ T cells. These results suggest that nonimmune cells actively control the status of immune cells including $CD4^+$ T cells, and dysregulation of this control can cause autoimmune diseases (Sawa et al. 2006).

Th17 cells have been shown to be involved in many autoimmune disease models and human diseases including rheumatoid arthritis. Consistent with this property, serum IL-17 levels and Th17 cells were both unusually high in aged F759 mice. When F759 mice were bred on an IL-17-deficient background, the arthritis was significantly suppressed. On the other hand, the overexpression of IL-17 in F759 mice accelerated arthritis development. Among the more than 30 kinds of cytokines and chemokines, only serum IL-6 and certain chemokines were upregulated after overexpression of IL-17 in vivo. Moreover, the serum concentration of IL-6 after IL-17 overexpression was significantly higher in F759 mice than in control mice, indicating a positive interaction between IL-17 and IL-6 signaling. In vitro investigation showed that a combination of IL-17 and IL-6 synergistically induces IL-6 and inflammatory chemokines in type I collagen+ nonimmune cells such as fibroblasts and endothelial cells (see bar graph in Fig. 2.3). This synergistic effect depends on NF-κB and STAT3. Thus, it was suggested that NF-κB activation by cytokines such as IL-17 stimulates a sufficiently minimal amount of IL-6 in nonimmune cells that in turn synergistically acts with IL-17 to induce more IL-6, which leads to the development of inflammation. In fact, this positive feedback loop of IL-6 signaling, or inflammation amplifier, has been found important for the pathogenesis of autoimmune diseases including F759 arthritis and experimental autoimmune encephalomyelitis (EAE), a disease that resembles multiple sclerosis in animal models (Fig. 2.3) (Ogura et al. 2008). In addition, graft rejection was significantly inhibited in mice with defective CCL2 expression caused by inflammation amplifier activation in the basement cells of tracheal epithelial cells. In this context, CCL2 expression is triggered by traumatic stress, which increases IL-6 and epidermal growth factor, followed by Th1 cell accumulation in the graft (Lee et al. 2012). Thus, the inflammation amplifier can be applied not only to autoimmunity disorders but also to other inflammatory disorders in vivo. These results suggest that the mechanisms behind the clinical efficacy of the anti-IL-6R antibody are likely to be explained by activation of the inflammation amplifier.

2.5.3 Inflammation Amplifier-Regulating Genes as Potential Therapeutic Targets for Human Diseases

Because the inhibition of IL-6 signaling by anti-IL-6R treatment has been successful in humans, the identification of genes that regulate or are regulated by inflammation amplifier activation may provide therapeutic targets against inflammatory diseases. We conducted a functional genome-wide screening by using a lentivirus library having 65500 shRNAs (16000 ORFs) and DNA microarray experiments to identify those genes related to activation of the inflammation amplifier, finding more than 1,000 (Murakami et al. 2013). When these genes were analyzed by using a public database of human genetic association studies of complex diseases and disorders (Genetic Association Database at NIH; http://geneticassociationdb.nih.gov/cgi-bin/index.cgi), a large number were found to be associated with human diseases and disorders. The enriched disease categories went well beyond autoimmune diseases to include metabolic syndromes, such as atherosclerosis and type 2 diabetes, and neurodegenerative diseases, such as Alzheimer's disease and Parkinson's disease. These results, then, could be interpreted to show that inflammation, which is induced by the inflammation amplifier, relates to various human diseases and disorders, and that IL-6 inhibitors could have wide-ranging therapeutic effects.

2.6 A Four-Step Model for MHC Class II-Associated Autoimmune Diseases

We previously demonstrated that the activation of the inflammation amplifier by IL-17 and IL-6 leads to arthritis in F759 mice. To explain this relationship, we recently proposed a four-step model for MHC class II-associated autoimmune diseases (Murakami and Hirano 2011; Murakami et al. 2011). It has been believed that cognate tissue specific-antigen recognition by autoreactive T cells is a key step for the development of tissue-specific autoimmune diseases, particularly those genetically associated with MHC class II genes such as rheumatoid arthritis (Marrack et al. 2001). Indeed, the development of arthritis in F759 mice is dependent on MHC class II and CD4 (Sawa et al. 2006). Therefore, we investigated the involvement of joint-specific antigens in the development of arthritis in F759 mice. We established F759 mice having a single T-cell receptor (TCR) that recognizes non-joint antigens. F759 mice were crossed with Rag2-deficient mice as well as OT-2 or P25 TCR-transgenic (Tg) mice whose T cells recognize MHC class II-restricted peptides from ovalbumin (OVA) or peptide 25 from *Mycobacterium tuberculosis*. Theoretically, no $CD4^+$ T cells recognize joint antigens in these offspring. Unexpectedly, $CD4^+$ T cells bearing a single TCR that recognizes antigens not related to joint tissues induce arthritis in Rag2-deficient mice that have the F759 mutation. We therefore concluded that cognate antigen recognition by effector $CD4^+$ T cells is not necessary for tissue specificity in F759 mice (Murakami et al. 2011). From this, we hypothesized that disease specificity may be determined by the tissue itself such that local events in the joint may determine and initiate the disease via the inflammation amplifier. If the Th17 cell transfer is done before inducing experimental microbleeding, only the bled leg will develop arthritis. Even Th17 cells derived from TCR transgenic mice induced arthritis in the microbleeding-induced leg of F759 mice. These findings are consistent with the idea that local events determine the disease specificity even if activation of tissue antigen-specific T cells does not occur. We further observed that T cells accumulated in the arthritic joint. This microbleeding-induced accumulation of Th17 cells is dependent on the production of CCL20, a target of the inflammation amplifier, in the joint. Disease induction requires IL-17A produced by T cells, IL-6, and enhanced STAT3 signaling in type I collagen-expressing cells (Murakami and Hirano 2011). Thus, local microbleeding facilitates IL-6- and IL-17-dependent arthritis in the absence of tissue antigen recognition by activated T cells (Murakami and Hirano 2011).

Based on these results, we proposed that certain MHC class-II associated autoimmune diseases such as rheumatoid arthritis arise through a series of at least four steps: (1) T-cell activation regardless of antigen specificity; (2) local events inducing a tissue-specific accumulation of activated T cells; (3) transient activation of the inflammation amplifier, which is triggered by $CD4^+$ T-cell-derived cytokines such as IL-17A; and (4) enhanced sensitivity to T-cell-derived cytokines and/or IL-6 in type 1 collagen+ cells in the target tissue just like F759 mutation (Murakami and Hirano 2011; Murakami et al. 2011). After these four steps, chronic activation of the inflammation amplifier followed by development of the autoimmune disease occurs. It is likely that each step interacts with the others, and the degree of the contribution of each to the pathogenesis varies with the disease. Our four-step model provides a plausible explanation why tissue-specific antigens recognized by activated $CD4^+$ T cells have not been identified in several autoimmune diseases, especially those associated with MHC class II molecules. It is likely that in diseases where tissue antigen-specific T cells play a role, tissue antigen-specific recognition by T cells bypasses the requirement of local events, even though these local events can still affect the accumulation of tissue antigen-specific T cells in the target tissue. Our four-step model, therefore, should be applicable to a wide range of autoimmune and other chronic inflammatory diseases.

2.7 A Possible Physiological Role of the Inflammation Amplifier

In the previous sections, we described a pathogenic role for the inflammation amplifier during chronic inflammation. Although aberrant amplification of IL-6 and chemokine production by uncontrolled activation of STAT3 and NF-κB trigger autoimmunity and graft rejection, we recently found that the inflammation amplifier also functions at the steady state. Activation of the inflammation amplifier has been detected in blood vessels adjacent to the central nervous system (CNS). In the course of disease development in EAE, disease-causing $CD4^+$ T cells infiltrate into the CNS, a well-known immune-privileged tissue that restricts the intrusion of immune cells into the bloodstream by the blood–brain barrier. The blood–brain barrier is a specific blood vessel structure that is mediated by tight junctions and tight liner sheets established by pericytes, astrocytes, and macrophages.

Where and how pathogenic $CD4^+$ T cells enter the CNS was unclear until recently. Sallusto et al. reported that mice lacking CCR6, a receptor for CCL20, are highly resistant to EAE, and that the choroid plexus, a specialized epithelial structure in the brain, expresses CCL20 constitutively, thereby potentially acting as an attractant for the first wave of $CCR6^+$ Th17 (Reboldi et al. 2009). In this study, however, the CNS disease was induced by means of complete Freund's adjuvant, which is widely used for active immunization in animals, but at the same time is also an inducer of strong inflammatory responses that potentially affect the pathophysiological status of the whole body including the brain and spinal cord. To reduce background inflammatory responses, particularly at the initiation stage of EAE, we utilized an adoptive transfer model in which Th17 cells obtained from MOG-immunized mice were infused into naïve recipients so that the quiescence

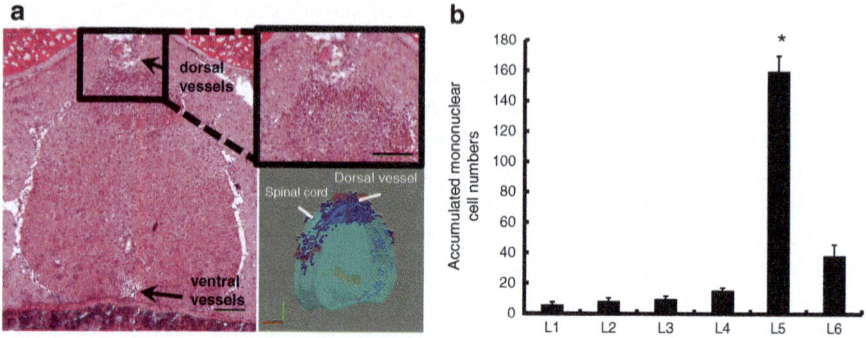

Fig. 2.4 The fifth lumbar cord is a gateway to the central nervous system (CNS). A cross section of the fifth lumbar (L5) cord (**a**) and actual cell numbers of mononuclear cells accumulated in each lumbar cord segment (**b**) at a preclinical phase of EAE. A magnified image around the dorsal vessel of L5 and a three-dimensional (3D) picture based on ten serial sections of L5 are shown on the *right* side of (**a**) (*top* and *bottom*, respectively)

of the CNS could be preserved. We found that Th17 cells preferentially accumulated in the fifth lumbar (L5) cord rather than the brain or other regions of the spinal cord at the preclinical phase of EAE (Fig. 2.4) (Arima et al. 2012), although we found pathogenic $CD4^+$ T-cell accumulation in the brain at a later stage after the T-cell transfer. CCL20 mRNA levels were highest in the dorsal vessel of L5 as compared with the other lumbar cords. Interestingly, even in naïve animals, CCL20 as well as many other chemokines were specifically upregulated in the dorsal vessel of L5. Additionally, NF-κB reporter mice showed that NF-κB activity is higher at L5 than L1 or cervical cords. Moreover, the elevated CCL20 levels at L5 were decreased in mice devoid of the inflammation amplifier such as IL-6-deficient mice (Fig. 2.5a). Even under normal conditions, it is known that some immune cells are present in the CNS, suggesting there exists a gateway to enter this restricted area. In this respect, it is tempting to speculate that low-grade activation of the inflammation amplifier at the L5 dorsal vessel in the steady state creates the gateway by inducing some level of chemokines, although direct evidence is needed to link low levels of inflammation amplifier activation and immune homeostasis, including immunological surveillance of the CNS.

2.8 Neuroimmune Interactions that Boost Inflammation Amplifier Activation

The answer to why the L5 specifically acts as the gateway comes from an unlikely source. The dorsal root ganglia (DRG) of the sensory neurons from the soleus muscle, which can be activated by a gravitation stimulus, are located beside L5 (Ohira et al. 2004). When mice were tail-suspended so that only the forelimbs could touch the ground and the hind legs were released from gravity stimuli, pathogenic Th17

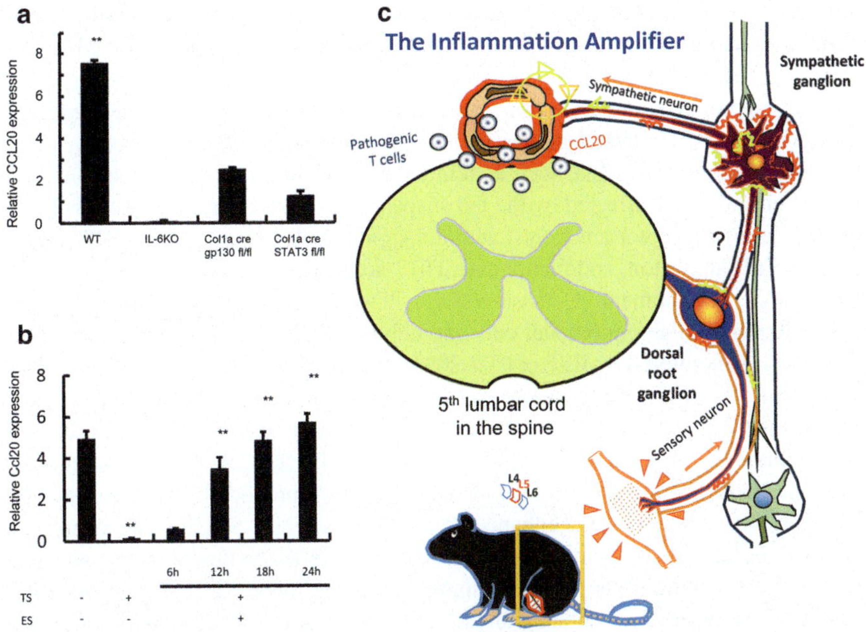

Fig. 2.5 Neural stimulation-mediated inflammation amplifier activation creates a gateway into the CNS by chemokine production. (**a**) CCL20 levels under steady state. Note that naïve mice with an inactive inflammation amplifier (*all three columns from right*) showed significantly reduced levels of CCL20, which implicates a physiological role of the inflammation amplifier in the steady state but not in the disease condition (see Fig. 2.3). (**b**) Absence of gravitational stimuli (tail suspension, TS) decreases CCL20 at the L5 dorsal vessel. Electric stimulation (ES) during TS restores the levels in a time-dependent manner. (**c**) Schematic representation of neural stimulation-mediated activation of the inflammation amplifier. Neural signals from gravitational stimuli in soleus muscles reach the L5 dorsal root ganglion. Subsequent activation of sympathetic nerves alters the status of L5 dorsal vessel endothelial cells to enhance the inflammation amplifier, which leads to the production of chemokines including CCL20. Norepinephrine is a mediator between the neural signal and inflammation amplifier activation. A neural network from the soleus muscle-derived sensory neurons to sympathetic neurons that reach L5 is not defined (depicted with a *question mark*)

cells negligibly accumulated at L5. Instead, these cells utilized a new gateway opened by gravity stimuli to the forelimbs and accumulated at the cervical cords. Consistent with this result, tail suspension significantly inhibited CCL20 expression in L5 dorsal blood vessels and decreased the expression of a neural activation marker, c-Fos, in the L5 DRG. In addition, when the soleus muscles of tail-suspended mice were artificially stimulated by electric pulses, CCL20 expression (Fig. 2.5b), pathogenic Th17 accumulation and c-Fos levels were all restored at L5. These data strongly suggest that neural activation via anti-gravitational responses by the soleus muscles plays a role in the activation of the inflammation amplifier and subsequent expression of chemokines including Th17-attracting CCL20 in L5 dorsal blood vessels before the development of EAE. How do afferent sensory neurons from the soleus muscle influence the status of the blood vessels at L5? Although a precise neural network remains unidentified, we demonstrated that sympathetic nerves are involved. Blood flow speed at the L5 dorsal vessel became slower when mice were tail-suspended, whereas electronic stimulation of the soleus muscles increased the speed, suggesting a contribution of automatic nerves including sympathetic ones. Importantly, blood flow speeds at other vessels, such as femoral vessels, brain surface vessels, and the portal vein, were not affected by the tail suspension. Furthermore, treatment with the norepinephrine receptor antagonist atenolol significantly suppressed CCL20 expression, NF-κB activation, and pathogenic Th17 accumulation in L5 vessels and abrogated EAE development. Consistent with these in vivo results, the addition of norepinephrine to an endothelial cell line culture enhanced inflammation amplifier activation as monitored by IL-6 or CCL20 expression. Thus, neural stimulation of the soleus muscles by gravity causes sympathetic nerve stimulation, which creates a gateway to the CNS at L5 vessels by activating the inflammation amplifier with secreted norepinephrine. MOG-specific, disease-causing Th17 cells exploit this gateway to infiltrate the CNS and induce local inflammation by producing cytokines such as IL-17 and IL-6, which further enhances the activation (Fig. 2.5c). Such neuroimmune interactions have also been reported by Tracey et al., in which they showed that vagus nerve stimulation inhibits proinflammatory cytokine release through the nicotinic acetylcholine receptor α7 subunit, and identified a subset of T cells that produce acetylcholine and can relay the neural signals (Borovikova et al. 2000); (Wang et al. 2003); (Rosas-Ballina et al. 2011). Therefore, interfering neuroimmune responses may be another promising approach for therapeutic interventions to inflammatory diseases.

2.9 Gate Theory

Gravitational stimuli upregulate various chemokines as a result of inflammation amplifier activation at L5 cord vessels via the sensory neurons of soleus muscles extending from the L5 DRG. This neural event creates a local gate for immune cells to enter the CNS. It turns out that stimulating other muscles can create a similar gate into blood vessels where DRG neurons are located. Electronic stimulation of the quadriceps or thigh muscles, which are known to be controlled by L3 DRG neurons, upregulated the expression of the inflammation amplifier target chemokine, CCL20, across L3 cord vessels in mice. Similarly, chemokine levels from the fifth cervical to fifth thoracic cord vessels were elevated by stimulations of epitrochlearis/triceps brachii and upper arm muscles regulated by neurons located at corresponding areas (Arima et al. 2012). In addition, we have found that the location and degree of pathogenic $CD4^+$ T-cell infiltration into the CNS can be changed in mice under mental stress (unpublished data). These observations led us to propose a gate theory in which the invasion of immune cells into target organs can be modulated by manipulating nerve activity. Although further studies are required to generalize this theory to organs, such a theory has promise for novel therapy against many chronic inflammatory diseases.

2.10 Conclusion

In this review, we summarized recent advances in understanding how IL-6 plays a central role in inflammation and various diseases. The success of anti-IL-6R therapy has shown the promise of targeting IL-6 signaling and production for treatment against inflammatory diseases. The inflammation amplifier is one prominent mechanism that produces a large amount of chemokines and IL-6 from nonimmune cells in vivo and is therefore a candidate worth considering for future targets. In addition to targeting traditional IL-6 signaling molecules such as JAK and STAT3, finding ways to inhibit the inflammation amplifier and a therapeutic method that can modulate regional neural activity that activates the amplifier may be beneficial to sufferers of chronic inflammatory diseases.

References

Akira S, Isshiki H, Sugita T, Tanabe O, Kinoshita S, Nishio Y, Nakajima T, Hirano T, Kishimoto T (1990) A nuclear factor for IL-6 expression (NF-IL6) is a member of a C/EBP family. EMBO J 9:1897–1906

Akira S, Nishio Y, Inoue M, Wang XJ, Wei S, Matsusaka T, Yoshida K, Sudo T, Naruto M, Kishimoto T (1994) Molecular cloning of APRF, a novel IFN-stimulated gene factor 3 p91-related transcription factor involved in the gp130-mediated signaling pathway. Cell 77:63–71

Arima Y, Harada M, Kamimura D, Park JH, Kawano F, Yull FE, Kawamoto T, Iwakura Y, Betz UA, Marquez G, Blackwell TS, Ohira Y, Hirano T, Murakami M (2012) Regional neural activation defines a gateway for autoreactive T cells to cross the blood–brain barrier. Cell 148:447–457

Atsumi T, Ishihara K, Kamimura D, Ikushima H, Ohtani T, Hirota S, Kobayashi H, Park SJ, Saeki Y, Kitamura Y, Hirano T (2002) A point mutation of Tyr-759 in interleukin 6 family cytokine receptor subunit gp130 causes autoimmune arthritis. J Exp Med 196:979–990

Atsumi T, Sato M, Kamimura D, Moroi A, Iwakura Y, Betz UA, Yoshimura A, Nishihara M, Hirano T, Murakami M (2009) IFN-gamma expression in CD8+ T cells regulated by IL-6 signal is involved in superantigen-mediated CD4+ T cell death. Int Immunol 21:73–80

Babon JJ, Kershaw NJ, Murphy JM, Varghese LN, Laktyushin A, Young SN, Lucet IS, Norton RS, Nicola NA (2012) Suppression of cytokine signaling by SOCS3: characterization of the mode of inhibition and the basis of its specificity. Immunity 36:239–250

Barr TA, Shen P, Brown S, Lampropoulou V, Roch T, Lawrie S, Fan B, O'Connor RA, Anderton SM, Bar-Or A, Fillatreau S, Gray D (2012) B cell depletion therapy ameliorates autoimmune disease through ablation of IL-6-producing B cells. J Exp Med 209:1001–1010

Borovikova LV, Ivanova S, Zhang M, Yang H, Botchkina GI, Watkins LR, Wang H, Abumrad N, Eaton JW, Tracey KJ (2000) Vagus nerve stimulation attenuates the systemic inflammatory response to endotoxin. Nature (Lond) 405:458–462

Boulanger MJ, Chow DC, Brevnova EE, Garcia KC (2003) Hexameric structure and assembly of the interleukin-6/IL-6 alpha-receptor/gp130 complex. Science 300:2101–2104

Calle EE, Rodriguez C, Walker-Thurmond K, Thun MJ (2003) Overweight, obesity, and mortality from cancer in a prospectively studied cohort of U.S. adults. N Engl J Med 348:1625–1638

Chalaris A, Garbers C, Rabe B, Rose-John S, Scheller J (2011) The soluble interleukin 6 receptor: generation and role in inflammation and cancer. Eur J Cell Biol 90:484–494

Chiang JJ, Eisenberger NI, Seeman TE, Taylor SE (2012) Negative and competitive social interactions are related to heightened proinflammatory cytokine activity. Proc Natl Acad Sci USA 109:1878–1882

Collison LW, Delgoffe GM, Guy CS, Vignali KM, Chaturvedi V, Fairweather D, Satoskar AR, Garcia KC, Hunter CA, Drake CG, Murray PJ, Vignali DA (2012) The composition and signaling of the IL-35 receptor are unconventional. Nat Immunol 13:290–299

Dai L, Liu D, Guo H, Wang Y, Bai Y (2012) Association between polymorphism in the promoter region of interleukin 6 (−174 G/C) and risk of Alzheimer's disease: a meta-analysis. J Neurol 259:414–419

Derouet D, Rousseau F, Alfonsi F, Froger J, Hermann J, Barbier F, Perret D, Diveu C, Guillet C, Preisser L, Dumont A, Barbado M, Morel A, deLapeyriere O, Gascan H, Chevalier S (2004) Neuropoietin, a new IL-6-related cytokine signaling through the ciliary neurotrophic factor receptor. Proc Natl Acad Sci USA 101:4827–4832

Dhamija S, Kuehne N, Winzen R, Doerrie A, Dittrich-Breiholz O, Thakur BK, Kracht M, Holtmann H (2011) Interleukin-1 activates synthesis of interleukin-6 by interfering with a KH-type splicing regulatory protein (KSRP)-dependent translational silencing mechanism. J Biol Chem 286:33279–33288

Drogemuller K, Helmuth U, Brunn A, Sakowicz-Burkiewicz M, Gutmann DH, Mueller W, Deckert M, Schluter D (2008) Astrocyte gp130 expression is critical for the control of *Toxoplasma* encephalitis. J Immunol 181:2683–2693

Ellingsgaard H, Hauselmann I, Schuler B, Habib AM, Baggio LL, Meier DT, Eppler E, Bouzakri K, Wueest S, Muller YD, Hansen AM, Reinecke M, Konrad D, Gassmann M, Reimann F, Halban PA, Gromada J, Drucker DJ, Gribble FM, Ehses JA, Donath MY (2011) Interleukin-6 enhances insulin secretion by increasing glucagon-like peptide-1 secretion from L cells and alpha cells. Nat Med 17:1481–1489

Ferreira MA, Matheson MC, Duffy DL, Marks GB, Hui J, Le Souef P, Danoy P, Baltic S, Nyholt DR, Jenkins M, Hayden C, Willemsen G, Ang W, Kuokkanen M, Beilby J, Cheah F, de Geus EJ, Ramasamy A, Vedantam S, Salomaa V, Madden PA, Heath AC, Hopper JL, Visscher PM, Musk B, Leeder SR, Jarvelin MR, Pennell C, Boomsma DI, Hirschhorn JN, Walters H, Martin NG, James A, Jones G, Abramson MJ, Robertson CF, Dharmage SC, Brown MA, Montgomery GW, Thompson PJ (2011) Identification of IL6R and chromosome 11q13.5 as risk loci for asthma. Lancet 378:1006–1014

Franke A, Balschun T, Karlsen TH, Hedderich J, May S, Lu T, Schuldt D, Nikolaus S, Rosenstiel P, Krawczak M, Schreiber S (2008) Replication of signals from recent studies of Crohn's disease identifies previously unknown disease loci for ulcerative colitis. Nat Genet 40:713–715

Ghoreschi K, Laurence A, Yang XP, Tato CM, McGeachy MJ, Konkel JE, Ramos HL, Wei L, Davidson TS, Bouladoux N, Grainger JR, Chen Q, Kanno Y, Watford WT, Sun HW, Eberl G, Shevach EM, Belkaid Y, Cua DJ, Chen W, O'Shea JJ (2010) Generation of pathogenic T(H)17 cells in the absence of TGF-beta signalling. Nature 467:967–971

Gilbert LA, Hemann MT (2010) DNA damage-mediated induction of a chemoresistant niche. Cell 143:355–366

Handel U, Brunn A, Drogemuller K, Muller W, Deckert M, Schluter D (2012) Neuronal gp130 expression is crucial to prevent neuronal loss, hyperinflammation, and lethal course of murine *Toxoplasma* encephalitis. Am J Pathol 181:163–173

Harker JA, Lewis GM, Mack L, Zuniga EI (2011) Late interleukin-6 escalates T follicular helper cell responses and controls a chronic viral infection. Science 334:825–829

Hatziapostolou M, Polytarchou C, Aggelidou E, Drakaki A, Poultsides GA, Jaeger SA, Ogata H, Karin M, Struhl K, Hadzopoulou-Cladaras M, Iliopoulos D (2011) An HNF4alpha-miRNA inflammatory feedback circuit regulates hepatocellular oncogenesis. Cell 147:1233–1247

Hibi M, Murakami M, Saito M, Hirano T, Taga T, Kishimoto T (1990) Molecular cloning and expression of an IL-6 signal transducer, gp130. Cell 63:1149–1157

Hingorani AD, Casas JP (2012) The interleukin-6 receptor as a target for prevention of coronary heart disease: a mendelian randomisation analysis. Lancet 379:1214–1224

Hirano T, Yasukawa K, Harada H, Taga T, Watanabe Y, Matsuda T, Kashiwamura S, Nakajima K, Koyama K, Iwamatsu A et al (1986) Complementary DNA for a novel human interleukin (BSF-2) that induces B lymphocytes to produce immunoglobulin. Nature (Lond) 324:73–76

Iliopoulos D, Hirsch HA, Struhl K (2009) An epigenetic switch involving NF-kappaB, Lin28, Let-7 MicroRNA, and IL6 links inflammation to cell transformation. Cell 139:693–706

Ivanov BS II, McKenzie L, Zhou CE, Tadokoro A, Lepelley JJ, Lafaille DJC, Littman DR (2006) The orphan nuclear receptor RORgamma directs the differentiation program of proinflammatory IL-17+ T helper cells. Cell 126:1121–1133

Jones MR, Quinton LJ, Blahna MT, Neilson JR, Fu S, Ivanov AR, Wolf DA, Mizgerd JP (2009) Zcchc11-dependent uridylation of microRNA directs cytokine expression. Nat Cell Biol 11:1157–1163

Jones SA, Scheller J, Rose-John S (2011) Therapeutic strategies for the clinical blockade of IL-6/gp130 signaling. J Clin Invest 121:3375–3383

Jostock T, Mullberg J, Ozbek S, Atreya R, Blinn G, Voltz N, Fischer M, Neurath MF, Rose-John S (2001) Soluble gp130 is the natural inhibitor of soluble interleukin-6 receptor transsignaling responses. Eur J Biochem 268:160–167

Kamimura D, Ishihara K, Hirano T (2003) IL-6 signal transduction and its physiological roles: the signal orchestration model. Rev Physiol Biochem Pharmacol 149:1–38

Karnowski A, Chevrier S, Belz GT, Mount A, Emslie D, D'Costa K, Tarlinton DM, Kallies A, Corcoran LM (2012) B and T cells collaborate in antiviral responses via IL-6, IL-21, and transcriptional activator and coactivator, Oct2 and OBF-1. J Exp Med 209:2049–2064

Kim MY, Oskarsson T, Acharyya S, Nguyen DX, Zhang XH, Norton L, Massague J (2009) Tumor self-seeding by circulating cancer cells. Cell 139:1315–1326

Kitamura H, Kamon H, Sawa S, Park SJ, Katunuma N, Ishihara K, Murakami M, Hirano T (2005) IL-6-STAT3 controls intracellular MHC class II alphabeta dimer level through cathepsin S activity in dendritic cells. Immunity 23:491–502

Lee J, Nakagiri T, Oto T, Harada M, Morii E, Shintani Y, Inoue M, Iwakura Y, Miyoshi S, Okumura M, Hirano T, Murakami M (2012) IL-6 amplifier, NF-kappaB-triggered positive feedback for IL-6 signaling, in grafts is involved in allogeneic rejection responses. J Immunol 189:1928–1936

Marrack P, Kappler J, Kotzin BL (2001) Autoimmune disease: why and where it occurs. Nat Med 7:899–905

Matsushita K, Takeuchi O, Standley DM, Kumagai Y, Kawagoe T, Miyake T, Satoh T, Kato H, Tsujimura T, Nakamura H, Akira S (2009) Zc3h12a is an RNase essential for controlling immune responses by regulating mRNA decay. Nature (Lond) 458:1185–1190

Minegishi Y, Saito M, Tsuchiya S, Tsuge I, Takada H, Hara T, Kawamura N, Ariga T, Pasic S, Stojkovic O, Metin A, Karasuyama H (2007) Dominant-negative mutations in the DNA-binding domain of STAT3 cause hyper-IgE syndrome. Nature (Lond) 448:1058–1062

Murakami M, Hirano T (2011) A four-step model for the IL-6 amplifier, a regulator of chronic inflammations in tissue-specific MHC class II-associated autoimmune diseases. Front Immunol 2:22

Murakami M, Narazaki M, Hibi M, Yawata H, Yasukawa K, Hamaguchi M, Taga T, Kishimoto T (1991) Critical cytoplasmic region of the interleukin 6 signal transducer gp130 is conserved in the cytokine receptor family. Proc Natl Acad Sci USA 88:11349–11353

Murakami M, Hibi M, Nakagawa N, Nakagawa T, Yasukawa K, Yamanishi K, Taga T, Kishimoto T (1993) IL-6-induced homodimerization of gp130 and associated activation of a tyrosine kinase. Science 260:1808–1810

Murakami M, Kamimura D, Hirano T (2004) New IL-6 (gp130) family cytokine members, CLC/NNT1/BSF3 and IL-27. Growth Factors 22:75–77

Murakami M, Okuyama Y, Ogura H, Asano S, Arima Y, Tsuruoka M, Harada M, Kanamoto M, Sawa Y, Iwakura Y, Takatsu K, Kamimura D, Hirano T (2011) Local microbleeding facilitates IL-6- and IL-17-dependent arthritis in the absence of tissue antigen recognition by activated T cells. J Exp Med 208:103–114

Murakami M, Harada M, Kamimura D, Ogura H, Okuyama Y, Kumai N, Okuyama A, Singh R, Jiang J-J, Atsumi T, Shiraya S, Nakatsuji Y, Kinoshita M, Kohsaka H, Nishida M, Sakoda S, Miyasaka N, Yamaguchi-Takihara K, Hirano T (2013) Disease-association analysis of an inflammation-related feedback loop. Cell Rep 3(3):946–959

Nishida K, Yoshida Y, Itoh M, Fukada T, Ohtani T, Shirogane T, Atsumi T, Takahashi-Tezuka M, Ishihara K, Hibi M, Hirano T (1999) Gab-family adapter proteins act downstream of cytokine and growth factor receptors and T- and B-cell antigen receptors. Blood 93:1809–1816

Nishihara M, Ogura H, Ueda N, Tsuruoka M, Kitabayashi C, Tsuji F, Aono H, Ishihara K, Huseby E, Betz UA, Murakami M, Hirano T (2007) IL-6-gp130-STAT3 in T cells directs the development of IL-17+ Th with a minimum effect on that of Treg in the steady state. Int Immunol 19:695–702

Nishimoto N, Kishimoto T, Yoshizaki K (2000) Anti-interleukin 6 receptor antibody treatment in rheumatic disease. Ann Rheum Dis 59(Suppl 1):i21–i27

Nurieva RI, Chung Y, Martinez GJ, Yang XO, Tanaka S, Matskevitch TD, Wang YH, Dong C (2009) Bcl6 mediates the development of T follicular helper cells. Science 325:1001–1005

O'Donovan A, Hughes BM, Slavich GM, Lynch L, Cronin MT, O'Farrelly C, Malone KM (2010) Clinical anxiety, cortisol and interleukin-6: evidence for specificity in emotion-biology relationships. Brain Behav Immun 24:1074–1077

Ogura H, Murakami M, Okuyama Y, Tsuruoka M, Kitabayashi C, Kanamoto M, Nishihara M, Iwakura Y, Hirano T (2008) Interleukin-17 promotes autoimmunity by triggering a positive-feedback loop via interleukin-6 induction. Immunity 29:628–636

Ohira Y, Kawano F, Stevens JL, Wang XD, Ishihara A (2004) Load-dependent regulation of neuromuscular system. J Gravit Physiol 11:P127–P128

Ohtani T, Ishihara K, Atsumi T, Nishida K, Kaneko Y, Miyata T, Itoh S, Narimatsu M, Maeda H, Fukada T, Itoh M, Okano H, Hibi M, Hirano T (2000) Dissection of signaling cascades through gp130 in vivo: reciprocal roles for STAT3- and SHP2-mediated signals in immune responses. Immunity 12:95–105

Park SJ, Nakagawa T, Kitamura H, Atsumi T, Kamon H, Sawa S, Kamimura D, Ueda N, Iwakura Y, Ishihara K, Murakami M, Hirano T (2004) IL-6 regulates in vivo dendritic cell differentiation through STAT3 activation. J Immunol 173:3844–3854

Park EJ, Lee JH, Yu GY, He G, Ali SR, Holzer RG, Osterreicher CH, Takahashi H, Karin M (2010) Dietary and genetic obesity promote liver inflammation and tumorigenesis by enhancing IL-6 and TNF expression. Cell 140:197–208

Pedersen BK, Steensberg A, Schjerling P (2001) Exercise and interleukin-6. Curr Opin Hematol 8:137–141

Pellegrini M, Calzascia T, Toe JG, Preston SP, Lin AE, Elford AR, Shahinian A, Lang PA, Lang KS, Morre M, Assouline B, Lahl K, Sparwasser T, Tedder TF, Paik JH, DePinho RA, Basta S, Ohashi PS, Mak TW (2011) IL-7 engages multiple mechanisms to overcome chronic viral infection and limit organ pathology. Cell 144:601–613

Pilati C, Amessou M, Bihl MP, Balabaud C, Nhieu JT, Paradis V, Nault JC, Izard T, Bioulac-Sage P, Couchy G, Poussin K, Zucman-Rossi J (2011) Somatic mutations activating STAT3 in human inflammatory hepatocellular adenomas. J Exp Med 208:1359–1366

Quintana A, Erta M, Ferrer B, Comes G, Giralt M, Hidalgo J (2013) Astrocyte-specific deficiency of interleukin-6 and its receptor reveal specific roles in survival, body weight and behavior. Brain Behav Immun 27(1):162–173

Reboldi A, Coisne C, Baumjohann D, Benvenuto F, Bottinelli D, Lira S, Uccelli A, Lanzavecchia A, Engelhardt B, Sallusto F (2009) C-C chemokine receptor 6-regulated entry of TH-17 cells into the CNS through the choroid plexus is required for the initiation of EAE. Nat Immunol 10:514–523

Rebouissou S, Amessou M, Couchy G, Poussin K, Imbeaud S, Pilati C, Izard T, Balabaud C, Bioulac-Sage P, Zucman-Rossi J (2009) Frequent in-frame somatic deletions activate gp130 in inflammatory hepatocellular tumours. Nature (Lond) 457:200–204

Rokavec M, Wu W, Luo JL (2012) IL6-mediated suppression of miR-200c directs constitutive activation of inflammatory signaling circuit driving transformation and tumorigenesis. Mol Cell 45:777–789

Rosas-Ballina M, Olofsson PS, Ochani M, Valdes-Ferrer SI, Levine YA, Reardon C, Tusche MW, Pavlov VA, Andersson U, Chavan S, Mak TW, Tracey KJ (2011) Acetylcholine-synthesizing T cells relay neural signals in a vagus nerve circuit. Science 334:98–101

Sarwar N, Butterworth AS, Freitag DF, Gregson J, Willeit P, Gorman DN, Gao P, Saleheen D, Rendon A, Nelson CP, Braund PS, Hall AS, Chasman DI, Tybjaerg-Hansen A, Chambers JC, Benjamin EJ, Franks PW, Clarke R, Wilde AA, Trip MD, Steri M, Witteman JC, Qi L, van der Schoot CE, de Faire U, Erdmann J, Stringham HM, Koenig W, Rader DJ, Melzer D, Reich D, Psaty BM, Kleber ME, Panagiotakos DB, Willeit J, Wennberg P, Woodward M, Adamovic S, Rimm EB, Meade TW, Gillum RF, Shaffer JA, Hofman A, Onat A, Sundstrom J, Wassertheil-Smoller S, Mellstrom D, Gallacher J, Cushman M, Tracy RP, Kauhanen J, Karlsson M, Salonen JT, Wilhelmsen L, Amouyel P, Cantin B, Best LG, Ben-Shlomo Y, Manson JE, Davey-Smith G, de Bakker PI, O'Donnell CJ, Wilson JF, Wilson AG, Assimes TL, Jansson JO, Ohlsson C, Tivesten A, Ljunggren O, Reilly MP, Hamsten A, Ingelsson E, Cambien F, Hung J, Thomas GN, Boehnke M, Schunkert H, Asselbergs FW, Kastelein JJ, Gudnason V, Salomaa V, Harris TB, Kooner JS, Allin KH, Nordestgaard BG, Hopewell JC, Goodall AH, Ridker PM, Holm H, Watkins H, Ouwehand WH, Samani NJ, Kaptoge S, Di Angelantonio E, Harari O, Danesh J (2012) Interleukin-6 receptor pathways in coronary heart disease: a collaborative meta-analysis of 82 studies. Lancet 379:1205–1213

Sawa S, Kamimura D, Jin GH, Morikawa H, Kamon H, Nishihara M, Ishihara K, Murakami M, Hirano T (2006) Autoimmune arthritis associated with mutated interleukin (IL)-6 receptor gp130 is driven by STAT3/IL-7-dependent homeostatic proliferation of CD4+ T cells. J Exp Med 203:1459–1470

Sawa Y, Arima Y, Ogura H, Kitabayashi C, Jiang JJ, Fukushima T, Kamimura D, Hirano T, Murakami M (2009) Hepatic interleukin-7 expression regulates T cell responses. Immunity 30:447–457

Skiniotis G, Boulanger MJ, Garcia KC, Walz T (2005) Signaling conformations of the tall cytokine receptor gp130 when in complex with IL-6 and IL-6 receptor. Nat Struct Mol Biol 12: 545–551

Srivastava S, Boyer JL (2010) Psychological stress is associated with relapse in type 1 autoimmune hepatitis. Liver Int 30:1439–1447

Stahl EA, Raychaudhuri S, Remmers EF, Xie G, Eyre S, Thomson BP, Li Y, Kurreeman FA, Zhernakova A, Hinks A, Guiducci C, Chen R, Alfredsson L, Amos CI, Ardlie KG, Barton A, Bowes J, Brouwer E, Burtt NP, Catanese JJ, Coblyn J, Coenen MJ, Costenbader KH, Criswell LA, Crusius JB, Cui J, de Bakker PI, De Jager PL, Ding B, Emery P, Flynn E, Harrison P, Hocking LJ, Huizinga TW, Kastner DL, Ke X, Lee AT, Liu X, Martin P, Morgan AW, Padyukov L, Posthumus MD, Radstake TR, Reid DM, Seielstad M, Seldin MF, Shadick NA, Steer S, Tak PP, Thomson W, van der Helm-van Mil AH, van der Horst-Bruinsma IE, van der Schoot CE, van Riel PL, Weinblatt ME, Wilson AG, Wolbink GJ, Wordsworth BP, Wijmenga C, Karlson EW, Toes RE, de Vries N, Begovich AB, Worthington J, Siminovitch KA, Gregersen PK, Klareskog L, Plenge RM (2010) Genome-wide association study meta-analysis identifies seven new rheumatoid arthritis risk loci. Nat Genet 42:508–514

Steptoe A, Hamer M, Chida Y (2007) The effects of acute psychological stress on circulating inflammatory factors in humans: a review and meta-analysis. Brain Behav Immun 21:901–912

Stumhofer JS, Tait ED, Quinn WJ 3rd, Hosken N, Spudy B, Goenka R, Fielding CA, O'Hara AC, Chen Y, Jones ML, Saris CJ, Rose-John S, Cua DJ, Jones SA, Elloso MM, Grotzinger J, Cancro MP, Levin SD, Hunter CA (2010) A role for IL-27p28 as an antagonist of gp130-mediated signaling. Nat Immunol 11:1119–1126

Suthaus J, Stuhlmann-Laeisz C, Tompkins VS, Rosean TR, Klapper W, Tosato G, Janz S, Scheller J, Rose-John S (2012) HHV-8-encoded viral IL-6 collaborates with mouse IL-6 in the development of multicentric Castleman disease in mice. Blood 119:5173–5181

Taga T, Kishimoto T (1997) Gp130 and the interleukin-6 family of cytokines. Annu Rev Immunol 15:797–819

Taga T, Hibi M, Hirata Y, Yamasaki K, Yasukawa K, Matsuda T, Hirano T, Kishimoto T (1989) Interleukin-6 triggers the association of its receptor with a possible signal transducer, gp130. Cell 58:573–581

Takahashi-Tezuka M, Yoshida Y, Fukada T, Ohtani T, Yamanaka Y, Nishida K, Nakajima K, Hibi M, Hirano T (1998) Gab1 acts as an adapter molecule linking the cytokine receptor gp130 to ERK mitogen-activated protein kinase. Mol Cell Biol 18:4109–4117

Tebbutt NC, Giraud AS, Inglese M, Jenkins B, Waring P, Clay FJ, Malki S, Alderman BM, Grail D, Hollande F, Heath JK, Ernst M (2002) Reciprocal regulation of gastrointestinal homeostasis by SHP2 and STAT-mediated trefoil gene activation in gp130 mutant mice. Nat Med 8:1089–1097

Veldhoen M, Hocking RJ, Flavell RA, Stockinger B (2006) Signals mediated by transforming growth factor-beta initiate autoimmune encephalomyelitis, but chronic inflammation is needed to sustain disease. Nat Immunol 7:1151–1156

Vgontzas AN, Bixler EO, Lin HM, Prolo P, Trakada G, Chrousos GP (2005) IL-6 and its circadian secretion in humans. Neuroimmunomodulation 12:131–140

Waetzig GH, Rose-John S (2012) Hitting a complex target: an update on interleukin-6 trans-signalling. Expert Opin Ther Targets 16:225–236

Wang J, Homer RJ, Chen Q, Elias JA (2000) Endogenous and exogenous IL-6 inhibit aeroallergen-induced Th2 inflammation. J Immunol 165:4051–4061

Wang H, Yu M, Ochani M, Amella CA, Tanovic M, Susarla S, Li JH, Yang H, Ulloa L, Al-Abed Y, Czura CJ, Tracey KJ (2003) Nicotinic acetylcholine receptor alpha7 subunit is an essential regulator of inflammation. Nature (Lond) 421:384–388

Yamasaki K, Taga T, Hirata Y, Yawata H, Kawanishi Y, Seed B, Taniguchi T, Hirano T, Kishimoto T (1988) Cloning and expression of the human interleukin-6 (BSF-2/IFN beta 2) receptor. Science 241:825–828

Yasukawa H, Ohishi M, Mori H, Murakami M, Chinen T, Aki D, Hanada T, Takeda K, Akira S, Hoshijima M, Hirano T, Chien KR, Yoshimura A (2003) IL-6 induces an anti-inflammatory response in the absence of SOCS3 in macrophages. Nat Immunol 4:551–556

Yin YW, Hu AM, Sun QQ, Liu HL, Wang Q, Zeng YH, Xu RJ, Hou ZZ, Zhang ZD, Zhang SJ, Zhou MQ (2012) Association between interleukin-6 gene −174 G/C polymorphism and the risk of coronary heart disease: a meta-analysis of 20 studies including 9619 cases and 10,919 controls. Gene (Amst) 503:25–30

Yoshimura A, Naka T, Kubo M (2007) SOCS proteins, cytokine signalling and immune regulation. Nat Rev Immunol 7:454–465

Zhang X, Yue P, Page BD, Li T, Zhao W, Namanja AT, Paladino D, Zhao J, Chen Y, Gunning PT, Turkson J (2012) Orally bioavailable small-molecule inhibitor of transcription factor Stat3 regresses human breast and lung cancer xenografts. Proc Natl Acad Sci USA 109:9623–9628

Zielinski CE, Mele F, Aschenbrenner D, Jarrossay D, Ronchi F, Gattorno M, Monticelli S, Lanzavecchia A, Sallusto F (2012) Pathogen-induced human TH17 cells produce IFN-gamma or IL-10 and are regulated by IL-1beta. Nature (Lond) 484:514–518

Chapter 3
The Roles of IL-17A and IL-17F in Infection and Inflammatory Disorders

Harumichi Ishigame and Susumu Nakae

Abstract Interleukin-17A (IL-17A) is the signature cytokine of the recently identified T helper 17 (Th17) cell subset. IL-17F shows the highest homology to IL-17A among the IL-17 family and binds to the same IL-17 receptor A (IL-17RA) and IL-17RC receptor complex as IL-17A. Recent studies have demonstrated that IL-17A and IL-17F are also produced by multiple lineages of immune cells, including γδ T cells, NKT cells, and innate lymphoid cells. These cytokines are involved in the development of inflammation, autoimmunity, allergy, and tumors, and also play important roles in host defenses against bacterial and fungal infections. However, it is now clear that IL-17A and IL-17F have overlapping—yet distinct—roles in innate and adoptive immune responses. Here, we describe the recent data on the roles of IL-17A and IL-17F in inflammatory disorders and host defense and discuss the potential of these molecules as therapeutic targets.

Keywords Allergy • Autoimmunity • Bacterial infection • Fungal infection • Th17 cell

---

H. Ishigame
Department of Immunobiology, Yale University School of Medicine, New Haven, CT 06520, USA

S. Nakae (✉)
Frontier Research Initiative, Laboratory of Systems Biology, Center for Experimental Medicine and Systems Biology, The Institute of Medical Science, University of Tokyo, 4-6-1 Shirokanedai, Minato, Tokyo 108-8639, Japan

Precursory Research for Embryonic Science and Technology (PRESTO), Japan Science and Technology Agency, Saitama, Japan
e-mail: snakae@ims.u-tokyo.ac.jp

T. Yoshimoto and T. Yoshimoto (eds.), *Cytokine Frontiers: Regulation of Immune Responses in Health and Disease*, DOI 10.1007/978-4-431-54442-5_3, © Springer Japan 2014

3.1 Introduction

Upon antigenic stimulation, naive CD4⁺ T cells differentiate into discrete functional T-cell subsets that have distinct cytokine production profiles. These different types of Th cells mediate distinct types of immune responses to protect the host against pathogens; however, their dysregulation can lead to autoimmune and allergic diseases (Zhou et al. 2009; Zhu et al. 2010). The identification of the T helper 17 (Th17) subset of CD4⁺ T cells was one of the most important advances in T-cell immunology since the discovery of Th1 and Th2 cells by Mosmann, Coffman, and their colleagues more than two decades ago (Mosmann et al. 1986). Interleukin-17A (IL-17A), commonly called IL-17, is produced by Th17 cells (Fig. 3.1). Five additional structurally related cytokines were recently identified, that is, IL-17B, IL-17C, IL-17D, IL-17E (also called IL-25), and IL-17F, which comprise the IL-17 family (Weaver et al. 2007; Iwakura et al. 2011). Th17 cells also produce IL-17F, IL-21, and IL-22, whereas Th1 and Th2 cells produce mainly interferon-γ (IFN-γ) and IL-4, respectively (Zhou et al. 2009; Zhu et al. 2010). Recent studies of IL-23, an IL-12 family cytokine, and Th17 cells have revealed important roles for IL-17A and IL-17F in the development of allergic and autoimmune diseases as well as in protective mechanisms against pathogens, functions that were previously believed to be mediated by Th1 or Th2 cells.

In this review, we summarize the recent progress of functional studies of IL-17A and IL-17F in inflammatory disorders and infectious diseases and then discuss areas for future research and the therapeutic potential of targeting IL-17A and IL-17F signaling in inflammatory disorders.

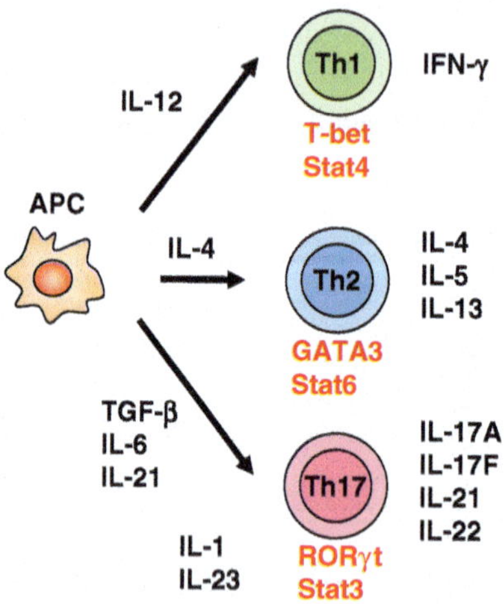

Fig. 3.1 T helper (Th)1, Th2, and Th17 cell differentiation by interleukin (IL-12) induces differentiation of naive CD4⁺ T cells into interferon (IFN)-γ-producing Th1 cells through Stat4 and T-bet activation. In contrast, IL-4 induces Stat6 activation, followed by expression of GATA-3. Th17 cell differentiation is induced by transforming growth factor (TGF)-β plus IL-6 or IL-21 by activating Stat3 and retinoic acid receptor-related orphan receptor-γt (RORγt). IL-1 and IL-23 are also required for differentiation, growth, and survival of Th17 cells. *APC* antigen-presenting cells

3.2 Biological Actions of IL-17A and IL-17F

IL-17A and IL-17F are highly homologous and bind to the same receptor complex. Furthermore, IL-17A and IL-17F can both be secreted as disulfide-linked homodimers or heterodimers. Thus, these two molecules are likely to have similar biological activities (reviewed by Chang and Dong 2011; Iwakura et al. 2011). Indeed, both IL-17A and IL-17F are involved in neutrophil recruitment and inflammation by inducing expression of genes encoding proinflammatory cytokines [tumor necrosis factor (TNF), IL-1, IL-6, granulocyte colony-stimulating factor (G-CSF), and granulocyte-macrophage colony-stimulating factor (GM-CSF)], chemokines (CXCL1, CXCL5, IL-8, CCL2, and CCL7), antimicrobial peptides (defensins and S100 proteins), and matrix metalloproteinases (MMP1, MMP3, and MMP13) in fibroblasts, endothelial cells, and epithelial cells (Fig. 3.2). IL-17A also promotes stem cell factor (SCF)- and G-CSF-mediated granulopoiesis and recruits neutrophils to inflammatory sites. It also induces expression of intercellular cell adhesion molecule 1 (ICAM-1) in keratinocytes, and iNOS and cyclooxygenase-2 in chondrocytes.

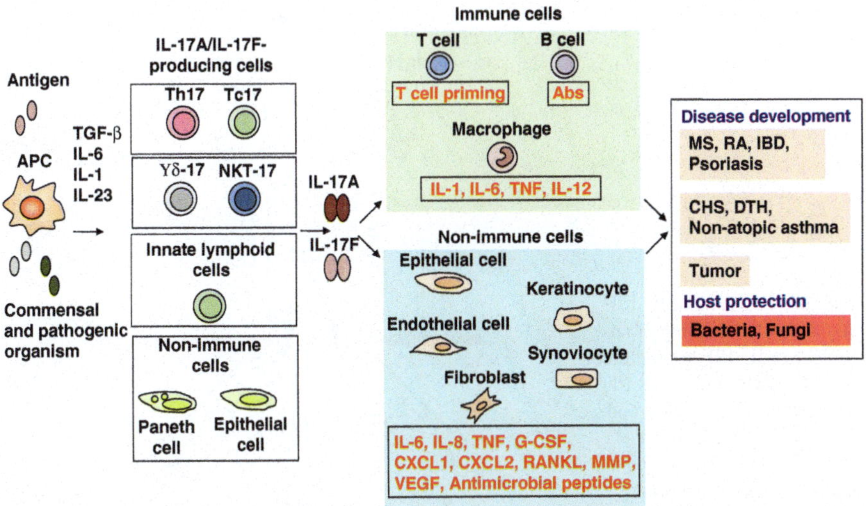

Fig. 3.2 The roles of IL-17A and IL-17F in inflammatory disease development and host defense against pathogens IL-17A and IL-17F are produced by various cell types, including T cells, innate immune cells, and nonlymphoid cells, in response to cytokines that are produced by antigen- and pathogen-stimulated antigen-presenting cells (APC). These cytokines activate lymphoid cells, such as T cells and B cells, to promote T-cell priming and antibody production, and activate non-lymphoid cells to produce many proinflammatory mediators and antimicrobial peptides. These mediators induce neutrophil recruitment at inflammatory sites, promote local tissue destruction, induce neovascularization in tumors, enhance osteoclastogenesis, and protect against pathogens. *CHS* contact hypersensitivity

3.3 Signaling Mechanisms of IL-17A and IL-17F

The IL-17-receptor (IL-17R) family includes six members (IL-17RA–IL-17RE) (reviewed by Gaffen 2009). Functional receptors for IL-17 family cytokines consist of homodimers or heterodimers, and IL-17RA is thought to be a shared receptor subunit for all family members (Fig. 3.3). The heterodimer consisting of IL-17RA and IL-17RC is a receptor for homodimers and heterodimers of IL-17A and IL-17F. However, the expression profiles of IL-17RA and IL-17RC differ among tissues and cell types: IL-17RA is expressed mostly by immune cells, and IL-17RC is preferentially expressed by nonimmune cells (Kuestner et al. 2007; Ishigame et al. 2009). In addition, the binding affinities of IL-17A and IL-17F for these receptors are different: IL-17A has higher affinity for IL-17RA and IL-17F has higher affinity for IL-17RC (Hymowitz et al. 2001; Kuestner et al. 2007; Wright et al. 2008). Therefore, the distinct functions of IL-17A and IL-17F may reflect, at least in part, differential expression of IL-17RA and IL-17RC, although the precise structures of the receptors for IL-17A and IL-17F remain to be elucidated. The SEF/IL-17R (SEFIR) domains of both IL-17RA and IL-17RC are required to activate NF-κB, MAPK, and C/EBP pathways in response to IL-17A or IL-17F (Maitra et al. 2007; Ho et al. 2010;

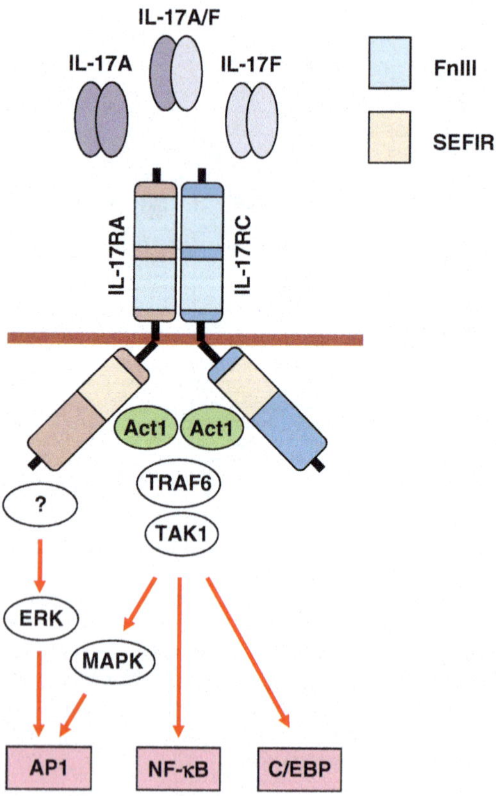

Fig. 3.3 IL-17A and IL-17F signaling after binding of an IL-17A or IL-17F homodimer or heterodimer to IL-17R (the heterodimer of IL-17RA and IL-17RC). The SEFIR domain-containing adapter molecules, Act1, associate with IL-17RA and/or IL-17RC through the SEFIR domains. Subsequently, the complex associates with TRAF6, leading to activation of NF-κB, AP-1, and/or C/EBP. Act1-independent ERK activation contributes to IL-17R signaling. *FnIII*, fibronectin III-like domain; *SEFIR*, SEF/IL-17R domain

Hu et al. 2010). The SEFIR domain-containing adaptor protein, Act1 (also called TRAF3IP3 and CIKS), associates directly with IL-17RA and IL-17RC via interaction with each SEFIR domain, resulting in recruitment of TRAF6 and TAK1 to activate NF-κB (Fig. 3.3).

3.4 IL-17A and IL-17F Producer Cells

Linking IL-17A- and IL-17F-producing $CD4^+$ T cells to IL-23 effector function led to the concept that Th17 cells belong to a distinct $CD4^+$ T-cell subset (Langrish et al. 2005; McGeachy and Cua 2008). Th17 cell differentiation from naive $CD4^+$ T cells is controlled by several cytokines, including transforming growth factor (TGF)-β, IL-6, and IL-21, which activate Stat3- and IRF4-dependent expression of retinoic acid receptor-related orphan receptor-γt (RORγt) [reviewed in (Korn et al. 2009; Zhou and Littman 2009; Hirahara et al. 2010)] (see Fig. 3.1). Other transcription factors, such as RORα, basic leucine zipper transcription factor (Batf), Runx1, and IκBζ, also regulate Th17 cell differentiation in cooperation with RORγt. Both IL-1 and IL-23 are also critical for Th17 cell differentiation, growth, survival, and effector functions. In humans, IL-1β, IL-21, IL-23, and TGF-β are required for development of Th17 cells expressing IL-17A, IL-17F, IL-22, and RORγt, although the requirement for TGF-β remains controversial (reviewed by Korn et al. 2009; Annunziato et al. 2012).

In addition to Th17 cells, a wide variety of T cells also produce IL-17A and IL-17F: these include cytotoxic $CD8^+$ T cells (Tc17) under conditions that are similar to those required by Th17 cells, but different from those required by IFN-γ-producing $CD8^+$ T cells (Tc1) (Ciric et al. 2009). Similarly, distinct populations of γδT (γδ-17) cells and NKT (NKT-17) cells produce IL-17A and IL-17F (reviewed by Cua and Tato 2010). IL-23 and IL-1 can directly induce γδ-17 cell development in the absence of IL-6 and T-cell receptor (TCR) ligation, because, in contrast to naïve $CD4^+$ and $CD8^+$ T cells, these cells constitutively express IL-23R, IL-1R, and RORγt (Sutton et al. 2009). Similarly, NKT cells produce IL-17A in the presence of IL-1 and IL-23 in combination with TCR stimulation (Rachitskaya et al. 2008). These two T-cell populations (γδ-17 and NKT-17) can rapidly produce IL-17A and IL-17F in response to proinflammatory cytokine stimulation and may therefore be important initial sources of these two cytokines.

More recently, innate immune cell populations, such as neutrophils, monocytes, natural killer cells, and innate lymphoid cells, have been shown capable of rapidly producing IL-17A and IL-17F (reviewed by Cua and Tato 2010). IL-17A is also produced by intestinal Paneth cells (Takahashi et al. 2008), whereas IL-17F mRNA, but not IL-17A mRNA, is expressed in colonic epithelial cells (Ishigame et al. 2009; Tong et al. 2012), suggesting that IL-17A and IL-17F from nonlymphoid cells may also regulate immune responses. Substantial efforts are underway to clarify the mechanisms that control IL-17A and IL-17F production in these cell types as well as the relative contributions of the resulting cytokines to immune responses.

3.5 IL-17A and IL-17F in Inflammatory Disorders

3.5.1 Multiple Sclerosis

Multiple sclerosis (MS) is a chronic inflammatory disease characterized by brain inflammation and demyelination. MS has long been believed to be a Th1 cell cytokine-mediated autoimmune disease. However, myelin oligodendrocyte glycoprotein (MOG)-induced experimental autoimmune encephalomyelitis (EAE) is much more severe in mice lacking IL-12 or interferon (IFN)-γ activity (Kastelein et al. 2007), arguing against the importance of Th1 cells in these diseases. Several studies reported that infiltration by Th17 cells, which show a proliferative response to myelin peptides, was found in the central nervous system (CNS) of MS patients (Durelli et al. 2009; Montes et al. 2009; Venken et al. 2010), suggesting involvement of Th17 cells. In fact, IL-23-, but not IL-12-, deficient mice are resistant to EAE development (Cua et al. 2003). EAE is also attenuated in *Il17a*$^{-/-}$ mice (Nakae et al. 2003; Komiyama et al. 2006; Yang et al. 2008; Ishigame et al. 2009). Use of IL-17R adaptor protein Act1 conditional knockout mice suggested that astrocytes are the main target of IL-17A signaling to recruit inflammatory cells into the CNS during the effector phase of EAE. However, in contrast to *Il23a*$^{-/-}$ mice, *Il17a*$^{-/-}$ mice had only slightly attenuated inflammation after MOG immunization (Nakae et al. 2003; Komiyama et al. 2006; Yang et al. 2008; Haak et al. 2009; Ishigame et al. 2009), suggesting that IL-17A is not the only critical mediator inducing EAE downstream of IL-23. IL-17F, however, is not required for EAE (Kreymborg et al. 2007; Yang et al. 2008; Haak et al. 2009; Ishigame et al. 2009). Recently, granulocyte-macrophage colony-stimulating factor (GM-CSF) produced by IL-23-polarized Th cells was reported to play a critical role in EAE development (Codarri et al. 2011; El-Behi et al. 2011). In addition to Th17 cells, $\gamma\delta$-17 cells were reported to be involved in the pathogenesis of EAE (Roark et al. 2007; Sutton et al. 2009; Petermann et al. 2010).

3.5.2 Rheumatoid Arthritis

Rheumatoid arthritis (RA) is a chronic systemic inflammatory disease that affects the synovial membranes of multiple joints. In humans, IL-17A is detected in synovial fluids from RA patients and induces proinflammatory cytokine production by synoviocytes (Chabaud et al. 1998). Collagen-induced arthritis (CIA) is a typical arthritis model that is induced by immunizing animals with type II collagen (IIC). Development of CIA is largely dependent on IL-23 and IL-17, because *Il23a*$^{-/-}$ and *Il17a*$^{-/-}$ mice, but not *Il12a*$^{-/-}$ mice, are resistant to induction (Murphy et al. 2003; Nakae et al. 2003; Ishigame et al. 2009). Both IL-17A and IL-17F activate synoviocytes, fibroblasts, and endothelial cells to produce various inflammatory cytokines and chemokines, including IL-1 and TNF (Fig. 3.1). IL-17A also directly promotes osteoclast differentiation by inducing RANKL in osteoblasts (Sato et al. 2006). On the other hand, sensitization of T cells following immunization with IIC as well as IIC-specific antibody production were significantly reduced in *Il17a*$^{-/-}$ mice (Nakae et al. 2003). These results suggest that IL-17A is involved in T-cell sensitization and antibody production, in addition to proinflammatory cytokine induction, in the effector phase. Because deficiency of IL-17A did not completely suppress CIA, involvement of IL-17F was suggested. However, CIA developed normally in *Il17f*$^{-/-}$ mice (Ishigame et al. 2009), indicating that IL-17A plays a more important role than IL-17F in this model. The importance of IL-17A has also been examined in several RA models, including IL-1 receptor antagonist-deficient (*Il1rn*$^{-/-}$) mice, HTLV-I transgenic mice (Iwakura et al. 2008), and K/BxN serum-induced arthritis (Jacobs et al. 2009). Furthermore, lack of IL-17A also suppressed arthritis development in mice carrying the Y759F mutation in the gp130 subunit of IL-6R, which disrupts SOCS3-mediated negative feedback (Ogura et al. 2008), and in SKG mice, which carry a mutation in ZAP70 of the TCR complex (Hirota et al. 2007). Interestingly, development of arthritis in *Il1rn*$^{-/-}$, K/BxN, and SKG mice depends on environmental factors, such as the indigenous microbe *Lactobacillus bifidus* (Abdollahi-Roodsaz et al. 2008), gut-residing segmented filamentous bacteria (SFB) (Wu et al. 2010), and fungi (Yoshitomi et al. 2005). These pathogens induce Th17 cell differentiation (see following), which results in arthritis, suggesting a link between innate immunity and autoimmune diseases.

3.5.3 *Inflammatory Bowel Disease*

Inflammatory bowel disease (IBD) is a chronic inflammatory condition of the gut, consisting of Crohn's disease and ulcerative colitis. Th17 cells are detected in the gut of Crohn's disease patients and some of these cells produce IFN-γ (Annunziato et al. 2007; Kleinschek et al. 2009). Transferring CD4^{+}CD45RB^{hi} T cells into lymphopenic mice provides a model of IBD. IL-23 is essential for colitis development in this model, whereas neither IL-17A nor IL-17F is required (Izcue et al. 2008; Leppkes et al. 2009). Transferring *Il17f*$^{-/-}$ CD4^{+}CD45RB^{hi} T cells together with anti-IL-17A Ab significantly reduced colitis (Leppkes et al. 2009), suggesting that IL-17A and IL-17F play redundant roles in colitis development. On the other hand, mice receiving *Il17a*$^{-/-}$ CD4^{+} T cells displayed accelerated wasting disease that was associated with increased Th1 cell-related cytokine production, suggesting a protective role for IL-17A (O'Connor et al. 2009). Because colitis was suppressed by adoptive transfer of CD4^{+}CD45RB^{hi} T cells that were deficient for IFN-γ (Ito and Fathman 1997), both Th1 and Th17 cells are involved in the pathogenesis of IBD, although their precise roles await further elucidation. The potential contribution of IL-17A to a dextran sodium sulfate (DSS)-induced acute colitis model is controversial. One study reported that *Il17a*$^{-/-}$ mice displayed a substantially reduced clinical score (Ito et al. 2008), whereas another study demonstrated that mice lacking IL-17A or given anti-IL-17A Ab showed severe weight loss and colonic epithelial damage (Ogawa et al. 2004; Yang et al. 2008). Conversely, *Il17f*$^{-/-}$ mice showed attenuated colonic inflammation, which was associated with reduced chemokine expression (Yang et al. 2008). Although the reason for the discrepancies in these studies is unclear at present, the intestinal microbial flora may have differed in these studies and affected the results. Recently, it was reported that *Helicobacter hepaticus-induced* chronic colitis is associated with IL-23- and IL-1-dependent production of IL-17A and IFN-γ in the colon by Thy1$^+$SCA-1$^+$RORγt$^+$IL-23R$^+$ innate lymphoid cells (Buonocore et al. 2010; Coccia et al. 2012). IL-17A-producing innate lymphoid cells are also found in patients with IBD (Buonocore et al. 2010), suggesting that both IL-17A-producing T cells and innate lymphoid cells contribute to IBD.

3.5.4 Psoriasis

Psoriasis is an inflammatory epidermal hyperproliferative skin disease. Psoriasis-like epidermal hyperplasia was induced in the ears of wild-type mice injected with IL-23, whereas little hyperplasia was observed in *Il17a*, *Il17ra*, or *Il22*$^{-/-}$ mice (Rizzo et al. 2010; Cai et al. 2011). In an imiquimod (Toll-like receptor 7/8 ligand)-induced psoriasis model, deletion of IL-17A and IL-17F signaling reduced the severity score of skin inflammation (van der Fits et al. 2009; Cai et al. 2011; Pantelyushin et al. 2012), but also resulted in increased numbers of IL-17A- and IL-22-producing CD4$^+$ and γδ T cells (El Malki et al. 2012). Similarly, T cell-specific Act1-deficient mice showed increased production of IL-17A and IL-22 and developed IL-23/IL-22-dependent spontaneous psoriasis-like skin inflammation (Wang et al. 2012), indicating a negative feedback role for IL-17A-Act1 signaling in Th17 responses. In contrast, psoriasiform dermatitis, which develops spontaneously in *Il1rn*$^{-/-}$ mice, was not dependent on either T cells or IL-17A (Nakajima et al. 2010), suggesting an autoinflammatory mechanism. Nonetheless, in humans, expression of pro-inflammatory cytokines, including IL-17A, IL-22, and IL-23, is elevated in psoriatic skin (Wilson et al. 2007; Lowes et al. 2008; Di Cesare et al. 2009). Importantly, recent clinical studies demonstrated that blockade of IL-17A signaling improved the clinical symptoms of psoriasis (Leonardi et al. 2012; Papp et al. 2012). Although no exact cellular source of IL-17A responsible for psoriasis development has been identified, γδ-17 cells and Tc17 cells have been suggested to be important for its pathogenesis (Res et al. 2010; Cai et al. 2011).

3.5.5 Systemic Lupus Erythematosus

Systemic lupus erythematosus (SLE) is a systemic autoimmune disease characterized by immune complex deposition caused by autoantibodies. IL-17A-producing

Th17 cells and $CD3^+CD4^-CD8^-$ cells are increased in patients with SLE (Crispin et al. 2008; Shah et al. 2010) and several mouse models of lupus (Hsu et al. 2008; Zhang et al. 2009). IL-17A acts directly on B cells and enhances survival, proliferation, and Ig class switching (Hsu et al. 2008; Doreau et al. 2009; Mitsdoerffer et al. 2010; Wu et al. 2010). Abrogation of IL-17A signaling by deleting IL-17RA reduced germinal center formation and humoral responses in recombinant inbred BXD2 mice (Hsu et al. 2008). However, another study, in mice lacking FcγRIIβ, found that IL-17A deficiency did not reduce autoantibody production or germinal center formation, but did protect against fetal glomerulonephritis by preventing recruitment of inflammatory myeloid-derived cells into the kidneys (Pisitkun et al. 2012). Thus, IL-17A may contribute to SLE pathogenesis by acting directly on B cells to enhance autoantibody production as well as by promoting local tissue inflammation.

3.5.6 Type 1 Diabetes

Type 1 diabetes (T1D) is an autoimmune disease characterized by destruction of the insulin-producing β cells in the pancreatic islets by autoreactive $CD4^+$ and $CD8^+$ T cells. In vitro differentiated Th17 cells induced T1D in nonobese diabetic (NOD)-severe combined immunodeficiency (scid) mice (Bending et al. 2009; Ciric et al. 2009; Martin-Orozco et al. 2009a). In that system, however, the development of diabetes was dependent on IFN-γ, but not IL-17A, suggesting that Th17 cells were reprogrammed into Th1 cells under the lymphopenic conditions. In contrast, adoptive transfer of IL-23-induced OTI Tc17 cells caused IL-17A- and IL-17F-dependent diabetes in recipients that expressed OVA in pancreatic β cells (Ciric et al. 2009). Progression of diabetes in NOD mice was inhibited by anti-IL-17A Ab (Emamaullee et al. 2009), but the incidence of hyperglycemia was comparable in $Il17a^{-/-}$ and $Il17a^{+/+}$ NOD mice (Komiyama et al. 2006). Thus, the precise roles of IL-17A and IL-17F in T1D remain to be elucidated.

3.5.7 Delayed-Type and Contact Hypersensitivity

Delayed-type hypersensitivity (DTH) and contact hypersensitivity (CHS), two T cell-mediated type IV hypersensitive responses, are attenuated in $Il17a^{-/-}$ mice, but not $Il17f^{-/-}$ mice, suggesting pathological roles for IL-17A in these allergic responses (Nakae et al. 2002; Ishigame et al. 2009). The antigen-specific T-cell population was reduced in $Il17a^{-/-}$ mice (Nakae et al. 2002; Ishigame et al. 2009), and Tc17 cell-derived IL-17A induced local inflammation during the elicitation phase of CHS (He et al. 2006).

3.5.8 Atopic Dermatitis

In patients with atopic dermatitis (AD), IL-17A expression is increased in local lesions, and the Th17 cell population is expanded in the peripheral blood (Koga et al. 2008). IL-17A may be beneficial in this setting because it is important for protection against *Staphylococcus aureus*, which aggravates AD (see following). Yet, IL-17A may also promote AD progression by facilitating local inflammation, as reported for mice lacking filaggrin, an important component of the skin barrier and a predisposing factor for atopic dermatitis (Oyoshi et al. 2009b).

3.5.9 Allergic Airway Inflammation

The pathogenesis of asthma is complex, as shown by its heterogeneity; for example, atopic versus non-atopic, eosinophilic versus neutrophilic, and steroid-effective mild asthma versus steroid-resistant severe asthma (Iwakura et al. 2008). Similar to their wild-type littermates, $Il17a^{-/-}$, $Il17f^{-/-}$, and $Il17a^{-/-}$ $Il17f^{-/-}$ mice were shown to develop OVA-induced airway eosinophilia (Nakae et al. 2002; Ishigame et al. 2009; Suzukawa et al. 2012). One study, however, suggested that IL-17A and IL-17F contribute positively and negatively, respectively, to this chronic allergic response (Yang et al. 2008). Alternatively, overexpression of IL-17A and IL-17F in the lungs causes increased pro-inflammatory cytokine and chemokine expression, resulting in inflammation associated with neutrophil infiltration (Park et al. 2005; Yang et al. 2008). IL-17A, but not IL-17F, is required for OVA-induced airway neutrophilia in DO11.10 and OTII mice (Nakae et al. 2007; Ishigame et al. 2009) and in mice injected with DO11.10 Th17 cells (Liang et al. 2007). Airway neutrophilia induced by house dust mite allergen, a major asthma allergen, was also dependent on IL-17A (Lajoie et al. 2010). The population of IL-17A-producing Th2 cells, in addition to those of Th2 cells and Th17 cells (Cosmi et al. 2010; Wang et al. 2010b), was preferentially increased in the lungs of patients with atopic asthma as well as of mice treated with certain protease allergens such as *Aspergillus*-derived proteases and papain, and resulted in inflammatory leukocyte influx and asthma exacerbation (Wang et al. 2010b). IL-17F, but not IL-17A, mediated airway neutrophilia after inhalation of *Aspergillus* proteases (Yang et al. 2008), suggesting involvement of innate immune cell-derived IL-17F in this response. Th17 cells also contribute to airway remodeling and neutrophilia during chronic airway inflammation (Wang et al. 2010a). Overall, these data suggest that IL-17A and IL-17F are involved in airway neutrophilia, but not eosinophilia, during allergic asthma. The detailed functional differences between IL-17A and IL-17F as well as the roles of Th17 and IL-17A-producing Th2 cells in asthma development remain to be elucidated.

3.6 IL-17A and IL-17F in Other Immune Cell-Mediated Diseases

IL-17A-induced neutrophil recruitment to inflamed sites contributes to various diseases. For example, IL-17A is crucial for neutrophil-mediated ischemia-reperfusion injury in the brain (Shichita et al. 2009). NKT cell-derived IL-17A plays a role in ozone-induced airway neutrophilia, a form of asthma independent of adaptive immunity (Pichavant et al. 2008). IL-17A, which may be produced by γδ T cells, is also involved in pulmonary fibrosis and neutrophilia induced by bleomycin (Wilson et al. 2010). Also, similar to many cytokines, IL-17A may play a role in sepsis (Flierl et al. 2008; Freitas et al. 2009).

Rodents treated with soluble IL-17RA showed reduced acute, but not chronic, rejection of cardiac or aortic transplants (Tang et al. 2001; Li et al. 2006), whereas *Il17a*$^{-/-}$ mice exhibited attenuated chronic, but not acute, rejection after cardiac transplantation (Gorbacheva et al. 2010; Itoh et al. 2010). This apparent discrepancy suggests that IL-17F and/or IL-17E may also be involved in these responses. IL-17A-mediated neutrophil recruitment promoted minor histocompatibility antigen-mismatched skin allograft rejection (Vokaer et al. 2010), yet IL-17A deficiency stimulated corneal allograft rejection by promoting Th2 cell responses (Yamada et al. 2009; Cunnusamy et al. 2010). The role of IL-17A in graft-versus-host disease (GVHD) is controversial: it has been shown to be protective (Yi et al. 2008; Oh et al. 2010), pathogenic (Kappel et al. 2009), and nonessential (Nakae et al. 2002), depending on the experimental protocol. IL-17F has not been examined in GVHD.

Obesity is now recognized as a risk factor for development of various diseases. The levels of blood IL-17A and IL-23 were higher in obese individuals than lean subjects (Sumarac-Dumanovic et al. 2009), and enhanced Th17 cell expansion and IL-17A production were observed in diet-induced obese mice, resulting in increased susceptibility to EAE and colitis (Winer et al. 2009). These observations suggest that obesity may accelerate induction of Th17 cell-mediated immune diseases. IL-17A-producing T cells, particularly γδ T cells, are detected in adipose tissues. IL-17A inhibited adipogenesis and lipid and glucose uptake by adipocytes (Goswami et al. 2009; Zuniga et al. 2010). Indeed, *Il17a*$^{-/-}$ mice showed enhanced obesity by diet (Zuniga et al. 2010). These observations suggest that, although obesity promotes Th17-mediated diseases, IL-17A and/or IL-17F regulate fat metabolism by inhibiting adipogenesis.

3.7 IL-17A and IL-17F in Tumor Development

Th17 cells infiltrate into tumor sites and draining lymph nodes in cancer patients (reviewed by Zou and Restifo 2010). Transplantation of IL-17A-overexpressing tumor cells into immunodeficient mice induced angiogenesis through induction of vascular endothelial growth factor (VEGF) expression, resulting in enhanced tumor growth (Numasaki et al. 2003). T cell-derived IL-17A also dramatically increased the release of angiogenic and chemoattractive IL-8 from tumor cells (Tartour et al. 1999). Growth of subcutaneously transplanted B16 melanoma cells and MB49 bladder carcinoma cells was delayed in *Il17a*$^{-/-}$ mice, whereas *Ifng*$^{-/-}$ mice showed accelerated growth and augmented IL-17A production in the tumors (Wang et al. 2009). Both IL-17A and IL-17F are also involved in the development of intestinal cancer in mouse models of familial adenomatous polyposis (Wu et al. 2009; Chae et al. 2010; Chae and Bothwell 2011; Grivennikov et al. 2012). In those models, commensal bacteria can promote IL-23/IL-17A-dependent colonic inflammation, resulting in accelerated tumor formation.

Th17 cells have also been shown to inhibit tumor development. Lung metastasis of B16-F10 melanoma was increased in *Il17a*$^{-/-}$ mice, whereas adoptive transfer of tumor-specific Th17 cells prevented tumor development associated with activation of tumor-specific CD8^{+} T cells (Martin-Orozco et al. 2009b). Tc17 cells also suppressed established B16-F10 melanomas (Hinrichs et al. 2009). *Il17a*$^{-/-}$ mice showed increased tumor growth and metastasis of MC38 cells in lung and subcutaneous tissues, which was associated with reduced IFN-γ-producing NK and CD8^{+} T cells (Kryczek et al. 2009). These results suggest that IL-17A indirectly stimulates anti-tumor immunity by promoting type 1 immune responses. Thus, IL-17A has both pro-tumor and anti-tumor activities, depending on the type and stage of tumors.

3.8 IL-17A and IL-17F in Host Defense Against Infection

3.8.1 Bacterial Infections

In contrast to the pathogenic roles of Th17 cytokines in autoimmune and allergic diseases, Th17 cytokines protect hosts from pathogens in epithelial and mucosal tissues, including the skin, lung, and intestine (Fig. 3.2). Indeed, studies using cytokine- and receptor-deficient mice showed that IL-17A and IL-17F were required for responses to such extracellular bacteria as *Klebsiella pneumoniae* in the lungs, *Citrobacter rodentium* in the colon, and *Staphylococcus aureus* in the skin (Aujla et al. 2008; Ishigame et al. 2009). IL-17A was also required for host defense mechanisms against intracellular pathogens in mice. Although *Il17ra*$^{-/-}$ mice were not more susceptible to *Mycobacterium tuberculosis* infection (Aujla et al. 2008), the IL-23/IL-17A pathway was required for Th1 cell-type immune responses that protected the host against the intracellular pathogen *Francisella tularensis* (Lin et al. 2009).

IL-17A is rapidly produced after microbial infection, and γδ T cells have been identified as a primary source during early infections by several types of bacteria, including *Listeria monocytogenes* and *Escherichia coli* (Shibata et al. 2007; Hamada et al. 2008). Those studies demonstrated that a lack of IL-17A resulted in increased bacterial burden, suggesting that γδ T-cell-derived IL-17A promotes neutrophil accumulation to eradicate bacteria. Although NKT cells, innate lymphoid cells, epithelial cells, and Paneth cells can also produce IL-17A and/or IL-17F, little is known regarding its role in antimicrobial immunity.

Th17 cells are enriched in the gastrointestinal tract under steady-state conditions, and intestinal Th17 cell development is largely dependent on a commensal microbe, SFB (Ivanov et al. 2008; Gaboriau-Routhiau et al. 2009; Ivanov et al. 2009; Wu et al. 2010). Notably, mice lacking Th17 cells because of the absence of SFB showed increased susceptibility to *C. rodentium* (Ivanov et al. 2009). Thus, intestinal commensal bacteria mediate the development of Th17 cells as well as other IL-17A- and IL-17F-producing cells, resulting in various effects on host immune responses.

3.8.2 Fungal Infections

Il23a$^{-/-}$ and *Il17ra*$^{-/-}$, but not *Il12a*$^{-/-}$, mice are highly susceptible to oral candidiasis as a result of defective neutrophil recruitment and antimicrobial peptide production (Conti et al. 2009). IL-17A plays a more important role than IL-17F in systemic *Candida albicans* infection, because only *Il17a*$^{-/-}$ mice, not *Il17f*$^{-/-}$ mice, show increased susceptibility (Saijo et al. 2010). Interestingly, fungal cell wall components induce Th17 cell differentiation in a dectin-1- and dectin-2-dependent manner. (LeibundGut-Landmann et al. 2007; Saijo et al. 2010).

In humans, hyper-IgE syndrome patients, in whom Th17 cell differentiation is suppressed by a *STAT3* mutation, are highly susceptible to *C. albicans* as well as *S. aureus and Streptococcus pneumoniae* infections, resulting in skin and lung inflammation (reviewed by Hernandez-Santos and Gaffen 2012). Similarly, people who have an IL-17RA autosomal recessive deficiency, an IL-17F autosomal dominant deficiency, or a dectin-1 or CARD9 deficiency show impaired Th17-type immunity and are susceptible to candidiasis (Ferwerda et al. 2009; Glocker et al. 2009; Liu et al. 2011). Furthermore, gain-of-function Stat1 mutations impair Th17-type immunity by enhancement of the Stat1-dependent responses to IL-17A inhibitors, that is, IFN-α/β, IFN-γ, and IL-27, and also underlie chronic mucocutaneous candidiasis (Liu et al. 2011). These observations suggest that Th17 responses are critical for host protection against *C. albicans*.

3.8.3 Viral Infections

Although it is established that Th1, Tc1, and type 1 IFN responses are critical for host defense against various viruses, IL-17A is also induced during viral infections. The contribution of IL-17A to the viral host defense mechanism is poorly understood. Published studies demonstrated that IL-17A signaling is not required for viral clearance during influenza A virus or herpes simplex type 1 virus (HSV-1) infection

(Molesworth-Kenyon et al. 2008; Crowe et al. 2009), whereas eradication of vaccine virus is partially dependent on IL-17A (Kohyama et al. 2007). Importantly, increased IL-17A production during influenza A virus or Theiler's murine encephalomyelitis virus infection in the lung, HSV-1 infection in the cornea, systemic HSV-2, and murine cytomegalovirus infection leads to neutrophil-dependent tissue injury, viral persistence, or mortality (Molesworth-Kenyon et al. 2008; Hou et al. 2009; Oyoshi et al. 2009a), suggesting that blockade of IL-17A signaling is beneficial in virus-induced immunopathology. By contrast, preferential depletion of Th17 cells in the intestine occurs during human immunodeficiency virus (HIV)-1 or simian immunodeficiency virus infection (Brenchley et al. 2008; Cecchinato et al. 2008), which dramatically increases the dissemination of bacteria from the gut, including nontyphoidal *Salmonella* serotypes (Raffatellu et al. 2008), suggesting that Th17 cells are essential for intestinal mucosal barrier function during HIV infection. The contributions of IL-17F to host defense against viral infections remain to be elucidated.

3.9 Concluding Remarks

Clinical studies have shown that blocking the activity of IL-12 and IL-23 (p40), IL-1, or IL-6, which are important for the development of Th17 cells and possibly innate IL-17A-producing cells, is effective for treatment of such inflammatory diseases as RA, MS, IBD, and psoriasis (Broderick et al. 2011; Elloso et al. 2012; Miossec and Kolls 2012). More recent clinical trials have demonstrated that anti-IL-17A or anti-IL-17RA therapies are also effective against RA and psoriasis (Genovese et al. 2010; Hueber et al. 2010; Leonardi et al. 2012; Papp et al. 2012). On the other hand, anti-IL-17A treatment failed to improve the clinical symptoms of Crohn's disease, and actually exacerbated the disease and local fungal infections (Hueber et al. 2012), indicating a protective role of IL-17A in this inflammatory process. Because IL-17A plays important roles in granulopoiesis and host defense against pathogens, anti-IL-17A treatments may increase the risk of opportunistic infections, as has been observed with therapies targeting other cytokines. Thus, clinical application of these approaches requires caution, with particular attention paid to staphylococcal and candida infections. Because IL-17F is often functionally redundant with IL-17A in host defenses against infections and shows relatively lower proinflammatory activity, anti-IL-17A treatment may be safer than other biological therapies, such as anti-IL-17RA treatment. Because other IL-17 family members, especially IL-17C and IL-17E, are pro-inflammatory and may be involved in host defense responses, an understanding of these cytokine functions will allow development of more effective treatments for allergic and autoimmune disorders and tumors, without compromising host defense. Finally, a growing body of evidence suggests that gut microbiota influence Th17 responses, which in turn impact host inflammatory responses. Elucidating the host–microbiota interaction may extend our understanding of the cellular mechanisms involved in the control of inflammatory disorders.

Acknowledgements This work was supported by PRESTO, Japan Science and Technology Agency (S.N.). H.I. is a Postdoctoral Fellow supported by a Postdoctoral Fellowship for Research Abroad from the Japan Society for the Promotion of Science.

References

Abdollahi-Roodsaz S, Joosten LA, Koenders MI, Devesa I, Roelofs MF, Radstake TR, Heuvelmans-Jacobs M, Akira S, Nicklin MJ, Ribeiro-Dias F, van den Berg WB (2008) Stimulation of TLR2 and TLR4 differentially skews the balance of T cells in a mouse model of arthritis. J Clin Invest 118:205–216

Annunziato F, Cosmi L, Santarlasci V, Maggi L, Liotta F, Mazzinghi B, Parente E, Fili L, Ferri S, Frosali F, Giudici F, Romagnani P, Parronchi P, Tonelli F, Maggi E, Romagnani S (2007) Phenotypic and functional features of human Th17 cells. J Exp Med 204:1849–1861

Annunziato F, Cosmi L, Liotta F, Maggi E, Romagnani S (2012) Defining the human T helper 17 cell phenotype. Trends Immunol 33:505–512

Aujla SJ, Chan YR, Zheng M, Fei M, Askew DJ, Pociask DA, Reinhart TA, McAllister F, Edeal J, Gaus K, Husain S, Kreindler JL, Dubin PJ, Pilewski JM, Myerburg MM, Mason CA, Iwakura Y, Kolls JK (2008) IL-22 mediates mucosal host defense against gram-negative bacterial pneumonia. Nat Med 14:275–281

Bending D, De La Pena H, Veldhoen M, Phillips JM, Uyttenhove C, Stockinger B, Cooke A (2009) Highly purified Th17 cells from BDC2.5NOD mice convert into Th1-like cells in NOD/SCID recipient mice. J Clin Invest 119:565–572

Brenchley JM, Paiardini M, Knox KS, Asher AI, Cervasi B, Asher TE, Scheinberg P, Price DA, Hage CA, Kholi LM, Khoruts A, Frank I, Else J, Schacker T, Silvestri G, Douek DC (2008) Differential Th17 CD4 T-cell depletion in pathogenic and nonpathogenic lentiviral infections. Blood 112:2826–2835

Broderick L, Tourangeau LM, Kavanaugh A, Wasserman SI (2011) Biologic modulators in allergic and autoinflammatory diseases. Curr Opin Allergy Clin Immunol 11:355–360

Buonocore S, Ahern PP, Uhlig HH, Ivanov II, Littman DR, Maloy KJ, Powrie F (2010) Innate lymphoid cells drive interleukin-23-dependent innate intestinal pathology. Nature (Lond) 464:1371–1375

Cai Y, Shen X, Ding C, Qi C, Li K, Li X, Jala VR, Zhang HG, Wang T, Zheng J, Yan J (2011) Pivotal role of dermal IL-17-producing gamma-delta T cells in skin inflammation. Immunity 35:596–610

Cecchinato V, Trindade CJ, Laurence A, Heraud JM, Brenchley JM, Ferrari MG, Zaffiri L, Tryniszewska E, Tsai WP, Vaccari M, Parks RW, Venzon D, Douek DC, O'Shea JJ, Franchini G (2008) Altered balance between Th17 and Th1 cells at mucosal sites predicts AIDS progression in simian immunodeficiency virus-infected macaques. Mucosal Immunol 1:279–288

Chabaud M, Fossiez F, Taupin JL, Miossec P (1998) Enhancing effect of IL-17 on IL-1-induced IL-6 and leukemia inhibitory factor production by rheumatoid arthritis synoviocytes and its regulation by Th2 cytokines. J Immunol 161:409–414

Chae WJ, Bothwell AL (2011) IL-17F deficiency inhibits small intestinal tumorigenesis in ApcMin/+ mice. Biochem Biophys Res Commun 414:31–36

Chae WJ, Gibson TF, Zelterman D, Hao L, Henegariu O, Bothwell AL (2010) Ablation of IL-17A abrogates progression of spontaneous intestinal tumorigenesis. Proc Natl Acad Sci USA 107:5540–5544

Chang SH, Dong C (2011) Signaling of interleukin-17 family cytokines in immunity and inflammation. Cell Signal 23:1069–1075

Ciric B, El-behi M, Cabrera R, Zhang GX, Rostami A (2009) IL-23 drives pathogenic IL-17-producing CD8+ T cells. J Immunol 182:5296–5305

Coccia M, Harrison OJ, Schiering C, Asquith MJ, Becher B, Powrie F, Maloy KJ (2012) IL-1beta mediates chronic intestinal inflammation by promoting the accumulation of IL-17A secreting innate lymphoid cells and CD4(+) Th17 cells. J Exp Med 209:1595–1609

Codarri L, Gyulveszi G, Tosevski V, Hesske L, Fontana A, Magnenat L, Suter T, Becher B (2011) RORgammat drives production of the cytokine GM-CSF in helper T cells, which is essential for the effector phase of autoimmune neuroinflammation. Nat Immunol 12:560–567

Conti HR, Shen F, Nayyar N, Stocum E, Sun JN, Lindemann MJ, Ho AW, Hai JH, Yu JJ, Jung JW, Filler SG, Masso-Welch P, Edgerton M, Gaffen SL (2009) Th17 cells and IL-17 receptor signaling are essential for mucosal host defense against oral candidiasis. J Exp Med 206: 299–311

Cosmi L, Maggi L, Santarlasci V, Capone M, Cardilicchia E, Frosali F, Querci V, Angeli R, Matucci A, Fambrini M, Liotta F, Parronchi P, Maggi E, Romagnani S, Annunziato F (2010) Identification of a novel subset of human circulating memory CD4+ T cells that produce both IL-17A and IL-4. J Allergy Clin Immunol 125(222–230):e221–e224

Crispin JC, Oukka M, Bayliss G, Cohen RA, Van Beek CA, Stillman IE, Kyttaris VC, Juang YT, Tsokos GC (2008) Expanded double negative T cells in patients with systemic lupus erythematosus produce IL-17 and infiltrate the kidneys. J Immunol 181:8761–8766

Crowe CR, Chen K, Pociask DA, Alcorn JF, Krivich C, Enelow RI, Ross TM, Witztum JL, Kolls JK (2009) Critical role of IL-17RA in immunopathology of influenza infection. J Immunol 183:5301–5310

Cua DJ, Tato CM (2010) Innate IL-17-producing cells: the sentinels of the immune system. Nat Rev Immunol 10:479–489

Cua DJ, Sherlock J, Chen Y, Murphy CA, Joyce B, Seymour B, Lucian L, To W, Kwan S, Churakova T, Zurawski S, Wiekowski M, Lira SA, Gorman D, Kastelein RA, Sedgwick JD (2003) Interleukin-23 rather than interleukin-12 is the critical cytokine for autoimmune inflammation of the brain. Nature (Lond) 421:744–748

Cunnusamy K, Chen PW, Niederkorn JY (2010) IL-17 promotes immune privilege of corneal allografts. J Immunol 185:4651–4658

Di Cesare A, Di Meglio P, Nestle FO (2009) The IL-23/Th17 axis in the immunopathogenesis of psoriasis. J Invest Dermatol 129:1339–1350

Doreau A, Belot A, Bastid J, Riche B, Trescol-Biemont MC, Ranchin B, Fabien N, Cochat P, Pouteil-Noble C, Trolliet P, Durieu I, Tebib J, Kassai B, Ansieau S, Puisieux A, Eliaou JF, Bonnefoy-Berard N (2009) Interleukin 17 acts in synergy with B cell-activating factor to influence B cell biology and the pathophysiology of systemic lupus erythematosus. Nat Immunol 10:778–785

Durelli L, Conti L, Clerico M, Boselli D, Contessa G, Ripellino P, Ferrero B, Eid P, Novelli F (2009) T-helper 17 cells expand in multiple sclerosis and are inhibited by interferon-beta. Ann Neurol 65:499–509

El Malki K, Karbach SH, Huppert J, Zayoud M, Reissig S, Schuler R, Nikolaev A, Karram K, Munzel T, Kuhlmann CR, Luhmann HJ, von Stebut E, Wortge S, Kurschus FC, Waisman A (2012) An alternative pathway of imiquimod-induced psoriasis-like skin inflammation in the absence of interleukin-17 receptor A signaling. J Invest Dermatol 133:441–451

El-Behi M, Ciric B, Dai H, Yan Y, Cullimore M, Safavi F, Zhang GX, Dittel BN, Rostami A (2011) The encephalitogenicity of TH17 cells is dependent on IL-1- and IL-23-induced production of the cytokine GM-CSF. Nat Immunol 12:568–575

Elloso MM, Gomez-Angelats M, Fourie AM (2012) Targeting the Th17 pathway in psoriasis. J Leukoc Biol 92:1187–1197

Emamaullee JA, Davis J, Merani S, Toso C, Elliott JF, Thiesen A, Shapiro AM (2009) Inhibition of Th17 cells regulates autoimmune diabetes in NOD mice. Diabetes 58:1302–1311

Ferwerda B, Ferwerda G, Plantinga TS, Willment JA, van Spriel AB, Venselaar H, Elbers CC, Johnson MD, Cambi A, Huysamen C, Jacobs L, Jansen T, Verheijen K, Masthoff L, Morre SA, Vriend G, Williams DL, Perfect JR, Joosten LA, Wijmenga C, van der Meer JW, Adema GJ, Kullberg BJ, Brown GD, Netea MG (2009) Human dectin-1 deficiency and mucocutaneous fungal infections. N Engl J Med 361:1760–1767

Flierl MA, Rittirsch D, Gao H, Hoesel LM, Nadeau BA, Day DE, Zetoune FS, Sarma JV, Huber-Lang MS, Ferrara JL, Ward PA (2008) Adverse functions of IL-17A in experimental sepsis. FASEB J 22:2198–2205

Freitas A, Alves-Filho JC, Victoni T, Secher T, Lemos HP, Sonego F, Cunha FQ, Ryffel B (2009) IL-17 receptor signaling is required to control polymicrobial sepsis. J Immunol 182:7846–7854

Gaboriau-Routhiau V, Rakotobe S, Lecuyer E, Mulder I, Lan A, Bridonneau C, Rochet V, Pisi A, De Paepe M, Brandi G, Eberl G, Snel J, Kelly D, Cerf-Bensussan N (2009) The key role of segmented filamentous bacteria in the coordinated maturation of gut helper T cell responses. Immunity 31:677–689

Gaffen SL (2009) Structure and signalling in the IL-17 receptor family. Nat Rev Immunol 9: 556–567

Genovese MC, Van den Bosch F, Roberson SA, Bojin S, Biagini IM, Ryan P, Sloan-Lancaster J (2010) LY2439821, a humanized anti-interleukin-17 monoclonal antibody, in the treatment of patients with rheumatoid arthritis: a phase I randomized, double-blind, placebo-controlled, proof-of-concept study. Arthritis Rheum 62:929–939

Glocker EO, Hennigs A, Nabavi M, Schaffer AA, Woellner C, Salzer U, Pfeifer D, Veelken H, Warnatz K, Tahami F, Jamal S, Manguiat A, Rezaei N, Amirzargar AA, Plebani A, Hannesschlager N, Gross O, Ruland J, Grimbacher B (2009) A homozygous CARD9 mutation in a family with susceptibility to fungal infections. N Engl J Med 361:1727–1735

Gorbacheva V, Fan R, Li X, Valujskikh A (2010) Interleukin-17 promotes early allograft inflammation. Am J Pathol 177:1265–1273

Goswami J, Hernandez-Santos N, Zuniga LA, Gaffen SL (2009) A bone-protective role for IL-17 receptor signaling in ovariectomy-induced bone loss. Eur J Immunol 39:2831–2839

Grivennikov SI, Wang K, Mucida D, Stewart CA, Schnabl B, Jauch D, Taniguchi K, Yu GY, Osterreicher CH, Hung KE, Datz C, Feng Y, Fearon ER, Oukka M, Tessarollo L, Coppola V, Yarovinsky F, Cheroutre H, Eckmann L, Trinchieri G, Karin M (2012) Adenoma-linked barrier defects and microbial products drive IL-23/IL-17-mediated tumour growth. Nature (Lond) 491:254–258

Haak S, Croxford AL, Kreymborg K, Heppner FL, Pouly S, Becher B, Waisman A (2009) IL-17A and IL-17F do not contribute vitally to autoimmune neuro-inflammation in mice. J Clin Invest 119:61–69

Hamada S, Umemura M, Shiono T, Tanaka K, Yahagi A, Begum MD, Oshiro K, Okamoto Y, Watanabe H, Kawakami K, Roark C, Born WK, O'Brien R, Ikuta K, Ishikawa H, Nakae S, Iwakura Y, Ohta T, Matsuzaki G (2008) IL-17A produced by gammadelta T cells plays a critical role in innate immunity against *Listeria monocytogenes* infection in the liver. J Immunol 181:3456–3463

He D, Wu L, Kim HK, Li H, Elmets CA, Xu H (2006) $CD8^+$ IL-17-producing T cells are important in effector functions for the elicitation of contact hypersensitivity responses. J Immunol 177:6852–6858

Hernandez-Santos N, Gaffen SL (2012) Th17 cells in immunity to *Candida albicans*. Cell Host Microbe 11:425–435

Hinrichs CS, Kaiser A, Paulos CM, Cassard L, Sanchez-Perez L, Heemskerk B, Wrzesinski C, Borman ZA, Muranski P, Restifo NP (2009) Type 17 CD8+ T cells display enhanced antitumor immunity. Blood 114:596–599

Hirahara K, Ghoreschi K, Laurence A, Yang XP, Kanno Y, O'Shea JJ (2010) Signal transduction pathways and transcriptional regulation in Th17 cell differentiation. Cytokine Growth Factor Rev 21:425–434

Hirota K, Hashimoto M, Yoshitomi H, Tanaka S, Nomura T, Yamaguchi T, Iwakura Y, Sakaguchi N, Sakaguchi S (2007) T cell self-reactivity forms a cytokine milieu for spontaneous development of $IL-17^+$ Th cells that cause autoimmune arthritis. J Exp Med 204:41–47

Ho AW, Shen F, Conti HR, Patel N, Childs EE, Peterson AC, Hernandez-Santos N, Kolls JK, Kane LP, Ouyang W, Gaffen SL (2010) IL-17RC is required for immune signaling via an extended SEF/IL-17R signaling domain in the cytoplasmic tail. J Immunol 185:1063–1070

Hou W, Kang HS, Kim BS (2009) Th17 cells enhance viral persistence and inhibit T cell cytotoxicity in a model of chronic virus infection. J Exp Med 206:313–328
Hsu HC, Yang P, Wang J, Wu Q, Myers R, Chen J, Yi J, Guentert T, Tousson A, Stanus AL, Le TV, Lorenz RG, Xu H, Kolls JK, Carter RH, Chaplin DD, Williams RW, Mountz JD (2008) Interleukin 17-producing T helper cells and interleukin 17 orchestrate autoreactive germinal center development in autoimmune BXD2 mice. Nat Immunol 9:166–175
Hu Y, Ota N, Peng I, Refino CJ, Danilenko DM, Caplazi P, Ouyang W (2010) IL-17RC is required for IL-17A- and IL-17F-dependent signaling and the pathogenesis of experimental autoimmune encephalomyelitis. J Immunol 184:4307–4316
Hueber W, Patel DD, Dryja T, Wright AM, Koroleva I, Bruin G, Antoni C, Draelos Z, Gold MH, Durez P, Tak PP, Gomez-Reino JJ, Foster CS, Kim RY, Samson CM, Falk NS, Chu DS, Callanan D, Nguyen QD, Rose K, Haider A, Di Padova F (2010) Effects of AIN457, a fully human antibody to interleukin-17A, on psoriasis, rheumatoid arthritis, and uveitis. Sci Transl Med 2:52ra72
Hueber W, Sands BE, Lewitzky S, Vandemeulebroecke M, Reinisch W, Higgins PD, Wehkamp J, Feagan BG, Yao MD, Karczewski M, Karczewski J, Pezous N, Bek S, Bruin G, Mellgard B, Berger C, Londei M, Bertolino AP, Tougas G, Travis SP (2012) Secukinumab, a human anti-IL-17A monoclonal antibody, for moderate to severe Crohn's disease: unexpected results of a randomised, double-blind placebo-controlled trial. Gut 61:1693–1700
Hymowitz SG, Filvaroff EH, Yin JP, Lee J, Cai L, Risser P, Maruoka M, Mao W, Foster J, Kelley RF, Pan G, Gurney AL, de Vos AM, Starovasnik MA (2001) IL-17s adopt a cystine knot fold: structure and activity of a novel cytokine, IL-17F, and implications for receptor binding. EMBO J 20:5332–5341
Ishigame H, Kakuta S, Nagai T, Kadoki M, Nambu A, Komiyama Y, Fujikado N, Tanahashi Y, Akitsu A, Kotaki H, Sudo K, Nakae S, Sasakawa C, Iwakura Y (2009) Differential roles of interleukin-17A and -17F in host defense against mucoepithelial bacterial infection and allergic responses. Immunity 30:108–119
Ito H, Fathman CG (1997) CD45RBhigh CD4+ T cells from IFN-gamma knockout mice do not induce wasting disease. J Autoimmun 10:455–459
Ito R, Kita M, Shin-Ya M, Kishida T, Urano A, Takada R, Sakagami J, Imanishi J, Iwakura Y, Okanoue T, Yoshikawa T, Kataoka K, Mazda O (2008) Involvement of IL-17A in the pathogenesis of DSS-induced colitis in mice. Biochem Biophys Res Commun 377:12–16
Itoh S, Nakae S, Axtell RC, Velotta JB, Kimura N, Kajiwara N, Iwakura Y, Saito H, Adachi H, Steinman L, Robbins RC, Fischbein MP (2010) IL-17 contributes to the development of chronic rejection in a murine heart transplant model. J Clin Immunol 30:235–240
Ivanov II, Frutos RL, Manel N, Yoshinaga K, Rifkin DB, Sartor RB, Finlay BB, Littman DR (2008) Specific microbiota direct the differentiation of IL-17-producing T-helper cells in the mucosa of the small intestine. Cell Host Microbe 4:337–349
Ivanov II, Atarashi K, Manel N, Brodie EL, Shima T, Karaoz U, Wei D, Goldfarb KC, Santee CA, Lynch SV, Tanoue T, Imaoka A, Itoh K, Takeda K, Umesaki Y, Honda K, Littman DR (2009) Induction of intestinal Th17 cells by segmented filamentous bacteria. Cell 139:485–498
Iwakura Y, Nakae S, Saijo S, Ishigame H (2008) The roles of IL-17A in inflammatory immune responses and host defense against pathogens. Immunol Rev 226:57–79
Iwakura Y, Ishigame H, Saijo S, Nakae S (2011) Functional specialization of interleukin-17 family members. Immunity 34:149–162
Izcue A, Hue S, Buonocore S, Arancibia-Carcamo CV, Ahern PP, Iwakura Y, Maloy KJ, Powrie F (2008) Interleukin-23 restrains regulatory T cell activity to drive T cell-dependent colitis. Immunity 28:559–570
Jacobs JP, Wu HJ, Benoist C, Mathis D (2009) IL-17-producing T cells can augment autoantibody-induced arthritis. Proc Natl Acad Sci USA 106:21789–21794
Kappel LW, Goldberg GL, King CG, Suh DY, Smith OM, Ligh C, Holland AM, Grubin J, Mark NM, Liu C, Iwakura Y, Heller G, van den Brink MR (2009) IL-17 contributes to CD4-mediated graft-versus-host disease. Blood 113:945–952

Kastelein RA, Hunter CA, Cua DJ (2007) Discovery and biology of IL-23 and IL-27: related but functionally distinct regulators of inflammation. Annu Rev Immunol 25:221–242

Kleinschek MA, Boniface K, Sadekova S, Grein J, Murphy EE, Turner SP, Raskin L, Desai B, Faubion WA, de Waal MR, Pierce RH, McClanahan T, Kastelein RA (2009) Circulating and gut-resident human Th17 cells express CD161 and promote intestinal inflammation. J Exp Med 206:525–534

Koga C, Kabashima K, Shiraishi N, Kobayashi M, Tokura Y (2008) Possible pathogenic role of Th17 cells for atopic dermatitis. J Invest Dermatol 128:2625–2630

Kohyama S, Ohno S, Isoda A, Moriya O, Belladonna ML, Hayashi H, Iwakura Y, Yoshimoto T, Akatsuka T, Matsui M (2007) IL-23 enhances host defense against vaccinia virus infection via a mechanism partly involving IL-17. J Immunol 179:3917–3925

Komiyama Y, Nakae S, Matsuki T, Nambu A, Ishigame H, Kakuta S, Sudo K, Iwakura Y (2006) IL-17 plays an important role in the development of experimental autoimmune encephalomyelitis. J Immunol 177:566–573

Korn T, Bettelli E, Oukka M, Kuchroo VK (2009) IL-17 and Th17 Cells. Annu Rev Immunol 27:485–517

Kreymborg K, Etzensperger R, Dumoutier L, Haak S, Rebollo A, Buch T, Heppner FL, Renauld JC, Becher B (2007) IL-22 is expressed by Th17 cells in an IL-23-dependent fashion, but not required for the development of autoimmune encephalomyelitis. J Immunol 179:8098–8104

Kryczek I, Wei S, Szeliga W, Vatan L, Zou W (2009) Endogenous IL-17 contributes to reduced tumor growth and metastasis. Blood 114:357–359

Kuestner RE, Taft DW, Haran A, Brandt CS, Brender T, Lum K, Harder B, Okada S, Ostrander CD, Kreindler JL, Aujla SJ, Reardon B, Moore M, Shea P, Schreckhise R, Bukowski TR, Presnell S, Guerra-Lewis P, Parrish-Novak J, Ellsworth JL, Jaspers S, Lewis KE, Appleby M, Kolls JK, Rixon M, West JW, Gao Z, Levin SD (2007) Identification of the IL-17 receptor related molecule IL-17RC as the receptor for IL-17F. J Immunol 179:5462–5473

Lajoie S, Lewkowich IP, Suzuki Y, Clark JR, Sproles AA, Dienger K, Budelsky AL, Wills-Karp M (2010) Complement-mediated regulation of the IL-17A axis is a central genetic determinant of the severity of experimental allergic asthma. Nat Immunol 11:928–935

Langrish CL, Chen Y, Blumenschein WM, Mattson J, Basham B, Sedgwick JD, McClanahan T, Kastelein RA, Cua DJ (2005) IL-23 drives a pathogenic T cell population that induces autoimmune inflammation. J Exp Med 201:233–240

Leibund Gut-Landmann S, Gross O, Robinson MJ, Osorio F, Slack EC, Tsoni SV, Schweighoffer E, Tybulewicz V, Brown GD, Ruland J, Reis e Sousa C (2007) Syk- and CARD9-dependent coupling of innate immunity to the induction of T helper cells that produce interleukin 17. Nat Immunol 8:630–638

Leonardi C, Matheson R, Zachariae C, Cameron G, Li L, Edson-Heredia E, Braun D, Banerjee S (2012) Anti-interleukin-17 monoclonal antibody ixekizumab in chronic plaque psoriasis. N Engl J Med 366:1190–1199

Leppkes M, Becker C, Ivanov II, Hirth S, Wirtz S, Neufert C, Pouly S, Murphy AJ, Valenzuela DM, Yancopoulos GD, Becher B, Littman DR, Neurath MF (2009) RORgamma-expressing Th17 cells induce murine chronic intestinal inflammation via redundant effects of IL-17A and IL-17F. Gastroenterology 136:257–267

Li J, Simeoni E, Fleury S, Dudler J, Fiorini E, Kappenberger L, von Segesser LK, Vassalli G (2006) Gene transfer of soluble interleukin-17 receptor prolongs cardiac allograft survival in a rat model. Eur J Cardiothorac Surg 29:779–783

Liang SC, Long AJ, Bennett F, Whitters MJ, Karim R, Collins M, Goldman SJ, Dunussi-Joannopoulos K, Williams CM, Wright JF, Fouser LA (2007) An IL-17F/A heterodimer protein is produced by mouse Th17 cells and induces airway neutrophil recruitment. J Immunol 179:7791–7799

Lin Y, Ritchea S, Logar A, Slight S, Messmer M, Rangel-Moreno J, Guglani L, Alcorn JF, Strawbridge H, Park SM, Onishi R, Nyugen N, Walter MJ, Pociask D, Randall TD, Gaffen SL, Iwakura Y, Kolls JK, Khader SA (2009) Interleukin-17 is required for T helper 1 cell immunity and host resistance to the intracellular pathogen *Francisella tularensis*. Immunity 31:799–810

Liu L, Okada S, Kong XF, Kreins AY, Cypowyj S, Abhyankar A, Toubiana J, Itan Y, Audry M, Nitschke P, Masson C, Toth B, Flatot J, Migaud M, Chrabieh M, Kochetkov T, Bolze A, Borghesi A, Toulon A, Hiller J, Eyerich S, Eyerich K, Gulacsy V, Chernyshova L, Chernyshov V, Bondarenko A, Grimaldo RM, Blancas-Galicia L, Beas IM, Roesler J, Magdorf K, Engelhard D, Thumerelle C, Burgel PR, Hoernes M, Drexel B, Seger R, Kusuma T, Jansson AF, Sawalle-Belohradsky J, Belohradsky B, Jouanguy E, Bustamante J, Bue M, Karin N, Wildbaum G, Bodemer C, Lortholary O, Fischer A, Blanche S, Al-Muhsen S, Reichenbach J, Kobayashi M, Rosales FE, Lozano CT, Kilic SS, Oleastro M, Etzioni A, Traidl-Hoffmann C, Renner ED, Abel L, Picard C, Marodi L, Boisson-Dupuis S, Puel A, Casanova JL (2011) Gain-of-function human STAT1 mutations impair IL-17 immunity and underlie chronic mucocutaneous candidiasis. J Exp Med 208:1635–1648

Lowes MA, Kikuchi T, Fuentes-Duculan J, Cardinale I, Zaba LC, Haider AS, Bowman EP, Krueger JG (2008) Psoriasis vulgaris lesions contain discrete populations of Th1 and Th17 T cells. J Invest Dermatol 128:1207–1211

Maitra A, Shen F, Hanel W, Mossman K, Tocker J, Swart D, Gaffen SL (2007) Distinct functional motifs within the IL-17 receptor regulate signal transduction and target gene expression. Proc Natl Acad Sci USA 104:7506–7511

Martin-Orozco N, Chung Y, Chang SH, Wang YH, Dong C (2009a) Th17 cells promote pancreatic inflammation but only induce diabetes efficiently in lymphopenic hosts after conversion into Th1 cells. Eur J Immunol 39:216–224

Martin-Orozco N, Muranski P, Chung Y, Yang XO, Yamazaki T, Lu S, Hwu P, Restifo NP, Overwijk WW, Dong C (2009b) T helper 17 cells promote cytotoxic T cell activation in tumor immunity. Immunity 31:787–798

McGeachy MJ, Cua DJ (2008) Th17 cell differentiation: the long and winding road. Immunity 28:445–453

Miossec P, Kolls JK (2012) Targeting IL-17 and T(H)17 cells in chronic inflammation. Nat Rev Drug Discov 11:763–776

Mitsdoerffer M, Lee Y, Jager A, Kim HJ, Korn T, Kolls JK, Cantor H, Bettelli E, Kuchroo VK (2010) Proinflammatory T helper type 17 cells are effective B-cell helpers. Proc Natl Acad Sci USA 107:14292–14297

Molesworth-Kenyon SJ, Yin R, Oakes JE, Lausch RN (2008) IL-17 receptor signaling influences virus-induced corneal inflammation. J Leukoc Biol 83:401–408

Montes M, Zhang X, Berthelot L, Laplaud DA, Brouard S, Jin J, Rogan S, Armao D, Jewells V, Soulillou JP, Markovic-Plese S (2009) Oligoclonal myelin-reactive T-cell infiltrates derived from multiple sclerosis lesions are enriched in Th17 cells. Clin Immunol 130:133–144

Mosmann TR, Cherwinski H, Bond MW, Giedlin MA, Coffman RL (1986) Two types of murine helper T cell clone. I. Definition according to profiles of lymphokine activities and secreted proteins. J Immunol 136:2348–2357

Murphy CA, Langrish CL, Chen Y, Blumenschein W, McClanahan T, Kastelein RA, Sedgwick JD, Cua DJ (2003) Divergent pro- and antiinflammatory roles for IL-23 and IL-12 in joint autoimmune inflammation. J Exp Med 198:1951–1957

Nakae S, Komiyama Y, Nambu A, Sudo K, Iwase M, Homma I, Sekikawa K, Asano M, Iwakura Y (2002) Antigen-specific T cell sensitization is impaired in IL-17-deficient mice, causing suppression of allergic cellular and humoral responses. Immunity 17:375–387

Nakae S, Nambu A, Sudo K, Iwakura Y (2003) Suppression of immune induction of collagen-induced arthritis in IL-17-deficient mice. J Immunol 171:6173–6177

Nakae S, Suto H, Berry GJ, Galli SJ (2007) Mast cell-derived TNF can promote Th17 cell-dependent neutrophil recruitment in ovalbumin-challenged OTII mice. Blood 109:3640–3648

Nakajima A, Matsuki T, Komine M, Asahina A, Horai R, Nakae S, Ishigame H, Kakuta S, Saijo S, Iwakura Y (2010) TNF, but not IL-6 and IL-17, is crucial for the development of T cell-independent psoriasis-like dermatitis in Il1rn–/– mice. J Immunol 185:1887–1893

Numasaki M, Fukushi J, Ono M, Narula SK, Zavodny PJ, Kudo T, Robbins PD, Tahara H, Lotze MT (2003) Interleukin-17 promotes angiogenesis and tumor growth. Blood 101:2620–2627

O'Connor W Jr, Kamanaka M, Booth CJ, Town T, Nakae S, Iwakura Y, Kolls JK, Flavell RA (2009) A protective function for interleukin 17A in T cell-mediated intestinal inflammation. Nat Immunol 10:603–609

Ogawa A, Andoh A, Araki Y, Bamba T, Fujiyama Y (2004) Neutralization of interleukin-17 aggravates dextran sulfate sodium-induced colitis in mice. Clin Immunol 110:55–62

Ogura H, Murakami M, Okuyama Y, Tsuruoka M, Kitabayashi C, Kanamoto M, Nishihara M, Iwakura Y, Hirano T (2008) Interleukin-17 promotes autoimmunity by triggering a positive-feedback loop via interleukin-6 induction. Immunity 29:628–636

Oh I, Ozaki K, Meguro A, Hatanaka K, Kadowaki M, Matsu H, Tatara R, Sato K, Iwakura Y, Nakae S, Sudo K, Teshima T, Leonard WJ, Ozawa K (2010) Altered effector CD4+ T cell function in IL-21R−/− CD4+ T cell-mediated graft-versus-host disease. J Immunol 185:1920–1926

Oyoshi MK, Elkhal A, Kumar L, Scott JE, Koduru S, He R, Leung DY, Howell MD, Oettgen HC, Murphy GF, Geha RS (2009a) Vaccinia virus inoculation in sites of allergic skin inflammation elicits a vigorous cutaneous IL-17 response. Proc Natl Acad Sci USA 106:14954–14959

Oyoshi MK, Murphy GF, Geha RS (2009b) Filaggrin-deficient mice exhibit TH17-dominated skin inflammation and permissiveness to epicutaneous sensitization with protein antigen. J Allergy Clin Immunol 124:485–493, 493 e481

Pantelyushin S, Haak S, Ingold B, Kulig P, Heppner FL, Navarini AA, Becher B (2012) Rorgammat+ innate lymphocytes and gammadelta T cells initiate psoriasiform plaque formation in mice. J Clin Invest 122:2252–2256

Papp KA, Leonardi C, Menter A, Ortonne JP, Krueger JG, Kricorian G, Aras G, Li J, Russell CB, Thompson EH, Baumgartner S (2012) Brodalumab, an anti-interleukin-17-receptor antibody for psoriasis. N Engl J Med 366:1181–1189

Park H, Li Z, Yang XO, Chang SH, Nurieva R, Wang YH, Wang Y, Hood L, Zhu Z, Tian Q, Dong C (2005) A distinct lineage of CD4 T cells regulates tissue inflammation by producing interleukin 17. Nat Immunol 6:1133–1141

Petermann F, Rothhammer V, Claussen MC, Haas JD, Blanco LR, Heink S, Prinz I, Hemmer B, Kuchroo VK, Oukka M, Korn T (2010) gd T cells enhance autoimmunity by restraining regulatory T cell responses via an interleukin-23-dependent mechanism. Immunity 33:351–363

Pichavant M, Goya S, Meyer EH, Johnston RA, Kim HY, Matangkasombut P, Zhu M, Iwakura Y, Savage PB, DeKruyff RH, Shore SA, Umetsu DT (2008) Ozone exposure in a mouse model induces airway hyperreactivity that requires the presence of natural killer T cells and IL-17. J Exp Med 205:385–393

Pisitkun P, Ha HL, Wang H, Claudio E, Tivy CC, Zhou H, Mayadas TN, Illei GG, Siebenlist U (2012) Interleukin-17 cytokines are critical in development of fatal lupus glomerulonephritis. Immunity 37:1104–1115

Rachitskaya AV, Hansen AM, Horai R, Li Z, Villasmil R, Luger D, Nussenblatt RB, Caspi RR (2008) Cutting edge: NKT cells constitutively express IL-23 receptor and RORgammat and rapidly produce IL-17 upon receptor ligation in an IL-6-independent fashion. J Immunol 180:5167–5171

Raffatellu M, Santos RL, Verhoeven DE, George MD, Wilson RP, Winter SE, Godinez I, Sankaran S, Paixao TA, Gordon MA, Kolls JK, Dandekar S, Baumler AJ (2008) Simian immunodeficiency virus-induced mucosal interleukin-17 deficiency promotes *Salmonella* dissemination from the gut. Nat Med 14:421–428

Res PC, Piskin G, de Boer OJ, van der Loos CM, Teeling P, Bos JD, Teunissen MB (2010) Overrepresentation of IL-17A and IL-22 producing CD8 T cells in lesional skin suggests their involvement in the pathogenesis of psoriasis. PLoS One 5:e14108

Rizzo HL, Kagami S, Phillips KG, Kurtz SE, Jacques SL, Blauvelt A (2010) IL-23-mediated psoriasis-like epidermal hyperplasia is dependent on IL-17A. J Immunol 186:1495–1502

Roark CL, French JD, Taylor MA, Bendele AM, Born WK, O'Brien RL (2007) Exacerbation of collagen-induced arthritis by oligoclonal, IL-17-producing gamma delta T cells. J Immunol 179:5576–5583

Saijo S, Ikeda S, Yamabe K, Kakuta S, Ishigame H, Akitsu A, Fujikado N, Kusaka T, Kubo S, Chung SH, Komatsu R, Miura N, Adachi Y, Ohno N, Shibuya K, Yamamoto N, Kawakami K,

Yamasaki S, Saito T, Akira S, Iwakura Y (2010) Dectin-2 recognition of alpha-mannans and induction of Th17 cell differentiation is essential for host defense against *Candida albicans*. Immunity 32:681–691

Sato K, Suematsu A, Okamoto K, Yamaguchi A, Morishita Y, Kadono Y, Tanaka S, Kodama T, Akira S, Iwakura Y, Cua DJ, Takayanagi H (2006) Th17 functions as an osteoclastogenic helper T cell subset that links T cell activation and bone destruction. J Exp Med 203: 2673–2682

Shah K, Lee WW, Lee SH, Kim SH, Kang SW, Craft J, Kang I (2010) Dysregulated balance of Th17 and Th1 cells in systemic lupus erythematosus. Arthritis Res Ther 12:R53

Shibata K, Yamada H, Hara H, Kishihara K, Yoshikai Y (2007) Resident Vdelta1$^+$ gammadelta T cells control early infiltration of neutrophils after *Escherichia coli* infection via IL-17 production. J Immunol 178:4466–4472

Shichita T, Sugiyama Y, Ooboshi H, Sugimori H, Nakagawa R, Takada I, Iwaki T, Okada Y, Iida M, Cua DJ, Iwakura Y, Yoshimura A (2009) Pivotal role of cerebral interleukin-17-producing gammadeltaT cells in the delayed phase of ischemic brain injury. Nat Med 15:946–950

Sumarac-Dumanovic M, Stevanovic D, Ljubic A, Jorga J, Simic M, Stamenkovic-Pejkovic D, Starcevic V, Trajkovic V, Micic D (2009) Increased activity of interleukin-23/interleukin-17 proinflammatory axis in obese women. Int J Obes (Lond) 33:151–156

Sutton CE, Lalor SJ, Sweeney CM, Brereton CF, Lavelle EC, Mills KH (2009) Interleukin-1 and IL-23 induce innate IL-17 production from gammadelta T cells, amplifying Th17 responses and autoimmunity. Immunity 31:331–341

Suzukawa M, Morita H, Nambu A, Arae K, Shimura E, Shibui A, Yamaguchi S, Suzukawa K, Nakanishi W, Oboki K, Kajiwara N, Ohno T, Ishii A, Korner H, Cua DJ, Suto H, Yoshimoto T, Iwakura Y, Yamasoba T, Ohta K, Sudo K, Saito H, Okumura K, Broide DH, Matsumoto K, Nakae S (2012) Epithelial cell-derived IL-25, but not Th17 cell-derived IL-17 or IL-17F, is crucial for murine asthma. J Immunol 189:3641–3652

Takahashi N, Vanlaere I, de Rycke R, Cauwels A, Joosten LA, Lubberts E, van den Berg WB, Libert C (2008) IL-17 produced by Paneth cells drives TNF-induced shock. J Exp Med 205:1755–1761

Tang JL, Subbotin VM, Antonysamy MA, Troutt AB, Rao AS, Thomson AW (2001) Interleukin-17 antagonism inhibits acute but not chronic vascular rejection. Transplantation 72:348–350

Tartour E, Fossiez F, Joyeux I, Galinha A, Gey A, Claret E, Sastre-Garau X, Couturier J, Mosseri V, Vives V, Banchereau J, Fridman WH, Wijdenes J, Lebecque S, Sautes-Fridman C (1999) Interleukin 17, a T-cell-derived cytokine, promotes tumorigenicity of human cervical tumors in nude mice. Cancer Res 59:3698–3704

Tong Z, Yang XO, Yan H, Liu W, Niu X, Shi Y, Fang W, Xiong B, Wan Y, Dong C (2012) A protective role by interleukin-17F in colon tumorigenesis. PLoS One 7:e34959

van der Fits L, Mourits S, Voerman JS, Kant M, Boon L, Laman JD, Cornelissen F, Mus AM, Florencia E, Prens EP, Lubberts E (2009) Imiquimod-induced psoriasis-like skin inflammation in mice is mediated via the IL-23/IL-17 axis. J Immunol 182:5836–5845

Venken K, Hellings N, Hensen K, Rummens JL, Stinissen P (2010) Memory CD4+CD127high T cells from patients with multiple sclerosis produce IL-17 in response to myelin antigens. J Neuroimmunol 226:185–191

Vokaer B, Van Rompaey N, Lemaitre PH, Lhomme F, Kubjak C, Benghiat FS, Iwakura Y, Petein M, Field KA, Goldman M, Le Moine A, Charbonnier LM (2010) Critical role of regulatory T cells in Th17-mediated minor antigen-disparate rejection. J Immunol 185:3417–3425

Wang L, Yi T, Kortylewski M, Pardoll DM, Zeng D, Yu H (2009) IL-17 can promote tumor growth through an IL-6-Stat3 signaling pathway. J Exp Med 206:1457–1464

Wang Q, Li H, Yao Y, Xia D, Zhou J (2010a) The overexpression of heparin-binding epidermal growth factor is responsible for Th17-induced airway remodeling in an experimental asthma model. J Immunol 185:834–841

Wang YH, Voo KS, Liu B, Chen CY, Uygungil B, Spoede W, Bernstein JA, Huston DP, Liu YJ (2010b) A novel subset of CD4+ TH2 memory/effector cells that produce inflammatory IL-17 cytokine and promote the exacerbation of chronic allergic asthma. J Exp Med 207:2479–2491

Wang C, Wu L, Bulek K, Martin BN, Zepp JA, Kang Z, Liu C, Herjan T, Misra S, Carman JA, Gao J, Dongre A, Han S, Bunting KD, Ko JS, Xiao H, Kuchroo VK, Ouyang W, Li X (2012) The psoriasis-associated D10N variant of the adaptor Act1 with impaired regulation by the molecular chaperone hsp90. Nat Immunol 14:72–81

Weaver CT, Hatton RD, Mangan PR, Harrington LE (2007) IL-17 family cytokines and the expanding diversity of effector T cell lineages. Annu Rev Immunol 25:821–852

Wilson NJ, Boniface K, Chan JR, McKenzie BS, Blumenschein WM, Mattson JD, Basham B, Smith K, Chen T, Morel F, Lecron JC, Kastelein RA, Cua DJ, McClanahan TK, Bowman EP, de Waal MR (2007) Development, cytokine profile and function of human interleukin 17-producing helper T cells. Nat Immunol 8:950–957

Wilson MS, Madala SK, Ramalingam TR, Gochuico BR, Rosas IO, Cheever AW, Wynn TA (2010) Bleomycin and IL-1beta-mediated pulmonary fibrosis is IL-17A dependent. J Exp Med 207:535–552

Winer S, Paltser G, Chan Y, Tsui H, Engleman E, Winer D, Dosch HM (2009) Obesity predisposes to Th17 bias. Eur J Immunol 39:2629–2635

Wright JF, Bennett F, Li B, Brooks J, Luxenberg DP, Whitters MJ, Tomkinson KN, Fitz LJ, Wolfman NM, Collins M, Dunussi-Joannopoulos K, Chatterjee-Kishore M, Carreno BM (2008) The human IL-17F/IL-17A heterodimeric cytokine signals through the IL-17RA/IL-17RC receptor complex. J Immunol 181:2799–2805

Wu S, Rhee KJ, Albesiano E, Rabizadeh S, Wu X, Yen HR, Huso DL, Brancati FL, Wick E, McAllister F, Housseau F, Pardoll DM, Sears CL (2009) A human colonic commensal promotes colon tumorigenesis via activation of T helper type 17 T cell responses. Nat Med 15:1016–1022

Wu HJ, Ivanov II, Darce J, Hattori K, Shima T, Umesaki Y, Littman DR, Benoist C, Mathis D (2010) Gut-residing segmented filamentous bacteria drive autoimmune arthritis via T helper 17 cells. Immunity 32:815–827

Yamada J, Hamuro J, Fukushima A, Ohteki T, Terai K, Iwakura Y, Yagita H, Kinoshita S (2009) MHC-matched corneal allograft rejection in an IFN-gamma/IL-17-independent manner in C57BL/6 mice. Invest Ophthalmol Vis Sci 50:2139–2146

Yang XO, Chang SH, Park H, Nurieva R, Shah B, Acero L, Wang YH, Schluns KS, Broaddus RR, Zhu Z, Dong C (2008) Regulation of inflammatory responses by IL-17F. J Exp Med 205:1063–1075

Yi T, Zhao D, Lin CL, Zhang C, Chen Y, Todorov I, LeBon T, Kandeel F, Forman S, Zeng D (2008) Absence of donor Th17 leads to augmented Th1 differentiation and exacerbated acute graft-versus-host disease. Blood 112:2101–2110

Yoshitomi H, Sakaguchi N, Kobayashi K, Brown GD, Tagami T, Sakihama T, Hirota K, Tanaka S, Nomura T, Miki I, Gordon S, Akira S, Nakamura T, Sakaguchi S (2005) A role for fungal beta-glucans and their receptor Dectin-1 in the induction of autoimmune arthritis in genetically susceptible mice. J Exp Med 201:949–960

Zhang Z, Kyttaris VC, Tsokos GC (2009) The role of IL-23/IL-17 axis in lupus nephritis. J Immunol 183:3160–3169

Zhou L, Littman DR (2009) Transcriptional regulatory networks in Th17 cell differentiation. Curr Opin Immunol 21:146–152

Zhou L, Chong MM, Littman DR (2009) Plasticity of CD4+ T cell lineage differentiation. Immunity 30:646–655

Zhu J, Yamane H, Paul WE (2010) Differentiation of effector CD4 T cell populations. Annu Rev Immunol 28:445–489

Zou W, Restifo NP (2010) TH17 cells in tumour immunity and immunotherapy. Nat Rev Immunol 10:248–256

Zuniga LA, Shen WJ, Joyce-Shaikh B, Pyatnova EA, Richards AG, Thom C, Andrade SM, Cua DJ, Kraemer FB, Butcher EC (2010) IL-17 regulates adipogenesis, glucose homeostasis, and obesity. J Immunol 185:6947–6959

Chapter 4
IL-18

Regulation and Physiological Roles of IL-18

Wen Li, Yoshimasa Tanaka, and Haruki Okamura

Abstract Interleukin-18 (IL-18) belongs to the IL-1 cytokine family, which has crucial roles in innate immunity and inflammation. These cytokines share similar molecular structures, receptor structures, and signal transduction pathways. Each cytokine, however, acts on distinct types of cells that express their specific receptor. The functional IL-18 receptor is specifically induced by cytokines such as IL-2 and IL-12. During innate immune responses, IL-1β acts primarily on Th17 cells, leading to the recruitment of neutrophils, whereas IL-18 in the presence of IL-2 or IL-12 induces Th1-type responses in natural killer (NK) and $CD8^+$ T cells, which induces reactive oxygen species and nitric oxide production in macrophages. IL-18 is constitutively expressed in hematopoietic cells and nonhematopoietic cells in tissues where cells are rapidly dividing. Similar to IL-1β, the inactive precursor of IL-18 is proteolytically converted to a mature form by caspase-1 in inflammasome complexes. Although the molecular mechanism by which IL-18 is activated by inflammasomes has been clarified, the regulatory mechanism underlying the expression of IL-18 receptors, the effect of signaling on organelles such as mitochondria, and the biological roles of IL-18 in nonhematopoietic cells remain to be elucidated. Accumulating evidence suggests that IL-18 is not merely a cytokine inducer, but a molecule involved in the viability and terminal differentiation of various cells.

Keywords Cytokine • IL-18 • Inflammasome • Inflammation • Innate immunity

W. Li • H. Okamura (✉)
Laboratory of Tumor Immunology and Cell Therapy, Hyogo College of Medicine,
1-1 Mukogawa-cho, Nishinomiya, Hyogo 663-8501, Japan
e-mail: haruoka@hyo-med.ac.jp

Y. Tanaka
Center for Innovation in Immunoregulative Technology and Therapeutics,
Graduate School of Medicine, Kyoto University, Kyoto, Japan

T. Yoshimoto and T. Yoshimoto (eds.), *Cytokine Frontiers: Regulation of Immune Responses in Health and Disease*, DOI 10.1007/978-4-431-54442-5_4, © Springer Japan 2014

4.1 Introduction

To maintain homeostasis, our body responds to physiological and injurious stresses such as ultraviolet (UV) radiation, reactive oxygen species (ROS), and infection by inducing local inflammation and systemic endocrine responses. Inflammation is necessary to sense and alleviate these stresses, repair wounds, and restore homeostasis. Various soluble mediators are involved in the promotion and suppression of inflammatory responses. Above all, IL-1 family cytokines play pivotal roles during inflammatory responses from initiation through to termination (Dinarello 2009; Sims and Smith 2010; Lukens et al. 2012; Boraschi et al. 2011). IL-1 family cytokines share similar molecular structures, receptor structures, and signal transduction pathways, but have different roles in inflammation because of the specific distribution of their receptors on target cells (Dinarello 2009; Sims and Smith 2010; Lukens et al. 2012; Boraschi et al. 2011).

Much progress has been made in understanding the regulatory roles of IL-1 family cytokines in inflammatory responses (Dinarello 2009; Sims and Smith 2010). Because dysregulated inflammation is still not successfully controlled by medical treatment, it is necessary to elucidate the precise functions and mechanisms of each IL-1 family member and the mutual relationship between these cytokines during inflammation. In this review, we focus on the mechanisms underlying the effect of IL-18 on lymphocytes and the biological and pathological roles of IL-18 in comparison with other IL-1 family members (Fig. 4.1). IL-1α and IL-1β, for instance, act on Th17 cells, endowing them with the ability to recruit and activate neutrophils (Sutton et al. 2006; Ben-Sasson et al. 2009; Chung et al. 2009). IL-33 is released from necrotic or damaged cells and may be involved in wound healing and fibrosis (Kurowska-Stolarska et al. 2011; Lopetuso et al. 2012). It is noteworthy that caspase-1 cleaves the inactive precursors to produce biologically active forms of IL-1β and IL-18, whereas caspase-1 inactivates IL-33.

Among the 11 IL-1 family members, IL-1β, IL-18, and possibly IL-37 are activated by inflammasomes. During inflammation, IL-1β regulates the induction of IL-17-producing cells, whereas IL-18 acts primarily on interferon (IFN)-γ-producing Th1-type lymphocytes in the presence of IL-2 or IL-12 to promote their survival and expansion, induce cytokine production, and enhance cytotoxic activity. However, little is known about the precise mechanism underlying IL-18 signaling in target cells and the roles of IL-18 in inflammation as a mediator bridging innate and acquired immunity (Nakanishi et al. 2001; Dinarello 2007; Arend et al. 2008). IL-18 may play pivotal roles in the onset of inflammation following the activation of inflammasomes. Although IL-1 family cytokines appear to be evolutionarily conserved and are ubiquitously present in a wide range of species from invertebrates to vertebrates, the roles of IL-18 in nonimmune cells remain poorly understood.

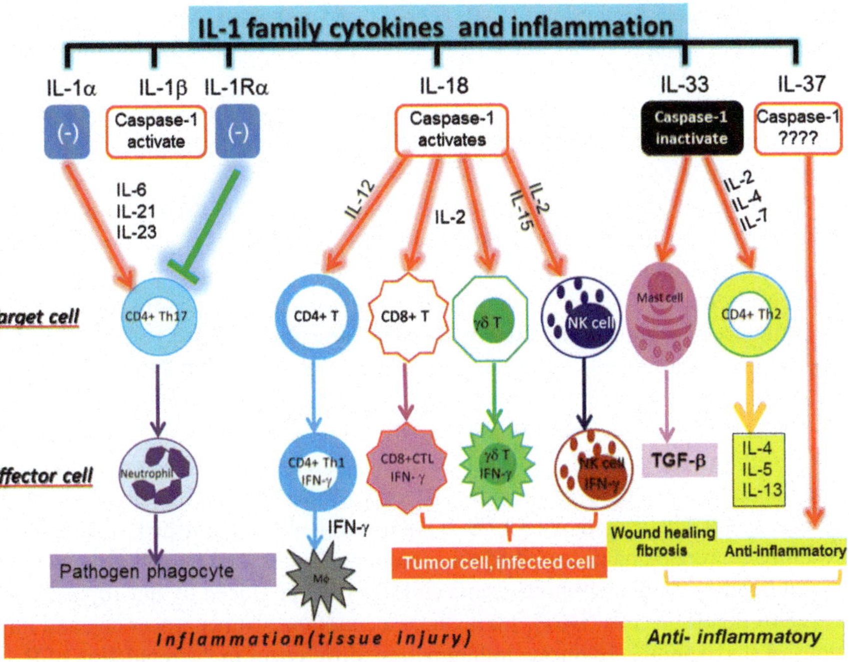

Fig. 4.1 Immunomodulatory roles of interleukin-1 (IL-1) family cytokines. Each IL-1 family member acts on effector cells committed to other stimuli. IL-12 and IL-18 promote Th1 precursor cells to proliferate and produce IFN-γ. IL-1 α/β stimulate Th17 cells in the presence of IL-6 and IL-21 to produce IL-17, and IL-33 in combination with IL-2 or IL-7 induces proliferation and cytokine production in Th2 cells and mast cells. IL-37 binds to IL-18Rα, inhibiting the recruitment of IL-18Rβ and regulating inflammation

4.2 Regulation of IL-18 Production by Inflammasomes

IL-18 was originally identified as an IFN-γ inducing factor in 1995. We previously identified and purified a factor in the blood of mice treated with killed *Propionibacterium acnes* followed by bacterial lipopolysaccharide (LPS). Subsequently, the gene was cloned and the gene product named IL-18 (Okamura et al. 1995). This cytokine lacks a signal sequence and requires IL-1β-converting enzyme (ICE, caspase-1) for its maturation (Gu et al. 1997). However, it took a long time to elucidate the mechanism of processing and activation of IL-18 by caspase-1. In parallel with progress in the study of IL-1 family cytokines, the mechanisms responsible for the recognition and response to different danger signals were clarified.

In 2002, Martinon et al. identified a caspase-1-activating protein complex composed of a pattern recognition receptor, NACHT, LRR, and PYD domain-containing protein 3 (NLRP3), an adaptor molecule, apoptosis-associated speck-like protein containing a CARD (ASC), and caspase-1 (Martinon et al. 2002). This protein complex was named the inflammasome. Activated inflammasomes convert pro-inflammatory cytokines IL-1β and IL-18 to their mature forms. Because many studies have shown that inflammasomes are critical in the pathogenesis of a wide spectrum of diseases, much attention has been paid to the regulation of inflammasome activation.

IL-1β promotes Th17 cells to recruit neutrophils and is used as a measure of inflammasome activation, whereas IL-18 has attracted less attention, probably because the physiological roles of IL-18 are obscure. Although both IL-1β and IL-18 are converted to biologically active forms by inflammasomes, pro-IL-1β itself needs to be induced by signals initiated by Toll-like receptors (TLRs) before being processed. In contrast, pro-IL-18 is constitutively expressed in cells (Puren et al. 1999), which strongly suggests that the cleavage and secretion of IL-18 precedes that of IL-1β during inflammatory responses. IL-18 is thus able to respond to alarm signals at the very initial stage of inflammation. In support of this, IL-18 was detected in the circulation a few hours earlier than IL-1β in mice that had been administered *P. acnes* and challenged with LPS (Okamura et al., unpublished observations). Notably, no reports have described the simultaneous conversion of IL-1β and IL-18 to their mature forms within the same cell. IL-1β is produced by macrophages and IL-18 by dendritic cells (DCs). Taking into account the distinct processing and functions of IL-1β and IL-18 during inflammation, differences in the regulation of IL-1β and IL-18 production at a cellular level should be examined more extensively.

The activation of inflammasomes is similar to that of the apoptosome in terms of the involvement of CARD (Munoz-Pinedo 2012). Apoptosomes are involved in cellular apoptosis, whereas the activation of inflammasomes leads to cell death called pyroptosis (Miao et al. 2011; Ceballos-Olvera et al. 2011; Schmidt and Lenz 2012). Although both apoptosis and pyroptosis describe cell death, the former is dependent on caspase-9 and caspase-3, and the latter on caspase-1. Pyroptosis shares unique characteristics with both necrosis and apoptosis. Apoptosis occurs in almost all types of cells, whereas pyroptosis is specifically observed in macrophages and DCs. Pyroptosis is considered a mechanism to eliminate intracellular pathogens. Cells bearing activated inflammasomes are thus committed to die. During pyroptosis, dying cells release cytosolic molecules including cytokines such as IL-1β and IL-18. It is noteworthy that some inflammasomes induce both cell death and the secretion of IL-1β and IL-18, whereas others secrete these cytokines in the absence of cell death (Miao et al. 2011; Ceballos-Olvera et al. 2011; Schmidt and Lenz 2012); this might be attributable to the different subcellular location of each inflammasome. In addition, caspse-11-deficient mice failed to secrete IL-1β in response to bacterial antigens, although IL-18 production was not impaired (Schmidt and Lenz 2012). Thus, the processing of IL-1β and IL-18 may be differentially regulated.

Members of IL-1 family cytokines can be divided into two groups based on their activation mechanism. One group includes IL-1β and IL-18, which strongly induce inflammation and require caspase-1 activation for their conversion to biologically active forms. The second group can also be hydrolyzed by proteases. However, these cytokines do not necessarily require proteolytic processing for their maturation. For example, caspase-1-mediated cleavage of IL-33 inactivates the cytokine (Cayrol and Girard 2009; Luthi et al. 2009).

Recently, a novel non-canonical inflammasome activation pathway dependent on caspase-11 was proposed (Kayagaki et al. 2011). Caspase-11-deficient macrophages infected with bacteria failed to activate caspase-1, whereas the canonical activation of inflammasomes by ATP or monosodium uric acid was not impaired. In addition, IL-11 was required for pyroptosis induced by the non-canonical inflammasome. Type I interferon (IFN) signaling may induce functional caspase-11 that activates NLRP3 inflammasome and causes cell death in the absence of caspase-1 involvement (Rathinam et al. 2012). Furthermore, caspase-11-deficient DCs failed to secrete IL-1β in response to non-canonical stimuli, whereas the secretion of IL-18 was not impaired (Schmidt and Lenz 2012). These observations strongly suggest that the three events—IL-1β production, IL-18 secretion, and pyroptosis—are likely to be mutually exclusive and differentially regulated.

4.3 Inflammasome-Independent Maturation of Pro-IL-18

IL-18 can be activated by proteases other than caspase-1, such as proteinase 3 (Sugawara et al. 2001), chymase (Omoto et al. 2006), and granzyme B (Omoto et al. 2010). In addition, a unique and unknown caspase-1-independent mechanism is likely to exist for the maturation of IL-18 (Tsutsui et al. 1999). The pathophysiological significance of these observations remains to be determined.

4.4 IL-18 in Nonhematopoietic Cells

IL-18 is constitutively expressed by hematopoietic cells such as macrophages and DCs and nonhematopoietic cells such as keratinocytes and intestinal epithelial cells. How IL-18 is activated and acts on nonhematopoietic cells is poorly understood. IL-18 is highly expressed in peripheral tissues such as skin, secretary glands, and the intestine, where the cells are rapidly dividing and require high levels of energy. It is also an evolutionarily ancient molecule. Because the biological roles of IL-18 in immune/inflammatory responses have not been fully elucidated, an increased understanding of the roles of IL-18 in nonhematopoietic cells will be of interest.

4.5 Molecular Mechanisms Underlying IL-18 Signaling

IL-18 exhibits its physiological functions in the context of its receptors IL-18Rα (also termed IL1R5 or IL-1Rrp1) and IL-18Rβ (also termed IL-1R7 or IL-1RAcPL), which belong to the IL-1-receptor family (Dinarello 2009; Sims and Smith 2010;

Lukens et al. 2012; Boraschi et al. 2011; O'Neill 2008). Recent analyses revealed novel roles or IL-18Rs in the regulation of inflammatory responses. IL-18Rs belong to the IL-1R/TLR family. The known IL-1R family receptors consist of ten members (Dinarello 2009; Sims and Smith 2010; Lukens et al. 2012; Boraschi et al. 2011; O'Neill 2008). IL-1R1, IL-18Rα, IL-1RrP2, and Ti/ST2 contain ligand-binding motifs, IL-1RAcP and IL-18Rβ are accessory receptors, and TIGIRR and IL-1RAPL are orphan receptors. IL-1R2 and TIR8/SIGIRR are decoy receptors lacking an intracellular TIR domain. Among these, TIR8/SIGIRR and T1/ST2 are involved in negative signaling.

Upon the binding of IL-18 to IL-18Rα, the complex recruits IL-18Rβ and transduces signals in lymphocytes. IL-18Rα has a TIR domain, which initiates the signaling cascade. Although IL-18Rα is constitutively expressed in lymphocytes such as NK cells and T cells, IL-18Rβ appears to be induced by activation stimuli (Nakanishi et al. 2001; Kawashima and Miossec. 2003; Li et al. 2007). IL-18 and other cytokines such as IL-2, IL-12, IL-15, and IL-21 synergistically induce IFN-γ, which may be responsible for the upregulation of IL-18Rs, especially the IL-18Rβ chain. IL-18Rs appear to be constitutively expressed in intestinal epithelial cells and brain cells (Kolinska et al. 2008; Alboni et al. 2011). Because IL-18 alone fails to induce cytokine production in resting lymphocytes and acts on lymphocytes synergistically with other factors, such as IFN-γ induction by IL-18 plus IL-12, IL-18Rs can be upregulated in activated lymphocytes. The regulation of IL-18Rs expression in different cell types should be investigated to elucidate further the roles of IL-18 in immune/inflammatory responses.

IL-18Rs are ubiquitously expressed on the surface of hematopoietic and nonhematopoietic cells (Dinarello 2009; Sims and Smith 2010; Lukens et al. 2012; Boraschi et al. 2011; O'Neill 2008), and their signaling pathways are well characterized. However, the regulation and location of IL-18Rs and their signaling pathways in nonhematopoietic cells such as keratinocytes and epithelial cells are unknown. Although IL-1α and IL-33 function as extracellular cytokines and intracellular nuclear factors regulating the transcription of genes, no evidence has suggested that IL-18 functions as a nuclear factor.

IL-1, IL-18, and IL-33 transduce signals upon binding to IL-1RI, IL-18Rα, and ST2, respectively, followed by the recruitment of the MyD88 adaptor molecule (O'Neill 2008). Although these receptors are expressed by different types of cells, the same intracellular signaling pathway appears to be used (Fig. 4.2). After binding of IL-18 to IL-18Rα, the complex recruits IL-18Rβ. The resulting triplet complex further recruits the adaptor molecule MyD88, leading to the successive activation of IRAK1, IRAK4, or IRAK2. Phosphorylated IRAK1 and IRAK4 interact with TRAF6, serving as a platform that recruits and activates the kinase TAK1, which in turn activates the IKK complex, leading to the degradation of IκB. The generation of free NF-κB that translocates from the cytosol to the nucleus results in the transcription of NF-κB-dependent genes involved in cell growth/survival and apoptosis. Besides the central MyD88-IRAK-TRAF6 signaling pathway, the p38 MAPK, PI3K/Akt, and ERK pathways are also activated downstream of IL-18Rs (Neumann et al. 2002; Yoo et al. 2005; Hirata et al. 2008; Fortin et al. 2009; Ahn et al. 2012).

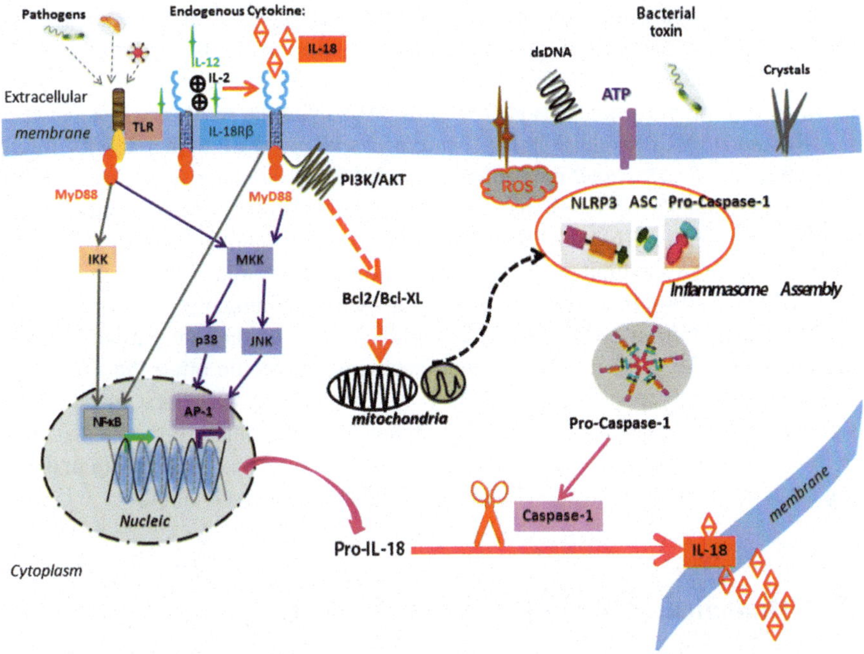

Fig. 4.2 IL-18 signaling may be transduced via diverse pathways. Because functional IL-18Rs are not constitutively expressed in lymphocytes and their expression is regulated by other environmental factors such as cytokines (IL-2, IL-12) and antigens (αCD3-TCR), it is likely that IL-18 signaling is co-transduced with signals downstream of these factors. Therefore, it is difficult to identify IL-18-specific signaling pathways. IL-18 may activate signaling pathways involved in the production of cytokines as well as cellular viability and differentiation by maintaining the integrity of mitochondrial functions

The activation of these pathways closely couples with the central pathway, but appears to be involved in independent functions of NF-κB. Because IL-18 signaling is enhanced by signals from other cytokines and environmental factors, it is difficult to identify signals that are unique to the IL-18R pathway.

It was previously demonstrated that IL-18 deficiency resulted in the attenuation of inflammatory responses, whereas IL-18Rs deficiency aggravated inflammatory responses (Lewis and Dinarello 2006; Gutcher et al. 2006; Nold-Petry et al. 2009; Nozaki et al. 2012). This apparently paradoxical observation demonstrates the complexity of IL-18R regulatory signaling during inflammation. Thus, it is possible to postulate the existence of an agonist in addition to IL-18 that can signal through IL-18Rs. Consistent with this hypothesis, the IL-1-like orphan molecule IL-1F7 (recently redesignated as IL-37) can bind to IL-18Rα- and IL-18-binding proteins (IL-18BP) (Boraschi et al. 2011; Kumar et al. 2002; Pan et al. 2001; Bufler et al. 2002). Although the physiological role of IL-37 has long been enigmatic, it was recently demonstrated to competitively inhibit IL-18 by interacting with IL-18Rs (Boraschi et al. 2011). IL-37 is produced by many cell types and is converted to its mature form most effectively by caspase-1 (Boraschi et al. 2011; Kumar et al. 2002). The functional orthologue of murine IL-37 has not yet been identified.

IL-37 associates with IL-18Rα and inhibits the recruitment of IL-18Rβ, leading to the disruption of signal transduction via IL-18Rs, similar to interactions between IL-1Rα and IL-1RI. In addition, IL-37 interacts with IL-18BP to increase its inhibitory effect on IL-18 signaling (Bufler et al. 2002). In contrast to IL-37, IL-18BP directly associates with IL-18 and interferes with its activity. Because the concentration of circulating IL-18BP is about 20 fold greater than that of IL-18, it seems unlikely that IL-18 exhibits systemic activity under the conditions of high levels of IL-18BP. Therefore, IL-18 may act locally at sites of inflammation.

It may be hypothesized that IL-37-bound IL-18Rα transduces negative signals in the presence of a currently unidentified accessory molecule other than IL-18Rβ. TIR8/SIGIRR, an orphan receptor of the IL-1R family, may transduce the inhibitory signal because the receptor is required to mediate the antagonistic effects of IL-36Rα in inflammatory responses (Costelloe et al. 2008). The precise mechanism underlying the inhibition of IL-18 signaling by IL-37 is not fully understood.

4.6 Biological Roles of IL-18 in Cellular Responses

As already mentioned, IL-18 activates p38 MAPK, PI3K/Akt, and ERK pathways, in addition to the central MyD88-IRAK-TRAF6 signaling pathway (Neumann et al. 2002; Yoo et al. 2005; Hirata et al. 2008; Fortin et al. 2009; Ahn et al. 2012). This finding indicates that IL-18 may induce various cellular functions such as cellular viability, growth, differentiation, and migration in addition to the induction of cytokines. Indeed, several reports have demonstrated the effect of IL-18 on cell viability and expansion. IL-18 activates signals related to cell survival such as cIAP, TRAF1, PI3K/Akt, Bcl-2, and Bcl-X_L (Hodge et al. 2006; Hosotani et al. 2008; Li et al. 2010), resulting in the prolonged survival of the activated cells. Thus, IL-18 can selectively increase the population size of lymphocytes that express a functional receptor. Although IL-18 was originally discovered as a cytokine inducer, it is unlikely that it only induces cytokine production without cell differentiation. Effective cytokine production is often achieved in parallel with the expansion of cytokine-producing cells. During inflammation, some IL-1 family cytokines promote the expansion and effector functions of activated cells. Therefore, it will be of great interest if the roles of constitutively expressed IL-18 by nonhematopoietic cells such as keratinocytes and epithelial cells are elucidated. Because these cells are located in tissues where cells are rapidly dividing, IL-18 may play a role in the terminal differentiation of cells.

IL-18 is activated by inflammasomes induced by mitochondrial dysfunction. It is intriguing that IL-18 activates signals involved in the integrity of mitochondria such as PI3K/Akt, Bcl-2, and Bcl-X_L (Li et al. 2007, 2010; Hodge et al. 2006; Hosotani et al. 2008). Therefore, it will be necessary to determine the roles of IL-18 in the regulation of mitochondrial functions.

4.7 Roles of IL-18 in Infection

It is well established that IL-18 and IL-1β contribute to host defense against infection by various pathogens (Dinarello 2009; Nakanishi et al. 2001; Dinarello and Fantuzzi 2003; Sahoo et al. 2011; van de Veerdonk et al. 2011). In the presence of IL-2, IL-12, and IL-15, IL-18 has a pivotal functional role in the activation of lymphocytes (Nakanishi et al. 2001). IFN-γ produced by activated lymphocytes stimulates monocytes or macrophages to produce high levels of reactive oxygen species (ROS) and NO, thus killing intracellular pathogens such as *Mycobacterium tuberculosis*. In contrast, IL-1β induces the development of Th17 cells, resulting in the recruitment of neutrophils that engulf and kill extracellular pathogens. In addition, IL-1β induces high levels of cytochrome *c* oxidase subunit II (COX2) and prostaglandin E_2 (PGE_2) production that suppress IFN-γ production (Sahoo et al. 2011; van de Veerdonk et al. 2011). IL-18 and IL-1β may thus have redundant and nonredundant roles in the eradication of microbial pathogens.

The activation of inflammasomes is indispensable for defense against infections as IL-18 and IL-1β are essential factors in the first line of defense. During inflammation, inflammasomes induce cytokine production and are critical for microbial elimination. Pyroptotic (and pyronecrotic) cell death is involved in the protection from bacterial infections (van de Veerdonk et al. 2011). Pyroptotic cell death may deprive the intracellular bacteria of a suitable environment for growth, or may expose intracellular bacteria to the extracellular immune system to induce humoral and cellular responses. Thus, inflammasomes play essential roles in defense against infections.

Upon virus infection, NOD-like receptor (NLRs) such as NLRP3, AIM2, and RIG-1 assemble to form caspase-1-activating inflammasomes, and the resulting production of IL-18 plays roles in host defense similar to infection with intracellular bacteria (Kanneganti 2010; Rathinam and Fitzgerald 2010; Gram et al. 2012). IL-18 activates and expands NK cells and cytotoxic T cells, which directly kill virally infected cells or secrete IFN-γ to indirectly eradicate infected cells (Nakanishi et al. 2001). Although bacteria and viruses have evolved tools to modulate the function of inflammasomes, such as bacterial virulence factor Yop and viral decoy proteins or proteases that inhibit IL-1β and IL-18 production, the precise mechanism of inhibition remains unknown (Lamkanfi and Dixit 2011; Jacobs and Damania 2012).

4.8 Antitumor Activity of IL-18

Inflammasomes exhibit both tumor-promoting and tumor-suppressing functions (Grace and Gabriel 2011; Laurence et al. 2012), and consequently IL-18 has dual effects on carcinogenesis and tumor progression (Nakanishi et al. 2001; Vidal-Vanaclocha et al. 2006; Park et al. 2007; Srivastava et al. 2010; Takashi et al. 2006).

Tumor cells are considered to arise in the microenvironmental niche during chronic inflammation. During inflammatory bowel disease, the production of ROS, growth factors, and cell survival-promoting factors, as well as DNA damage, may be closely linked to tumorigenesis. IL-18 may enhance the production of these factors, and may thus promote tumorigenesis. In addition, IL-18 may directly act on tumor cells and enhance their proliferation and migration (Vidal-Vanaclocha et al. 2006; Park et al. 2007; Srivastava et al. 2010). Because tumor cells often produce IL-18 or express IL-18Rs, IL-18 may decrease the susceptibility of tumor cells to cytotoxicity by immune cells (Vidal-Vanaclocha et al. 2006).

However, IL-18 was initially demonstrated to have strong antitumor activity (Nakanishi et al. 2001; Vidal-Vanaclocha et al. 2006). Inflammasomes upstream of IL-18 were involved in the inhibition of tumor growth (Grace and Gabriel 2011; Laurence et al. 2012). In support of this, mice deficient in inflammasome components or IL-18 exhibited augmented tumorigenesis in mice (Irving et al. 2010; Grace et al. 2011; Rosalba et al. 2010). Because IL-18 itself fails to directly suppress tumorigenicity, the cytokine most likely activates immune cells such as NK cells and cytotoxic T cells, resulting in antitumor activity. It is noteworthy that IL-18 may mediate multiple effector functions against tumors with different malignant potential, because IL-18 signals to various IL-18R-bearing cells in concert with other cytokines and environmental factors.

4.9 Pathological Roles of IL-18

Before the discovery of inflammasomes, IL-18 was shown to be involved in various diseases including autoimmune diseases, inflammatory diseases, cancer, metabolic diseases, cardiovascular diseases, and mental diseases. These diseases occur in various organs and tissues such as brain, heart, lung, liver, kidney, and skin. Similarly, a number of studies demonstrated the importance of dysregulated inflammasomes in acute and chronic diseases in addition to their defensive roles (Martinon et al. 2009; Conforti-Andreoni et al. 2011; Menu and Vince 2011; Strowig et al. 2012; Lamkanfi and Dixit 2012; Maritinon and Tschopp 2012).

Although IL-18 and upstream inflammasomes appear to be involved in these diseases (Dinarello et al. 2009; Sims and Smith 2010; Lukens et al. 2012; Arend et al. 2008; Maritinon et al. 2009; Conforti-Andreoni et al. 2011; Menu and Vince 2011; Strowig et al. 2012; Lamkanfi and Dixit 2012; Maritinon and Tschopp 2012), the effects of IL-18 do not completely overlap with those of inflammasomes. For instance, IL-18-deficient Apo $E^{-/-}$ mice exhibit less severe atherosclerosis (Elhage et al. 2003) compared with Apo $E^{-/-}$ mice deficient for inflammasome components (Menu et al. 2011). Thus, it is necessary to determine which inflammasomes are activated and which cytokines are produced in specific disease models to elucidate the roles of IL-18 and inflammasomes. The regulation of receptor expression for each cytokine may also influence the physiological and clinical course of disease.

4.9.1 Rheumatoid Arthritis

IL-18 is expressed in synovial cells from patients with rheumatoid arthritis (RA) (Gracie et al. 1999; Yokota et al. 2012). In an animal model of RA, IL-18 administration aggravated RA and the administration of neutralizing anti-IL-18 antibodies or IL-18BP alleviated pathological damage such as cartilage destruction (Plater-Zyberk et al. 2001). However, IL-18-deficient mice developed antigen-induced arthritis (Santos et al. 2006). Regarding inflammasomes, high levels of inflammasome components were detected in synovial cells in RA (Kolly et al. 2010). Polymorphisms in inflammasome component genes were associated with RA in one study (Kastbom et al. 2010), whereas another study reported the opposite result (Ben Hamad et al. 2012). Thus, a consensus has not been reached for the role of IL-18 and inflammasomes in RA.

4.9.2 Multiple Sclerosis

IL-18 was suggested to have a role during the onset of multiple sclerosis (MS) because high levels of IL-18 were detected in the circulation and cerebrospinal fluid of MS patients (Fassbender et al. 1999). However, in an adoptive transfer model of experimental autoimmune encephalomyelitis, the transfer of antigen-specific T cells pretreated with IL-18 caused more aggressive pathogenic changes in normal mice (Ito et al. 2003). In addition, IL-18 deficiency decreased disease severity in an animal model. IL-18R deficiency, however, enhanced pathological damage (Gutcher et al. 2006). Thus, it is difficult to determine the roles of IL-18 in MS. The roles of NLRP3 inflammasome in experimental autoimmune encephalomyelitis (EAE) are also rather confusing. High levels of caspase-1 are expressed in lesions of MS patients, and the activation of caspase-1 appears to be crucial for the development of EAE (Shaw et al. 2011). Although some reports indicated a critical role for NLRP3 inflammasome in the progression of EAE and the development of Th1 and Th17 cells, other studies concluded there was no causative association between EAE and inflammasomes (Shaw et al. 2011).

4.9.3 Type 1 Diabetes

Because IL-18 strongly augments the development of Th1 cells and induces excessive production of ROS via IFN-γ induction, the involvement of IL-18 was suggested to be involved in the destruction of β cells in the islets of nonobese diabetic mice (Rothe and Kolb 1998). Similar to MS, previous studies have yielded conflicting results regarding the effect of IL-18 polymorphisms in the development of type 1 diabetes (T1D) (Thompson and Humphries 2007). Thus, a consensus has not been reached for the roles of IL-18 during the onset and pathophysiological course of T1D, as observed by single nucleotide polymorphism (SNP) analyses of inflammasome genes (Shaw et al. 2011; Rothe and Kolb 1998; Thompson and Humphries 2007).

4.9.4 Inflammatory Bowel Disease (IBD)

In patients with Crohn's disease, ulcerative colitis, and celiac disease, high levels of IL-18 were observed in the circulation, and IL-18Rs and caspase-1 were observed in inflamed mucosa (Ludwiczek et al. 2005; Wiercinska-Drapalo et al. 2005; Salvati et al. 2002). In a mouse model of IBD, IL-18 deficiency reduced clinical severity, whereas IL-18 transgene expression exacerbated pathological changes (Ishikura et al. 2003). NLRP3-deficient mice treated with dextran sodium sulfate (DSS) exhibited attenuated colitis, compared to wild-type mice (Bauer et al. 2012). In this model, bacterial flora were suggested to be involved in susceptibility to disease.

4.9.5 Cardiovascular Diseases and Metabolic Syndrome

Many reports have suggested that the level of circulating IL-18 is associated with a risk of death from cardiovascular disease or myocardial dysfunction (Blankenberg et al. 2002; Jefferis et al. 2011; Badimon 2012). Accumulating evidence suggests that biologically active IL-18 is involved in the development of atherosclerosis or plaque instability possibly through the induction of IFN-γ (Mallat et al. 2001). The active form of IL-18, together with mature caspase-1, is highly expressed in atherosclerotic lesions (Lee et al. 2012; Usui et al. 2012). In animal models, IL-18-deficient $ApoE^{-/-}$ mice developed reduced disease severity (Elhage et al. 2003). However, SNPs of the IL-18 gene indicated no association between IL-18 polymorphisms and cardiovascular diseases (Thompson and Humphries 2007). Some reports demonstrated the roles of IL-18 and inflammasomes in metabolic syndromes (Zirlik et al. 2007; Zilverschoon et al. 2008; Troseid et al. 2009, 2010; Vandanmagsar et al. 2011; Yamaoka-Tojo et al. 2011). Interestingly, IL-18 failed to transduce signals in patients with type 2 diabetes, although the mechanism of IL-18 tolerance remains to be elucidated (Troseid et al. 2009). Recent studies suggested the effect of inflammasomes on the development of cardiovascular and myocardial diseases (Takahashi 2011; Garg 2011; Gage et al. 2012), and the roles of IL-18 and inflammasomes are currently being examined.

4.9.6 Psychiatric Diseases: Depression

Severe, long-term, and repeated stress can cause depression often accompanied with disorders in various tissues and organs. Psychiatric diseases are associated with cytokines including IL-18 (Spalletta et al. 2006; Nishida et al. 2009; Bossu et al. 2010; Matsunaga et al. 2011; Haastrup et al. 2012). Recent studies demonstrated the linkage between psychological stresses and inflammasomes (Shelton et al. 2011; Iwata et al. 2013; Maslanik et al. 2013). It is hypothesized that psychological stresses can influence systemic disease through inflammasomes. Because depression is a leading cause of disability worldwide, increasing attention will be paid to the association between depression and inflammasomes in future studies.

4.9.7 *Other Diseases*

High levels of IL-18 expression were observed in the circulation and lesions of various diseases including acute graft-versus-host disease (aGVHD) (Fujimori et al. 2000; Nakamura et al. 2000; Reddy et al. 2001; Min et al. 2004), asthma (Shaw et al. 2011), Sjogren's syndrome (Sakai et al. 2008), juvenile idiopathic arthritis (Jelusic et al. 2007), and systemic lupus erythematosis (Kahlenberg et al. 2011). Although the effect of IL-18 on the onset of these diseases has been suggested, the role of IL-18 in disease pathogenesis has not been fully elucidated. Although IL-18 was considered to exacerbate pathogenic manifestations, the administration of IL-18 in a murine model of aGVHD significantly ameliorated disease severity. Furthermore, IL-18-deficient mice exhibited marked susceptibility to aGVHD (Reddy et al. 2001; Min et al. 2004). The regulation and mechanisms underlying these two opposite effects of IL-18 remains unknown. To clarify these contradictory results is an important step toward elucidating the physiological roles of IL-18 and IL-18Rs.

4.10 Roles of IL-18 in Wound Healing and Fibrosis

From a pathological point of view, IL-18 appears to be a pro-fibrotic, pro-angiogenetic, remodeling, and pro-hypertrophic factor. The increased synthesis and deposition of collagen resulted in organ fibrosis, which impaired organ functions. IL-18 induces interstitial myocardial fibrosis in the heart (Platis et al. 2008). In addition, felodipine, a calcium channel inhibitor, ameliorated perivascular fibrosis induced by fructose burden, probably through the inhibition of inflammasomes (Xing et al. 2008). In support of these observations, IL-18 was shown to induce the proliferation of fibroblasts and collagen deposition (Fix et al. 2011).

The roles of IL-18 in angiogenesis and vascular remodeling have long been examined. As for other diseases, IL-18 exhibited dual functions, modulating the effects both positively (Cao et al. 1999; Amin et al. 2010) and negatively (Qiao et al. 2007; Zheng et al. 2010). IL-18 has been regarded as a hypertrophy- and migration-promoting factor through the activation of the PI3K/Akt pathway(Chandrasekar et al. 2005; Valente et al. 2012). Thus, IL-18 and inflammasomes play dichotomous roles in immune and inflammatory responses and wound repair (Fix et al. 2011; Kang et al. 2012; Kampfer et al. 2000; Ishida et al. 2004; Artlett 2013).

The signals following the activation of IL-18 and inflammasomes appear to be transduced along diverse pathways, leading to tissue injury and the induction of wound healing. Because IL-18 functions under the influence of other cytokines or environmental factors, IL-18 by itself is unlikely to regulate the expansion and function of effector cells or determine the clinical and pathological features of disease. IL-18 may play roles in the protection and survival of differentiating or rapidly dividing cells exposed to various stresses. Therefore, it is essential to determine the roles of IL-18 in nonhematopoietic cells, which would help further understanding of the apparently confusing roles of IL-18 in immune/inflammatory responses. The difficulty in developing a therapeutic strategy targeting IL-18 or inflammasomes may stem from the multifaceted nature of IL-18.

4.11 Conclusions

Pro-IL-18 is converted to its mature form by inflammasomes activated by a variety of cellular alarm signals such as oxidative stress, endoplasmic reticulum stress, and ATP leakage. During inflammation, IL-18 may be involved in the maintenance of homeostasis of cells and tissues by facilitating the lymphocyte functions. Thus, IL-18 is not a mere cytokine inducer. Although IL-18 is likely to play pivotal roles in the proliferation, differentiation, and survival of lymphocytes, its intrinsic function remains unknown. Because IL-18 augments the expression of Bcl-X_L, it may be involved in energy metabolism by maintaining mitochondrial integrity. Elucidation of the roles of IL-18 in nonhematopoietic cells may help us understand the physiological functions of this cytokine. IL-18Rs are not constitutively expressed by lymphocytes, and their expression is regulated by other factors. In addition, IL-18 signals are co-transduced with signals delivered through other receptors. Therefore, it is difficult to identify the pathways unique to IL-18Rs and to elucidate the mechanisms underlying the regulation of IL-18Rs signaling.

References

Ahn IE, Ju JH, Lee SY, Park JS, Oh HJ, Kim HR, Lee SH, Park SH, Kim HY, Cho ML (2012) Upregulation of stromal cell-derived factor by IL-17 and IL-18 via a phosphatidylinositol 3-kinase-dependent pathway. Scand J Immunol 76:433–439

Alboni S, Montanari C, Benatti C, Blom JM, Simone ML, Brunello N, Caggia F, Guidotti G, Marcondes MC, Sanchez-Alavez M et al (2011) Constitutive and LPS-regulated expression of interleukin-18 receptor beta variants in the mouse brain. Brain Behav Immun 25:483–493

Allen IC, TeKippe EM, Woodford RM, Uronis JM, Holl EK, Rogers AB, Herfarth HH, Jobin C, Ting JP (2010) The NLRP3 inflammasome functions as a negative regulator of tumorigenesis during colitis-associated cancer. J Exp Med 207:1045–1056

Amin MA, Rabquer BJ, Mansfield PJ, Ruth JH, Marotte H, Haas CS, Reamer EN, Koch AE (2010) Interleukin 18 induces angiogenesis in vitro and in vivo via Src and Jnk kinases. Ann Rheum Dis 69:2204–2212

Arend WP, Palmer G, Gabay C (2008) IL-1, IL-18, and IL-33 families of cytokines. Immunol Rev 223:20–38

Artlett CM (2013) Inflammasomes in wound healing and fibrosis. J Pathol 229:157–167

Badimon L (2012) Interleukin-18: a potent pro-inflammatory cytokine in atherosclerosis. Cardiovasc Res 96:172–175, discussion 176–180

Bauer C, Duewell P, Lehr HA, Endres S, Schnurr M (2012) Protective and aggravating effects of Nlrp3 inflammasome activation in IBD models: influence of genetic and environmental factors. Dig Dis 30(suppl 1):82–90

Ben Hamad M, Cornelis F, Marzouk S, Chabchoub G, Bahloul Z, Rebai A, Fakhfakh F, Ayadi H, Petit-Teixeira E, Maalej A (2012) Association study of CARD8 (p.C10X) and NLRP3 (p.Q705K) variants with rheumatoid arthritis in French and Tunisian populations. Int J Immunogenet 39:131–136

Ben-Sasson SZ, Hu-Li J, Quiel J, Cauchetaux S, Ratner M, Shapira I, Dinarello CA, Paul WE (2009) IL-1 acts directly on CD4 T cells to enhance their antigen-driven expansion and differentiation. Proc Natl Acad Sci USA 106:7119–7124

Blankenberg S, Tiret L, Bickel C, Peetz D, Cambien F, Meyer J, Rupprecht HJ, AtheroGene I (2002) Interleukin-18 is a strong predictor of cardiovascular death in stable and unstable angina. Circulation 106:24–30

Boraschi D, Lucchesi D, Hainzl S, Leitner M, Maier E, Mangelberger D, Oostingh GJ, Pfaller T, Pixner C, Posselt G et al (2011) IL-37: a new anti-inflammatory cytokine of the IL-1 family. Eur Cytokine Netw 22:127–147

Bossu P, Ciaramella A, Salani F, Vanni D, Palladino I, Caltagirone C, Scapigliati G (2010) Interleukin-18, from neuroinflammation to Alzheimer's disease. Curr Pharm Des 16: 4213–4224

Bufler P, Azam T, Gamboni-Robertson F, Reznikov LL, Kumar S, Dinarello CA, Kim SH (2002) A complex of the IL-1 homologue IL-1F7b and IL-18-binding protein reduces IL-18 activity. Proc Natl Acad Sci USA 99:13723–13728

Cao R, Farnebo J, Kurimoto M, Cao Y (1999) Interleukin-18 acts as an angiogenesis and tumor suppressor. FASEB J 13:2195–2202

Cayrol C, Girard JP (2009) The IL-1-like cytokine IL-33 is inactivated after maturation by caspase-1. Proc Natl Acad Sci USA 106:9021–9026

Ceballos-Olvera I, Sahoo M, Miller MA, Del Barrio L, Re F (2011) Inflammasome-dependent pyroptosis and IL-18 protect against *Burkholderia pseudomallei* lung infection while IL-1beta is deleterious. PLoS Pathog 7:e1002452

Chandrasekar B, Mummidi S, Claycomb WC, Mestril R, Nemer M (2005) Interleukin-18 is a pro-hypertrophic cytokine that acts through a phosphatidylinositol 3-kinase-phosphoinositide-dependent kinase-1-Akt-GATA4 signaling pathway in cardiomyocytes. J Biol Chem 280:4553–4567

Chen GY, Nunez G (2011) Inflammasomes in intestinal inflammation and cancer. Gastroenterology 141:1986–1999

Chen GY, Liu M, Wang F, Bertin J, Nunez G (2011) A functional role for Nlrp6 in intestinal inflammation and tumorigenesis. J Immunol 186:7187–7194

Chung Y, Chang SH, Martinez GJ, Yang XO, Nurieva R, Kang HS, Ma L, Watowich SS, Jetten AM, Tian Q, Dong C (2009) Critical regulation of early Th17 cell differentiation by interleukin-1 signaling. Immunity 30:576–587

Conforti-Andreoni C, Ricciardi-Castagnoli P, Mortellaro A (2011) The inflammasomes in health and disease: from genetics to molecular mechanisms of autoinflammation and beyond. Cell Mol Immunol 8:135–145

Costelloe C, Watson M, Murphy A, McQuillan K, Loscher C, Armstrong ME, Garlanda C, Mantovani A, O'Neill LA, Mills KH, Lynch MA (2008) IL-1F5 mediates anti-inflammatory

activity in the brain through induction of IL-4 following interaction with SIGIRR/TIR8. J Neurochem 105:1960–1969
Dinarello CA (2007) Interleukin-18 and the pathogenesis of inflammatory diseases. Semin Nephrol 27:98–114
Dinarello CA (2009) Immunological and inflammatory functions of the interleukin-1 family. Annu Rev Immunol 27:519–550
Dinarello CA, Fantuzzi G (2003) Interleukin-18 and host defense against infection. J Infect Dis 187(suppl 2):S370–S384
Elhage R, Jawien J, Rudling M, Ljunggren HG, Takeda K, Akira S, Bayard F, Hansson GK (2003) Reduced atherosclerosis in interleukin-18 deficient apolipoprotein E-knockout mice. Cardiovasc Res 59:234–240
Fassbender K, Mielke O, Bertsch T, Muehlhauser F, Hennerici M, Kurimoto M, Rossol S (1999) Interferon-gamma-inducing factor (IL-18) and interferon-gamma in inflammatory CNS diseases. Neurology 53:1104–1106
Fix C, Bingham K, Carver W (2011) Effects of interleukin-18 on cardiac fibroblast function and gene expression. Cytokine 53:19–28
Fortin CF, Ear T, McDonald PP (2009) Autocrine role of endogenous interleukin-18 on inflammatory cytokine generation by human neutrophils. FASEB J 23:194–203
Fujimori Y, Takatsuka H, Takemoto Y, Hara H, Okamura H, Nakanishi K, Kakishita E (2000) Elevated interleukin (IL)-18 levels during acute graft-versus-host disease after allogeneic bone marrow transplantation. Br J Haematol 109:652–657
Gage J, Hasu M, Thabet M, Whitman SC (2012) Caspase-1 deficiency decreases atherosclerosis in apolipoprotein E-null mice. Can J Cardiol 28:222–229
Garg NJ (2011) Inflammasomes in cardiovascular diseases. Am J Cardiovasc Dis 1:244–254
Gracie JA, Forsey RJ, Chan WL, Gilmour A, Leung BP, Greer MR, Kennedy K, Carter R, Wei XQ, Xu D et al (1999) A proinflammatory role for IL-18 in rheumatoid arthritis. J Clin Invest 104:1393–1401
Gram AM, Frenkel J, Ressing ME (2012) Inflammasomes and viruses: cellular defence versus viral offence. J Gen Virol 93:2063–2075
Gu Y, Kuida K, Tsutsui H, Ku G, Hsiao K, Fleming MA, Hayashi N, Higashino K, Okamura H, Nakanishi K et al (1997) Activation of interferon-gamma inducing factor mediated by interleukin-1beta converting enzyme. Science 275:206–209
Gutcher I, Urich E, Wolter K, Prinz M, Becher B (2006) Interleukin 18-independent engagement of interleukin 18 receptor-alpha is required for autoimmune inflammation. Nat Immunol 7:946–953
Haastrup E, Bukh JD, Bock C, Vinberg M, Thorner LW, Hansen T, Werge T, Kessing LV, Ullum H (2012) Promoter variants in IL18 are associated with onset of depression in patients previously exposed to stressful-life events. J Affect Disord 136:134–138
Hirata J, Kotani J, Aoyama M, Kashiwamura S, Ueda H, Kuroda Y, Usami M, Okamura H, Marukawa S (2008) A role for IL-18 in human neutrophil apoptosis. Shock 30:628–633
Hodge DL, Subleski JJ, Reynolds DA, Buschman MD, Schill WB, Burkett MW, Malyguine AM, Young HA (2006) The proinflammatory cytokine interleukin-18 alters multiple signaling pathways to inhibit natural killer cell death. J Interferon Cytokine Res 26:706–718
Hosotani Y, Kashiwamura S, Kimura-Shimmyo A, Sekiyama A, Ueda H, Ikeda T, Mimura O, Okamura H (2008) Interleukin-18 prevents apoptosis via PI3K/Akt pathway in normal human keratinocytes. J Dermatol 35:514–524
Ishida Y, Kondo T, Takayasu T, Iwakura Y, Mukaida N (2004) The essential involvement of cross-talk between IFN-gamma and TGF-beta in the skin wound-healing process. J Immunol 172:1848–1855
Ishikura T, Kanai T, Uraushihara K, Iiyama R, Makita S, Totsuka T, Yamazaki M, Sawada T, Nakamura T, Miyata T et al (2003) Interleukin-18 overproduction exacerbates the development of colitis with markedly infiltrated macrophages in interleukin-18 transgenic mice. J Gastroenterol Hepatol 18:960–969

Ito A, Matejuk A, Hopke C, Drought H, Dwyer J, Zamora A, Subramanian S, Vandenbark AA, Offner H (2003) Transfer of severe experimental autoimmune encephalomyelitis by IL-12- and IL-18-potentiated T cells is estrogen sensitive. J Immunol 170:4802–4809

Iwata M, Ota KT, Duman RS (2013) The inflammasome: pathways linking psychological stress, depression, and systemic illnesses. Brain Behav Immun 31C:105–114

Jacobs SR, Damania B (2012) NLRs, inflammasomes, and viral infection. J Leukoc Biol 92: 469–477

Jefferis BJ, Papacosta O, Owen CG, Wannamethee SG, Humphries SE, Woodward M, Lennon LT, Thomson A, Welsh P, Rumley A et al (2011) Interleukin 18 and coronary heart disease: prospective study and systematic review. Atherosclerosis 217:227–233

Jelusic M, Lukic IK, Tambic-Bukovac L, Dubravcic K, Malcic I, Rudan I, Batinic D (2007) Interleukin-18 as a mediator of systemic juvenile idiopathic arthritis. Clin Rheumatol 26: 1332–1334

Kahlenberg JM, Thacker SG, Berthier CC, Cohen CD, Kretzler M, Kaplan MJ (2011) Inflammasome activation of IL-18 results in endothelial progenitor cell dysfunction in systemic lupus erythematosus. J Immunol 187:6143–6156

Kampfer H, Muhl H, Manderscheid M, Kalina U, Kauschat D, Pfeilschifter J, Frank S (2000) Regulation of interleukin-18 (IL-18) expression in keratinocytes (HaCaT): implications for early wound healing. Eur Cytokine Netw 11:626–633

Kang MJ, Choi JM, Kim BH, Lee CM, Cho WK, Choe G, Kim DH, Lee CG, Elias JA (2012) IL-18 induces emphysema and airway and vascular remodeling via IFN-gamma, IL-17A, and IL-13. Am J Respir Crit Care Med 185:1205–1217

Kanneganti TD (2010) Central roles of NLRs and inflammasomes in viral infection. Nat Rev Immunol 10:688–698

Kastbom A, Johansson M, Verma D, Soderkvist P, Rantapaa-Dahlqvist S (2010) CARD8 p.C10X polymorphism is associated with inflammatory activity in early rheumatoid arthritis. Ann Rheum Dis 69:723–726

Kawashima M, Miossec P (2003) Heterogeneity of response of rheumatoid synovium cell subsets to interleukin-18 in relation to differential interleukin-18 receptor expression. Arthritis Rheum 48:631–637

Kayagaki N, Warming S, Lamkanfi M, Vande Walle L, Louie S, Dong J, Newton K, Qu Y, Liu J, Heldens S et al (2011) Non-canonical inflammasome activation targets caspase-11. Nature 479:117–121

Kolinska J, Lisa V, Clark JA, Kozakova H, Zakostelecka M, Khailova L, Sinkora M, Kitanovicova A, Dvorak B (2008) Constitutive expression of IL-18 and IL-18R in differentiated IEC-6 cells: effect of TNF-alpha and IFN-gamma treatment. J Interferon Cytokine Res 28:287–296

Kolly L, Busso N, Palmer G, Talabot-Ayer D, Chobaz V, So A (2010) Expression and function of the NALP3 inflammasome in rheumatoid synovium. Immunology 129:178–185

Kumar S, Hanning CR, Brigham-Burke MR, Rieman DJ, Lehr R, Khandekar S, Kirkpatrick RB, Scott GF, Lee JC, Lynch FJ et al (2002) Interleukin-1F7B (IL-1H4/IL-1F7) is processed by caspase-1 and mature IL-1F7B binds to the IL-18 receptor but does not induce IFN-gamma production. Cytokine 18:61–71

Kurowska-Stolarska M, Hueber A, Stolarski B, McInnes IB (2011) Interleukin-33: a novel mediator with a role in distinct disease pathologies. J Intern Med 269:29–35

Lamkanfi M, Dixit VM (2011) Modulation of inflammasome pathways by bacterial and viral pathogens. J Immunol 187:597–602

Lamkanfi M, Dixit VM (2012) Inflammasomes and their roles in health and disease. Annu Rev Cell Dev Biol 28:137–161

Lee Y, Schulte DJ, Shimada K, Chen S, Crother TR, Chiba N, Fishbein MC, Lehman TJ, Arditi M (2012) Interleukin-1beta is crucial for the induction of coronary artery inflammation in a mouse model of Kawasaki disease. Circulation 125:1542–1550

Lewis EC, Dinarello CA (2006) Responses of IL-18- and IL-18 receptor-deficient pancreatic islets with convergence of positive and negative signals for the IL-18 receptor. Proc Natl Acad Sci USA 103:16852–16857

Li W, Kashiwamura S, Ueda H, Sekiyama A, Okamura H (2007) Protection of CD8+ T cells from activation-induced cell death by IL-18. J Leukoc Biol 82:142–151

Li W, Kubo S, Okuda A, Yamamoto H, Ueda H, Tanaka T, Nakamura H, Yamanishi H, Terada N, Okamura H (2010) Effect of IL-18 on expansion of gammadelta T cells stimulated by zoledronate and IL-2. J Immunother 33:287–296

Lopetuso LR, Scaldaferri F, Pizarro TT (2012) Emerging role of the interleukin (IL)-33/ST2 axis in gut mucosal wound healing and fibrosis. Fibrogenesis Tissue Repair 5:18

Ludwiczek O, Kaser A, Novick D, Dinarello CA, Rubinstein M, Tilg H (2005) Elevated systemic levels of free interleukin-18 (IL-18) in patients with Crohn's disease. Eur Cytokine Netw 16:27–33

Lukens JR, Gross JM, Kanneganti TD (2012) IL-1 family cytokines trigger sterile inflammatory disease. Front Immunol 3:315

Luthi AU, Cullen SP, McNeela EA, Duriez PJ, Afonina IS, Sheridan C, Brumatti G, Taylor RC, Kersse K, Vandenabeele P et al (2009) Suppression of interleukin-33 bioactivity through proteolysis by apoptotic caspases. Immunity 31:84–98

Majima T, Ichikura T, Chochi K, Kawabata T, Tsujimoto H, Sugasawa H, Kuranaga N, Takayama E, Kinoshita M, Hiraide H et al (2006) Exploitation of interleukin-18 by gastric cancers for their growth and evasion of host immunity. Int J Cancer 118:388–395

Mallat Z, Corbaz A, Scoazec A, Besnard S, Leseche G, Chvatchko Y, Tedgui A (2001) Expression of interleukin-18 in human atherosclerotic plaques and relation to plaque instability. Circulation 104:1598–1603

Martinon F, Burns K, Tschopp J (2002) The inflammasome: a molecular platform triggering activation of inflammatory caspases and processing of proIL-beta. Mol Cell 10:417–426

Martinon F, Mayor A, Tschopp J (2009) The inflammasomes: guardians of the body. Annu Rev Immunol 27:229–265

Maslanik T, Mahaffey L, Tannura K, Beninson L, Greenwood BN, Fleshner M (2013) The inflammasome and danger associated molecular patterns (DAMPs) are implicated in cytokine and chemokine responses following stressor exposure. Brain Behav Immun 28:54–62

Matsunaga H, Hokari R, Ueda T, Kurihara C, Hozumi H, Higashiyama M, Okada Y, Watanabe C, Komoto S, Nakamura M et al (2011) Physiological stress exacerbates murine colitis by enhancing proinflammatory cytokine expression that is dependent on IL-18. Am J Physiol Gastrointest Liver Physiol 301:G555–G564

Menu P, Vince JE (2011) The NLRP3 inflammasome in health and disease: the good, the bad and the ugly. Clin Exp Immunol 166:1–15

Menu P, Pellegrin M, Aubert JF, Bouzourene K, Tardivel A, Mazzolai L, Tschopp J (2011) Atherosclerosis in ApoE-deficient mice progresses independently of the NLRP3 inflammasome. Cell Death Dis 2:e137

Miao EA, Rajan JV, Aderem A (2011) Caspase-1-induced pyroptotic cell death. Immunol Rev 243:206–214

Min CK, Maeda Y, Lowler K, Liu C, Clouthier S, Lofthus D, Weisiger E, Ferrara JL, Reddy P (2004) Paradoxical effects of interleukin-18 on the severity of acute graft-versus-host disease mediated by CD4+ and CD8+ T-cell subsets after experimental allogeneic bone marrow transplantation. Blood 104:3393–3399

Munoz-Pinedo C (2012) Signaling pathways that regulate life and cell death: evolution of apoptosis in the context of self-defense. Adv Exp Med Biol 738:124–143

Nakamura H, Komatsu K, Ayaki M, Kawamoto S, Murakami M, Uoshima N, Yagi T, Hasegawa T, Yasumi M, Karasuno T et al (2000) Serum levels of soluble IL-2 receptor, IL-12, IL-18, and IFN-gamma in patients with acute graft-versus-host disease after allogeneic bone marrow transplantation. J Allergy Clin Immunol 106:S45–S50

Nakanishi K, Yoshimoto T, Tsutsui H, Okamura H (2001) Interleukin-18 regulates both Th1 and Th2 responses. Annu Rev Immunol 19:423–474

Neumann D, Lienenklaus S, Rosati O, Martin MU (2002) IL-1beta-induced phosphorylation of PKB/Akt depends on the presence of IRAK-1. Eur J Immunol 32:3689–3698

Nishida K, Kamizato M, Kawai T, Masuda K, Takeo K, Teshima-Kondo S, Tanahashi T, Rokutan K (2009) Interleukin-18 is a crucial determinant of vulnerability of the mouse rectum to psychosocial stress. FASEB J 23:1797–1805

Nold-Petry CA, Nold MF, Nielsen JW, Bustamante A, Zepp JA, Storm KA, Hong JW, Kim SH, Dinarello CA (2009) Increased cytokine production in interleukin-18 receptor alpha-deficient cells is associated with dysregulation of suppressors of cytokine signaling. J Biol Chem 284: 25900–25911

Nozaki Y, Kinoshita K, Yano T, Asato K, Shiga T, Hino S, Niki K, Nagare Y, Kishimoto K, Shimazu H et al (2012) Signaling through the interleukin-18 receptor alpha attenuates inflammation in cisplatin-induced acute kidney injury. Kidney Int 82:892–902

O'Neill LA (2008) The interleukin-1 receptor/Toll-like receptor superfamily: 10 years of progress. Immunol Rev 226:10–18

Okamura H, Tsutsi H, Komatsu T, Yutsudo M, Hakura A, Tanimoto T, Torigoe K, Okura T, Nukada Y, Hattori K et al (1995) Cloning of a new cytokine that induces IFN-gamma production by T cells. Nature 378:88–91

Omoto Y, Tokime K, Yamanaka K, Habe K, Morioka T, Kurokawa I, Tsutsui H, Yamanishi K, Nakanishi K, Mizutani H (2006) Human mast cell chymase cleaves pro-IL-18 and generates a novel and biologically active IL-18 fragment. J Immunol 177:8315–8319

Omoto Y, Yamanaka K, Tokime K, Kitano S, Kakeda M, Akeda T, Kurokawa I, Gabazza EC, Tsutsui H, Katayama N et al (2010) Granzyme B is a novel interleukin-18 converting enzyme. J Dermatol Sci 59:129–135

Pan G, Risser P, Mao W, Baldwin DT, Zhong AW, Filvaroff E, Yansura D, Lewis L, Eigenbrot C, Henzel WJ, Vandlen R (2001) IL-1H, an interleukin 1-related protein that binds IL-18 receptor/ IL-1Rrp. Cytokine 13:1–7

Park S, Cheon S, Cho D (2007) The dual effects of interleukin-18 in tumor progression. Cell Mol Immunol 4:329–335

Plater-Zyberk C, Joosten LA, Helsen MM, Sattonnet-Roche P, Siegfried C, Alouani S, van De Loo FA, Graber P, Aloni S, Cirillo R et al (2001) Therapeutic effect of neutralizing endogenous IL-18 activity in the collagen-induced model of arthritis. J Clin Invest 108:1825–1832

Platis A, Yu Q, Moore D, Khojeini E, Tsau P, Larson D (2008) The effect of daily administration of IL-18 on cardiac structure and function. Perfusion 23:237–242

Puren AJ, Fantuzzi G, Dinarello CA (1999) Gene expression, synthesis, and secretion of interleukin 18 and interleukin 1beta are differentially regulated in human blood mononuclear cells and mouse spleen cells. Proc Natl Acad Sci USA 96:2256–2261

Qiao H, Sonoda KH, Ikeda Y, Yoshimura T, Hijioka K, Jo YJ, Sassa Y, Tsutsumi-Miyahara C, Hata Y, Akira S, Ishibashi T (2007) Interleukin-18 regulates pathological intraocular neovascularization. J Leukoc Biol 81:1012–1021

Rathinam VA, Fitzgerald KA (2010) Inflammasomes and anti-viral immunity. J Clin Immunol 30:632–637

Rathinam VA, Vanaja SK, Waggoner L, Sokolovska A, Becker C, Stuart LM, Leong JM, Fitzgerald KA (2012) TRIF licenses caspase-11-dependent NLRP3 inflammasome activation by gram-negative bacteria. Cell 150:606–619

Reddy P, Teshima T, Kukuruga M, Ordemann R, Liu C, Lowler K, Ferrara JL (2001) Interleukin-18 regulates acute graft-versus-host disease by enhancing Fas-mediated donor T cell apoptosis. J Exp Med 194:1433–1440

Rothe H, Kolb H (1998) The APC1 concept of type I diabetes. Autoimmunity 27:179–184

Sahoo M, Ceballos-Olvera I, del Barrio L, Re F (2011) Role of the inflammasome, IL-1beta, and IL-18 in bacterial infections. Scientific World Journal 11:2037–2050

Sakai A, Sugawara Y, Kuroishi T, Sasano T, Sugawara S (2008) Identification of IL-18 and Th17 cells in salivary glands of patients with Sjogren's syndrome, and amplification of IL-17-mediated secretion of inflammatory cytokines from salivary gland cells by IL-18. J Immunol 181:2898–2906

Salcedo R, Worschech A, Cardone M, Jones Y, Gyulai Z, Dai RM, Wang E, Ma W, Haines D, O'Huigin C et al (2010) MyD88-mediated signaling prevents development of adenocarcinomas of the colon: role of interleukin 18. J Exp Med 207:1625–1636

Salvati VM, MacDonald TT, Bajaj-Elliott M, Borrelli M, Staiano A, Auricchio S, Troncone R, Monteleone G (2002) Interleukin 18 and associated markers of T helper cell type 1 activity in coeliac disease. Gut 50:186–190

Santos LL, Milenkovski GP, Hall PH, Leech M, Sharma L, Takeda K, Akira S, Kitching AR, Morand EF (2006) IL-18 is redundant in T-cell responses and in joint inflammation in antigen-induced arthritis. Immunol Cell Biol 84:166–173

Schmidt RL, Lenz LL (2012) Distinct licensing of IL-18 and IL-1beta secretion in response to NLRP3 inflammasome activation. PLoS One 7:e45186

Shaw PJ, McDermott MF, Kanneganti TD (2011) Inflammasomes and autoimmunity. Trends Mol Med 17:57–64

Shelton RC, Claiborne J, Sidoryk-Wegrzynowicz M, Reddy R, Aschner M, Lewis DA, Mirnics K (2011) Altered expression of genes involved in inflammation and apoptosis in frontal cortex in major depression. Mol Psychiatry 16:751–762

Sims JE, Smith DE (2010) The IL-1 family: regulators of immunity. Nat Rev Immunol 10:89–102

Spalletta G, Bossu P, Ciaramella A, Bria P, Caltagirone C, Robinson RG (2006) The etiology of poststroke depression: a review of the literature and a new hypothesis involving inflammatory cytokines. Mol Psychiatry 11:984–991

Srivastava S, Salim N, Robertson MJ (2010) Interleukin-18: biology and role in the immunotherapy of cancer. Curr Med Chem 17:3353–3357

Strowig T, Henao-Mejia J, Elinav E, Flavell R (2012) Inflammasomes in health and disease. Nature 481:278–286

Sugawara S, Uehara A, Nochi T, Yamaguchi T, Ueda H, Sugiyama A, Hanzawa K, Kumagai K, Okamura H, Takada H (2001) Neutrophil proteinase 3-mediated induction of bioactive IL-18 secretion by human oral epithelial cells. J Immunol 167:6568–6575

Sutton C, Brereton C, Keogh B, Mills KH, Lavelle EC (2006) A crucial role for interleukin (IL)-1 in the induction of IL-17-producing T cells that mediate autoimmune encephalomyelitis. J Exp Med 203:1685–1691

Takahashi M (2011) Role of the inflammasome in myocardial infarction. Trends Cardiovasc Med 21:37–41

Thompson SR, Humphries SE (2007) Interleukin-18 genetics and inflammatory disease susceptibility. Genes Immun 8:91–99

Troseid M, Seljeflot I, Hjerkinn EM, Arnesen H (2009) Interleukin-18 is a strong predictor of cardiovascular events in elderly men with the metabolic syndrome: synergistic effect of inflammation and hyperglycemia. Diabetes Care 32:486–492

Troseid M, Seljeflot I, Arnesen H (2010) The role of interleukin-18 in the metabolic syndrome. Cardiovasc Diabetol 9:11

Tsutsui H, Kayagaki N, Kuida K, Nakano H, Hayashi N, Takeda K, Matsui K, Kashiwamura S, Hada T, Akira S et al (1999) Caspase-1-independent, Fas/Fas ligand-mediated IL-18 secretion from macrophages causes acute liver injury in mice. Immunity 11:359–367

Usui F, Shirasuna K, Kimura H, Tatsumi K, Kawashima A, Karasawa T, Hida S, Sagara J, Taniguchi S, Takahashi M (2012) Critical role of caspase-1 in vascular inflammation and development of atherosclerosis in Western diet-fed apolipoprotein E-deficient mice. Biochem Biophys Res Commun 425:162–168

Valente AJ, Yoshida T, Murthy SN, Sakamuri SS, Katsuyama M, Clark RA, Delafontaine P, Chandrasekar B (2012) Angiotensin II enhances AT1-Nox1 binding and stimulates arterial smooth muscle cell migration and proliferation through AT1, Nox1, and interleukin-18. Am J Physiol Heart Circ Physiol 303:H282–H296

van de Veerdonk FL, Netea MG, Dinarello CA, Joosten LA (2011) Inflammasome activation and IL-1beta and IL-18 processing during infection. Trends Immunol 32:110–116

Vandanmagsar B, Youm YH, Ravussin A, Galgani JE, Stadler K, Mynatt RL, Ravussin E, Stephens JM, Dixit VD (2011) The NLRP3 inflammasome instigates obesity-induced inflammation and insulin resistance. Nat Med 17:179–188

Vidal-Vanaclocha F, Mendoza L, Telleria N, Salado C, Valcarcel M, Gallot N, Carrascal T, Egilegor E, Beaskoetxea J, Dinarello CA (2006) Clinical and experimental approaches to the pathophysiology of interleukin-18 in cancer progression. Cancer Metastasis Rev 25:417–434

Wiercinska-Drapalo A, Flisiak R, Jaroszewicz J, Prokopowicz D (2005) Plasma interleukin-18 reflects severity of ulcerative colitis. World J Gastroenterol 11:605–608

Xing SS, Tan HW, Bi XP, Zhong M, Zhang Y, Zhang W (2008) Felodipine reduces cardiac expression of IL-18 and perivascular fibrosis in fructose-fed rats. Mol Med 14:395–402

Yamaoka-Tojo M, Tojo T, Wakaume K, Kameda R, Nemoto S, Takahira N, Masuda T, Izumi T (2011) Circulating interleukin-18: a specific biomarker for atherosclerosis-prone patients with metabolic syndrome. Nutr Metab (Lond) 8:3

Yokota K, Miyazaki T, Hemmatazad H, Gay RE, Kolling C, Fearon U, Suzuki H, Mimura T, Gay S, Ospelt C (2012) The pattern-recognition receptor nucleotide-binding oligomerization domain – containing protein 1 promotes production of inflammatory mediators in rheumatoid arthritis synovial fibroblasts. Arthritis Rheum 64:1329–1337

Yoo JK, Kwon H, Khil LY, Zhang L, Jun HS, Yoon JW (2005) IL-18 induces monocyte chemotactic protein-1 production in macrophages through the phosphatidylinositol 3-kinase/Akt and MEK/ERK1/2 pathways. J Immunol 175:8280–8286

Zheng JN, Pei DS, Mao LJ, Liu XY, Sun FH, Zhang BF, Liu YQ, Liu JJ, Li W, Han D (2010) Oncolytic adenovirus expressing interleukin-18 induces significant antitumor effects against melanoma in mice through inhibition of angiogenesis. Cancer Gene Ther 17:28–36

Zilverschoon GR, Tack CJ, Joosten LA, Kullberg BJ, van der Meer JW, Netea MG (2008) Interleukin-18 resistance in patients with obesity and type 2 diabetes mellitus. Int J Obes (Lond) 32:1407–1414

Zirlik A, Abdullah SM, Gerdes N, MacFarlane L, Schonbeck U, Khera A, McGuire DK, Vega GL, Grundy S, Libby P, de Lemos JA (2007) Interleukin-18, the metabolic syndrome, and subclinical atherosclerosis: results from the Dallas Heart Study. Arterioscler Thromb Vasc Biol 27: 2043–2049

Zitvogel L, Kepp O, Galluzzi L, Kroemer G (2012) Inflammasomes in carcinogenesis and anticancer immune responses. Nat Immunol 13:343–351

Chapter 5
Interleukin-21: A Pleiotropic Mediator of Immunity and Inflammation with Broad Therapeutic Potential

Rosanne Spolski and Warren J. Leonard

Abstract Interleukin-21 (IL-21) is a member of the family of cytokines that share the common cytokine receptor γ-chain, γ_c. IL-21 plays critical roles in the differentiation and function of multiple lymphoid and myeloid lineages. In this chapter, we review the inflammatory and immunosuppressive activities of this cytokine and the specific mechanisms by which this cytokine regulates both the innate and adaptive immune responses. Despite the fact that IL-21 and its receptor were only identified in 2000, a tremendous amount has been learned already, leading to the development of potential therapeutic approaches to cancer as well as potentially allowing for the specific manipulation of IL-21 effects in a spectrum of autoimmune diseases, allergic responses, and viral infections.

Keywords Antitumor • Autoimmune disease • IL-21 • Plasma cell

5.1 Introduction

Interleukin-21 (IL-21) is the most recently discovered member of a family of type 1 cytokines that play critical roles in the regulation of immune function (Spolski and Leonard 2008). Together with IL-2, IL-4, IL-7, IL-9, and IL-15, IL-21 shares as a receptor component the common cytokine receptor γ-chain, γ_c, which is mutated in humans with XSCID (Noguchi et al. 1993), a disease in which T cells and natural killer (NK) cells are profoundly diminished and B cells are nonfunctional (Leonard 2001). Since IL-21 and its receptor were discovered in 2000, a tremendous amount has been learned about the production and regulation of this cytokine as well as its role in the immune system (Spolski and Leonard 2008). Although IL-21 is primarily

R. Spolski • W.J. Leonard (✉)
Laboratory of Molecular Immunology, National Heart, Lung, and Blood Institute, 9000 Rockville Pike (Bldg. 10/7B05), Bethesda, MD 20892, USA
e-mail: spolskir@nhlbi.nih.gov; wjl@helix.nih.gov

T. Yoshimoto and T. Yoshimoto (eds.), *Cytokine Frontiers: Regulation of Immune Responses in Health and Disease*, DOI 10.1007/978-4-431-54442-5_5,

produced by subsets of $CD4^+$ T cells and natural killer (NK)T cells, the IL-21 receptor is found on most hematopoietic cells, both lymphoid and myeloid, and therefore has the capacity to positively or negatively regulate the function of both innate and adaptive immune responses to pathogens and tumors. Additionally, however, it has become clear that IL-21 can also cause pathogenic effects in diverse inflammatory responses, exacerbating autoimmune and allergic disease (Spolski and Leonard 2008). Given these pleiotropic actions of IL-21, it is important to understand the mechanisms involved in its regulation of specific immune processes, and this evolving knowledge already is allowing therapeutic manipulation of the actions of this cytokine. In this chapter, we review the biology and signaling mechanisms of IL-21, its role in disease processes, and the therapeutic potential of augmenting or inhibiting the actions of this cytokine.

5.2 IL-21 and IL-21 Receptor Structure and Signaling

The IL-21 receptor (IL-21R) was initially identified in a genomic sequencing project on the basis of its shared properties of type 1 cytokine receptors, and the predicted IL-21R amino acid sequence was most similar to that of the IL-2 receptor β-chain (Ozaki et al. 2000; Parrish-Novak et al. 2000). The ligand IL-21 was also identified (Parrish-Novak et al. 2000) and was found to signal through a heterodimeric receptor composed of IL-21R and γ_c (Asao et al. 2001). The affinity of IL-21 for IL-21R is much stronger ($K_d \sim 70$ pM) than its affinity for γ_c ($K_d \sim 160$ μM), but the latter is also required for signaling (Asao et al. 2001; Kang et al. 2010). Based on nuclear magnetic resonance (NMR) spectroscopy, the three-dimensional structure of human IL-21 is a typical up–up-down–down four-α-helical bundle type 1 cytokine (Bondensgaard et al. 2007). The amino acid residues involved in the interaction of IL-21 with IL-21R and γ_c were putatively identified by molecular homology modeling, and based on these predictions, IL-21 mutants retaining high-affinity binding for IL-21R but negligible binding to γ_c were generated (Kang et al. 2010) and shown to function as potent antagonists of IL-21-induced proliferation of human cells. X-ray crystallization-based analysis of the IL-21–IL-21R complex expressed in mammalian cells with partial glycosylation has revealed homology to other γ_c-family cytokines, with stabilization of the domain structure of IL-21R by sugar chains bridging the two fibronectin domains (Hamming et al. 2012).

As already discussed, IL-21 is produced by $CD4^+$ T cells, most abundantly by Th17 and T follicular helper (Tfh) cells, as well as by NKT cells (Spolski and Leonard 2008). IL-21 is induced both by signals to the antigen receptor on $CD4^+$ T cells (Kim et al. 2005) and by mycobacterial antigens in NKT cells (Harada et al. 2006), implicating the role of this cytokine in both early innate responses to pathogens as well as in later adaptive responses. IL-21R is expressed on B cells, $CD4^+$ T cells, $CD8^+$ T cells, NK cells, NKT cells, dendritic cells, macrophages, keratinocytes, and epithelial cells, consistent with the broad range of actions of IL-21

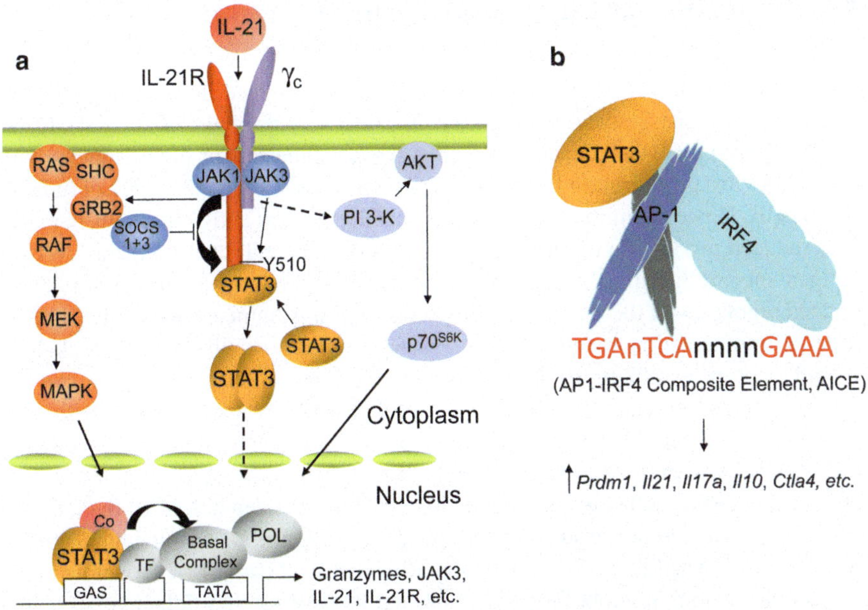

Fig. 5.1 Interleukin (IL)-21 signals through multiple pathways. **(a)** IL-21 activates the JAK/STAT pathway, using JAK1 and JAK3, with STAT3 as the dominant STAT protein. It also activates and signals via the MAP kinase and the PI-3-kinase pathways. **(b)** In activated CD4$^+$ T cells and Th17 cells, STAT3 is often part of a complex with AP-1 and IRF4 transcription factors that can bind to AP-1/IRF4 consensus elements (AICEs) to modulate expression of a range of target genes

(Spolski and Leonard 2008). Surface IL-21R is expressed throughout B-cell development, with high-level expression on mature follicular B cells, but then markedly decreasing as B cells terminally differentiate into plasma cells (Good et al. 2006). IL-21R is expressed on maturing thymocytes even though its expression is not essential for thymic differentiation. Low levels of IL-21R are found on naïve CD4$^+$ and CD8$^+$ T cells, with increased IL-21R expression induced by either antigen receptor signals or stimulation with IL-21 (Jin et al. 2004).

Analogous to other γ_c-family cytokines, IL-21 signals via the JAK-STAT pathway (Fig. 5.1a). Binding of IL-21 to its receptor activates the Janus family tyrosine kinases JAK1 and JAK3, leading to a rapid and sustained activation of STAT3, somewhat weaker activation of STAT1, and weaker and more transient activation of STAT5 (Zeng et al. 2007). IL-21-mediated gene activation in CD4$^+$ T cells often involves a complex composed of STAT3, IRF4, and AP1 family members JUN and BATF (Fig. 5.1b) (Ciofani et al. 2012; Glasmacher et al. 2012; Li et al. 2012). Based on studies using inhibitors, the PI 3-kinase and mitogen-activated protein (MAP) kinase pathways are also required for IL-21 signaling (Zeng et al. 2007). It is possible that these different signaling pathways are differentially important in different cell types or the stage of development.

5.3 Regulation of B-Cell Function by IL-21

The effect of IL-21 on B-cell function was initially assessed in vitro as well as in vivo using *Il21* transgenic as well as *Il21r* KO mice (Ozaki et al. 2002, 2004). IL-21 is not required for B-cell development in either bone marrow or the periphery, with normal in vitro proliferation in response to B-cell-receptor signals and IL-4 in *Il21r* KO B cells. However, immunization with T-cell-dependent antigens results in markedly reduced production of antigen-specific IgG1 and significantly higher levels of antigen-specific IgE in *Il21r* KO mice, an unexpected result given the usually coordinated regulation of these immunoglobulin isotypes. Administration of IL-21 at the time of immunization reduces IgE responses in vivo, and IL-21 can block IgE-specific isotype switching in B cells in vitro (Suto et al. 2002). However, in vitro experiments with human B cells have shown that IL-21, when combined with anti-CD40 and IL-4, could elevate IgE levels by human B cells, indicating potential differences between IL-21-mediated regulation of IgE in mouse and human B cells (Wood et al. 2004). The elevated IgE in *Il21r* KO mice is dependent on IL-4, as *Il4/Il21r* DKO mice do not produce IgE, but surprisingly, the DKO mice have greatly reduced levels of all Ig isotypes, demonstrating that IL-21 and IL-4 cooperate in the regulation of immunoglobulin production and suggesting that defective signaling by these cytokines explains the B-cell defect in X-SCID (Ozaki et al. 2002).

5.4 IL-21 Promotes B-Cell Apoptosis

A surprising finding in light of the significant role for IL-21 in the regulation of Ig production is that IL-21 can also potently induce apoptosis of B cells. It was initially observed that IL-21, when combined with a CD40 signal, enhances B-cell proliferation (Parrish-Novak et al. 2000), but subsequently IL-21 was shown to reduce B-cell proliferation in the context of an Toll-like receptor (TLR) signal (Jin et al. 2004; Mehta et al. 2003; Ozaki et al. 2004). This latter effect is caused by a strong pro-apoptotic signal from IL-21 in the context of lipopolysaccharide (LPS) or CpG, which activate TLRs, even though IL-21 enhances proliferation of B cells in the context of B-cell receptor (BCR) signaling or T-cell costimulatory signals (Jin et al. 2004). IL-21-mediated apoptosis is dependent on caspase and the induction of the pro-apoptotic mitochondrial protein BIM, as apoptosis is eliminated in B cells from *Bim* knockout (KO) mice (Jin et al. 2004). Elevated IL-21 levels in transgenic mice or after hydrodynamic injection of mice with an IL-21 expression vector results in potent in vivo apoptosis of naïve B cells (Ozaki et al. 2004). These results indicate that the B-cell response to IL-21 is dictated by the context of costimulatory signals present at the time of antigen encounter. The apoptotic effect of IL-21 may serve to eliminate incompletely activated, potentially auto-reactive B cells, potentially serving a role similar role for B cells to that served by IL-2 in T cells during the process of activation-induced cell death (AICD).

Other γ_c-family cytokines can provide anti-apoptotic costimulatory signals for IL-21. For example, IL-4 prevents LPS+IL-21-induced apoptosis in B cells via its induction of the anti-apoptotic protein, BCL-XL (Tsuruoka et al. 2007). In addition, IL-2 has been shown to modulate the pro-apoptotic effects of IL-21 in vivo via down-regulation of IL-21(Tortola et al. 2010). Immunization with *Streptococcus pneumoniae* phosphorylcholine, an antigen that targets a T-independent marginal zone B-cell response, induces anti-PC IgM levels in wild-type (WT) mice, but the response is much lower than that observed in *Il2ra* knockout (KO) mice. This decrease in IgM presumably results from the tenfold increase in expression of IL-21 in the *Il2ra* KO mice, with an increased death of activated marginal-zone B cells (Tortola et al. 2010). Normal responses to PC are induced in *Il2ra*/*Il21r* DKO mice, presumably because the elevated IL-21 can no longer signal to mediate B-cell apoptosis.

5.5 IL-21 Induces Plasma Cell Differentiation

Although elevated levels of IL-21 lead to the apoptosis of B cells in both IL-21 transgenic mice as well as WT mice subjected to hydrodynamic injection of IL-21 plasmid DNA, these mice also accumulate increased numbers of splenic B cells. Consistent with the increased serum Ig levels in these mice, they have increased numbers of plasma cells (Ozaki et al. 2004), and the combination of IL-21 and anti-IgM directly induces the differentiation of splenic B cells, resulting in the accumulation of Syndecan-1$^+$ plasma cells. This effect of IL-21 results from its ability to potently induce the expression of B-lymphocyte-induced maturation protein 1 (BLIMP1) (Kwon et al. 2009; Ozaki et al. 2004), a transcription factor that functions as a master switch for the plasma cell program in B cells (Calame et al. 2003). Importantly, IL-21 in combination with BCR or CD40 signals can also drive human naïve cord blood B cells and post-switch memory cells to express BLIMP1 as well as activation-induced cytidine deaminase (AID) and to terminally differentiate into plasma cells (Ettinger et al. 2005). In combination with BAFF/BLys, a tumor necrosis factor (TNF) family cytokine produced by dendritic cells in the marginal zone, IL-21 also induces BLIMP1 and AID in marginal zone memory B cells, leading to the rapid antigen-independent induction of plasma cells (Ettinger et al. 2007). So far, IL-21 is the most potent cytokine for driving plasma cell differentiation.

5.6 IL-21 Promotes Germinal Center Development and Function

Following antigen stimulation in the presence of T-cell help, B cells can differentiate along an extrafollicular path or can enter follicles where they undergo affinity maturation in response to IL-21 signals from Tfh cells (King et al. 2008). Tfh cells are a population of antigen-specific CD4$^+$ T helper cells that develop in lymphoid

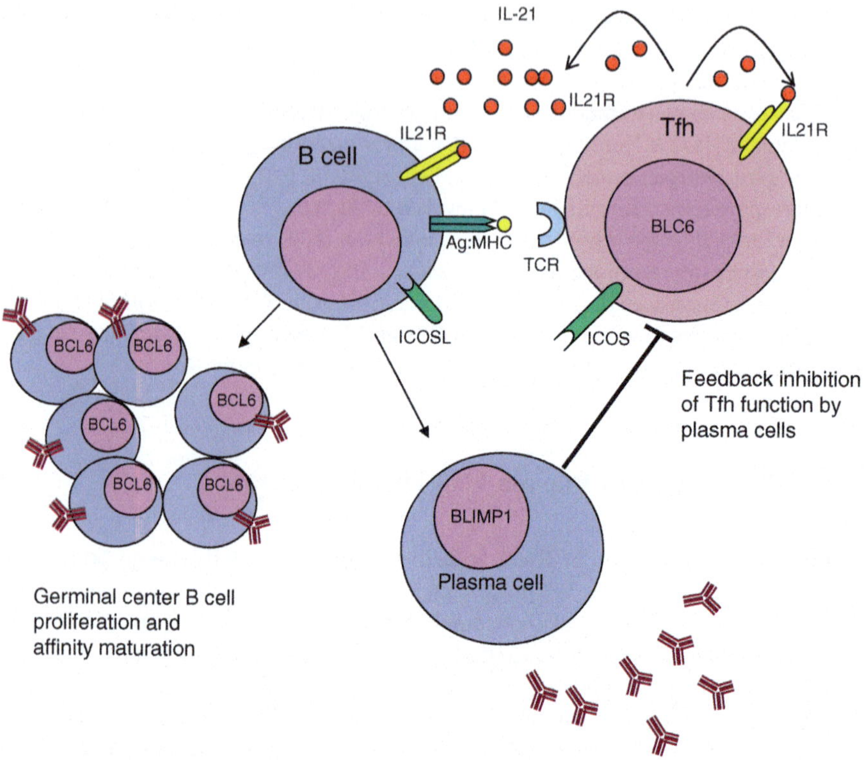

Fig. 5.2 IL-21 plays a critical role in germinal center (GC) function. IL-21 is produced by Tfh cells in the follicle and acts on both Tfh and GC B cells. IL-21 induces GC B-cell proliferation and affinity maturation through its induction of BCL6 and drives terminal B-cell differentiation into plasma cells via its induction of BLIMP1. Plasma cells can present antigen to Tfh cells and mediate inhibition of Tfh function

organs in response to antigen presented by dendritic cells (Fig. 5.2). The subsequent induction of the transcription factor BCL6 mediates expression of the chemokine receptor, CXCR5, which promotes migration to the B-cell zone. Based on the use of IL-21 reporter mice, Tfh cells were shown to express high levels of IL-21 as well as interferon (IFN)-γ, IL-2, and IL-4 (Luthje et al. 2012). The role for IL-21 in the formation of the germinal center (GC) has been investigated using *Il21* KO mice. After immunization with T-dependent antigens, GCs form in these *Il21* KO mice, but the number of GC B cells ($GL7^{+}CD95^{+}$) is reduced, and these cells do not persist as long as in the WT mice (Linterman et al. 2010; Zotos et al. 2010). Interestingly, Tfh cells develop and proliferate in the *Il21* KO GCs, but their numbers decline faster than in wild-type (WT) mice. Immunization studies using mixed bone marrow chimeras formed from WT/*Il21r* KO mice have revealed that IL-21 acts directly on B cells to induce BCL6 to orchestrate the formation and maintenance of GCs (Linterman et al. 2010). Tfh cells not only require IL-21R expression for their optimal maintenance, but IL-21 is also required for the normal development of Tfh cells (King et al. 2008). Analysis of the Ag-specific Ig repertoire in memory B cells that develop in GCs in the absence of IL-21 signaling has revealed a significant reduction in somatic mutation of *Vh* genes required for affinity maturation, which indicates an important effect of IL-21 on memory B-cell formation (Zotos et al. 2010). IL-6 and IL-21 partially share signaling pathways in CD4$^+$ T cells, and relatively normal Tfh generation can occur in the absence of either cytokine, but mice lacking both IL-6 and IL-21 have greatly diminished ability to generate Tfh cells in vivo in response to viral infection (Eto et al. 2011). These studies identified follicular B cells in the draining lymph node as a source for IL-6 that then can induce IL-21 production by CD4$^+$ T cells, leading to an IL-6/IL-21 synergistic effect on Tfh cell development.

Tfh and plasma cell differentiation are controlled, respectively, by BCL6 and BLIMP1, transcription factors that are known to negatively regulate each other (Calame et al. 2003). Interestingly, antigen-activated plasma cells, which express low levels of major histocompatibility complex (MHC) class II and other surface proteins required for antigen presentation, can downregulate the expression of both BCL6 and IL-21 in Tfh cells, demonstrating the presence of a negative feedback loop for these two populations of cells such that high levels of plasma cells can diminish the generation of additional GC memory B cells (Pelletier et al. 2010).

5.7 IL-21 and CD4$^+$ T-Cell Differentiation

Interestingly, *Il21r* KO mice have normal development of thymic and peripheral CD4$^+$ T cells. IL-21 is produced by multiple subsets of CD4$^+$ T cells, including Th1, Th2, Th9, Th17, and Tfh cells, with Th17 and Tfh cells producing the highest levels. IL-6 and IL-21 are essential for the maximal production of IL-21 by Th17 cells (Zhou et al. 2007; Dienz et al. 2009). At the time of the initial discovery of IL-21, only Th1 and Th2 CD4$^+$ T-cell subsets and function had been characterized, and the ability of IL-21 to decrease IFN-γ expression in Th1 cells at the time of their priming was observed (Suto et al. 2006), but other studies showed that IL-21 could induce the expression of Th1-associated genes (Strengell et al. 2002). In fact, in vivo immunization in Th1 delayed-type hypersensitivity models reveals higher inflammatory responses in *Il21r* KO mice, accompanied by higher IFN-γ production (Wurster et al. 2002), and infection of *Il21r* KO mice with *Schistosoma mansoni* results in a reduced Th2 response and faster clearance of lung granulomas (Pesce et al. 2006), indicating that IL-21 is required for these inflammatory responses.

IL-21 plays a key role in the differentiation of Th17 cells, as it induces IL-23R expression on naïve CD4$^+$ T cells, thus allowing Th17 cell expansion. IL-21 also upregulates expression of RORγt, a transcription factor that is required for the differentiation of Th17 cells (Ivanov et al. 2007). IL-17 production was found to be markedly lower in *Il21r* KO CD4$^+$ T cells stimulated in vitro with transforming growth factor (TGF)-β and IL-6 than in corresponding WT cells, demonstrating that the induction of IL-21 by IL-6 leads to the production of IL-21 and the amplification of this pathway (Korn et al. 2007; Nurieva et al. 2007; Zhou et al. 2007). IL-21-induced Th17 cell differentiation plays a role in the development of autoimmune diseases, as discussed below. Overall, IL-21 has effects on differentiation or cytokine production of multiple Th populations.

5.8 IL-21 Controls Regulatory T- and B-Cell Function

IL-21 plays an indirect role in the generation of regulatory T (Treg) cells. *Il6* KO mice have reduced levels of IL-21, and in response to in vivo antigen signaling these mice produce an expanded population of $FoxP3^+$ Treg cells instead of Th17 cells (Korn et al. 2007). Correspondingly, IL-21 can inhibit Treg generation in vitro (Korn et al. 2007). The inhibitory effects of IL-21 on Treg differentiation are further indicated by the increased population of $FoxP3^+$ $CD4^+$ T cells in *Il21* KO mice (Attridge et al. 2012) (Fig. 5.3a). This effect of IL-21 on Treg cells is indirect and results from decreased IL-2 production by $CD4^+$ T cells, which leads to reduced availability of IL-2 for maintaining the viability of Treg cells in vivo (Attridge et al. 2012) (Fig. 5.3b).

In addition to its indirect regulation of Treg function, IL-21 is immunosuppressive via its induction of IL-10 in several subsets of lymphoid cells (Fig. 5.3c). IL-10 is one of the most immunosuppressive cytokines and is produced by multiple hematopoietic lineages, including T cells, B cells, NK cells, dendritic cells, macrophages, neutrophils, and mast cells (Saraiva and O'Garra 2010). A relationship between IL-21 and IL-10 was initially suggested by their coexpression in the serum of autoimmune BXSB-*Yaa* mice with system lupus erythematosus (SLE) (Ozaki et al. 2004). Indeed, *Il21r* KO T cells have decreased levels of IL-10 whereas *Il21* TG T cells had increased levels of IL-10, and IL-21 induces IL-10 in both naïve $CD4^+$ and $CD8^+$ T cells in vitro (Spolski et al. 2009). Additionally, polarization of CD4 and $CD8^+$ T cells to Th17 or Tc17 populations is associated with lower production of IL-10 if IL-21 signaling is absent, and IL-21 can also induce IL-10 in Tc1 and Tc17 polarized cells (Spolski et al. 2009).

The Tr1 population of immunosuppressive $CD4^+$ T cells lacks FoxP3 expression, and these cells produce IL-10 in response to IL-27. IL-27, which is produced by dendritic cells, was found to initiate the induction of ICOS and the transcription factor c-Maf, which then induces IL-21 expression (Pot et al. 2009). Thus, IL-27 indirectly mediates induction of IL-10 via an IL-27 to IL-21 to IL-10 cytokine cascade to effect the induction of immunosuppressive $CD4^+$ T cells (Fig. 5.3d).

A subpopulation of B cells (B10 cells) produces IL-10 and has strong immunoregulatory activity in several autoimmune diseases, including experimental autoimmune encephalitis (EAE), colitis, type I diabetes, and arthritis (Yanaba et al. 2008). IL-21 is required not only for B10 cell expansion, but also for IL-10 secretion and the immunosuppressive activity of these cells in vivo (Yoshizaki et al. 2012) (Fig. 5.3e). IL-21 is also a critical component of the in vitro cocktail of cytokines that can greatly expand this B10 population, which allowed therapeutic use of B10 cells in adoptive immunotherapy of autoimmune EAE (Yoshizaki et al. 2012).

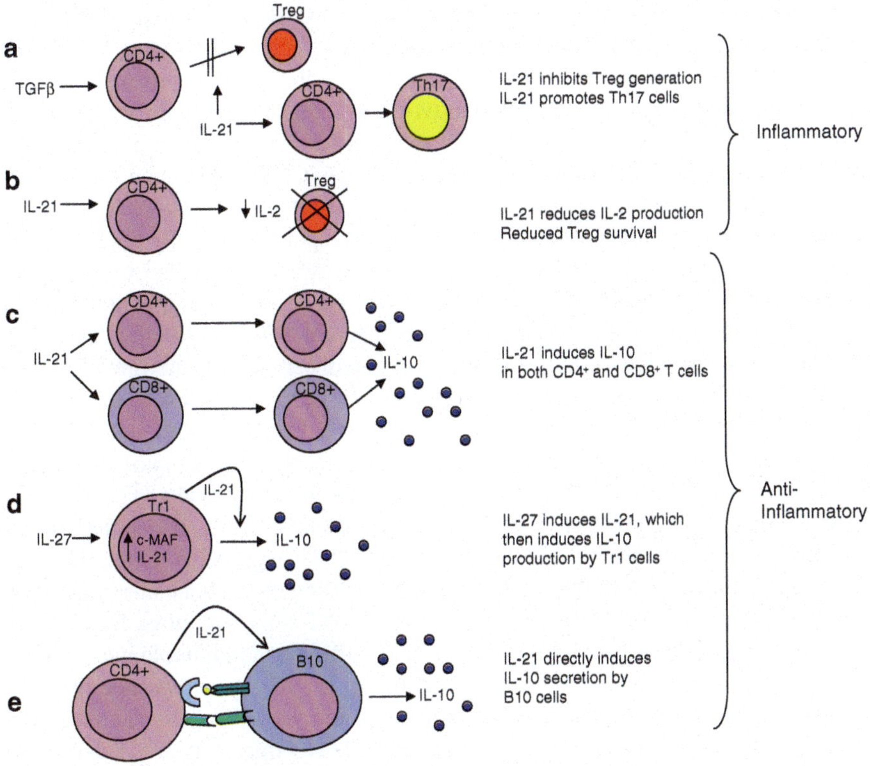

Fig. 5.3 IL-21 has both inflammatory and anti-inflammatory roles based on its effects on T and B regulatory cells. (**a**) IL-21 inhibits Treg differentiation and promotes the induction of inflammatory Th17 cells. (**b**) IL-21 indirectly inhibits Treg survival via its downregulation of IL-2 production by CD4⁺ T cells. (**c**) IL-21 directly induces IL-10 in T cells, leading to immunosuppressive activities. (**d**) IL-27 stimulates IL-21 production by Tr1 regulatory cells, and IL-21 then induces production of IL-10. (**e**) IL-21 is required for the differentiation and effector function of B10 regulatory cells through its induction of IL-10

5.9 IL-21 Is Required for Optimal $CD8^+$ T-Cell Proliferation and Function

$CD8^+$ T-cell numbers in the thymus and periphery of *Il21r* KO mice are normal, but IL-21 nevertheless has profound proliferative and functional effects on these cells. IL-21 alone can only minimally induce $CD8^+$ T-cell proliferation in vitro, but it has a strong synergistic effect when combined with either IL-7 or IL-15, cytokines that play major roles in $CD8^+$ expansion (Zeng et al. 2005). Similarly, although IL-21 alone has no effect on IFN-γ expression by $CD8^+$ T cells, it enhances expression induced by IL-15 (Zeng et al. 2005).

IL-21 has distinctive actions on antigen-stimulated $CD8^+$ T-cell function, as compared to IL-2 or IL-15. IL-2 and IL-15 each induce a mature effector phenotype characterized by elevated CD44, Eomes, granzyme B, and increased cytolytic activity, whereas IL-21 induces a less activated effector phenotype, with lower levels of CD44, Eomes, granzyme B, increased CD62L, and decreased cytolytic activity (Hinrichs et al. 2008). Importantly, $CD8^+$ T cells primed with antigen in the presence of IL-21 have memory cell characteristics and enhanced persistence in vivo (Hinrichs et al. 2008). IL-21 also upregulates CD28 expression on $CD8^+$ T cells and thereby inhibits senescence (Alves et al. 2005).

$CD8^+$ T-cell responses are critical in cytotoxic responses to virally infected cells. When mice were immunized with vaccinia virus expressing human immunodeficiency virus (HIV) gp160 antigen, *Il21r* KO mice displayed significantly lower $CD8^+$ T-cell expansion and cytotoxic activity than did WT mice (Zeng et al. 2005). This lower activity is in contrast to the situation observed in acute infection with influenza, where *Il21r* KO mice have similar initial $CD8^+$ T-cell responses and viral clearance to those observed in WT mice (Frohlich et al. 2009). Although *Il21r* KO mice and WT mice respond similarly during the acute phase of lymphocytic choriomeningitis virus (LCMV) infection, chronic LCMV infection results in severe functional "exhaustion" in *Il21* KO $CD8^+$ T cells, with lower IL-2, IFN-γ, and tumor necrosis factor (TNF)-α but elevated inhibitory PD-1 expression (Elsaesser et al. 2009; Frohlich et al. 2009; Yi et al. 2009). Studies using mixed bone marrow chimeric mice have revealed that IL-21 signaling in $CD8^+$ T cells is required for effective viral responses and that *Il21r* KO $CD8^+$ T cells suffer from immunological exhaustion and are deleted. $CD8^+$ T cells responding to LCMV in the absence of IL-21 produce significantly less IL-2 during both the primary effector phase and memory response to virus (Yi et al. 2010). Treatment of *Il21* KO mice with IL-21 during the first week of infection restores the antiviral activity of $CD8^+$ T cells but leads to immunopathology related to the activated $CD8^+$ T-cell response (Yi et al. 2009). Collectively, these studies demonstrate that although the $CD8^+$ T-cell response to acute viral infections does not necessarily require IL-21, the presence of IL-21 during a primary response to virus significantly affects the quality and persistence of the $CD8^+$ T-cell antiviral response during the chronic phase of infection.

5.10 IL-21 Regulation of NK and NKT Cell Function

IL-21 is not required for NK cell development but can affect mature NK function and proliferation (Kasaian et al. 2002; Ozaki et al. 2002). Although IL-21 alone does not augment proliferation of NK cells, it can amplify NK proliferative responses to IL-2 or IL-15 (Toomey et al. 2003). IL-21-treated NK cells display enhanced cytolytic function and IFN-γ production (Kasaian et al. 2002), while decreasing expression of the inhibitory Ly49D receptor (Brady et al. 2004) and the NKG2D receptors on NK cells (Burgess et al. 2006). IL-21 induces high levels of IL-10 in populations of NK cells, and the early production of IL-10 by NK cells in the in vivo response to mouse cytomegalovirus infection is important for limiting $CD8^+$ T-cell-dependent pathology in this disease (Lee et al. 2009).

Natural killer T cells (NKT) are a population of T cells that have a restricted TCR repertoire, with specificity for glycolipid antigens when presented by CD1d. IL-21 can enhance the proliferation of NKT cells in response to TCR signals but only in combination with either IL-2 or IL-15 (Coquet et al. 2007). NKT cells respond to IL-21, but importantly they also are major producers of IL-21 during early innate responses to pathogens and hence can regulate later adaptive responses by B and T cells (Coquet et al. 2007). After immunization with the glycolipid α-GalCer, a subpopulation of NKT cells appeared that had acquired the Tfh markers CXCR5 and PD-1. These NKTfh cells can provide help for B cells that is dependent on IL-21 (Chang et al. 2012); these cells depend on BCL6 expression and are more efficient than $CD4^+$ Tfh cells in promoting the development of GC B cells, although less effective in promoting affinity maturation and plasma cell differentiation (Chang et al. 2012).

5.11 IL-21 Has Inhibitory Effects on Dendritic Cells

Although many of the effects of IL-21 on lymphoid cells involve proliferative responses or enhanced effector functions, IL-21 primarily has inhibitory effects on dendritic cells. For example, although IL-15 can induce maturation of dendritic cells (DCs) with enhanced antigen-presenting ability, IL-21-treated DCs have reduced ability to upregulate MHC class II, CD86, or CD80 costimulatory molecules (Brandt et al. 2003). Importantly, it has become clear that IL-21 induces apoptosis of conventional dendritic cells (cDCs) through a STAT3- and BIM-dependent mechanism (Wan et al. 2013), helping to explain the negative regulatory actions of IL-21 on immune responses.

5.12 IL-21 Has Potent Antitumor Effects

The ability of IL-21 to control effector function of $CD8^+$ T cells, NK cells, and NKT cells makes it an attractive candidate therapeutic agent in cancer. Early studies showed that systemic expression of IL-21 can inhibit growth of large established melanomas (Wang et al. 2003), an effect that was predominantly mediated by NK cells. Importantly, there was no substantial toxicity of IL-21 even at high doses. The killing of tumor cells by IL-21-activated NK cells appears to require the presence of inhibitory NKG2D ligands on tumor targets (Takaki et al. 2005). When mice with large established melanomas are treated with adoptively transferred tumor-specific $CD8^+$ T cells followed by treatment with IL-2, IL-15, IL-21, or combinations of these cytokines, either IL-15 or IL-21 can induce partial tumor regression, but the combination of these cytokines can result in complete regression of some of the tumors as well as increased survival of most of the treated mice (Zeng et al. 2005). Interestingly, IL-21 confers a distinct differentiation program on $CD8^+$ T cells as compared to that conferred by IL-2 (Hinrichs et al. 2008). When tumor-specific $CD8^+$ T cells were primed in vitro with antigen and either IL-21 or IL-2, the IL-21-primed cells expressed less granzyme B, CD44, and IL-2Rα than did cells primed in the presence of IL-2. Surprisingly, in spite of less effector function in vitro, the IL-21-primed $CD8^+$ T cells display enhanced antitumor activity upon adoptive transfer into mice with large established melanomas. These IL-21-primed $CD8^+$ T cells have characteristics of memory cells, including increased levels of CD62L and enhanced longevity in vivo after the tumors had regressed (Hinrichs et al. 2008).

These preclinical studies indicate that IL-21 has significant antitumor activity involving both innate and adaptive immune components. IL-21 is currently in phase I and phase II clinical trials for melanoma, renal carcinoma, and certain other tumors (Petrella et al. 2012). Consistent with studies in mice, favorable therapeutic responses have been observed, without major adverse events.

5.13 Role of IL-21 in Autoimmune Disease

Earlier, we noted that IL-21 promotes Ig production and terminal B-cell differentiation as well as having effects on T cells. These characteristics are consistent with the possibility that IL-21 might promote autoimmune disease. Indeed, genome-wide association studies have identified risk variants in the *Il21* gene for celiac disease (van Heel et al. 2007), psoriasis, and psoriatic arthritis (Liu et al. 2008), lupus (Hughes et al. 2011), inflammatory bowel disease (Marquez et al. 2009), and type 1 diabetes (Cooper et al. 2008). Work in animal models discussed here has begun to elucidate the mechanisms by which IL-21 can exacerbate these autoimmune diseases.

5.13.1 Systemic Lupus Erythematosus

Systemic lupus erythematosus (SLE) is a chronic autoimmune disease characterized by the pathological accumulation of auto-reactive antibodies. In the BXSB-*Yaa* mouse model of SLE, serum IL-21 levels increase with age and correlate with the severity of disease, consistent with the known role for IL-21 in plasma cell differentiation and Ig production (Ozaki et al. 2004). When BXSB-*Yaa* mice were crossed onto the *Il21r* KO background, there was a significant reduction in levels of anti-nuclear antibodies accompanied by an absence of kidney disease and increased survival (Bubier et al. 2009). These experiments suggest that blocking IL-21 could potentially be used therapeutically to inhibit disease progression. Indeed, treatment of the lupus-prone MRL-Fas/*lpr* mouse with IL-21R/Fc fusion proteins as an IL-21 blocking agent decreases Ig levels and kidney pathology (Herber et al. 2007).

Importantly, in addition to the mouse studies, studies in patients with SLE have shown elevated levels of serum IL-21, and this is accompanied by elevated numbers of $CD4^+CXCR5^+BCL6^+$ Tfh cells as well as increased circulating germinal center B cells (Terrier et al. 2012). These studies suggest that targeting IL-21 in these patients could potentially alleviate disease progression.

5.13.2 Experimental Autoimmune Encephalitis

IL-21 plays a role in experimental autoimmune encephalitis (EAE), a mouse model of human multiple sclerosis, which is induced by immunization with myelin with adjuvants. Administration of IL-21 before disease induction increases severity of disease, correlating with the presence of inflammatory NK cells whose depletion prevents the effects of IL-21 (Vollmer et al. 2005). Interestingly, *Il21r* KO mice have significantly reduced numbers of IL-17-producing cells and reduced EAE disease progression (Korn et al. 2007; Nurieva et al. 2007), indicating that IL-21 is required for the normal induction and expansion of inflammatory Th17 cells. When B regulatory cells (B10 cells), which produce IL-10 in response to IL-21, are expanded in vitro and then transferred into mice in which EAE disease has been initiated, there is a marked decrease in the severity of the disease even after symptoms have begun to appear (Yoshizaki et al. 2012). These experiments suggest that in EAE initiation and progression IL-21 can play both positive and negative roles via its effects on Th17 differentiation and B10 cell production, respectively.

5.13.3 Experimental Autoimmune Uveitis

Experimental autoimmune uveitis (EAU) shares pathological features with human uveitis, a group of inflammatory diseases that compromise intraocular function. EAU is induced in mice by immunization with retinal protein and is associated with the accumulation of inflammatory Th17 cells. IL-21 reporter transgenic mice were used to demonstrate that $CD4^+$ T cells expressing IL-21 accumulate in the retina (Wang et al. 2011). Moreover, *Il21r* KO mice are resistant to the development of EAU, and these mice have diminished IL-17A and IL-1β expression in draining lymph nodes. When lymph node (LN) cells from WT and *Il21r* KO EAE mice were adoptively transferred into WT recipients, there was significantly less disease development in mice that received KO LN cells. These studies suggest that IL-21 promotes the development of uveitis via its effects on inflammatory cells and mediators.

5.13.4 Autoimmune Diabetes

Autoimmune (type 1) diabetes is a T-cell-mediated disease characterized by immune cell infiltration into the pancreas followed by destruction of the insulin-producing β cells in the islets. The non-obese diabetic mouse (NOD) is a valuable animal model for studies of the progression of type 1 diabetes. One genetic locus associated with disease development is *Idd3*, which contains the genes encoding IL-2 and IL-21. IL-21 levels are elevated in the NOD mouse (King et al. 2004), and disease is ablated in NOD mice lacking the IL-21R (Spolski et al. 2008; Sutherland et al. 2009). Inflammatory Th17 cells whose differentiation is augmented by IL-21 have been implicated in diabetes development. Antigen-presenting cells from NOD mice are more efficient than WT APCs at the induction of Th17 $CD4^+$ cells, and this effect requires IL-21R on the APCs, suggesting that IL-21 acts directly on these cells to induce secretion of mediators including prostaglandin E_2 and IL-6, which cause more efficient Th17 induction (Liu et al. 2011). APCs in *Il21r* KO mice have an antigen-presenting defect characterized by lower levels of class II MHC and CD86 (Van Belle et al. 2012). In addition, *Il21r* KO APCs express decreased levels of the chemokine receptor CCR7, resulting in defective infiltration into the pancreatic LNs, with inefficient priming of $CD4^+$ T cells and failure to help $CD8^+$ T cells to infiltrate pancreatic islets. Thus, IL-21 is essential for the development of type 1 diabetes in the NOD mouse model system. Correspondingly, studies using an IL-21R-Fc fusion protein have revealed that IL-21 neutralization can reduce lymphoid infiltration into the islets at an early stage of the disease but that treatment of newly diabetic mice cannot block further diabetes development (McGuire et al. 2011). However, the combination of IL-21 blocking with syngeneic islet transplant was able to reverse diabetes, even when the IL-21 blockade was short term (McGuire et al. 2011). These studies suggest great promise for therapeutic strategies aimed at blocking IL-21 function at the time of islet transplant in humans with type I diabetes.

5.14 Inflammatory Bowel Disease and Colorectal Cancer

Inflammatory bowel disease (IBD) in humans involves enhanced inflammatory responses to the microbiota present in the gut lumen, leading to the extensive tissue damage seen in Crohn's disease or ulcerative colitis (Bouma and Strober 2003). Analysis of biopsies from patients with either Crohn's disease or ulcerative colitis has revealed higher levels of IL-21 protein expression in mucosal samples compared to controls (Fantini et al. 2008; Monteleone et al. 2005). In studies of the role of IL-21 in gut inflammation in the dextran sodium sulfate (DSS) mouse model for colitis (Fina et al. 2008), treatment with DSS results in elevated levels of IL-21 as well as increased levels of IL-17 expression in gut tissue, whereas *Il21* KO mice have diminished pathology, with reduced infiltration of inflammatory cells into the gut and lower IL-17 production.

In addition to IL-21 potentially promoting colitis by the induction of IL-17 production in $CD4^+$ T cells, epithelial cells in the gut are also targets for IL-21. Intestinal epithelial cells express IL-21R and responded to IL-21 with the production of MIP-3α, a T-cell chemoattractant for cells such as Th1 and Th17 that express CCR6 (Caruso et al. 2007). IL-21 also promotes the secretion of a number of matrix metalloproteases that are involved in tissue damage in the gut (Monteleone et al. 2006).

The chronic inflammation that accompanies IBD increases the risk of colon cancer. An increased expression of IL-21 in IBD patients led to an examination of whether IL-21 and its associated inflammatory response were responsible for the onset of colon cancer. IL-21 mRNA and protein were indeed overexpressed in human colon cancer samples (Stolfi et al. 2011). In a mouse model for colitis-associated colon cancer, *Il21* KO mice were resistant to the development of colon cancer, with decreased production of IL-6, IL-17A, and alternatively activated macrophage recruitment to the tumors. IL-21 inhibition also reduced the number of tumors in the colon tissue (Stolfi et al. 2011). Thus, although IL-21 has potent antitumor actions, it also plays a role in chronic inflammation and can thereby enhance tumor growth.

5.15 Role for IL-21 in Allergic Disease

Given that IL-21 has a critical role in B-cell immunoglobulin production, it was expected that it would exert effects on allergic inflammation. Both positive and negative regulatory effects of IL-21 on airway allergic responses have been reported. In one study, administering IL-21 during the sensitization phase of the response resulted in both reduced eosinophilia and lower IgE levels after the airway challenge phase (Suto et al. 2002). However, *Il21r* KO mice are more resistant than WT mice to eosinophilic airway inflammation (Frohlich et al. 2007). A study in a mouse model of IgE-mediated allergic cutaneous reactions showed that if IL-21 is administered during the sensitization phase, there is suppression of both the cutaneous response and of IgE levels (Tamagawa-Mineoka et al. 2011). However, when IL-21 is administered after sensitization was complete, IgE levels remained the same but mast cell degranulation in the skin is suppressed. Mast cells express IL-21R, and IL-21 can block the release of histamine and other inflammatory mediators by mast cells when these are activated by the IgE signal.

The role for IL-21 in cutaneous inflammatory responses has been confirmed in a human psoriasis SCID mouse model, in which blocking IL-21 led to significant reduction in skin thickening and $CD4^+$ T-cell infiltration (Caruso et al. 2009). In this case, IL-21 promoted the recruitment of $CD4^+$ T cells and induced the production of IFN-γ by these cells, which was ablated when the $CD4^+$ T cells lacked IL-21R (Sarra et al. 2011). These studies in pulmonary and skin allergic responses suggest that an understanding of the role of IL-21 involves integrating data concerning its effects on multiple immune populations at different stages of development of these responses.

5.16 Role of IL-21 in Pulmonary Viral Infection

Although IL-21 has been shown to have little effect on the $CD8^+$ T-cell primary responses to systemic viruses such as LCMV, this contrasts to its effect on the response to respiratory infections that elicit strong inflammatory responses. Pneumonia virus of mouse (PVM) is a natural rodent pathogen that initiates a respiratory infection in mice which resembles severe human respiratory syncytial virus (RSV) disease. Studies using IL-21 reporter mice have revealed that IL-21 is expressed by CD4+ T cells in the lung following infection of animals with PVM (Spolski et al. 2012), with PVM-infected *Il21r* KO mice having less lung infiltration by neutrophils than WT mice. Although viral titer was similar in WT and *Il21r* KO mice, the absence of IL-21 signaling confers a survival advantage to *Il21r* KO mice, indicating that IL-21 can contribute to a detrimental inflammatory state in the lung. Consistent with this, blocking IL-21 activity in WT mice leads to increased survival after PVM infection (Spolski et al. 2012). These studies suggest that therapeutic manipulation of IL-21 signaling, such as IL-21 receptor blockade, may represent an immunomodulatory strategy for averting the pathology of this and potentially other respiratory virus infections as well.

5.17 Conclusion

Since IL-21 was discovered in 2000, a tremendous amount of knowledge has been generated about this cytokine. IL-21 exerts broad actions on a range of target cell types including lymphoid and myeloid populations. These actions include positive regulation of B-cell Ig production and plasma cell differentiation as a result of its direct effects on B cells as well as effects on Tfh differentiation, but IL-21 also limits immune responses based on its induction of apoptosis of B cells and DCs and its induction of IL-10-mediated suppression by T and B regulatory cells. The discovery of the antitumor activity of IL-21 mediated by its actions on CD8+ T cells and NK cells in various animal models led to rapid translation of this work into the clinical setting, with IL-21 showing promise in the treatment of several cancers in phase I and phase II clinical trials. However, IL-21 is also known to either induce or exacerbate a range of autoimmune diseases, as described herein. Accordingly, early clinical trials are now in progress to evaluate anti-IL-21 in the treatment of lupus and rheumatoid arthritis (see www.clinicaltrials.gov). A major challenge is to understand all the mechanisms by which IL-21 exerts both its positive and deleterious effects to be able to therapeutically modulate these without tipping the balance of immune homeostasis.

Acknowledgments R.S. and W.J.L. are supported by the Division of Intramural Research, National Heart, Lung, and Blood Institute, NIH. R.S. and W.J.L. are inventors on patents related to IL-21. We thank Dr. Jian-Xin Lin for valuable suggestions and critical comments.

References

Alves NL, Arosa FA, van Lier RA (2005) IL-21 sustains CD28 expression on IL-15-activated human naive CD8+ T cells. J Immunol 175:755–762

Asao H, Okuyama C, Kumaki S, Ishii N, Tsuchiya S, Foster D, Sugamura K (2001) Cutting edge: the common gamma-chain is an indispensable subunit of the IL-21 receptor complex. J Immunol 167:1–5

Attridge K, Wang CJ, Wardzinski L, Kenefeck R, Chamberlain JL, Manzotti C, Kopf M, Walker LS (2012) IL-21 inhibits T cell IL-2 production and impairs Treg homeostasis. Blood 119: 4656–4664

Bondensgaard K, Breinholt J, Madsen D, Omkvist DH, Kang L, Worsaae A, Becker P, Schiodt CB, Hjorth SA (2007) The existence of multiple conformers of interleukin-21 directs engineering of a superpotent analogue. J Biol Chem 282:23326–23336

Bouma G, Strober W (2003) The immunological and genetic basis of inflammatory bowel disease. Nat Rev Immunol 3:521–533

Brady J, Hayakawa Y, Smyth MJ, Nutt SL (2004) IL-21 induces the functional maturation of murine NK cells. J Immunol 172:2048–2058

Brandt K, Bulfone-Paus S, Foster DC, Ruckert R (2003) Interleukin-21 inhibits dendritic cell activation and maturation. Blood 102:4090–4098

Bubier JA, Sproule TJ, Foreman O, Spolski R, Shaffer DJ, Morse HC 3rd, Leonard WJ, Roopenian DC (2009) A critical role for IL-21 receptor signaling in the pathogenesis of systemic lupus erythematosus in BXSB-Yaa mice. Proc Natl Acad Sci USA 106:1518–1523

Burgess SJ, Marusina AI, Pathmanathan I, Borrego F, Coligan JE (2006) IL-21 down-regulates NKG2D/DAP10 expression on human NK and CD8+ T cells. J Immunol 176:1490–1497

Calame KL, Lin KI, Tunyaplin C (2003) Regulatory mechanisms that determine the development and function of plasma cells. Annu Rev Immunol 21:205–230

Caruso R, Fina D, Peluso I, Stolfi C, Fantini MC, Gioia V, Caprioli F, Del Vecchio Blanco G, Paoluzi OA, Macdonald TT, Pallone F, Monteleone G (2007) A functional role for interleukin-21 in promoting the synthesis of the T-cell chemoattractant, MIP-3alpha, by gut epithelial cells. Gastroenterology 132:166–175

Caruso R, Botti E, Sarra M, Esposito M, Stolfi C, Diluvio L, Giustizieri ML, Pacciani V, Mazzotta A, Campione E, Macdonald TT, Chimenti S, Pallone F, Costanzo A, Monteleone G (2009) Involvement of interleukin-21 in the epidermal hyperplasia of psoriasis. Nat Med 15: 1013–1015

Chang PP, Barral P, Fitch J, Pratama A, Ma CS, Kallies A, Hogan JJ, Cerundolo V, Tangye SG, Bittman R, Nutt SL, Brink R, Godfrey DI, Batista FD, Vinuesa CG (2012) Identification of Bcl-6-dependent follicular helper NKT cells that provide cognate help for B cell responses. Nat Immunol 13:35–43

Ciofani M, Madar A, Galan C, Sellars M, Mace K, Pauli F, Agarwal A, Huang W, Parkurst CN, Muratet M, Newberry KM, Meadows S, Greenfield A, Yang Y, Jain P, Kirigin FK, Birchmeier C, Wagner EF, Murphy KM, Myers RM, Bonneau R, Littman DR (2012) A validated regulatory network for Th17 cell specification. Cell 151:289–303

Cooper JD, Smyth DJ, Smiles AM, Plagnol V, Walker NM, Allen JE, Downes K, Barrett JC, Healy BC, Mychaleckyj JC, Warram JH, Todd JA (2008) Meta-analysis of genome-wide association study data identifies additional type 1 diabetes risk loci. Nat Genet 40:1399–1401

Coquet JM, Kyparissoudis K, Pellicci DG, Besra G, Berzins SP, Smyth MJ, Godfrey DI (2007) IL-21 is produced by NKT cells and modulates NKT cell activation and cytokine production. J Immunol 178:2827–2834

Dienz O, Eaton SM, Bond JP, Neveu W, Moquin D, Noubade R, Briso EM, Charland C, Leonard WJ, Ciliberto G, Teuscher C, Haynes L, Rincon M (2009) The induction of antibody production by IL-6 is indirectly mediated by IL-21 produced by CD4+ T cells. J Exp Med 206: 69–78

Elsaesser H, Sauer K, Brooks DG (2009) IL-21 is required to control chronic viral infection. Science 324:1569–1572

Eto D, Lao C, DiToro D, Barnett B, Escobar TC, Kageyama R, Yusuf I, Crotty S (2011) IL-21 and IL-6 are critical for different aspects of B cell immunity and redundantly induce optimal follicular helper CD4 T cell (Tfh) differentiation. PLoS One 6:e17739

Ettinger R, Sims GP, Fairhurst AM, Robbins R, da Silva YS, Spolski R, Leonard WJ, Lipsky PE (2005) IL-21 induces differentiation of human naive and memory B cells into antibody-secreting plasma cells. J Immunol 175:7867–7879

Ettinger R, Sims GP, Robbins R, Withers D, Fischer RT, Grammer AC, Kuchen S, Lipsky PE (2007) IL-21 and BAFF/BLyS synergize in stimulating plasma cell differentiation from a unique population of human splenic memory B cells. J Immunol 178:2872–2882

Fantini MC, Monteleone G, MacDonald TT (2008) IL-21 comes of age as a regulator of effector T cells in the gut. Mucosal Immunol 1:110–115

Fina D, Sarra M, Fantini MC, Rizzo A, Caruso R, Caprioli F, Stolfi C, Cardolini I, Dottori M, Boirivant M, Pallone F, Macdonald TT, Monteleone G (2008) Regulation of gut inflammation and th17 cell response by interleukin-21. Gastroenterology 134:1038–1048

Frohlich A, Marsland BJ, Sonderegger I, Kurrer M, Hodge MR, Harris NL, Kopf M (2007) IL-21 receptor signaling is integral to the development of Th2 effector responses in vivo. Blood 109:2023–2031

Frohlich A, Kisielow J, Schmitz I, Freigang S, Shamshiev AT, Weber J, Marsland BJ, Oxenius A, Kopf M (2009) IL-21R on T cells is critical for sustained functionality and control of chronic viral infection. Science 324:1576–1580

Glasmacher E, Agrawal S, Chang AB, Murphy TL, Zeng W, Vander Lugt B, Khan AA, Ciofani M, Spooner CJ, Rutz S, Hackney J, Nurieva R, Escalante CR, Ouyang W, Littman DR, Murphy KM, Singh H (2012) A genomic regulatory element that directs assembly and function of immune-specific AP-1-IRF complexes. Science 338:975–980

Good KL, Bryant VL, Tangye SG (2006) Kinetics of human B cell behavior and amplification of proliferative responses following stimulation with IL-21. J Immunol 177:5236–5247

Hamming OJ, Kang L, Svensson A, Karlsen JL, Rahbek-Nielsen H, Paludan SR, Hjorth SA, Bondensgaard K, Hartmann R (2012) Crystal structure of interleukin-21 receptor (IL-21R) bound to IL-21 reveals that sugar chain interacting with WSXWS motif is integral part of IL-21R. J Biol Chem 287:9454–9460

Harada M, Magara-Koyanagi K, Watarai H, Nagata Y, Ishii Y, Kojo S, Horiguchi S, Okamoto Y, Nakayama T, Suzuki N, Yeh WC, Akira S, Kitamura H, Ohara O, Seino K, Taniguchi M (2006) IL-21-induced Bepsilon cell apoptosis mediated by natural killer T cells suppresses IgE responses. J Exp Med 203:2929–2937

Herber D, Brown TP, Liang S, Young DA, Collins M, Dunussi-Joannopoulos K (2007) IL-21 has a pathogenic role in a lupus-prone mouse model and its blockade with IL-21R.Fc reduces disease progression. J Immunol 178:3822–3830

Hinrichs CS, Spolski R, Paulos CM, Gattinoni L, Kerstann KW, Palmer DC, Klebanoff CA, Rosenberg SA, Leonard WJ, Restifo NP (2008) IL-2 and IL-21 confer opposing differentiation programs to CD8+ T cells for adoptive immunotherapy. Blood 111:5326–5333

Hughes T, Kim-Howard X, Kelly JA, Kaufman KM, Langefeld CD, Ziegler J, Sanchez E, Kimberly RP, Edberg JC, Ramsey-Goldman R, Petri M, Reveille JD, Martin J, Brown EE, Vila LM, Alarcon GS, James JA, Gilkeson GS, Moser KL, Gaffney PM, Merrill JT, Vyse TJ, Alarcon-Riquelme ME, Nath SK, Harley JB, Sawalha AH (2011) Fine-mapping and transethnic genotyping establish IL2/IL21 genetic association with lupus and localize this genetic effect to IL21. Arthritis Rheum 63:1689–1697

Ivanov II, Zhou L, Littman DR (2007) Transcriptional regulation of Th17 cell differentiation. Semin Immunol 19:409–417

Jin H, Carrio R, Yu A, Malek TR (2004) Distinct activation signals determine whether IL-21 induces B cell costimulation, growth arrest, or Bim-dependent apoptosis. J Immunol 173: 657–665

Kang L, Bondensgaard K, Li T, Hartmann R, Hjorth SA (2010) Rational design of interleukin-21 antagonist through selective elimination of the gamma C binding epitope. J Biol Chem 285: 12223–12231

Kasaian MT, Whitters MJ, Carter LL, Lowe LD, Jussif JM, Deng B, Johnson KA, Witek JS, Senices M, Konz RF, Wurster AL, Donaldson DD, Collins M, Young DA, Grusby MJ (2002) IL-21 limits NK cell responses and promotes antigen-specific T cell activation: a mediator of the transition from innate to adaptive immunity. Immunity 16:559–569

Kim HP, Korn LL, Gamero AM, Leonard WJ (2005) Calcium-dependent activation of interleukin-21 gene expression in T cells. J Biol Chem 280:25291–25297

King C, Ilic A, Koelsch K, Sarvetnick N (2004) Homeostatic expansion of T cells during immune insufficiency generates autoimmunity. Cell 117:265–277

King C, Tangye SG, Mackay CR (2008) T follicular helper (TFH) cells in normal and dysregulated immune responses. Annu Rev Immunol 26:741–766

Korn T, Bettelli E, Gao W, Awasthi A, Jager A, Strom TB, Oukka M, Kuchroo VK (2007) IL-21 initiates an alternative pathway to induce proinflammatory T(H)17 cells. Nature (Lond) 448:484–487

Kwon H, Thierry-Mieg D, Thierry-Mieg J, Kim HP, Oh J, Tunyaplin C, Carotta S, Donovan CE, Goldman ML, Tailor P, Ozato K, Levy DE, Nutt SL, Calame K, Leonard WJ (2009) Analysis of interleukin-21-induced Prdm1 gene regulation reveals functional cooperation of STAT3 and IRF4 transcription factors. Immunity 31:941–952

Lee SH, Kim KS, Fodil-Cornu N, Vidal SM, Biron CA (2009) Activating receptors promote NK cell expansion for maintenance, IL-10 production, and CD8 T cell regulation during viral infection. J Exp Med 206:2235–2251

Leonard WJ (2001) Cytokines and immunodeficiency diseases. Nat Rev Immunol 1:200–208

Li P, Spolski R, Liao W, Wang L, Murphy TL, Murphy KM, Leonard WJ (2012) BATF-JUN is critical for IRF4-mediated transcription in T cells. Nature (Lond) 490:543–546

Linterman MA, Beaton L, Yu D, Ramiscal RR, Srivastava M, Hogan JJ, Verma NK, Smyth MJ, Rigby RJ, Vinuesa CG (2010) IL-21 acts directly on B cells to regulate Bcl-6 expression and germinal center responses. J Exp Med 207:353–363

Liu Y, Helms C, Liao W, Zaba LC, Duan S, Gardner J, Wise C, Miner A, Malloy MJ, Pullinger CR, Kane JP, Saccone S, Worthington J, Bruce I, Kwok PY, Menter A, Krueger J, Barton A, Saccone NL, Bowcock AM (2008) A genome-wide association study of psoriasis and psoriatic arthritis identifies new disease loci. PLoS Genet 4:e1000041

Liu SM, Lee DH, Sullivan JM, Chung D, Jager A, Shum BO, Sarvetnick NE, Anderson AC, Kuchroo VK (2011) Differential IL-21 signaling in APCs leads to disparate Th17 differentiation in diabetes-susceptible NOD and diabetes-resistant NOD.Idd3 mice. J Clin Invest 121:4303–4310

Luthje K, Kallies A, Shimohakamada Y, Belz GT, Light A, Tarlinton DM, Nutt SL (2012) The development and fate of follicular helper T cells defined by an IL-21 reporter mouse. Nat Immunol 13:491–498

Marquez A, Orozco G, Martinez A, Palomino-Morales R, Fernandez-Arquero M, Mendoza JL, Taxonera C, Diaz-Rubio M, Gomez-Garcia M, Nieto A, Lopez-Nevot MA, de la Concha EG, Martin J, Urcelay E (2009) Novel association of the interleukin 2-interleukin 21 region with inflammatory bowel disease. Am J Gastroenterol 104:1968–1975

McGuire HM, Walters S, Vogelzang A, Lee CM, Webster KE, Sprent J, Christ D, Grey S, King C (2011) Interleukin-21 is critically required in autoimmune and allogeneic responses to islet tissue in murine models. Diabetes 60:867–875

Mehta DS, Wurster AL, Whitters MJ, Young DA, Collins M, Grusby MJ (2003) IL-21 induces the apoptosis of resting and activated primary B cells. J Immunol 170:4111–4118

Monteleone G, Monteleone I, Fina D, Vavassori P, Del Vecchio Blanco G, Caruso R, Tersigni R, Alessandroni L, Biancone L, Naccari GC, MacDonald TT, Pallone F (2005) Interleukin-21 enhances T-helper cell type I signaling and interferon-gamma production in Crohn's disease. Gastroenterology 128:687–694

Monteleone G, Caruso R, Fina D, Peluso I, Gioia V, Stolfi C, Fantini MC, Caprioli F, Tersigni R, Alessandroni L, MacDonald TT, Pallone F (2006) Control of matrix metalloproteinase production in human intestinal fibroblasts by interleukin 21. Gut 55:1774–1780

Noguchi M, Yi H, Rosenblatt HM, Filipovich AH, Adelstein S, Modi WS, McBride OW, Leonard WJ (1993) Interleukin-2 receptor gamma chain mutation results in X-linked severe combined immunodeficiency in humans. Cell 73:147–157

Nurieva R, Yang XO, Martinez G, Zhang Y, Panopoulos AD, Ma L, Schluns K, Tian Q, Watowich SS, Jetten AM, Dong C (2007) Essential autocrine regulation by IL-21 in the generation of inflammatory T cells. Nature (Lond) 448:480–483

Ozaki K, Kikly K, Michalovich D, Young PR, Leonard WJ (2000) Cloning of a type I cytokine receptor most related to the IL-2 receptor beta chain. Proc Natl Acad Sci USA 97: 11439–11444

Ozaki K, Spolski R, Feng CG, Qi CF, Cheng J, Sher A, Morse HC 3rd, Liu C, Schwartzberg PL, Leonard WJ (2002) A critical role for IL-21 in regulating immunoglobulin production. Science 298:1630–1634

Ozaki K, Spolski R, Ettinger R, Kim HP, Wang G, Qi CF, Hwu P, Shaffer DJ, Akilesh S, Roopenian DC, Morse HC 3rd, Lipsky PE, Leonard WJ (2004) Regulation of B cell differentiation and plasma cell generation by IL-21, a novel inducer of Blimp-1 and Bcl-6. J Immunol 173: 5361–5371

Parrish-Novak J, Dillon SR, Nelson A, Hammond A, Sprecher C, Gross JA, Johnston J, Madden K, Xu W, West J, Schrader S, Burkhead S, Heipel M, Brandt C, Kuijper JL, Kramer J, Conklin D, Presnell SR, Berry J, Shiota F, Bort S, Hambly K, Mudri S, Clegg C, Moore M, Grant FJ, Lofton-Day C, Gilbert T, Rayond F, Ching A, Yao L, Smith D, Webster P, Whitmore T, Maurer M, Kaushansky K, Holly RD, Foster D (2000) Interleukin 21 and its receptor are involved in NK cell expansion and regulation of lymphocyte function. Nature (Lond) 408:57–63

Pelletier N, McHeyzer-Williams LJ, Wong KA, Urich E, Fazilleau N, McHeyzer-Williams MG (2010) Plasma cells negatively regulate the follicular helper T cell program. Nat Immunol 11:1110–1118

Pesce J, Kaviratne M, Ramalingam TR, Thompson RW, Urban JF Jr, Cheever AW, Young DA, Collins M, Grusby MJ, Wynn TA (2006) The IL-21 receptor augments Th2 effector function and alternative macrophage activation. J Clin Invest 116:2044–2055

Petrella TM, Tozer R, Belanger K, Savage KJ, Wong R, Smylie M, Kamel-Reid S, Tron V, Chen BE, Hunder NN, Hagerman L, Walsh W, Eisenhauer EA (2012) Interleukin-21 has activity in patients with metastatic melanoma: a phase II study. J Clin Oncol 30:3396–3401

Pot C, Jin H, Awasthi A, Liu SM, Lai CY, Madan R, Sharpe AH, Karp CL, Miaw SC, Ho IC, Kuchroo VK (2009) Cutting edge: IL-27 induces the transcription factor c-Maf, cytokine IL-21, and the costimulatory receptor ICOS that coordinately act together to promote differentiation of IL-10-producing Tr1 cells. J Immunol 183:797–801

Saraiva M, O'Garra A (2010) The regulation of IL-10 production by immune cells. Nat Rev Immunol 10:170–181

Sarra M, Caruso R, Cupi ML, Monteleone I, Stolfi C, Campione E, Diluvio L, Mazzotta A, Botti E, Chimenti S, Costanzo A, MacDonald TT, Pallone F, Monteleone G (2011) IL-21 promotes skin recruitment of CD4(+) cells and drives IFN-gamma-dependent epidermal hyperplasia. J Immunol 186:5435–5442

Spolski R, Leonard WJ (2008) Interleukin-21: basic biology and implications for cancer and autoimmunity. Annu Rev Immunol 26:57–79

Spolski R, Kashyap M, Robinson C, Yu Z, Leonard WJ (2008) IL-21 signaling is critical for the development of type I diabetes in the NOD mouse. Proc Natl Acad Sci USA 105:14028–14033

Spolski R, Kim HP, Zhu W, Levy DE, Leonard WJ (2009) IL-21 mediates suppressive effects via its induction of IL-10. J Immunol 182:2859–2867

Spolski R, Wang L, Wan CK, Bonville CA, Domachowske JB, Kim HP, Yu Z, Leonard WJ (2012) IL-21 promotes the pathologic immune response to pneumovirus infection. J Immunol 188:1924–1932

Stolfi C, Rizzo A, Franze E, Rotondi A, Fantini MC, Sarra M, Caruso R, Monteleone I, Sileri P, Franceschilli L, Caprioli F, Ferrero S, MacDonald TT, Pallone F, Monteleone G (2011) Involvement of interleukin-21 in the regulation of colitis-associated colon cancer. J Exp Med 208:2279–2290

Strengell M, Sareneva T, Foster D, Julkunen I, Matikainen S (2002) IL-21 up-regulates the expression of genes associated with innate immunity and Th1 response. J Immunol 169:3600–3605

Sutherland AP, Van Belle T, Wurster AL, Suto A, Michaud M, Zhang D, Grusby MJ, von Herrath M (2009) Interleukin-21 is required for the development of type 1 diabetes in NOD mice. Diabetes 58:1144–1155

Suto A, Nakajima H, Hirose K, Suzuki K, Kagami S, Seto Y, Hoshimoto A, Saito Y, Foster DC, Iwamoto I (2002) Interleukin 21 prevents antigen-induced IgE production by inhibiting germ line C(epsilon) transcription of IL-4-stimulated B cells. Blood 100:4565–4573

Suto A, Wurster AL, Reiner SL, Grusby MJ (2006) IL-21 inhibits IFN-gamma production in developing Th1 cells through the repression of Eomesodermin expression. J Immunol 177:3721–3727

Takaki R, Hayakawa Y, Nelson A, Sivakumar PV, Hughes S, Smyth MJ, Lanier LL (2005) IL-21 enhances tumor rejection through a NKG2D-dependent mechanism. J Immunol 175:2167–2173

Tamagawa-Mineoka R, Kishida T, Mazda O, Katoh N (2011) IL-21 reduces immediate hypersensitivity reactions in mouse skin by suppressing mast cell activation or IgE production. J Invest Dermatol 131:1513–1520

Terrier B, Costedoat-Chalumeau N, Garrido M, Geri G, Rosenzwajg M, Musset L, Klatzmann D, Saadoun D, Cacoub P (2012) Interleukin 21 correlates with T cell and B cell subset alterations in systemic lupus erythematosus. J Rheumatol 39:1819–1828

Toomey JA, Gays F, Foster D, Brooks CG (2003) Cytokine requirements for the growth and development of mouse NK cells in vitro. J Leukoc Biol 74:233–242

Tortola L, Yadava K, Bachmann MF, Muller C, Kisielow J, Kopf M (2010) IL-21 induces death of marginal zone B cells during chronic inflammation. Blood 116:5200–5207

Tsuruoka N, Arima M, Arguni E, Saito T, Kitayama D, Sakamoto A, Hatano M, Tokuhisa T (2007) Bcl6 is required for the IL-4-mediated rescue of the B cells from apoptosis induced by IL-21. Immunol Lett 110:145–151

Van Belle TL, Nierkens S, Arens R, von Herrath MG (2012) Interleukin-21 receptor-mediated signals control autoreactive T cell infiltration in pancreatic islets. Immunity 36:1060–1072

van Heel DA, Franke L, Hunt KA, Gwilliam R, Zhernakova A, Inouye M, Wapenaar MC, Barnardo MC, Bethel G, Holmes GK, Feighery C, Jewell D, Kelleher D, Kumar P, Travis S, Walters JR, Sanders DS, Howdle P, Swift J, Playford RJ, McLaren WM, Mearin ML, Mulder CJ, McManus R, McGinnis R, Cardon LR, Deloukas P, Wijmenga C (2007) A genome-wide association study for celiac disease identifies risk variants in the region harboring IL2 and IL21. Nat Genet 39:827–829

Vollmer TL, Liu R, Price M, Rhodes S, La Cava A, Shi FD (2005) Differential effects of IL-21 during initiation and progression of autoimmunity against neuroantigen. J Immunol 174: 2696–2701

Wan CK, Oh J, Li P, West EE, Wong EA, Andraski AB, Spolski R, Yu ZX, He J, Kelsall BL, Leonard WJ (2013) The cytokines IL-21 and GM-CSF have opposing regulatory roles in the apoptosis of conventional dendritic cells. Immunity 38:514–527

Wang G, Tschoi M, Spolski R, Lou Y, Ozaki K, Feng C, Kim G, Leonard WJ, Hwu P (2003) In vivo antitumor activity of interleukin 21 mediated by natural killer cells. Cancer Res 63:9016–9022

Wang L, Yu CR, Kim HP, Liao W, Telford WG, Egwuagu CE, Leonard WJ (2011) Key role for IL-21 in experimental autoimmune uveitis. Proc Natl Acad Sci USA 108:9542–9547

Wood N, Bourque K, Donaldson DD, Collins M, Vercelli D, Goldman SJ, Kasaian MT (2004) IL-21 effects on human IgE production in response to IL-4 or IL-13. Cell Immunol 231: 133–145

Wurster AL, Rodgers VL, Satoskar AR, Whitters MJ, Young DA, Collins M, Grusby MJ (2002) Interleukin 21 is a T helper (Th) cell 2 cytokine that specifically inhibits the differentiation of naive Th cells into interferon gamma-producing Th1 cells. J Exp Med 196:969–977

Yanaba K, Bouaziz JD, Matsushita T, Magro CM, St. Clair EW, Tedder TF (2008) B-lymphocyte contributions to human autoimmune disease. Immunol Rev 223:284–299

Yi JS, Du M, Zajac AJ (2009) A vital role for interleukin-21 in the control of a chronic viral infection. Science 324:1572–1576

Yi JS, Ingram JT, Zajac AJ (2010) IL-21 deficiency influences CD8 T cell quality and recall responses following an acute viral infection. J Immunol 185:4835–4845

Yoshizaki A, Miyagaki T, DiLillo DJ, Matsushita T, Horikawa M, Kountikov EI, Spolski R, Poe JC, Leonard WJ, Tedder TF (2012) Regulatory B cells control T-cell autoimmunity through IL-21-dependent cognate interactions. Nature (Lond) 491:264–268

Zeng R, Spolski R, Finkelstein SE, Oh S, Kovanen PE, Hinrichs CS, Pise-Masison CA, Radonovich MF, Brady JN, Restifo NP, Berzofsky JA, Leonard WJ (2005) Synergy of IL-21 and IL-15 in regulating CD8+ T cell expansion and function. J Exp Med 201:139–148

Zeng R, Spolski R, Casas E, Zhu W, Levy DE, Leonard WJ (2007) The molecular basis of IL-21-mediated proliferation. Blood 109:4135–4142

Zhou L, Ivanov II, Spolski R, Min R, Shenderov K, Egawa T, Levy DE, Leonard WJ, Littman DR (2007) IL-6 programs T(H)-17 cell differentiation by promoting sequential engagement of the IL-21 and IL-23 pathways. Nat Immunol 8:967–974

Zotos D, Coquet JM, Zhang Y, Light A, D'Costa K, Kallies A, Corcoran LM, Godfrey DI, Toellner KM, Smyth MJ, Nutt SL, Tarlinton DM (2010) IL-21 regulates germinal center B cell differentiation and proliferation through a B cell-intrinsic mechanism. J Exp Med 207:365–378

Chapter 6
Interleukin-22: A Bridge Between Epithelial Innate Host Defense and Immune Cells

Xiaoting Wang and Wenjun Ouyang

Abstract Interleukin-22 (IL-22), an IL-10 family cytokine, is produced by various leukocytes. The receptor of IL-22, however, is preferentially detected on peripheral tissue epithelial cells. IL-22 functions as a unique messenger from immune system to tissue epithelial cells and to regulate homeostasis of epithelia. IL-22 is able to directly enhance antimicrobial defense mechanisms in epithelial cells and to facilitate epithelial barrier repair and wound healing process. It, therefore, possesses an irreplaceable role in host defense against certain pathogens that specifically invade epithelial cells. In addition, IL-22 can help to preserve the integrity and homeostasis of various epithelial organs during infection or inflammation. The importance of its tissue-protective function is manifested in many inflammatory situations such as inflammatory bowel diseases (IBD) and hepatitis. On the other hand, as a cytokine, IL-22 is capable of induction of proinflammatory responses, especially in synergy with other cytokines. Consequently, IL-22 contributes to pathogenesis of certain inflammatory diseases for example psoriasis.

Keywords Autoimmune diseases • Epithelial homeostasis • IBD • IL-22 • Psoriasis

6.1 IL-22 and IL-22 Receptors

IL-22 was originally cloned from IL-9-treated mouse T cells in 2000 and named IL-TIF (IL-10 related T-cell-derived inducible factor) (Dumoutier et al. 2000a). The human orthologue was subsequently identified with 79 % identity to mouse IL-22 and 25 % identity to human IL-10 (Dumoutier et al. 2000c; Xie et al. 2000). Because of the weak but significant homology with IL-10, IL-22 was classified as a member

X. Wang • W. Ouyang (✉)
Department of Immunology, Genentech, Inc, 1 DNA Way, Mail Stop 34, South San Francisco, CA 94080, USA
e-mail: wang.xiaoting@gene.com; ouyang.wenjun@gene.com

T. Yoshimoto and T. Yoshimoto (eds.), *Cytokine Frontiers: Regulation of Immune Responses in Health and Disease*, DOI 10.1007/978-4-431-54442-5_6, © Springer Japan 2014

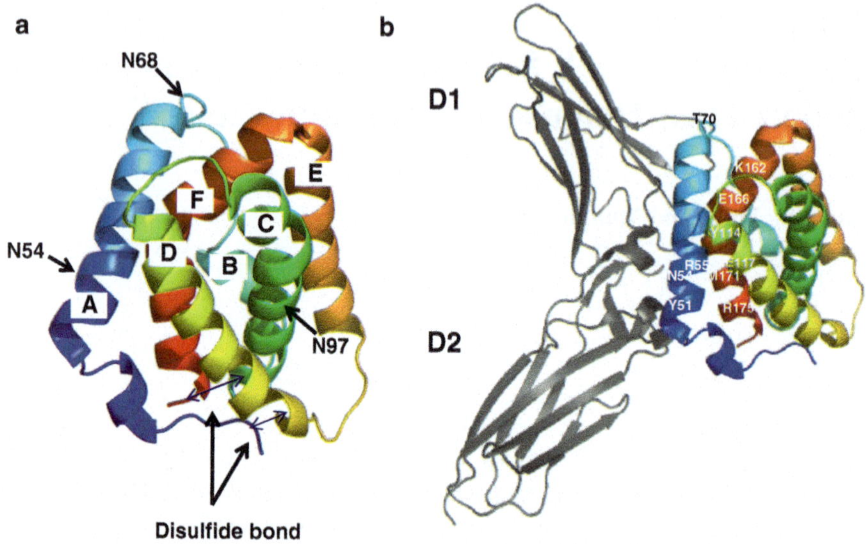

Fig. 6.1 Structure of human interleukin (IL)-22 and IL-22R1. (**a**) Crystalline structure of IL-22 from IL-22–IL-22R1 complex (Protein Data Bank ID code 3DGC). Six helices (*A–F*) are colored from the N-terminus (*blue*) to C-terminus (*red*). Three glycosylation sites (N54, N68, N97) are labeled with *arrowed lines*. Two disulfide bonds (C40–C132, C89–C178) are shown with *blue double-arrow lines*. (**b**) Complex of IL-22–IL-22R1. IL-22R1 (*dark grey*) contains D1 and D2 domains. Critical amino acids for IL-22R1 and IL-10R2 binding on IL-22 are labeled

in the IL-10 cytokine family, along with IL-10, IL-19, IL-20, IL-24, IL-26, IL-28A, IL-28B, and IL-29 (Ouyang et al. 2011). Cytokines from the IL-10 family share 20–30 % amino acid identity and also a weaker homology with cytokines in the interferon (IFN) family. Together, IL-10 and IFN cytokine families form the class II cytokine superfamily.

6.1.1 Gene and Structure of IL-22

The human IL-22 gene consists of five introns and six exons and is located on chromosome 12q15, adjacent to genes of IFN-γ and IL-26. The isolated cDNA encodes a 179-amino-acid-long protein with a 33-amino-acid signal peptide (Dumoutier et al. 2000b). Sharing structural similarity with other IL-10 family members, IL-22 contains six α-helices (A–F), forming a bundle structure with a compact hydrophobic core inside (Fig. 6.1a). In contrast to IL-10, which forms an intertwined dimer for receptor binding, IL-22 is a monomer (Nagem et al. 2002). IL-22 is heavily glycosylated with three N-linked glycosylation sites (Asn-X-Ser/Thr) at asparagines (Asn) -54, -68 and -97) (Fig. 6.1a) (Xu et al. 2004). However, the biological function of IL-22 is not fully dependent on glycosylation. The *Escherichia coli-derived* protein has normal biological activity without glycosylation. In fact, comparison of the IL-22 structures purified from *Drosophila* and *E. coli* reveals that the glycosylation only causes minor structural changes (Xu et al. 2005). There are four cysteines within IL-22 to create two disulfide linkages (Cys40–Cys130, Cys89–Cys178). The first disulfide bond holds together the N-terminal coil and the DE loop, and the second disulfide bond stabilizes the N-terminus of helix C and the C-terminus of helix F (Fig. 6.1a) (Nagem et al. 2002).

IL-22 interacts with the heterodimeric membrane-spanning receptor complex, IL-22R1 (also known as IL-22Ra1, CRF2-9, and zcytor11), and IL-10R2. Both subunits belong to the class II cytokine receptor or interferon receptor family with signature fibronectin type III domains (Ouyang et al. 2011). The gene of IL-22R1 is located on chromosome 1 and encodes a 574-amino-acid protein that contains an extracellular ligand-binding domain, a membrane-spanning helix, and an intracellular domain (Xie et al. 2000). *Il-10r2* is located on chromosome 21 and shares a similar organization with a much shorter cytoplasmic tail (Kotenko et al. 1997). IL-10R2 is a shared common receptor chain for several other IL-10 family cytokines including IL-10, IL-26, IL-28A, IL-28B, and IL-29 (Ouyang et al. 2011). The finding of receptor sharing within the same family is not uncommon, as the common γ-chain is shared by multiple cytokines. Similarly, IL-22R1 can also pair with IL-20R2 to engage the binding of IL-20 and IL-24. Although IL-10R2 is ubiquitously expressed, IL-22R1 is primarily detected on cells with epithelial origin.

6.1.2 Receptor Binding

As a monomer, IL-22 can form a 1:1 complex with the soluble IL-22R1 extracellular domain in solution. The two tandem fibronectin domains (D1 and D2) of IL-22R1 rotate to an angle to form an L-shape, where the interface comes in contact with IL-22 residues on helix A, the AB loop, and helix F (Fig. 6.1b). Several residues that are critical for IL-22R1 binding are located in helix F (Lys-162, Glu-166, Met-172, Arg-175) and the AB loop (Thr-70), and create either hydrogen bonds or van der Waals interactions between IL-22 and IL-22R1 (Jones et al. 2008). The specific residues in contact surfaces ensure the specificity between IL-22-IL-22R1 interactions and are distinct from other cytokine–receptor complexes in the family. IL-10R2 comes in contact with IL-22/IL-22R1 complex at three distinct surface sites. Helices A and B in IL-22 are predominantly engaged in binding to two of three sites on IL-10R2 whereas the third site interacts with IL-22R1. In contrast to IL-22R1-binding sites where most of the residues locate toward the C-terminus of IL-22, IL-10R2-binding spots on IL-22 are mainly located on the N-terminal end of the helix, including residues Tyr-51, Asn-54, Arg-55, Tyr-114, and Glu-117 (Fig. 6.1b) (Logsdon et al. 2004). The formation of the IL-22–IL-22R1–IL-10R2 complex is sequential. Surface plasmon resonance studies revealed high-affinity binding of IL-22 to IL-22R1 (20 nM) but very weak affinity (120 μM) to IL-10R2 (Logsdon et al. 2004). In fact, IL-10R2 displays a diminished affinity to IL-22 with a substantially increased affinity to the IL-22/IL–22R1 complex (Logsdon et al. 2002).

6.1.3 IL-22-Binding Protein

A protein that shares 34 % sequence identity with the extracellular region of IL-22R1 was discovered by screening genomic DNA databases and named IL-22-binding protein (IL-22BP, also known as IL-22Ra2, CRF2-10, CRF2-s1, and ZcytoR16). IL-22BP lacks transmembrane and intracellular domains and functions as a naturally occurring IL-22 antagonist (Gruenberg et al. 2001; Kotenko et al. 2001; Logsdon et al. 2002; Weiss et al. 2004; Xu et al. 2001). Soluble receptors or membrane-bound decoy receptors that serve as negative regulators for cytokines have previously been found for other cytokines, such as IL-1 and IL-18, but IL-22BP is the only naturally expressed soluble receptor discovered in the class II cytokine receptor family. IL-22BP is encoded by a distinct gene on chromosome 6, in a proximal distance of the class II cytokine receptor cluster containing Il-20r1 and ifngr1. Although IL-22BP shares weaker homology with other receptors of the IL-10 family cytokine such as IL-10R1 (29 %), it only specifically blocks IL-22 signaling. IL-22BP is expressed in a wide range of tissues including lymphoid and nonlymphoid tissues with three mRNA variants produced by alternative splicing, and the variant-2 with 231 amino acids neutralizes IL-22 activity. IL-22BP structurally resembles IL-22R1 with two disulfide bonds stabilizing D1 and D2 domains, and the interaction interface to IL-22 is located within IL-22 helix A, the AB loop, and helix F. In contrast to IL-22R1, IL-22BP forms five completely different salt bridges or hydrogen bonds, and a hydrophobic cluster with Phe-57, Phe-171, and Met-172 on IL-22. The binding affinity of IL-22BP to IL-22 (~1 pM) is significantly higher than IL-22–IL-22R1 (~nM), illuminating the tight regulation of IL-22BP on IL-22 activity (Jones et al. 2008).

6.1.4 Signal Transduction

Upon binding to the receptor complex, IL-22 activates Janus Kinase (Jak) and Signal Transducer and Activator of Transcription (Stat) signaling pathways, particularly, phosphorylation of tyrosine kinases Jak1 and Tyk2 (Fig. 6.2). Subsequently, transcription factor Stat3 is activated, and in some cells phosphorylation of Stat1 and Stat5 may also occur (Dumoutier et al. 2000a, 2003; Lejeune et al. 2002; Xie et al. 2000). A serine residue of Stat3 (Ser 727) can be phosphorylated by IL-22 stimulation, and this serine phosphorylation is required to achieve maximal transactivation (Lejeune et al. 2002). In the intracellular tail of the IL-22R1 chain, there are in total eight tyrosine residues that can be phosphorylated by activated Jak1 or Tyk2 and serve as potential docking sites for Stat molecules upon IL-22 stimulation. Interestingly, although these Tyr residuals are necessary for the binding of Stat5 and Stat1 to IL-22R1, they are dispensable for Stat3 activation (Dumoutier et al. 2009). Instead, the coiled-coil domain of Stat3 can bind directly to the C-terminal tail of the IL-22R1 chain (Fig. 6.2). The biological significant of this finding is unclear at

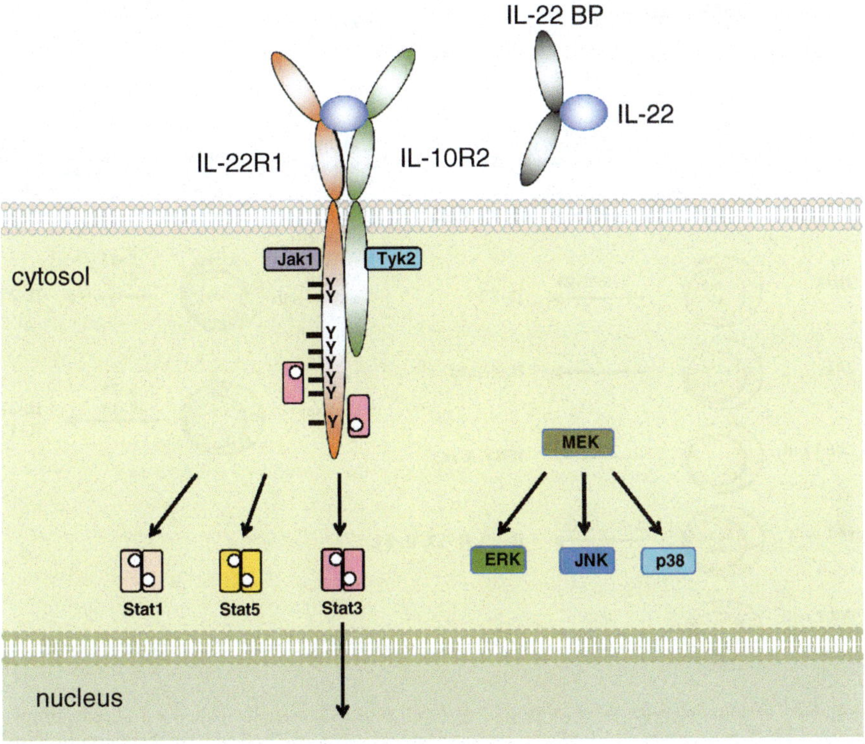

Fig. 6.2 IL-22 binds to receptors and activates downstream signaling transduction. The engagement of IL-22 to the IL-22R1–IL-10R2 complex activates Jak/Stat pathways and primarily signals through Stat3. IL-22R can associate with Stat3 in a tyrosine-independent manner. Stat1, Stat5, and MAP kinase pathways can be activated under certain conditions. *BP*, binding protein

present. In vivo, Stat3 is directly activated by IL-22 in epithelia and is necessary in mediating the downstream biological effects of IL-22 (Pickert et al. 2009; Sugimoto et al. 2008; Zheng et al. 2007). In addition to Stat molecules, IL-22 is able to activate key kinases MEK1/2, ERK1/2, JNK, and p38 of MAP kinase pathways, as well as Akt, SOCS-3, and transcription factors NF-κB and AP-1 in different cell types (Andoh et al. 2005; Brand et al. 2006).

6.2 Cellular Sources and Regulation of IL-22

IL-22 is broadly expressed by many types of leukocytes, including $CD4^+$ T cells, $CD8^+$ T cells, and γδ T cells, as well as various innate lymphocytes (ILCs) (Fig. 6.3) (Ouyang et al. 2011; Rutz and Ouyang 2011). Given the essential roles of IL-22 in tissue protection under various homeostatic, infectious, and inflammatory states as discussed herein, these different cellular sources are not redundant for the biological

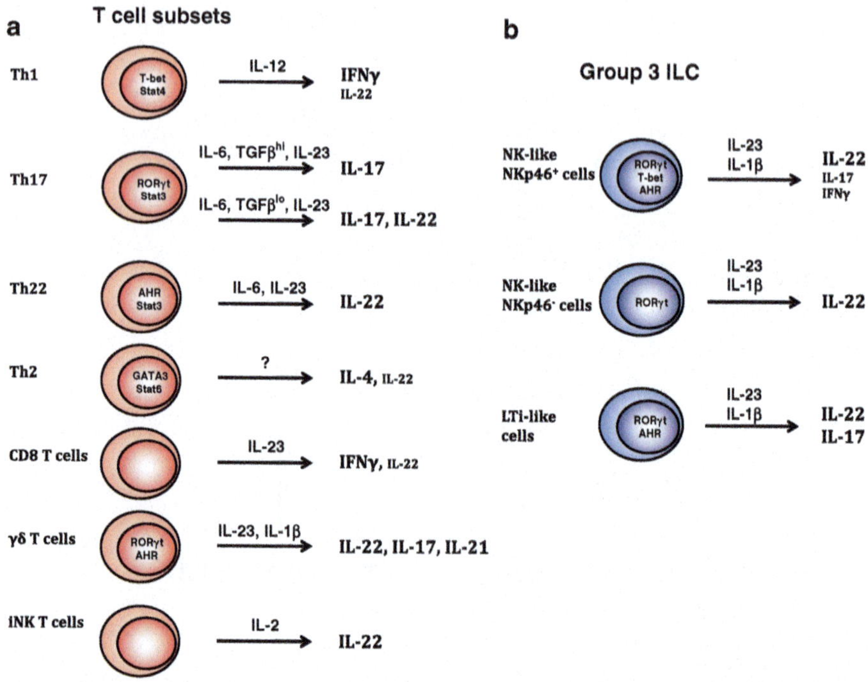

Fig. 6.3 Various cell types produce IL-22. (**a**) T-cell subsets secrete IL-22 under different stimulations. CD4 T cells are major producers and are categorized into Th17 and Th22 subgroups based on their cytokine profiles. (**b**) Innate lymphocytes (ILCs) are innate sources of IL-22. LTi-like and NK-like ILCs, although defined with different surface markers, are able to produce IL-22 under similar conditions

functions of IL-22 in vivo. Especially, the activation signals required for IL-22 production in these cells and homing properties of these cells into different organs are all different.

6.2.1 CD4 T Helper Cells

IL-22 was originally identified as a cytokine produced from activated murine CD4$^+$ T cells and human T cells (Dumoutier et al. 2000a, c; Wolk et al. 2002). In various CD4$^+$ T helper cell subsets, Th1 cells produce a higher level of IL-22 whereas Th2 cells express a very low, but detectable, level of IL-22 (Gurney 2004; Wolk et al. 2002). A new T helper subset, Th17 cells, was described in 2005 (Harrington et al. 2005; Park et al. 2005). Th17 cells, which primarily express the signature cytokine IL-17, are critical mediators of pathogenesis of inflammatory disorders and antibacterial host defense (Ouyang et al. 2008). Shortly after the discovery of Th17 cells, IL-22 was identified as another functionally important cytokine that helped to define this T helper lineage (Chung et al. 2006; Liang et al. 2006; Zheng et al. 2007). IL-22 production from Th17 cells is significantly higher than that from Th1 cells. Transforming growth factor (TGF)-β and IL-6 are necessary and sufficient to induce Th17 cell differentiation from naïve CD4⁺ T cells (Bettelli et al. 2006; Mangan et al. 2006; Veldhoen et al. 2006). Surprisingly, we noticed three different 'flavors' of Th17 cells depending on the concentration of TGF-β present in the culture (Fig. 6.3a) (Zheng et al. 2007). With a high dose of TGF-β, T cells produce only IL-17 but not much IL-22. These IL-17-only-producing T cells have been identified in human ulcerative colitis (UC) patients, and their presence is correlated with active inflammation in the intestine (Broadhurst et al. 2010). A small amount of TGF-β in combination with IL-6 can promote Th17 cells producing both IL-17 and IL-22. IL-6 alone without addition of TGF-β is not able to induce IL-17 production from T cells. Under this condition, however, T cells produce the highest level of IL-22, suggesting TGF-β is differentially required for IL-17 and IL-22 regulation. The IL-22-only-producing $CD4^+$ T helper cells were discovered in human peripheral blood mononuclear cells (PBMC) and are named Th22 cells (Duhen et al. 2009; Eyerich et al. 2009; Trifari et al. 2009). The Th22 cells not only express CCR6, a cell-surface marker for Th17 cells, but also the skin-homing receptors CCR4 and CCR10. Consistently, Th22 cells were found in epidermis of human inflammatory skin disorders, and expressed genes involved in tissue remodeling, suggesting an important role in skin homeostasis. Additional to αβ $CD4^+$ T cells, both $CD8^+$ T cells and γδ T cells produce IL-22 under the stimulation of IL-23 (Fig. 6.3a) (Billerbeck et al. 2010; Ciric et al. 2009; Martin et al. 2009; Sutton et al. 2009; Zheng et al. 2007). Mouse unconventional γδ T cells were found to coexpress IL-22 with IL-17 and IL-21 in response to IL-23 and IL-1β without T-cell-receptor engagement (Billerbeck et al. 2010; Ciric et al. 2009; Martin et al. 2009; Sutton et al. 2009). These cells express CCR6, orphan nuclear receptor RORγt, aryl hydrocarbon receptor (AHR), IL-23R, and receptors for pathogen products, and mediate antibacterial host defense and pathogenesis of experimental autoimmune encephalomyelitis (EAE), a murine model for multiple sclerosis (MS). Invariant natural killer (NK) T cells, another unconventional T-cell type that express both NK and T-cell markers, are capable of releasing IL-22 upon T-cell receptor (TCR) stimulation and protecting liver from inflammatory damage in a murine acute hepatitis model (Goto et al. 2009; Wahl et al. 2009).

6.2.2 Innate Lymphocytes

NK cells in human PBMC were reported to express a low level of IL-22, suggesting that IL-22 may also be produced by innate leukocytes (Wolk et al. 2002). Indeed, the induction of IL-22 in the colon with *Citrobactor rodentium* infection remained intact in Rag2-deficient mice, supporting a non-T-cell source of IL-22 (Zheng et al. 2008). A NK-like ILC subset was subsequently discovered by several groups as one of the major innate sources of IL-22 (Fig. 6.3b) (Cella et al. 2009;

Satoh-Takayama et al. 2008). The NK-like cells express NK-specific marker NKp46 and an intermediate level of NK1.1, but have low or no expression of stimulatory or inhibitory NK receptors and possess no effector function of NK cells. Moreover, these cells express CD117, CD127, and a high level of RORγt. Developmentally, $NKp46^{+}CD3^{-}$ cells are dependent on RORγt, but not IL-15, which is required for NK cells. They produce abundant IL-22 but little IL-17 in response to commensal microbiota. A similar IL-22-producing counterpart was identified from human mucosa-associated lymphoid tissues such as the tonsil and Peyer's patches (Cella et al. 2009). Instead of NKp46, the human cells express NKp44, as well as CD56 and CD127 (Cupedo et al. 2009). The $NKp44^{+}$ cells were named NK-22 because of their ability to produce IL-22 and respond to IL-23. These NK-22 cells are important sources of IL-22 and regulate the epithelial host defense mechanism and maintenance of dynamic equilibrium between microbial flora and immune surveillance. It is noteworthy that IL-22-producing NK-like ILCs are not only associated with mucosal tissues and restricted to NKp46 expression but are also identified in other organs such as skin dermis, suggesting they may broadly participate in host defense through IL-22 production (Dhiman et al. 2009; Hughes et al. 2009; Satoh-Takayama et al. 2009).

In parallel, a previously defined ILC subset, lymphoid tissue inducer (LTi) cells, was noticed to be able to produce IL-22 upon IL-23 stimulation (Fig. 6.3b) (Takatori et al. 2009). Splenocytes from $Rag2^{-/-}$ mice that are lacking both T and NK T cells still produce a large amount of IL-22 in reaction to IL-23, whereas in $Rag2^{-/-}$ common $\gamma^{-/-}$ splenocytes, IL-22 production is completely abrogated. A group of LTi-like cells, which express CCR6, RORγt, AHR, and IL-23R, are identified as the sources of IL-22 (Takatori et al. 2009; Veldhoen et al. 2008). LTi is essential for the formation of secondary lymphoid organs during embryogenesis (Mebius 2003). Both inhibitor of differentiation 2 (Id2) and RORγt are required for LTi development (Sun et al. 2000; Yokota et al. 1999). The LTi cells phenotypically lack surface markers for T cells, B cells, and myeloid cells, but express lymphotoxin-α (LT-α), LT-β, several chemokine receptors, and cytokine receptors including CD127 and CD117. Additionally, LTi cells express CD4 in the mouse (Spits and Di Santo 2011). Because LTi-like cells are able to secrete IL-22 upon activation and preferentially reside in mucosal-associated lymphoid tissues, they may play essential functions in maintaining mucosal homeostasis (Cupedo et al. 2009; Takatori et al. 2009).

A study suggested that LTi-like cells but not the NK-like cells were the major sources of IL-22 during *C. rodentium* infection in the colon (Sonnenberg et al. 2011). However, given many common features shared by LTi-like cells and NK-like ILCs, these cells may belong to the same ILC group with slight differences in the expression of certain cell-surface markers. The connection between LTi-like and NK-like cells has been built not only on their functions but also in lineage development (Sawa et al. 2010). The LTi-like cells could upregulate NK cell-surface marker CD56 and acquire a low level of cytolytic activity when culturing under conditions favoring NK cell differentiation (Crellin et al. 2010; Cupedo et al. 2009). Despite a suggestion that NK-like $IL22^{+}$ cells may come from IL-1β-stimulated immature NK cells, both LTi-like and NK-like IL-22-producing cells remain their RORc expression throughout the culture and represent a stable RORc^{+} lineage that is functionally and developmentally distinct from conventional NK cells (Crellin et al. 2010; Hughes et al. 2009, 2010; Satoh-Takayama et al. 2010). Although initially IL-22-producing ILCs have been given different names by different groups, such as NK-22, NCR22, and ILC-22, the nomenclature has been made uniform and all IL-22-producing ILCs are grouped into group 3 ILCs (Fig. 6.3b) (Spits et al. 2013).

6.2.3 Regulation

Similar to other cytokines, the production of IL-22 from leukocytes is tightly controlled (Ouyang et al. 2011; Rutz and Ouyang 2011). The regulation of IL-22 is best studied in CD4^{+} T helper cells. Because L-22 is a signature cytokine of Th17 cells, many pathways and factors that are essential for Th17 cell development and IL-17 production also control the expression of IL-22 (Korn et al. 2009; Ouyang et al. 2008). IL-6, IL-21, and IL-23 can all activate Stat3 in T cells. All three cytokines play critical roles in regulation of Th17 development and IL-17 and IL-22 production. The downstream transcription factor Stat3 thus also has an indispensable function in induction of both IL-17 and IL-22 from T cells (Yang et al. 2007). Other pro-inflammatory cytokines including IL-1β and tumor necrosis factor (TNF)-α can augment IL-17 and IL-22 expression. As discussed earlier, IL-23 not only induces IL-22 production from T-cell subsets but also is indispensable for IL-22 expression from ILCs, especially in vivo (Zheng et al. 2007, 2008). The orphan nuclear receptors RORc in human, RORγt, and RORα in mouse are considered as master transcription factors for Th17 lineage formation, and these factors also participate in regulation of IL-22 production in Th17 cells (Ivanov et al. 2006; Volpe et al. 2009; Yang et al. 2008). The AHR, a ligand-dependent transcription factor sensing xenobiotic metabolites, promotes Th17 cell differentiation and drives IL-22 expression (Quintana et al. 2008; Veldhoen et al. 2008). AHR ligand stimulation of human PBMC strongly upregulates IL-22 production in Th17 cells, and Notch signaling enhances IL-22 secretion indirectly through activating the AHR (Alam et al. 2010; Brembilla et al. 2011; Ramirez et al. 2010).

Several pathways, including TGF-β, inducible costimulator (ICOS), and IL-27, negatively regulate IL-22 production in T cells (Liu and Rohowsky-Kochan 2011; Paulos et al. 2010; Rutz et al. 2011; Volpe et al. 2009; Zheng et al. 2007). We recently identified c-Maf as a potential downstream transcription repressor that mediated the suppression of IL-22 in T cells by TGF-β. Because TGF-β is required for IL-17 induction in T cells, it controls the balance of IL-17 and IL-22 in T cells. IL-22 shares similar biological functions with IL-17 for induction of antimicrobial responses, chemokines, and other pro-inflammatory genes from epithelial cells. During pathogen invasion, coexpression of IL-22 and IL-17 may enhance the host defense mechanism through the synergistic effects of both cytokines (Aujla et al. 2008). Consistently, in inflammatory diseases such as psoriasis and rheumatoid arthritis, Th17 cells coexpress IL-17 and IL-22 in inflamed tissues and may contribute to the pathogenesis (Pene et al. 2008). On the other hand, IL-22 has strong tissue-protective functions (Zenewicz et al. 2008; Zheng et al. 2007). As discussed herein, IL-22-only-producing Th22 cells may have distinct biological roles from Th17 cells in both infectious and autoimmune diseases.

6.3 Biological Functions of IL-22

Carrying the name of "interleukin," IL-22 was first searched within the lymphopoietic system for its biological function. However, from freshly isolated human monocytes, T cells, NK cells, and B cells, there was no detectable IL-22R1 expression even with activation (Wolk et al. 2004). Consistently, IL-22 induced neither any cytokine production from PBMC nor upregulation of activation markers. IL-22 is secreted by various leukocyte subsets, whereas the expression of IL-22R1 is restricted to epithelial cells and the cells originated from epithelium (Aggarwal et al. 2001; Dumoutier et al. 2000c). Therefore, IL-22 is a key cytokine to mediate the crosstalk between immune systems and peripheral tissues. In contrast to IL-10, which performs primarily regulatory functions during inflammation, IL-22 specifically upregulates the production of antimicrobial peptides, pro-inflammatory chemokines, and cytokines, and directly stimulates epithelial cell proliferation and the self-repair process (Ouyang et al. 2011). IL-22 can elicit three major biological functions from various epithelial tissues: (1) antimicrobial host defense mechanisms, (2) tissue-protective effects, and (3) pro-inflammatory functions.

6.3.1 Epithelial Host Defense

6.3.1.1 IL-22 Induces the Production of Antimicrobial Peptides from Epithelial Cells

To enhance innate defense, IL-22 drives the production of a broad spectrum of antimicrobial proteins from epithelial cells in tissues that are directly exposed to external pathogens. IL-22 acts on skin keratinocytes and induces S100 family proteins containing S100A7 (psoriasis), S100A8, and S100A9, and defensins including β-defensin 2 and β-defensin 3 (Liang et al. 2006; Sa et al. 2007; Wolk et al. 2004, 2006). In the lung, bronchial epithelial cells preferentially produce β-defensin 2, S100A7, S100A12 (calgranulin), and lipocalin-2 upon stimulation of IL-22 (Aujla et al. 2008). IL-22 can also stimulate colonic mucosal epithelium to express S100A8, S100A9, RegIIIβ, RegIIIγ, haptoglobin, SAA3, and lactotransferrin (Zheng et al. 2008). Furthermore, IL-22 was reported to induce hepatic lipopolysaccharide (LPS)-binding protein (LBP) to the serum concentration known to neutralize LPS, a major outer membrane component of gram-negative bacteria (Wolk et al. 2007). All these antimicrobial peptides exert essential roles in host defense (Kolls et al. 2008).

6.3.1.2 Role of IL-22 in Epithelial Defense Against Bacterial Infection

It has been speculated since its discovery that IL-22 might have an important role in host defense (Aggarwal et al. 2001; Dumoutier et al. 2000c; Wolk et al. 2004). Early effort with in vivo infection of an intracellular pathogen, *Listeria monocytogenes*, indicates that IL-22 is not essential for both the innate and adaptive immunity against the infection (Zenewicz et al. 2007). Because IL-22R1 is preferentially expressed on various epithelial cells and IL-22 induces the expression of many antimicrobial peptides (Sa et al. 2007), we hypothesized that the major function of IL-22 may be in control of extracellular bacteria, especially on the mucosal surface. Previously we have demonstrated that IL-22 is a downstream cytokine of IL-23 that can induce the expression of IL-22 from various lymphocytes (Zheng et al. 2007). A study published at the same time from Weaver's group established an indispensable role of IL-23 in host defense against *C. rodentium*, a gram-negative bacterium that infects murine colon epithelial cells and induces transient infectious colitis (Mangan et al. 2006). We speculated that IL-22 could also contribute significantly in this model of mucosal infection. Indeed, IL-22 expression is augmented very quickly in the colon in an IL-23-dependent manner upon *C. rodentium* infection (Zheng et al. 2008). IL-22 is both necessary and sufficient for controlling systemic bacteria dissemination and mortality during the early stage of the infection. Mice deficient in IL-22 succumb within the second week after infection. IL-22 can directly induce many antimicrobial peptides including the Reg family and S100 family of peptides from intestinal epithelial cells (IECs). The induction of Reg family peptides such as RegIIIβ and RegIIIγ by *C. rodentium* is abolished in the IL-22-deficient mice. Recombinant RegIIIγ can partially rescue the IL-22 deficient mice, suggesting RegIIIγ is one of the downstream host defense molecules induced by IL-22. Reg family peptides can directly kill *gram*$^{+}$ bacteria and induce aggregation of *gram*$^{-}$ bacteria (Cash et al. 2006; Iovanna et al. 1991).

More interestingly, in this model, ILCs, but not T cells, are the major early IL-22 producing cells in the intestine (Cella et al. 2009; Satoh-Takayama et al. 2008; Zheng et al. 2008). T cells, however, are still required for full control of *C. rodentium* infection, because the development of the anti-*C. rodentium* IgG response that eventually clears the bacteria from colon and CD4 T cells is required for this response (Bry and Brenner 2004; Maaser et al. 2004). Furthermore, a second wave of IL-22 production is also detected after *C. rodentium* inoculation in mice (Basu et al. 2012), which provides more complete host protection in the early stage of infection. This second wave of IL-22 is produced by Th22 cells in an IL-6-dependent, but IL-23-independent, manner.

In addition to controlling invading pathogens in the intestine, IL-22 can help to maintain the homeostasis of the mucosal barrier and commensal bacteria. A recent study shows that depletion of IL-22-producing ILCs leads to peripheral dissemination of commensal bacteria, especially *Alcaligenes* species, and systemic inflammation (Sonnenberg et al. 2012). Exogenous IL-22 is able to prevent the systemic dissemination of commensal bacteria and control the inflammation. Furthermore, IL-22 controls segmented filamentous bacteria (SFB) (Ivanov et al. 2009;

Upadhyay et al. 2012), which initiate and augment Th17 responses in the mouse small intestine. IL-22 not only reduces the amount of SFB detected in the small intestine, also regulates host defense against bacterial infections in other organs, for example, *Klebsiella pneumoniae* infection in the lung (Aujla et al. 2008). In this model, IL-22 cooperates with IL-17 to induce the expression of proinflammatory cytokines and chemokines, as well as lipocalin-2, that kills *K. pneumoniae* in the lung.

In summary, IL-22 is one of the essential cytokines that regulate epithelial defense function and protect epithelial integrity. The functions of IL-22 in host defense can be categorized in several areas. First, IL-22 directly induces innate antimicrobial mechanisms via promoting the expression of bactericidal peptides. Second, IL-22 preserves the epithelial integrity and promotes the survival of epithelial cells during infection and inflammation. Third, IL-22 also augments wound-healing responses and acts on stem cells to promote recovery of the epithelial organs from damage caused by infection. Finally, IL-22 can synergize with other pro-inflammatory cytokines to induce the expression of chemokines and cytokines that recruit and activate other leukocytes.

6.3.1.3 Potential Functions of IL-22 in Viral Infection

The type III interferon (IFN) (or IFN-λ family) cytokines, IL-28A, IL-28B, and IL-29 (also called IFN-λ2, IFN-λ3, and IFN-λ1, respectively), exert important antiviral functions (Ouyang et al. 2011). Interestingly, these cytokines structurally are much more closely related to IL-22 than type I IFNs. In addition, type III IFNs share the same IL-10R2 chain with IL-22 as a common receptor subunit. It is tempting, therefore, to speculate a role of IL-22 in antiviral responses. However, although type III IFNs elicit a typical IFN-stimulated gene factor 3 (ISGF3) signal pathway in antiviral response, IL-22 has not been reported to directly induce ISGF3 complex and antiviral activities in cells. It is still possible that IL-22 can enhance or synergize with IFNs during antiviral responses.

Alternatively, as IL-22 promotes tissue-protective functions, it may help to alleviate tissue damage caused by viral infection. IL-22 can also enhance antibacterial activity from the tissue and reduce the chance of secondary bacterial infections after viral invasion. Recent studies with various viral infection models support these hypotheses. In a chronic lymphocytic choriomeningitis virus (LCMV) infection model with clone 13, IL-7 limited organ pathology and damage (Pellegrini et al. 2011) . One of the downstream factors induced by IL-7 in this model is IL-22, which provides an important cytoprotective function and reduces the severity of hepatitis during infection. Consistently, IL-22 is elevated in the liver during chronic hepatitis B virus infection in humans (Feng et al. 2012a). The increased expression of IL-22 in the liver is correlated with the grade of the inflammation and the proliferation of liver progenitor cells. Both in vitro and in vivo, IL-22 is able to promote the proliferation of liver progenitor cells, suggesting a role of IL-22 in liver protection during hepatitis B virus (HBV) infection. A similar tissue protection role of

IL-22 is also observed in the lung during influenza infection (Kumar et al. 2013; Paget et al. 2012). IL-22-deficient mice infected with influenza develop much more severe damage of the tracheal epithelial cells than that in wild-type (WT) mice. IL-22 knockout mice also do not recover their body weight lost during the infection. In the gut, the major cellular sources of IL-22 are T cells and innate lymphoid cells. In patients with human immunodeficiency virus (HIV) infection, there is a significant reduction of IL-22-producing T helper subsets, which is correlated with compromised epithelial integrity and increased bacterial translocation (Kim et al. 2012). Long-term antiretroviral therapy is able to increase IL-22-producing T cells and reverse the tissue damage in the gut. These data support an important role of IL-22 in preventing HIV-induced mucosal immunopathogenesis in the gut.

All together these data support a critical function of IL-22 in protecting various epithelial tissues from virus- and inflammation-caused destruction. Further studies are needed to investigate whether there is a direct role of IL-22 in controlling certain viral infections in epithelial cells. On the other hand, IL-22 can synergize with other cytokines to enhance inflammation under certain conditions. It has been reported that IL-22-deficient mice are more resistant to the lethal West Nile virus (WNV) encephalitis. In this model, IL-22 promotes chemokine pathways to recruit neutrophils into the central nervous system (CNS) (Wang et al. 2012).

6.3.1.4 Role of IL-22 in Yeast Infection

Indispensable functional roles of Th17 and IL-17 in host defense against yeast infection have been established (Ouyang et al. 2008). Patients with mutations in either IL-17 receptor or IL-17F develop chronic mucocutaneous candidiasis (CMC) (Puel et al. 2011). Interestingly, in patients with CMC, their PBMCs not only produce less IL-17A but also less IL-22 upon stimulation with *Candida albicans* (Eyerich et al. 2008). In addition, antibodies against IL-17A, IL-17F, and IL-22 have been observed from patients with autoimmune polyendocrine syndrome type I (APS-I)/autoimmune polyendocrinopathy candidiasis ectodermal dystrophy (APECED) (Kisand et al. 2010; Puel et al. 2010). Dampened IL-22 response as well as IL-17 response is considered to associate with CMC in these patients. However, a direct causal relationship remains to be established (Zelante et al. 2011).

The role of IL-22 in preclinical yeast infection has been examined in recent years. During infection of oropharyngeal candidiasis, IL-22 seems to be dispensable though Th17 and IL-17 pathways are required for host defense (Conti et al. 2009). Similarly, IL-22 is not required for optimal host defense when skin is infected with *C. albicans* (Kagami et al. 2010a). However, IL-22 is required to prevent *C. albicans* from disseminating into stomach and kidney when the yeast is infected introgastrically (De Luca et al. 2010). In this case, IL-22 induces various antimicrobial peptides such as S100A8 and RegIII-γ. Furthermore, during *Aspergillus fumigatus* infection in the lung, IL-22 is induced through a Dectin-1- and IL-23-dependent pathway and contributes to the early innate host defense in the lung (Gessner et al. 2012). These data support that IL-22, similar to IL-17, participates in host defense of yeast infection.

6.3.2 *Maintenance of Epithelial Homeostasis*

The antimicrobial functions elicited by IL-22 help to prevent tissue damage from invading pathogens. In addition, IL-22 directly promotes tissue protection and regeneration. First, IL-22 has anti-apoptotic functions and prevents cell damage and cell death caused by inflammation (Cella et al. 2009; Radaeva et al. 2004; Zenewicz et al. 2007). Second, IL-22 promotes wound repair through increasing cell migration and proliferation of keratinocytes, fibroblasts, and IECs (Boniface et al. 2005; Brand et al. 2006; McGee et al. 2013). Stat3 activation by IL-22 is essential for epithelial protection and mucosal wound healing in murine colitis triggered by chemical epithelial damage, inflammatory T cells, or bacterial infection (Pickert et al. 2009; Sugimoto et al. 2008; Zenewicz et al. 2008; Zheng et al. 2008). Third, IL-22 helps to maintain mucosal barrier functions through the restoration of a mucus layer and tight junction. Mucus-associated proteins (MUC), structural components of mucus, form a static external barrier along the epithelial cell surface and are critical for epithelial protection and the wound-healing process (Ho et al. 2006; Van der Sluis et al. 2006). MUC1, MUC3, MUC10 and MUC13 were upregulated from colonic epithelial cells upon IL-22 stimulation (Sugimoto et al. 2008). IL-22 mediates enterocyte homeostasis and restores expression and redistribution of tight junction protein zonula occludens (OZ)-1 (Kim et al. 2012; Klatt et al. 2012). Finally, IL-22 is able to act on epithelial stem cells and progenitor cells and promote tissue regeneration. Given these unique functions of IL-22 in tissue protection, IL-22 may contribute to regulation of the pathogenesis in many diseases.

6.3.2.1 Tissue-Protective Role of IL-22 in Inflammation

The IL-22 receptor is expressed on many other cells with epithelial origins including hepatocytes in the liver and acinar cells in the pancreas. IL-22 plays an indispensable role in protecting these cells during inflammation. IL-22 expression is induced in liver during concanavalin A (ConA)-induced hepatitis. Blocking the IL-22 pathway by a neutralizing antibody or using IL-22-deficient mice exacerbates hepatitis as indicated by increased aspartate aminotransferase (AST) and alanine aminotransferase (ALT) levels (Radaeva et al. 2004; Zenewicz et al. 2007). Overexpression of IL-22 can protect liver from injury caused by an agonist antibody targeting Fas, ConA, or carbon tetrachloride (Pan et al. 2004). In addition, treatment with exogenous IL-22 reduces alcoholic liver injury in mice receiving chronic-binge ethanol feeding (Ki et al. 2010). In these models, IL-22 exerts anti-apoptotic effects, induces the expression of antioxidative products, and promotes the proliferation of the hepatocytes. More interestingly, IL-22 is also involved in liver regeneration. The expression of IL-22R on hepatocytes is increased after a partial hepatectomy in mice. Blocking the IL-22 pathway in this case reduces the proliferation of hepatocytes (Ren et al. 2010). Pancreas is another organ that is targeted by IL-22. IL-22R1 is expressed at the highest level in pancreas compared to other human tissues, and acinar cells are the predominant IL-22R1⁺ cell population (Aggarwal et al. 2001). Both freshly isolated murine acinar cells and a transformed cell line were able to respond to IL-22 and upregulate pancreatitis-associated protein (PAP1)/regenerating gene (Reg)2 and osteopontin (OPN). Moreover, the induction was immediately observed after in vivo IL-22 administration in an animal model. PAP/Reg proteins belong to the C-type lectin (calcium-dependent lectin) gene superfamily, which also includes RegIII genes discussed previously (Zhang et al. 2003). They are not expressed in normal pancreas but are greatly elevated during pancreatic inflammation and directly stimulate β-cell proliferation, facilitate pancreatic islet regeneration, and ameliorate diabetic syndrome (Keim et al. 1984, 1992; Orelle et al. 1992; Watanabe et al. 1994). A report has showed that both β-cells and α-cells may express some level of IL-22R1, but there is no direct evidence to show that they respond to IL-22 (Shioya et al. 2008). In a cerulein-induced chronic pancreatitis model, exogenous IL-22 is able to reduce the release of digestive enzymes, pancreatic cell death, and inflammation, suggesting that IL-22 may have therapeutic value in this disease (Feng et al. 2012b).

IL-22R1 expression was found in airway epithelial cells, indicating IL-22 as a regulator in lung inflammation (Besnard et al. 2011). In asthmatic patients, and patients with interstitial lung diseases such as idiopathic pulmonary fibrosis (IPF), acute respiratory distress syndrome (ARDS), and pulmonary sarcoidosis, IL-22 expression is altered comparing to healthy lungs, suggesting IL-22 has biological function in pulmonary diseases (Whittington et al. 2004). As already discussed, IL-22 is essential for maintenance of lung integrity during bacterial, fungal, and viral infection (Aujla et al. 2008; Gessner et al. 2012; Kumar et al. 2013) and promotes both the pro-inflammatory antimicrobial responses and tissue-protective functions in the lung. Interestingly, dependent on the environmental factors, the functions of IL-22 can be beneficial or detrimental to the lung. In a bleomycin-induced lung inflammation model, IL-22 exerts a protective function when the IL-17A pathway is blocked. However, in the presence of the IL-17 pathway, IL-22 can synergize with IL-17 to amplify the inflammatory response (Sonnenberg et al. 2010). On the other hand, there is evidence that IL-22 limits the Th2 inflammation in an allergic airway model (Takahashi et al. 2011; Taube et al. 2011), represses IL-25 production from lung epithelial cells, and reduces eosinophil infiltration in the airway (Takahashi et al. 2011).

In addition to its role in promoting liver regeneration, IL-22 also facilitates the regeneration of thymus (Dudakov et al. 2012). IL-22-deficient mice displayed significant delayed thymic regeneration after sublethal total body irradiation. Furthermore, IL-22 functions in maintaining intestinal stem cells (ISCs), which reside within intestinal crypts and could generate the entire crypt–villus structure (Barker et al. 2007; Hanash et al. 2012; Medema and Vermeulen 2011; Simons and Clevers 2011). In a murine graft-versus-host disease model, blocking the IL-22 pathway results in the loss of the ISCs and the increase of crypt apoptosis. Although the IL-22 receptor is not expressed on leukocytes, a role of IL-22 in retaining lymphoid structure has been revealed recently. In the murine *C. rodentium* infection model, IL-22 is downstream of the lymphoid toxin pathway. IL-22 is required for the maintenance of the isolated lymphoid follicles and colonic patches in the colon during infection (Ota et al. 2011). In this case, IL-22 does not directly act on leukocytes; rather, it helps to preserve the epithelial architecture and chemokines that required for the organization of the lymphoid follicles.

6.3.2.2 IL-22 May Be Beneficial for IBD Through Its Roles Both in Tissue Protection and Antimicrobial Functions

Inflammatory bowel diseases can be classified into either ulcerative colitis (UC) or Crohn's disease (CD), both of which manifest chronic inflammation in the gastrointestinal tract. Although UC mainly affects only the colon with continuous superficial inflammation, CD involves the entire intestinal track with patchy, sometimes transmural inflammation. Both environmental factors and genetic predispositions contribute to the pathogenesis of the IBD, although the exact triggers of the disease remain unclear. Recent genome-wide association studies (GWAS) in IBD and data from preclinical animal models suggest that maintenance of the homeostasis of the intestinal microflora and mucosal epithelial barrier may be important.

The IL-22 pathway is not linked with IBD in GWAS analysis, but significantly increased expression of IL-22 is reported in inflamed intestinal mucosa biopsies from IBD patients, and the level of IL-22 is correlated with disease severity (Andoh et al. 2005; Brand et al. 2006; Schmechel et al. 2008). As discussed, IL-23 is an essential upstream cytokine that induces the expression of IL-22 in the intestine. Both IL-23R and IL-12/IL-23p40 subunit are associated with IBD genetically. Interestingly, there is a correlation between serum IL-22 levels and IL-23R minor alleles in CD, in which lower IL-22 expression is linked with protective single nucleotide polymorphisms (SNPs) of IL-23 (Schmechel et al. 2008). As discussed previously, IL-22 effectively induces or synergizes the secretion of pro-inflammatory cytokines and chemokines (IL-6, IL-8, IL-11, leukemia inhibitory factor), pro-angiogenic mediators (MMP-1, MMP-3, MMP-10), growth factor (amphiregulin), and immune suppressor (IL-10, SOCS3) in colon (Andoh et al. 2005; Brand et al. 2006; Nagalakshmi et al. 2004). IL-22 facilitates IFN-γ for inducible nitric oxide synthase (iNOS) production in human colon carcinoma cells, and iNOS is associated with inflammation and immune activation (Ziesche et al. 2007).

IL-22 is upregulated and has a beneficial role in several preclinical colitis models (Neufert et al. 2010; Pickert et al. 2009; Sugimoto et al. 2008; Zenewicz et al. 2008). Although IL-22 does not directly repress inflammatory responses, it may help to dampen the symptoms in IBD in several ways (Fig. 6.4). First, IL-22 induces the production of antimicrobial peptides, such as Reg family of C-type letins, defensins, and cathelicidins, from intestinal epithelial cells and Paneth cells (Pickert et al. 2009; Sugimoto et al. 2008; Zenewicz et al. 2008; Zheng et al. 2008). Second, IL-22 preserves the integrity of the mucosal epithelial layers. IL-22 not only may reduce epithelial apoptosis but also helps to retain the tight junctions in epithelial cells during infection and inflammation (Cella et al. 2009; Kim et al. 2012). IL-22 also promotes epithelial proliferation and triggers wound-healing responses (Brand et al. 2006;

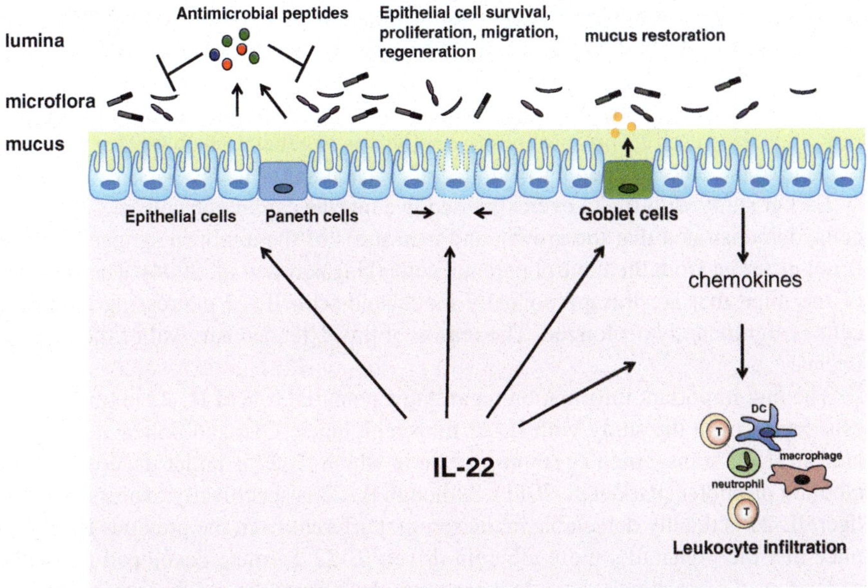

Fig. 6.4 IL-22 protects intestinal epithelia in inflammatory bowel disease (IBD). IL-22 restores epithelial integrity through four major biological functions: promoting antimicrobial peptide secretion from epithelial cells; directly stimulating epithelial cell survival, proliferation, migration, and regeneration; increasing the production of mucous proteins by goblet cells, and triggering leukocyte infiltration via activating the chemokine production from epithelial cells

Neufert et al. 2010; Pickert et al. 2009). In addition, IL-22 can direct act on intestinal epithelial stem cells and enhance the regeneration (Hanash et al. 2012). Third, IL-22 augments the production of MUC proteins, which can sequester luminal flora from directly interacting with epithelial cells (Sugimoto et al. 2008). Finally, IL-22 stimulates chemokine production from epithelial cells to boost leukocyte infiltration.

Based on its functions, IL-22 may offer certain therapeutic value in IBD, especially when epithelial damage and changes of luminal flora are major triggers of intestinal inflammation. As IL-22 does not suppress inflammation directly, additional anti-inflammatory treatments may be needed to cooperate with the effects of IL-22. In fact, IL-22 by itself can exacerbate intestinal inflammation under certain conditions. In colitis induced by transferring CD45Rb^{low} CD4^{+} memory cells that are deficient in IL-10 pathway, IL-22 amplifies the intestinal inflammation (Kamanaka et al. 2011). A similar pathogenic role of IL-22 is again revealed in a small intestinal inflammation model triggered by infection with *Toxoplasma gondii* (Munoz et al. 2009).

6.3.2.3 Role of IL-22 in Oncogenesis

Stat3 is a known oncogene and has an important role in many different types of cancer (Hodge et al. 2005). Cytokines, such as IL-6, that activate Stat3, have been connected with tumorigenesis under chronic inflammatory conditions (Naugler et al. 2007). Because IL-22 directly targets epithelial cells and induces Stat3 activation, it is possible that IL-22 may promote certain epithelial-derived tumors. Under homeostatic states, IL-22 can induce epithelial proliferation and elicit wound-healing responses. However, its pro-proliferative activity is in general much weaker than that of some of the other epithelial growth factors such as EGF and KGF (Sa et al. 2007).

Earlier study with IL-22 overexpressed in a murine carcinoma cell line, colon 26 cells, demonstrated that the growth and metastasis of the tumor in syngeneic mice is not different from the control parental cells (Nagakawa et al. 2004). The survival of the mice that are intraperitoneally inoculated with IL-22-expressing colon 26 cells is significantly prolonged. The reason of this extended survival of the mice is unclear.

The first important information regarding a potential role of IL-22 in tumorigenesis comes from the study with IL-22 transgenic mice. Gao and colleagues generated an IL-22 transgenic (Tg) mouse strain in which IL-22 is under the control of a albumin promoter (Park et al. 2011). Although IL-22 is specifically expressed in the liver, IL-22 is readily detectable in the serum. Different from the previous IL-22 Tg mice that die prenatally, these albumin-driven IL-22 Tg mice developed normally and survived more than 2 years. In these mice, the authors did not find any spontaneously developed tumors. Interestingly, IL-22 Tg mice are more susceptible to diethylnitrosamine, a liver carcinogen that induces hepatocarcinoma, supporting a role of IL-22 in promoting the growth of existing cancers. IL-22 is overexpressed in adipose tissue under the control of aP2 promoter in a Tg line (Wang et al. 2011). Although no obvious metabolic phenotypes are observed in these Tg mice, all Tg mice developed spontaneous liposarcomas after feeding with a high-fat diet (HFD). Interestingly, none of the wild-type (WT) mice fed with HFD or IL-22 Tg mice fed with a normal diet develop similar tumors. In both cases, it is unclear what the local IL-22 concentration is in the targeted tissue, which may be significantly higher than normal endogenous level of IL-22.

The potential role of physiological IL-22 level in tumorigenesis has been recently examined in colon cancer models using IL-22, IL-22BP knockout mice, and IL-22 neutralizing antibody (Huber et al. 2012). In the carcinogen-induced, colitis-associated colon cancer model, IL-22BP knockout mice have increased tumor burden, which is dependent on IL-22. Similar results are also observed in the $APC^{min/+}$ model of colon cancer. These data again support a role of IL-22 in promoting existing colon cancer. On the other hand, the development of colon cancer is associated with colitis, and IL-22 can alleviate colitis in preclinical models. IL-22 may prevent the initiation of colon cancer by repression of colitis. Indeed, anti-IL-22 treatment can increase tumor burden if treated early in the model, supporting a protective role of IL-22 in colon tumorigenesis at the early stage.

Although IL-22 promotes the existing colon tumors at the late stage, the long-term outcome of these tumors has not been examined. Stat3 was reported to have a dual role in the development of colon cancer in $APC^{min/+}$ mice carrying IEC-specific deletion of Stat3 (Musteanu et al. 2010). In the absence of Stat3, the number of intestinal epithelial adenoma is reduced at an early stage, suggesting the

Stat3 pathway may play a role in promoting tumor growth. Interesting, at the later stage, there is significantly increased tumor burden in both small intestine and colon without Stat3. Loss of Stat3 not only enhances the tumor progression and invasiveness but also significantly reduces the survival of $APC^{min/+}$ mice. Together, the data suggest that activated Stat3 may help control the progression of intestinal cancers. It is unclear whether there is a similar role of IL-22, given its function in activation of Stat3.

6.3.3 *Inflammation and Autoimmune Diseases*

Similar to many other cytokines including IL-6, TNF-α, and IL-1β, IL-22 is able to promote inflammatory responses. Three major functions of IL-22 contribute to the inflammatory cascade. First, many of the antimicrobial peptides such as S100 family proteins, induced by IL-22, can recruit and activate leukocytes. Second, IL-22 stimulates the expression of a large group of chemokines from epithelial cells. Third, IL-22 can directly act on liver and other organs to trigger the release of acute-phase proteins, whose plasma concentration changes in response to inflammation to restrain infection, damp local inflammation, and rebuild tissue homeostasis (Baumann and Gauldie 1994; Steel and Whitehead 1994). As discussed earlier, however, despite these pro-inflammatory effects, the net functional outcome of IL-22 in inflammation and infection could be an alleviated inflammatory response (Radaeva et al. 2004; Sugimoto et al. 2008; Zenewicz et al. 2007, 2008; Zheng et al. 2008). Under other conditions, IL-22 elicits strong pro-inflammatory effects and contributes to the pathogenesis of many inflammatory diseases (Kamanaka et al. 2011; Munoz et al. 2009; Zheng et al. 2007). These contradictory effects of IL-22 are likely caused by the shifted balance between the tissue-protective effects and pro-inflammatory functions of IL-22. Thus, pro-inflammatory functions of IL-22 are largely dictated by the context of tissue microenvironment, type of infection, or other pro-inflammatory mediators.

6.3.3.1 IL-22 Induces Expression of Chemokines and Triggers Acute-Phase Reaction

Despite the minimal effect of IL-22 on cytokine secretion, IL-22 strongly upregulates many chemokines in various organs and contributes to immune regulation by recruiting leukocyte infiltration. IL-22 directly upregulates CXCL10 (IP-10), CCL2 (MCP-1), IL-8, and CXCL1 in hepatocytes, resulting in the recruitment of monocytes, effector T cells, NK cells, and neutrophils (Donnelly et al. 2004; Liang et al. 2010). Similarly, chemoattractants IL-8, CXCL1, CXCL2 (MIP2-a), CXCL3 (MIP2-b), CXCL6, and CCL7 were positively regulated by IL-22 in human colonic subepithelial myofibroblasts and were elevated in inflamed colonic lesions of IBD patients (Andoh et al. 2005; Brand et al. 2006). There was an even greater variety of chemokines induced by IL-22 in human keratinocytes including IL-8, CXCL1, CXCL7, CXCL9 (MIG), CXCL10, CXCL11 (I-TAC), CCL2, CCL5 (RANTES), CCL20 (MIP3-a), and CCL26 (Eotaxin-3). All these chemokines not only activate chemotaxis in various leukocyte populations (monocytes, macrophages, dendritic cells, activated T cells, eosinophils, basophils, and neutrophils), but also stimulate other processes such as mitogenesis, synthesis of extracellular matrix, glucose metabolism, and phagocytosis (Eyerich et al. 2009; Sa et al. 2007). Similar observations were found in pulmonary diseases and rheumatoid arthritis (RA) (Aujla et al. 2008; Besnard et al. 2011; Ikeuchi et al. 2005). In preclinical studies, the infiltration of leukocytes into inflammation sites was significantly reduced in IL-22-deficient animals or those treated with IL-22-neutralizing antibody.

The liver is the predominant organ of pro-inflammatory response. IL-22 was initially characterized as a hepatocyte-stimulating factor for its ability to upregulate acute-phase proteins, such as serum amyloid A (SAA), a1-antichymotrypsin, and haptoglobin in several hepatoma cell lines (Dumoutier et al. 2000c). In addition, the production of both fibrinogen and LBP by hepatocytes was enhanced after IL-22 administration (Liang et al. 2010; Wolk et al. 2007). With sustained exposure to IL-22 delivered by the adenoviral system or liver-specific transgene expression, circulating blood cells were altered with increased blood platelets and decreased red blood cells. Furthermore, body weight loss, thymic atrophy, and renal proximal tubule metabolic activity associated with acute inflammation were also detected (Liang et al. 2010). Besides acting on the vital organs, IL-22 is expressed in RA synovial tissues and mononuclear cells in synovial fluid, and contributes to joint inflammation primarily through directly promoting fibroblast proliferation and monocyte chemoattractant protein 1 (MCP-1) (Ikeuchi et al. 2005).

6.3.3.2 Pathogenic Role of IL-22 in Psoriasis

The possible link between IL-22 and psoriasis was first revealed by studying the closely related family member IL-20 (Blumberg et al. 2001). Overexpressing IL-20 in Tg mice results in abnormal skin phenotypes, including wrinkled and shiny skin with a thickened epidermis. Histological analysis further identifies epidermal hyperplasia and compact stratum corneum, consistent with some of the histological features observed in psoriatic skin, in these mice. Interestingly, similar phenotypes are also observed in IL-22 and IL-24 Tg mice, suggesting common biological functions of this group of cytokines in the skin (He and Liang 2010; Wolk et al. 2009). IL-22 receptors are highly expressed by keratinocytes (Sa et al. 2007). Detailed in vitro studies with human primary epidermis demonstrate that IL-22 as well as other family members including IL-19, IL-20, and IL-24 can directly induce many downstream biological features observed in psoriatic skin. For example, IL-22 activates p-STAT3 in human keratinocytes and induces the expression of several pro-inflammatory proteins including members of S100 family of calcium-binding proteins, β-defensins, and matrix metalloproteinases (MMP) (Boniface et al. 2005; Ma et al. 2008; Sa et al. 2007; Wolk et al. 2004). Specifically, S100A7, S100A8, and

S100A9 are produced promptly when normal human epidermal keratinocytes are treated with IL-22, and they function as chemoattractants of innate immune cells (Roth et al. 2003; Watson et al. 1998). β-Defensin 2 and -3 are induced by IL-22 in human primary keratinocytes, and contribute to inflammation through activation and degranulation of mast cells and CCR6-mediated chemotaxis for immature DCs and CD4 memory T cells (Niyonsaba et al. 2001; Yang et al. 1999). MMPs, also called matrixins, have an essential role in cell mobility and tissue remodeling (Nagase and Woessner 1999). In addition, IL-22 promotes epidermal keratinocyte proliferation and abnormal differentiation similar to those observed in psoriatic skin. Among all its family members, IL-22 appears to be the most potent in induction of these biological effects from keratinocytes (Sa et al. 2007).

In humans, elevated IL-22 has been detected in lesional skin and serum from psoriasis patients (Boniface et al. 2005, 2007; Kagami et al. 2010b; Pene et al. 2008; Wolk et al. 2004). In addition, IL-22-producing T helper subsets, including Th17 and Th22 cells, have been isolated from psoriatic skin. Th17 cells produce both IL-22 and IL-17. IL-17 also exerts essential pathogenic functions in psoriasis. It induces many similar genes as those by IL-22 from keratinocytes (Liang et al. 2006; Tohyama et al. 2009). Blocking the IL-17 pathway ameliorates skin inflammation in many preclinical psoriatic models (van der Fits et al. 2009). More importantly, in clinic, blocking the antibody targeting IL-17 or the IL-17 receptor results in very impressive efficacy in treatment of psoriasis (Hueber et al. 2010; Leonardi et al. 2012; Papp et al. 2012). IL-22 is able to synergize with IL-17 and amplify inflammatory responses in the skin.

Human genetic studies in psoriasis also support an important role of IL-22 in the pathogenesis. Both IL-12/IL-23 p40 subunit and IL-23 receptor have been linked to psoriasis in GWAS (Capon et al. 2007). A protective SNP identified in IL-23R leads to reduced IL-23R expression and decreased IL-22 production upon IL-23 stimulation in T cells (Pidasheva et al. 2011). IL-23 has important functions in Th17 cell development in vivo (McGeachy and Cua 2007) and is indispensable for IL-22 induction from both T cells and ILCs (Ouyang et al. 2011). In clinic, the antibody-neutralizing IL-12/IL-23 p40 subunit succeeds in treatment of psoriasis (Krueger et al. 2007). The efficacy of anti-IL-23 could result partly from the block of IL-22 induction by IL-23. Indeed, in a preclinical mouse model, IL-23 is able to directly induce skin inflammation and acanthosis, and these phenotypes are alleviated in IL-22-deficient mice (Zheng et al. 2007). Blocking IL-22 is efficacious in several other preclinical models of skin inflammation (Ma et al. 2008; Van Belle et al. 2012).

Another protein, Act1, is identified to associate with psoriasis by three independent GWAS (Ellinghaus et al. 2010; Huffmeier et al. 2010; Strange et al. 2010). Act1 is an important adaptor molecule recruited to the receptor of IL-17 and is essential for the function of IL-17 (Chang et al. 2006; Qian et al. 2007). The variant of Act1 identified in psoriasis replaces asparagine for the aspartic acid at the position 10. This loss-of-function mutation results in abolished IL-17-dependent functions. Given an important pathogenic function of IL-17 as already discussed, it is paradoxical that this Act1 mutation is associated with increased risk of development of psoriasis. A recent study, however, revealed that in the mouse this mutation leads to a significantly increased production of IL-22 from T cells (Wang et al. 2013). In addition, these mice spontaneously develop skin inflammation that can be blocked by IL-22-neutralizing antibody.

Taken together, these data support an important role of IL-22 in the pathogenesis of psoriasis. However, it is unclear whether blocking IL-22 by itself will provide sufficient therapeutic benefit in psoriasis. First, several other IL-10 family members, IL-19, IL-20, and IL-24 are also elevated in human psoriatic skin and can induce similar downstream biological effects in the skin. Second, although IL-22 induces many unique patterns and genes from keratinocytes in psoriasis, it is a relative weak cytokine in promoting inflammatory responses in comparison with other cytokines such as IL-17 and TNF-α. IL-22 usually synergizes with these cytokines to further amplify inflammation. Further clinical studies may be necessary to fully assess the role of IL-22 in psoriasis.

References

Aggarwal S, Xie MH, Maruoka M, Foster J, Gurney AL (2001) Acinar cells of the pancreas are a target of interleukin-22. J Interferon Cytokine Res 21:1047–1053

Alam MS, Maekawa Y, Kitamura A, Tanigaki K, Yoshimoto T, Kishihara K, Yasutomo K (2010) Notch signaling drives IL-22 secretion in CD4+ T cells by stimulating the aryl hydrocarbon receptor. Proc Natl Acad Sci USA 107:5943–5948

Andoh A, Zhang Z, Inatomi O, Fujino S, Deguchi Y, Araki Y, Tsujikawa T, Kitoh K, Kim-Mitsuyama S, Takayanagi A et al (2005) Interleukin-22, a member of the IL-10 subfamily, induces inflammatory responses in colonic subepithelial myofibroblasts. Gastroenterology 129:969–984

Aujla SJ, Chan YR, Zheng M, Fei M, Askew DJ, Pociask DA, Reinhart TA, McAllister F, Edeal J, Gaus K et al (2008) IL-22 mediates mucosal host defense against gram-negative bacterial pneumonia. Nat Med 14:275–281

Barker N, van Es JH, Kuipers J, Kujala P, van den Born M, Cozijnsen M, Haegebarth A, Korving J, Begthel H, Peters PJ, Clevers H (2007) Identification of stem cells in small intestine and colon by marker gene Lgr5. Nature (Lond) 449:1003–1007

Basu R, O'Quinn DB, Silberger DJ, Schoeb TR, Fouser L, Ouyang W, Hatton RD, Weaver CT (2012) Th22 cells are an important source of IL-22 for host protection against enteropathogenic bacteria. Immunity 37:1061–1075

Baumann H, Gauldie J (1994) The acute phase response. Immunol Today 15:74–80

Besnard AG, Sabat R, Dumoutier L, Renauld JC, Willart M, Lambrecht B, Teixeira MM, Charron S, Fick L, Erard F et al (2011) Dual role of IL-22 in allergic airway inflammation and its cross-talk with IL-17A. Am J Respir Crit Care Med 183:1153–1163

Bettelli E, Carrier Y, Gao W, Korn T, Strom TB, Oukka M, Weiner HL, Kuchroo VK (2006) Reciprocal developmental pathways for the generation of pathogenic effector TH17 and regulatory T cells. Nature (Lond) 441:235–238

Billerbeck E, Kang YH, Walker L, Lockstone H, Grafmueller S, Fleming V, Flint J, Willberg CB, Bengsch B, Seigel B et al (2010) Analysis of CD161 expression on human CD8+ T cells defines a distinct functional subset with tissue-homing properties. Proc Natl Acad Sci USA 107:3006–3011

Blumberg H, Conklin D, Xu WF, Grossmann A, Brender T, Carollo S, Eagan M, Foster D, Haldeman BA, Hammond A et al (2001) Interleukin 20: discovery, receptor identification, and role in epidermal function. Cell 104:9–19

Boniface K, Bernard FX, Garcia M, Gurney AL, Lecron JC, Morel F (2005) IL-22 inhibits epidermal differentiation and induces proinflammatory gene expression and migration of human keratinocytes. J Immunol 174:3695–3702

Boniface K, Guignouard E, Pedretti N, Garcia M, Delwail A, Bernard FX, Nau F, Guillet G, Dagregorio G, Yssel H et al (2007) A role for T cell-derived interleukin 22 in psoriatic skin inflammation. Clin Exp Immunol 150:407–415

Brand S, Beigel F, Olszak T, Zitzmann K, Eichhorst ST, Otte JM, Diepolder H, Marquardt A, Jagla W, Popp A et al (2006) IL-22 is increased in active Crohn's disease and promotes proinflammatory gene expression and intestinal epithelial cell migration. Am J Physiol Gastrointest Liver Physiol 290:G827–G838

Brembilla NC, Ramirez JM, Chicheportiche R, Sorg O, Saurat JH, Chizzolini C (2011) In vivo dioxin favors interleukin-22 production by human CD4+ T cells in an aryl hydrocarbon receptor (AhR)-dependent manner. PLoS One 6:e18741

Broadhurst MJ, Leung JM, Kashyap V, McCune JM, Mahadevan U, McKerrow JH, Loke P (2010) IL-22+ CD4+ T cells are associated with therapeutic *Trichuris trichiura* infection in an ulcerative colitis patient. Sci Transl Med 2:60ra88

Bry L, Brenner MB (2004) Critical role of T cell-dependent serum antibody, but not the gut-associated lymphoid tissue, for surviving acute mucosal infection with *Citrobacter rodentium*, an attaching and effacing pathogen. J Immunol 172:433–441

Capon F, Di Meglio P, Szaub J, Prescott NJ, Dunster C, Baumber L, Timms K, Gutin A, Abkevic V, Burden AD et al (2007) Sequence variants in the genes for the interleukin-23 receptor (IL23R) and its ligand (IL12B) confer protection against psoriasis. Hum Genet 122:201–206

Cash HL, Whitham CV, Behrendt CL, Hooper LV (2006) Symbiotic bacteria direct expression of an intestinal bactericidal lectin. Science 313:1126–1130

Cella M, Fuchs A, Vermi W, Facchetti F, Otero K, Lennerz JK, Doherty JM, Mills JC, Colonna M (2009) A human natural killer cell subset provides an innate source of IL-22 for mucosal immunity. Nature (Lond) 457:722–725

Chang SH, Park H, Dong C (2006) Act1 adaptor protein is an immediate and essential signaling component of interleukin-17 receptor. J Biol Chem 281:35603–35607

Chung Y, Yang X, Chang SH, Ma L, Tian Q, Dong C (2006) Expression and regulation of IL-22 in the IL-17-producing CD4+ T lymphocytes. Cell Res 16:902–907

Ciric B, El-behi M, Cabrera R, Zhang GX, Rostami A (2009) IL-23 drives pathogenic IL-17-producing CD8+ T cells. J Immunol 182:5296–5305

Conti HR, Shen F, Nayyar N, Stocum E, Sun JN, Lindemann MJ, Ho AW, Hai JH, Yu JJ, Jung JW et al (2009) Th17 cells and IL-17 receptor signaling are essential for mucosal host defense against oral candidiasis. J Exp Med 206:299–311

Crellin NK, Trifari S, Kaplan CD, Cupedo T, Spits H (2010) Human NKp44+IL-22+ cells and LTi-like cells constitute a stable RORC+ lineage distinct from conventional natural killer cells. J Exp Med 207:281–290

Cupedo T, Crellin NK, Papazian N, Rombouts EJ, Weijer K, Grogan JL, Fibbe WE, Cornelissen JJ, Spits H (2009) Human fetal lymphoid tissue-inducer cells are interleukin 17-producing precursors to RORC+ CD127+ natural killer-like cells. Nat Immunol 10:66–74

De Luca A, Zelante T, D'Angelo C, Zagarella S, Fallarino F, Spreca A, Iannitti RG, Bonifazi P, Renauld JC, Bistoni F et al (2010) IL-22 defines a novel immune pathway of antifungal resistance. Mucosal Immunol 3:361–373

Dhiman R, Indramohan M, Barnes PF, Nayak RC, Paidipally P, Rao LV, Vankayalapati R (2009) IL-22 produced by human NK cells inhibits growth of *Mycobacterium tuberculosis* by enhancing phagolysosomal fusion. J Immunol 183:6639–6645

Donnelly RP, Sheikh F, Kotenko SV, Dickensheets H (2004) The expanded family of class II cytokines that share the IL-10 receptor-2 (IL-10R2) chain. J Leukoc Biol 76:314–321

Dudakov JA, Hanash AM, Jenq RR, Young LF, Ghosh A, Singer NV, West ML, Smith OM, Holland AM, Tsai JJ et al (2012) Interleukin-22 drives endogenous thymic regeneration in mice. Science 336:91–95

Duhen T, Geiger R, Jarrossay D, Lanzavecchia A, Sallusto F (2009) Production of interleukin 22 but not interleukin 17 by a subset of human skin-homing memory T cells. Nat Immunol 10: 857–863

Dumoutier L, Louahed J, Renauld JC (2000a) Cloning and characterization of IL-10-related T cell-derived inducible factor (IL-TIF), a novel cytokine structurally related to IL-10 and inducible by IL-9. J Immunol 164:1814–1819

Dumoutier L, Van Roost E, Ameye G, Michaux L, Renauld JC (2000b) IL-TIF/IL-22: genomic organization and mapping of the human and mouse genes. Genes Immun 1:488–494

Dumoutier L, Van Roost E, Colau D, Renauld JC (2000c) Human interleukin-10-related T cell-derived inducible factor: molecular cloning and functional characterization as an hepatocyte-stimulating factor. Proc Natl Acad Sci USA 97:10144–10149

Dumoutier L, Lejeune D, Hor S, Fickenscher H, Renauld JC (2003) Cloning of a new type II cytokine receptor activating signal transducer and activator of transcription (STAT)1, STAT2 and STAT3. Biochem J 370:391–396

Dumoutier L, de Meester C, Tavernier J, Renauld JC (2009) New activation modus of STAT3: a tyrosine-less region of the interleukin-22 receptor recruits STAT3 by interacting with its coiled-coil domain. J Biol Chem 284:26377–26384

Ellinghaus E, Ellinghaus D, Stuart PE, Nair RP, Debrus S, Raelson JV, Belouchi M, Fournier H, Reinhard C, Ding J et al (2010) Genome-wide association study identifies a psoriasis susceptibility locus at TRAF3IP2. Nat Genet 42:991–995

Eyerich K, Foerster S, Rombold S, Seidl HP, Behrendt H, Hofmann H, Ring J, Traidl-Hoffmann C (2008) Patients with chronic mucocutaneous candidiasis exhibit reduced production of Th17-associated cytokines IL-17 and IL-22. J Invest Dermatol 128:2640–2645

Eyerich S, Eyerich K, Pennino D, Carbone T, Nasorri F, Pallotta S, Cianfarani F, Odorisio T, Traidl-Hoffmann C, Behrendt H et al (2009) Th22 cells represent a distinct human T cell subset involved in epidermal immunity and remodeling. J Clin Invest 119:3573–3585

Feng D, Kong X, Weng H, Park O, Wang H, Dooley S, Gershwin ME, Gao B (2012a) Interleukin-22 promotes proliferation of liver stem/progenitor cells in mice and patients with chronic hepatitis B virus infection. Gastroenterology 143(188–198):e187

Feng D, Park O, Radaeva S, Wang H, Yin S, Kong X, Zheng M, Zakhari S, Kolls JK, Gao B (2012b) Interleukin-22 ameliorates cerulein-induced pancreatitis in mice by inhibiting the autophagic pathway. Int J Biol Sci 8:249–257

Gessner MA, Werner JL, Lilly LM, Nelson MP, Metz AE, Dunaway CW, Chan YR, Ouyang W, Brown GD, Weaver CT, Steele C (2012) Dectin-1-dependent interleukin-22 contributes to early innate lung defense against *Aspergillus fumigatus*. Infect Immun 80:410–417

Goto M, Murakawa M, Kadoshima-Yamaoka K, Tanaka Y, Nagahira K, Fukuda Y, Nishimura T (2009) Murine NKT cells produce Th17 cytokine interleukin-22. Cell Immunol 254:81–84

Gruenberg BH, Schoenemeyer A, Weiss B, Toschi L, Kunz S, Wolk K, Asadullah K, Sabat R (2001) A novel, soluble homologue of the human IL-10 receptor with preferential expression in placenta. Genes Immun 2:329–334

Gurney AL (2004) IL-22, a Th1 cytokine that targets the pancreas and select other peripheral tissues. Int Immunopharmacol 4:669–677

Hanash AM, Dudakov JA, Hua G, O'Connor MH, Young LF, Singer NV, West ML, Jenq RR, Holland AM, Kappel LW et al (2012) Interleukin-22 protects intestinal stem cells from immune-mediated tissue damage and regulates sensitivity to graft versus host disease. Immunity 37:339–350

Harrington LE, Hatton RD, Mangan PR, Turner H, Murphy TL, Murphy KM, Weaver CT (2005) Interleukin 17-producing CD4+ effector T cells develop via a lineage distinct from the T helper type 1 and 2 lineages. Nat Immunol 6:1123–1132

He M, Liang P (2010) IL-24 transgenic mice: in vivo evidence of overlapping functions for IL-20, IL-22, and IL-24 in the epidermis. J Immunol 184:1793–1798

Ho SB, Dvorak LA, Moor RE, Jacobson AC, Frey MR, Corredor J, Polk DB, Shekels LL (2006) Cysteine-rich domains of muc3 intestinal mucin promote cell migration, inhibit apoptosis, and accelerate wound healing. Gastroenterology 131:1501–1517

Hodge DR, Hurt EM, Farrar WL (2005) The role of IL-6 and STAT3 in inflammation and cancer. Eur J Cancer 41:2502–2512

Huber S, Gagliani N, Zenewicz LA, Huber FJ, Bosurgi L, Hu B, Hedl M, Zhang W, O'Connor W Jr, Murphy AJ et al (2012) IL-22BP is regulated by the inflammasome and modulates tumorigenesis in the intestine. Nature (Lond) 491:259–263

Hueber W, Patel DD, Dryja T, Wright AM, Koroleva I, Bruin G, Antoni C, Draelos Z, Gold MH, Durez P et al (2010) Effects of AIN457, a fully human antibody to interleukin-17A, on psoriasis, rheumatoid arthritis, and uveitis. Sci Transl Med 2:52ra72

Huffmeier U, Uebe S, Ekici AB, Bowes J, Giardina E, Korendowych E, Juneblad K, Apel M, McManus R, Ho P et al (2010) Common variants at TRAF3IP2 are associated with susceptibility to psoriatic arthritis and psoriasis. Nat Genet 42:996–999

Hughes T, Becknell B, McClory S, Briercheck E, Freud AG, Zhang X, Mao H, Nuovo G, Yu J, Caligiuri MA (2009) Stage 3 immature human natural killer cells found in secondary lymphoid tissue constitutively and selectively express the TH 17 cytokine interleukin-22. Blood 113: 4008–4010

Hughes T, Becknell B, Freud AG, McClory S, Briercheck E, Yu J, Mao C, Giovenzana C, Nuovo G, Wei L et al (2010) Interleukin-1beta selectively expands and sustains interleukin-22+ immature human natural killer cells in secondary lymphoid tissue. Immunity 32:803–814

Ikeuchi H, Kuroiwa T, Hiramatsu N, Kaneko Y, Hiromura K, Ueki K, Nojima Y (2005) Expression of interleukin-22 in rheumatoid arthritis: potential role as a proinflammatory cytokine. Arthritis Rheum 52:1037–1046

Iovanna J, Orelle B, Keim V, Dagorn JC (1991) Messenger RNA sequence and expression of rat pancreatitis-associated protein, a lectin-related protein overexpressed during acute experimental pancreatitis. J Biol Chem 266:24664–24669

Ivanov II, McKenzie BS, Zhou L, Tadokoro CE, Lepelley A, Lafaille JJ, Cua DJ, Littman DR (2006) The orphan nuclear receptor RORgammat directs the differentiation program of proinflammatory IL-17+ T helper cells. Cell 126:1121–1133

Ivanov II, Atarashi K, Manel N, Brodie EL, Shima T, Karaoz U, Wei D, Goldfarb KC, Santee CA, Lynch SV et al (2009) Induction of intestinal Th17 cells by segmented filamentous bacteria. Cell 139:485–498

Jones BC, Logsdon NJ, Walter MR (2008) Structure of IL-22 bound to its high-affinity IL-22R1 chain. Structure 16:1333–1344

Kagami S, Rizzo HL, Kurtz SE, Miller LS, Blauvelt A (2010a) IL-23 and IL-17A, but not IL-12 and IL-22, are required for optimal skin host defense against Candida albicans. J Immunol 185:5453–5462

Kagami S, Rizzo HL, Lee JJ, Koguchi Y, Blauvelt A (2010b) Circulating Th17, Th22, and Th1 cells are increased in psoriasis. J Invest Dermatol 130:1373–1383

Kamanaka M, Huber S, Zenewicz LA, Gagliani N, Rathinam C, O'Connor W Jr, Wan YY, Nakae S, Iwakura Y, Hao L, Flavell RA (2011) Memory/effector (CD45RB(lo)) CD4 T cells are controlled directly by IL-10 and cause IL-22-dependent intestinal pathology. J Exp Med 208:1027–1040

Keim V, Rohr G, Stockert HG, Haberich FJ (1984) An additional secretory protein in the rat pancreas. Digestion 29:242–249

Keim V, Iovanna JL, Orelle B, Verdier JM, Busing M, Hopt U, Dagorn JC (1992) A novel exocrine protein associated with pancreas transplantation in humans. Gastroenterology 103:248–254

Ki SH, Park O, Zheng M, Morales-Ibanez O, Kolls JK, Bataller R, Gao B (2010) Interleukin-22 treatment ameliorates alcoholic liver injury in a murine model of chronic-binge ethanol feeding: role of signal transducer and activator of transcription 3. Hepatology 52:1291–1300

Kim CJ, Nazli A, Rojas OL, Chege D, Alidina Z, Huibner S, Mujib S, Benko E, Kovacs C, Shin LY et al (2012) A role for mucosal IL-22 production and Th22 cells in HIV-associated mucosal immunopathogenesis. Mucosal Immunol 5:670–680

Kisand K, Boe Wolff AS, Podkrajsek KT, Tserel L, Link M, Kisand KV, Ersvaer E, Perheentupa J, Erichsen MM, Bratanic N et al (2010) Chronic mucocutaneous candidiasis in APECED or

thymoma patients correlates with autoimmunity to Th17-associated cytokines. J Exp Med 207:299–308
Klatt NR, Estes JD, Sun X, Ortiz AM, Barber JS, Harris LD, Cervasi B, Yokomizo LK, Pan L, Vinton CL et al (2012) Loss of mucosal CD103+ DCs and IL-17+ and IL-22+ lymphocytes is associated with mucosal damage in SIV infection. Mucosal Immunol 5:646–657
Kolls JK, McCray PB Jr, Chan YR (2008) Cytokine-mediated regulation of antimicrobial proteins. Nat Rev Immunol 8:829–835
Korn T, Bettelli E, Oukka M, Kuchroo VK (2009) IL-17 and Th17 Cells. Annu Rev Immunol 27:485–517
Kotenko SV, Krause CD, Izotova LS, Pollack BP, Wu W, Pestka S (1997) Identification and functional characterization of a second chain of the interleukin-10 receptor complex. EMBO J 16:5894–5903
Kotenko SV, Izotova LS, Mirochnitchenko OV, Esterova E, Dickensheets H, Donnelly RP, Pestka S (2001) Identification, cloning, and characterization of a novel soluble receptor that binds IL-22 and neutralizes its activity. J Immunol 166:7096–7103
Krueger GG, Langley RG, Leonardi C, Yeilding N, Guzzo C, Wang Y, Dooley LT, Lebwohl M (2007) A human interleukin-12/23 monoclonal antibody for the treatment of psoriasis. N Engl J Med 356:580–592
Kumar P, Thakar MS, Ouyang W, Malarkannan S (2013) IL-22 from conventional NK cells is epithelial regenerative and inflammation protective during influenza infection. Mucosal Immunol 6:69–82
Lejeune D, Dumoutier L, Constantinescu S, Kruijer W, Schuringa JJ, Renauld JC (2002) Interleukin-22 (IL-22) activates the JAK/STAT, ERK, JNK, and p38 MAP kinase pathways in a rat hepatoma cell line. Pathways that are shared with and distinct from IL-10. J Biol Chem 277:33676–33682
Leonardi C, Matheson R, Zachariae C, Cameron G, Li L, Edson-Heredia E, Braun D, Banerjee S (2012) Anti-interleukin-17 monoclonal antibody ixekizumab in chronic plaque psoriasis. N Engl J Med 366:1190–1199
Liang SC, Tan XY, Luxenberg DP, Karim R, Dunussi-Joannopoulos K, Collins M, Fouser LA (2006) Interleukin (IL)-22 and IL-17 are coexpressed by Th17 cells and cooperatively enhance expression of antimicrobial peptides. J Exp Med 203:2271–2279
Liang SC, Nickerson-Nutter C, Pittman DD, Carrier Y, Goodwin DG, Shields KM, Lambert AJ, Schelling SH, Medley QG, Ma HL et al (2010) IL-22 induces an acute-phase response. J Immunol 185:5531–5538
Liu H, Rohowsky-Kochan C (2011) Interleukin-27-mediated suppression of human Th17 cells is associated with activation of STAT1 and suppressor of cytokine signaling protein 1. J Interferon Cytokine Res 31:459–469
Logsdon NJ, Jones BC, Josephson K, Cook J, Walter MR (2002) Comparison of interleukin-22 and interleukin-10 soluble receptor complexes. J Interferon Cytokine Res 22:1099–1112
Logsdon NJ, Jones BC, Allman JC, Izotova L, Schwartz B, Pestka S, Walter MR (2004) The IL-10R2 binding hot spot on IL-22 is located on the N-terminal helix and is dependent on N-linked glycosylation. J Mol Biol 342:503–514
Ma HL, Liang S, Li J, Napierata L, Brown T, Benoit S, Senices M, Gill D, Dunussi-Joannopoulos K, Collins M et al (2008) IL-22 is required for Th17 cell-mediated pathology in a mouse model of psoriasis-like skin inflammation. J Clin Invest 118:597–607
Maaser C, Housley MP, Iimura M, Smith JR, Vallance BA, Finlay BB, Schreiber JR, Varki NM, Kagnoff MF, Eckmann L (2004) Clearance of *Citrobacter rodentium* requires B cells but not secretory immunoglobulin A (IgA) or IgM antibodies. Infect Immun 72:3315–3324
Mangan PR, Harrington LE, O'Quinn DB, Helms WS, Bullard DC, Elson CO, Hatton RD, Wahl SM, Schoeb TR, Weaver CT (2006) Transforming growth factor-beta induces development of the T(H)17 lineage. Nature (Lond) 441:231–234
Martin B, Hirota K, Cua DJ, Stockinger B, Veldhoen M (2009) Interleukin-17-producing gammadelta T cells selectively expand in response to pathogen products and environmental signals. Immunity 31:321–330

McGeachy MJ, Cua DJ (2007) The link between IL-23 and Th17 cell-mediated immune pathologies. Semin Immunol 19:372–376

McGee HM, Schmidt BA, Booth CJ, Yancopoulos GD, Valenzuela DM, Murphy AJ, Stevens S, Flavell RA, Horsley V (2013) IL-22 promotes fibroblast-mediated wound repair in the skin. J Invest Dermatol 133(5):1321–1329

Mebius RE (2003) Organogenesis of lymphoid tissues. Nat Rev Immunol 3:292–303

Medema JP, Vermeulen L (2011) Microenvironmental regulation of stem cells in intestinal homeostasis and cancer. Nature (Lond) 474:318–326

Munoz M, Heimesaat MM, Danker K, Struck D, Lohmann U, Plickert R, Bereswill S, Fischer A, Dunay IR, Wolk K et al (2009) Interleukin (IL)-23 mediates *Toxoplasma gondii*-induced immunopathology in the gut via matrix metalloproteinase-2 and IL-22 but independent of IL-17. J Exp Med 206:3047–3059

Musteanu M, Blaas L, Mair M, Schlederer M, Bilban M, Tauber S, Esterbauer H, Mueller M, Casanova E, Kenner L et al (2010) Stat3 is a negative regulator of intestinal tumor progression in Apc(Min) mice. Gastroenterology 138(1003–1011):e1001–e1005

Nagakawa H, Shimozato O, Yu L, Takiguchi Y, Tatsumi K, Kuriyama T, Tagawa M (2004) Expression of interleukin-22 in murine carcinoma cells did not influence tumour growth in vivo but did improve survival of the inoculated hosts. Scand J Immunol 60:449–454

Nagalakshmi ML, Rascle A, Zurawski S, Menon S, de Waal Malefyt R (2004) Interleukin-22 activates STAT3 and induces IL-10 by colon epithelial cells. Int Immunopharmacol 4:679–691

Nagase H, Woessner JF Jr (1999) Matrix metalloproteinases. J Biol Chem 274:21491–21494

Nagem RA, Colau D, Dumoutier L, Renauld JC, Ogata C, Polikarpov I (2002) Crystal structure of recombinant human interleukin-22. Structure 10:1051–1062

Naugler WE, Sakurai T, Kim S, Maeda S, Kim K, Elsharkawy AM, Karin M (2007) Gender disparity in liver cancer due to sex differences in MyD88-dependent IL-6 production. Science 317:121–124

Neufert C, Pickert G, Zheng Y, Wittkopf N, Warntjen M, Nikolaev A, Ouyang W, Neurath MF, Becker C (2010) Activation of epithelial STAT3 regulates intestinal homeostasis. Cell Cycle 9:652–655

Niyonsaba F, Someya A, Hirata M, Ogawa H, Nagaoka I (2001) Evaluation of the effects of peptide antibiotics human beta-defensins-1/-2 and LL-37 on histamine release and prostaglandin D(2) production from mast cells. Eur J Immunol 31:1066–1075

Orelle B, Keim V, Masciotra L, Dagorn JC, Iovanna JL (1992) Human pancreatitis-associated protein. Messenger RNA cloning and expression in pancreatic diseases. J Clin Invest 90:2284–2291

Ota N, Wong K, Valdez PA, Zheng Y, Crellin NK, Diehl L, Ouyang W (2011) IL-22 bridges the lymphotoxin pathway with the maintenance of colonic lymphoid structures during infection with Citrobacter rodentium. Nat Immunol 12:941–948

Ouyang W, Kolls JK, Zheng Y (2008) The biological functions of T helper 17 cell effector cytokines in inflammation. Immunity 28:454–467

Ouyang W, Rutz S, Crellin NK, Valdez PA, Hymowitz SG (2011) Regulation and functions of the IL-10 family of cytokines in inflammation and disease. Annu Rev Immunol 29:71–109

Paget C, Ivanov S, Fontaine J, Renneson J, Blanc F, Pichavant M, Dumoutier L, Ryffel B, Renauld JC, Gosset P et al (2012) Interleukin-22 is produced by invariant natural killer T lymphocytes during influenza A virus infection: potential role in protection against lung epithelial damages. J Biol Chem 287:8816–8829

Pan H, Hong F, Radaeva S, Gao B (2004) Hydrodynamic gene delivery of interleukin-22 protects the mouse liver from concanavalin A-, carbon tetrachloride-, and Fas ligand-induced injury via activation of STAT3. Cell Mol Immunol 1:43–49

Papp KA, Leonardi C, Menter A, Ortonne JP, Krueger JG, Kricorian G, Aras G, Li J, Russell CB, Thompson EH, Baumgartner S (2012) Brodalumab, an anti-interleukin-17-receptor antibody for psoriasis. N Engl J Med 366:1181–1189

Park H, Li Z, Yang XO, Chang SH, Nurieva R, Wang YH, Wang Y, Hood L, Zhu Z, Tian Q, Dong C (2005) A distinct lineage of CD4 T cells regulates tissue inflammation by producing interleukin 17. Nat Immunol 6:1133–1141

Park O, Wang H, Weng H, Feigenbaum L, Li H, Yin S, Ki SH, Yoo SH, Dooley S, Wang FS et al (2011) In vivo consequences of liver-specific interleukin-22 expression in mice: Implications for human liver disease progression. Hepatology 54:252–261

Paulos CM, Carpenito C, Plesa G, Suhoski MM, Varela-Rohena A, Golovina TN, Carroll RG, Riley JL, June CH (2010) The inducible costimulator (ICOS) is critical for the development of human T(H)17 cells. Sci Transl Med 2:55ra78

Pellegrini M, Calzascia T, Toe JG, Preston SP, Lin AE, Elford AR, Shahinian A, Lang PA, Lang KS, Morre M et al (2011) IL-7 engages multiple mechanisms to overcome chronic viral infection and limit organ pathology. Cell 144:601–613

Pene J, Chevalier S, Preisser L, Venereau E, Guilleux MH, Ghannam S, Moles JP, Danger Y, Ravon E, Lesaux S et al (2008) Chronically inflamed human tissues are infiltrated by highly differentiated Th17 lymphocytes. J Immunol 180:7423–7430

Pickert G, Neufert C, Leppkes M, Zheng Y, Wittkopf N, Warntjen M, Lehr HA, Hirth S, Weigmann B, Wirtz S et al (2009) STAT3 links IL-22 signaling in intestinal epithelial cells to mucosal wound healing. J Exp Med 206:1465–1472

Pidasheva S, Trifari S, Phillips A, Hackney JA, Ma Y, Smith A, Sohn SJ, Spits H, Little RD, Behrens TW et al (2011) Functional studies on the IBD susceptibility gene IL23R implicate reduced receptor function in the protective genetic variant R381Q. PLoS One 6:e25038

Puel A, Doffinger R, Natividad A, Chrabieh M, Barcenas-Morales G, Picard C, Cobat A, Ouachee-Chardin M, Toulon A, Bustamante J et al (2010) Autoantibodies against IL-17A, IL-17F, and IL-22 in patients with chronic mucocutaneous candidiasis and autoimmune polyendocrine syndrome type I. J Exp Med 207:291–297

Puel A, Cypowyj S, Bustamante J, Wright JF, Liu L, Lim HK, Migaud M, Israel L, Chrabieh M, Audry M et al (2011) Chronic mucocutaneous candidiasis in humans with inborn errors of interleukin-17 immunity. Science 332:65–68

Qian Y, Liu C, Hartupee J, Altuntas CZ, Gulen MF, Jane-Wit D, Xiao J, Lu Y, Giltiay N, Liu J et al (2007) The adaptor Act1 is required for interleukin 17-dependent signaling associated with autoimmune and inflammatory disease. Nat Immunol 8:247–256

Quintana FJ, Basso AS, Iglesias AH, Korn T, Farez MF, Bettelli E, Caccamo M, Oukka M, Weiner HL (2008) Control of T(reg) and T(H)17 cell differentiation by the aryl hydrocarbon receptor. Nature 453:65–71

Radaeva S, Sun R, Pan HN, Hong F, Gao B (2004) Interleukin 22 (IL-22) plays a protective role in T cell-mediated murine hepatitis: IL-22 is a survival factor for hepatocytes via STAT3 activation. Hepatology 39:1332–1342

Ramirez JM, Brembilla NC, Sorg O, Chicheportiche R, Matthes T, Dayer JM, Saurat JH, Roosnek E, Chizzolini C (2010) Activation of the aryl hydrocarbon receptor reveals distinct requirements for IL-22 and IL-17 production by human T helper cells. Eur J Immunol 40:2450–2459

Ren X, Hu B, Colletti LM (2010) IL-22 is involved in liver regeneration after hepatectomy. Am J Physiol Gastrointest Liver Physiol 298:G74–G80

Roth J, Vogl T, Sorg C, Sunderkotter C (2003) Phagocyte-specific S100 proteins: a novel group of proinflammatory molecules. Trends Immunol 24:155–158

Rutz S, Ouyang W (2011) Regulation of interleukin-10 and interleukin-22 expression in T helper cells. Curr Opin Immunol 23:605–612

Rutz S, Noubade R, Eidenschenk C, Ota N, Zeng W, Zheng Y, Hackney J, Ding J, Singh H, Ouyang W (2011) Transcription factor c-Maf mediates the TGF-beta-dependent suppression of IL-22 production in T(H)17 cells. Nat Immunol 12:1238–1245

Sa SM, Valdez PA, Wu J, Jung K, Zhong F, Hall L, Kasman I, Winer J, Modrusan Z, Danilenko DM, Ouyang W (2007) The effects of IL-20 subfamily cytokines on reconstituted human epidermis suggest potential roles in cutaneous innate defense and pathogenic adaptive immunity in psoriasis. J Immunol 178:2229–2240

Satoh-Takayama N, Vosshenrich CA, Lesjean-Pottier S, Sawa S, Lochner M, Rattis F, Mention JJ, Thiam K, Cerf-Bensussan N, Mandelboim O et al (2008) Microbial flora drives interleukin 22 production in intestinal NKp46+ cells that provide innate mucosal immune defense. Immunity 29:958–970

Satoh-Takayama N, Dumoutier L, Lesjean-Pottier S, Ribeiro VS, Mandelboim O, Renauld JC, Vosshenrich CA, Di Santo JP (2009) The natural cytotoxicity receptor NKp46 is dispensable for IL-22-mediated innate intestinal immune defense against *Citrobacter rodentium*. J Immunol 183:6579–6587

Satoh-Takayama N, Lesjean-Pottier S, Vieira P, Sawa S, Eberl G, Vosshenrich CA, Di Santo JP (2010) IL-7 and IL-15 independently program the differentiation of intestinal CD3-NKp46+ cell subsets from Id2-dependent precursors. J Exp Med 207:273–280

Sawa S, Cherrier M, Lochner M, Satoh-Takayama N, Fehling HJ, Langa F, Di Santo JP, Eberl G (2010) Lineage relationship analysis of RORgammat+ innate lymphoid cells. Science 330:665–669

Schmechel S, Konrad A, Diegelmann J, Glas J, Wetzke M, Paschos E, Lohse P, Goke B, Brand S (2008) Linking genetic susceptibility to Crohn's disease with Th17 cell function: IL-22 serum levels are increased in Crohn's disease and correlate with disease activity and IL23R genotype status. Inflamm Bowel Dis 14:204–212

Shioya M, Andoh A, Kakinoki S, Nishida A, Fujiyama Y (2008) Interleukin 22 receptor 1 expression in pancreas islets. Pancreas 36:197–199

Simons BD, Clevers H (2011) Stem cell self-renewal in intestinal crypt. Exp Cell Res 317:2719–2724

Sonnenberg GF, Nair MG, Kirn TJ, Zaph C, Fouser LA, Artis D (2010) Pathological versus protective functions of IL-22 in airway inflammation are regulated by IL-17A. J Exp Med 207: 1293–1305

Sonnenberg GF, Monticelli LA, Elloso MM, Fouser LA, Artis D (2011) CD4(+) lymphoid tissue-inducer cells promote innate immunity in the gut. Immunity 34:122–134

Sonnenberg GF, Monticelli LA, Alenghat T, Fung TC, Hutnick NA, Kunisawa J, Shibata N, Grunberg S, Sinha R, Zahm AM et al (2012) Innate lymphoid cells promote anatomical containment of lymphoid-resident commensal bacteria. Science 336:1321–1325

Spits H, Di Santo JP (2011) The expanding family of innate lymphoid cells: regulators and effectors of immunity and tissue remodeling. Nat Immunol 12:21–27

Spits H, Artis D, Colonna M, Diefenbach A, Di Santo JP, Eberl G, Koyasu S, Locksley RM, McKenzie AN, Mebius RE et al (2013) Innate lymphoid cells: a proposal for uniform nomenclature. Nat Rev Immunol 13:145–149

Steel DM, Whitehead AS (1994) The major acute phase reactants: C-reactive protein, serum amyloid P component and serum amyloid A protein. Immunol Today 15:81–88

Strange A, Capon F, Spencer CC, Knight J, Weale ME, Allen MH, Barton A, Band G, Bellenguez C, Bergboer JG et al (2010) A genome-wide association study identifies new psoriasis susceptibility loci and an interaction between HLA-C and ERAP1. Nat Genet 42:985–990

Sugimoto K, Ogawa A, Mizoguchi E, Shimomura Y, Andoh A, Bhan AK, Blumberg RS, Xavier RJ, Mizoguchi A (2008) IL-22 ameliorates intestinal inflammation in a mouse model of ulcerative colitis. J Clin Invest 118:534–544

Sun Z, Unutmaz D, Zou YR, Sunshine MJ, Pierani A, Brenner-Morton S, Mebius RE, Littman DR (2000) Requirement for RORgamma in thymocyte survival and lymphoid organ development. Science 288:2369–2373

Sutton CE, Lalor SJ, Sweeney CM, Brereton CF, Lavelle EC, Mills KH (2009) Interleukin-1 and IL-23 induce innate IL-17 production from gammadelta T cells, amplifying Th17 responses and autoimmunity. Immunity 31:331–341

Takahashi K, Hirose K, Kawashima S, Niwa Y, Wakashin H, Iwata A, Tokoyoda K, Renauld JC, Iwamoto I, Nakayama T, Nakajima H (2011) IL-22 attenuates IL-25 production by lung epithelial cells and inhibits antigen-induced eosinophilic airway inflammation. J Allergy Clin Immunol 128(1067–1076):e1061–e1066

Takatori H, Kanno Y, Watford WT, Tato CM, Weiss G, Ivanov II, Littman DR, O'Shea JJ (2009) Lymphoid tissue inducer-like cells are an innate source of IL-17 and IL-22. J Exp Med 206:35–41

Taube C, Tertilt C, Gyulveszi G, Dehzad N, Kreymborg K, Schneeweiss K, Michel E, Reuter S, Renauld JC, Arnold-Schild D et al (2011) IL-22 is produced by innate lymphoid cells and limits inflammation in allergic airway disease. PLoS One 6:e21799

Tohyama M, Hanakawa Y, Shirakata Y, Dai X, Yang L, Hirakawa S, Tokumaru S, Okazaki H, Sayama K, Hashimoto K (2009) IL-17 and IL-22 mediate IL-20 subfamily cytokine production in cultured keratinocytes via increased IL-22 receptor expression. Eur J Immunol 39: 2779–2788

Trifari S, Kaplan CD, Tran EH, Crellin NK, Spits H (2009) Identification of a human helper T cell population that has abundant production of interleukin 22 and is distinct from T(H)-17, T(H)1 and T(H)2 cells. Nat Immunol 10:864–871

Upadhyay V, Poroyko V, Kim TJ, Devkota S, Fu S, Liu D, Tumanov AV, Koroleva EP, Deng L, Nagler C et al (2012) Lymphotoxin regulates commensal responses to enable diet-induced obesity. Nat Immunol 13:947–953

Van Belle AB, de Heusch M, Lemaire MM, Hendrickx E, Warnier G, Dunussi-Joannopoulos K, Fouser LA, Renauld JC, Dumoutier L (2012) IL-22 is required for imiquimod-induced psoriasiform skin inflammation in mice. J Immunol 188:462–469

van der Fits L, Mourits S, Voerman JS, Kant M, Boon L, Laman JD, Cornelissen F, Mus AM, Florencia E, Prens EP, Lubberts E (2009) Imiquimod-induced psoriasis-like skin inflammation in mice is mediated via the IL-23/IL-17 axis. J Immunol 182:5836–5845

Van der Sluis M, De Koning BA, De Bruijn AC, Velcich A, Meijerink JP, Van Goudoever JB, Buller HA, Dekker J, Van Seuningen I, Renes IB, Einerhand AW (2006) Muc2-deficient mice spontaneously develop colitis, indicating that MUC2 is critical for colonic protection. Gastroenterology 131:117–129

Veldhoen M, Hocking RJ, Atkins CJ, Locksley RM, Stockinger B (2006) TGFbeta in the context of an inflammatory cytokine milieu supports de novo differentiation of IL-17-producing T cells. Immunity 24:179–189

Veldhoen M, Hirota K, Westendorf AM, Buer J, Dumoutier L, Renauld JC, Stockinger B (2008) The aryl hydrocarbon receptor links TH17-cell-mediated autoimmunity to environmental toxins. Nature 453:106–109

Volpe E, Touzot M, Servant N, Marloie-Provost MA, Hupe P, Barillot E, Soumelis V (2009) Multiparametric analysis of cytokine-driven human Th17 differentiation reveals a differential regulation of IL-17 and IL-22 production. Blood 114:3610–3614

Wahl C, Wegenka UM, Leithauser F, Schirmbeck R, Reimann J (2009) IL-22-dependent attenuation of T cell-dependent (ConA) hepatitis in herpes virus entry mediator deficiency. J Immunol 182:4521–4528

Wang Z, Yang L, Jiang Y, Ling ZQ, Li Z, Cheng Y, Huang H, Wang L, Pan Y, Yan X, Chen Y (2011) High fat diet induces formation of spontaneous liposarcoma in mouse adipose tissue with overexpression of interleukin 22. PLoS One 6:e23737

Wang P, Bai F, Zenewicz LA, Dai J, Gate D, Cheng G, Yang L, Qian F, Yuan X, Montgomery RR et al (2012) IL-22 signaling contributes to West Nile encephalitis pathogenesis. PLoS One 7:e44153

Wang C, Wu L, Bulek K, Martin BN, Zepp JA, Kang Z, Liu C, Herjan T, Misra S, Carman JA et al (2013) The psoriasis-associated D10N variant of the adaptor Act1 with impaired regulation by the molecular chaperone hsp90. Nat Immunol 14:72–81

Watanabe T, Yonemura Y, Yonekura H, Suzuki Y, Miyashita H, Sugiyama K, Moriizumi S, Unno M, Tanaka O, Kondo H et al (1994) Pancreatic beta-cell replication and amelioration of surgical diabetes by Reg protein. Proc Natl Acad Sci USA 91:3589–3592

Watson PH, Leygue ER, Murphy LC (1998) Psoriasin (S100A7). Int J Biochem Cell Biol 30:567–571

Weiss B, Wolk K, Grunberg BH, Volk HD, Sterry W, Asadullah K, Sabat R (2004) Cloning of murine IL-22 receptor alpha 2 and comparison with its human counterpart. Genes Immun 5:330–336

Whittington HA, Armstrong L, Uppington KM, Millar AB (2004) Interleukin-22: a potential immunomodulatory molecule in the lung. Am J Respir Cell Mol Biol 31:220–226

Wolk K, Kunz S, Asadullah K, Sabat R (2002) Cutting edge: immune cells as sources and targets of the IL-10 family members? J Immunol 168:5397–5402
Wolk K, Kunz S, Witte E, Friedrich M, Asadullah K, Sabat R (2004) IL-22 increases the innate immunity of tissues. Immunity 21:241–254
Wolk K, Witte E, Wallace E, Docke WD, Kunz S, Asadullah K, Volk HD, Sterry W, Sabat R (2006) IL-22 regulates the expression of genes responsible for antimicrobial defense, cellular differentiation, and mobility in keratinocytes: a potential role in psoriasis. Eur J Immunol 36:1309–1323
Wolk K, Witte E, Hoffmann U, Doecke WD, Endesfelder S, Asadullah K, Sterry W, Volk HD, Wittig BM, Sabat R (2007) IL-22 induces lipopolysaccharide-binding protein in hepatocytes: a potential systemic role of IL-22 in Crohn's disease. J Immunol 178:5973–5981
Wolk K, Haugen HS, Xu W, Witte E, Waggie K, Anderson M, Vom Baur E, Witte K, Warszawska K, Philipp S et al (2009) IL-22 and IL-20 are key mediators of the epidermal alterations in psoriasis while IL-17 and IFN-gamma are not. J Mol Med 87:523–536
Xie MH, Aggarwal S, Ho WH, Foster J, Zhang Z, Stinson J, Wood WI, Goddard AD, Gurney AL (2000) Interleukin (IL)-22, a novel human cytokine that signals through the interferon receptor-related proteins CRF2-4 and IL-22R. J Biol Chem 275:31335–31339
Xu W, Presnell SR, Parrish-Novak J, Kindsvogel W, Jaspers S, Chen Z, Dillon SR, Gao Z, Gilbert T, Madden K et al (2001) A soluble class II cytokine receptor, IL-22RA2, is a naturally occurring IL-22 antagonist. Proc Natl Acad Sci USA 98:9511–9516
Xu T, Logsdon NJ, Walter MR (2004) Crystallization and X-ray diffraction analysis of insect-cell-derived IL-22. Acta Crystallogr D Biol Crystallogr 60:1295–1298
Xu T, Logsdon NJ, Walter MR (2005) Structure of insect-cell-derived IL-22. Acta Crystallogr D Biol Crystallogr 61:942–950
Yang D, Chertov O, Bykovskaia SN, Chen Q, Buffo MJ, Shogan J, Anderson M, Schroder JM, Wang JM, Howard OM, Oppenheim JJ (1999) Beta-defensins: linking innate and adaptive immunity through dendritic and T cell CCR6. Science 286:525–528
Yang XO, Panopoulos AD, Nurieva R, Chang SH, Wang D, Watowich SS, Dong C (2007) STAT3 regulates cytokine-mediated generation of inflammatory helper T cells. J Biol Chem 282: 9358–9363
Yang XO, Pappu BP, Nurieva R, Akimzhanov A, Kang HS, Chung Y, Ma L, Shah B, Panopoulos AD, Schluns KS et al (2008) T helper 17 lineage differentiation is programmed by orphan nuclear receptors ROR alpha and ROR gamma. Immunity 28:29–39
Yokota Y, Mansouri A, Mori S, Sugawara S, Adachi S, Nishikawa S, Gruss P (1999) Development of peripheral lymphoid organs and natural killer cells depends on the helix-loop-helix inhibitor Id2. Nature (Lond) 397:702–706
Zelante T, Iannitti R, De Luca A, Romani L (2011) IL-22 in antifungal immunity. Eur J Immunol 41:270–275
Zenewicz LA, Yancopoulos GD, Valenzuela DM, Murphy AJ, Karow M, Flavell RA (2007) Interleukin-22 but not interleukin-17 provides protection to hepatocytes during acute liver inflammation. Immunity 27:647–659
Zenewicz LA, Yancopoulos GD, Valenzuela DM, Murphy AJ, Stevens S, Flavell RA (2008) Innate and adaptive interleukin-22 protects mice from inflammatory bowel disease. Immunity 29: 947–957
Zhang YW, Ding LS, Lai MD (2003) Reg gene family and human diseases. World J Gastroenterol 9:2635–2641
Zheng Y, Danilenko DM, Valdez P, Kasman I, Eastham-Anderson J, Wu J, Ouyang W (2007) Interleukin-22, a T(H)17 cytokine, mediates IL-23-induced dermal inflammation and acanthosis. Nature (Lond) 445:648–651
Zheng Y, Valdez PA, Danilenko DM, Hu Y, Sa SM, Gong Q, Abbas AR, Modrusan Z, Ghilardi N, de Sauvage FJ, Ouyang W (2008) Interleukin-22 mediates early host defense against attaching and effacing bacterial pathogens. Nat Med 14:282–289
Ziesche E, Bachmann M, Kleinert H, Pfeilschifter J, Muhl H (2007) The interleukin-22/STAT3 pathway potentiates expression of inducible nitric-oxide synthase in human colon carcinoma cells. J Biol Chem 282:16006–16015

Chapter 7
IL-23 in Health and Disease

Jonathan P. Sherlock, Luis A. Zuniga, and Daniel J. Cua

Abstract The discovery of IL-23 and characterization of the biology governed by this cytokine has led to dramatic new insights within immunology. IL-23 is central to the pathogenesis of many immune-mediated inflammatory disorders and was shown to act via a novel IL-17-producing T helper subset, the Th17 cell. Many innate cells also respond to IL-23 and are important in both resistance to infection and in mediating autoimmune pathology. These cells are characterized by expression of RORγt transcription regulator and include subsets of γδ T cells, natural killer (NK) cells, innate lymphoid cells in the intestine and a recently described entheseal resident cell that resides at a key site for rheumatic illness. Stimulation of these various cells by IL-23 induces severe inflammation, much of which is mediated by the signature cytokines, including IL-17 and IL-22. Exploration of the biology of IL-23 has thus led to the identification of novel immune cell subsets as well as new paradigms for disease localization based upon the distribution of IL-23-responder populations. These new concepts are at the core of novel therapeutic strategies based upon neutralization of IL-23 or IL-17, which are showing encouraging results for treatment of psoriasis, multiple sclerosis, rheumatoid arthritis, and ankylosing spondylitis.

Keywords Autoimmunity • Inflammation • Th17 cells

J.P. Sherlock
Centre for Translational Inflammation Research, University of Birmingham, Birmingham, 15 2TT UK

Nuffield Department of Orthopaedics, Rheumatology and Musculoskeletal Sciences, University of Oxford, Oxford, 3 7HE UK

L.A. Zuniga • D.J. Cua (✉)
Merck Research Labs, 901 S. California Ave, Palo Alto, CA 94304, USA
e-mail: daniel.cua@merck.com

T. Yoshimoto and T. Yoshimoto (eds.), *Cytokine Frontiers: Regulation of Immune Responses in Health and Disease*, DOI 10.1007/978-4-431-54442-5_7, © Springer Japan 2014

7.1 Discovery and Immune Function of IL-23

7.1.1 Historical Perspective: A Case of Mistaken Identity

The discovery of IL-23 and its influence on T-cell biology led to a substantial revision in our understanding of the key cytokines and T-lymphocyte subsets in immune-mediated inflammation. Before the turn of the millennium, classical thought in immunology held that two major divisions of T helper (Th) lymphocytes existed: Th1 and Th2 cells (Mosmann et al. 1986). Although Th2 cells were known to play an important role in promoting allergic responses and combating extracellular bacterial and parasitic infection, Th1 cells were demonstrated to be critical in the response to intracellular viral and bacterial infections through their characteristic production of interferon (IFN)-γ and were also proposed to be the central players in autoimmune inflammation. However, despite this prevailing paradigm, it was well recognized that IFN-γ and the IFN-γ receptor (IFN-γR) are not necessary for development of experimental autoimmune encephalomyelitis (EAE), a prototypical model of neurodegenerative autoimmune disease (Ferber et al. 1996; Willenborg et al. 1996). Indeed, IFN-γR deficiency or IFN-γ neutralization in mice leads to significantly worse EAE pathology (Willenborg et al. 1996; Billiau et al. 1988; Lublin et al. 1993; Duong et al. 1992; Heremans et al. 1996). Conversely, administration of recombinant IFN-γ could suppress the development of EAE (Voorthuis et al. 1990). Nevertheless, an importance for Th1 cells continued to be implied by experiments that demonstrated that the p40 subunit of IL-12 is necessary for the development of EAE. As IL-12 is the key differentiation factor for Th1 cells, it was assumed that these cells were fundamental to EAE pathogenesis. These conclusions appeared to be supported by experiments demonstrating that genetic absence or antibody-mediated antagonism of IL-12p40 blocked development of EAE (Segal et al. 1998). A few years later, however, it was demonstrated that although IL-12p40 is essential for development of EAE, deletion of the other IL-12 subunit, p35, does not prevent this disease (Becher et al. 2002). These findings were extended to show that deletion of a chain of the receptor for IL-12, IL-12Rβ2, is also unable to block the pathology (Zhang et al. 2003). Together these studies questioned whether Th1 cells and their associated cytokines are truly critical in the pathogenesis of EAE and, more broadly, whether these factors are fundamental to autoimmune inflammation in these settings.

7.1.2 IL-23 Is the Critical Factor in Autoimmune Pathology

The puzzling findings regarding the discordant roles of IL-12p40 and p35 in EAE were explained only when our understanding of the IL-12 family of cytokines was expanded. Systematic searching of sequence databases for additional members of the IL-6 helical cytokine family led to the discovery of a novel protein with a molecular weight of approximately 19 kDa (Oppmann et al. 2000). This protein, termed "p19," was found to be closely related in sequence to IL-12p35, IL-6, and granulocyte colony-stimulating factor (G-CSF), yet it was poorly secreted when expressed alone in cells. By analogy with IL-12, which is composed of two chains, p35 and p40, it was hypothesized that p19 might also pair with another protein. Coexpression studies revealed that the p40 subunit of IL-12, but not other related proteins, could pair with p19 to form a secreted p19–p40 heterodimer. Additionally, natural p19–p40 heterodimers were found to be secreted by activated dendritic cells. This novel heterodimer was named IL-23 (Oppmann et al. 2000).

Because IL-23 shares the p40 subunit with IL-12, the possibility emerged that it is actually IL-23, and not IL-12, which is critically important in the development of autoimmune inflammation in the models studied, as the neutralization experiments targeting p40 would block both IL-12 and IL-23. This hypothesis was directly tested by producing mice specifically deficient in IL-23 through deletion of the p19 subunit. With these mice it was confirmed that IL-23, and not IL-12, is the essential cytokine in EAE (Cua et al. 2003). These findings were extended to show that IL-23 is also essential in animal models of inflammatory arthritis (Murphy et al. 2003) and inflammatory bowel disease (Yen et al. 2006), emphasizing its central role in various immune-mediated pathologies.

7.1.3 The Cellular Responders to IL-23: Th17 Cells

Given the known importance of IL-12 in the induction of Th1 cells and the homology of IL-12 and IL-23, it seemed logical to test whether IL-23 could act to induce the polarization of a distinct T-cell subset critical for mediating autoimmune inflammation. Indeed, an early report demonstrated that IL-23 stimulation results in production of IL-17 by memory T lymphocytes (Aggarwal et al. 2003). At the same time, we showed that IL-23p19-deficient mice lacked IL-17-producing cells in a model of inflammatory arthritis (Murphy et al. 2003), suggesting that IL-17$^+$ cells are selectively induced by IL-23. However, it is important to note that IL-23 in this system was found to act preferentially on *memory* cells, which is consistent with the original description of this cytokine in stimulating proliferation of these mature cells (Oppmann et al. 2000). Moreover, the observation that IL-23R is present on memory cells but not naive T cells (Parham et al. 2002; Awasthi et al. 2009) suggests that IL-23 cannot act at the very earliest stages of T-cell polarization. As expected, IL-23 was shown to be dispensable for polarization of naive cells toward IL-17-producing T cells (Th17), which develop instead in response to IL-6 and transforming growth factor (TGF)-β stimulation (Bettelli et al. 2006; Veldhoen et al. 2006a), with IL-6 inducing expression of IL-23R (Mangan et al. 2006). As TGF-β was known to also induce expression of Foxp3 and differentiation of Treg cells, attention focused on how IL-6 might inhibit this effect. Experiments demonstrated that IL-6 plays a critical role by antagonizing Foxp3 expression (Bettelli et al. 2006) and can induce Th17 cells through its induction of STAT3 and the Th17 lineage-determining transcription factor RORγt (Yang et al. 2007) (Fig. 7.1). Intestinal lamina propria cells from IL-6-deficient mice are thus deficient in expression of key Th17 cell factors, including IL-17F, RORγt, and IL-23R (Ivanov et al. 2006).

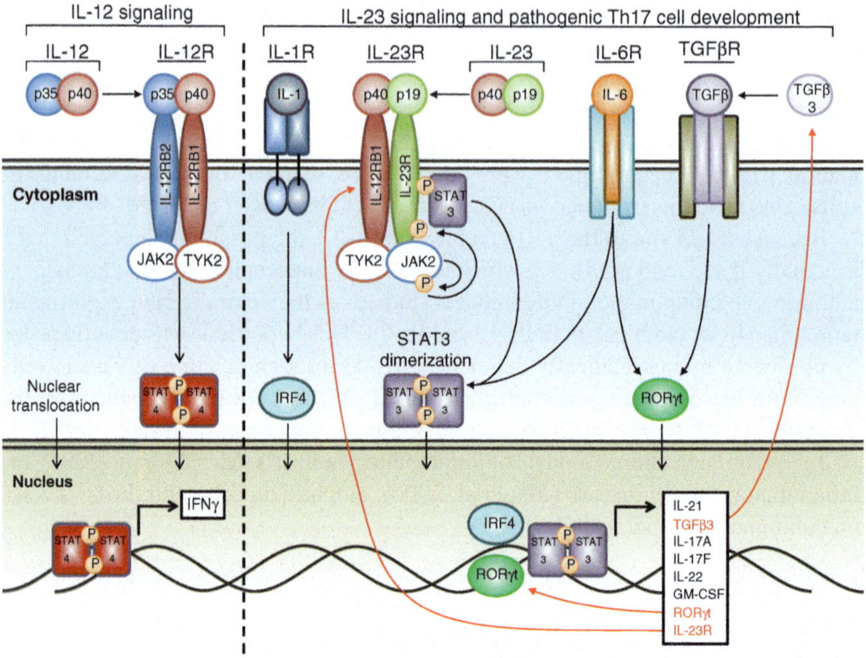

Fig. 7.1 Interleukin (IL)-12 and IL-23 signaling and factors contributing to pathogenic Th17 cell development. Th17 cell polarization occurs in the presence of IL-6 and transforming growth factor (TGF)-β, which activate STAT3 and RORγt, a lineage-specific transcription regulator. Both RORγt and STAT3 translocate to the nucleus to promote expression of Th17-specific genes, including IL-17 and IL-23R. If IL-23 is present, engagement with its receptor on a newly polarized Th17 cell allows TYK2 and JAK2 recruitment to the IL-23 receptor. Phosphorylation of JAK2 allows autophosphorylation of the receptor, which then facilitates STAT3 recruitment and activation (pSTAT3). pSTAT3 homodimerizes and translocates to the nucleus, thus amplifying expression of pathogenic Th17 cell genes in a feed-forward loop (*red arrows*) and stabilizing pathogenic Th17 cell effector commitment. IL-1 can contribute to pathogenic Th17 cell polarization via activation of IRF4, which can also participate in Th17 gene expression. It is likely that IRF4, RORγt, and STAT3 act in a molecular complex to regulate pathogenic Th17 genes. In contrast to IL-23R signaling, IL-12 signaling leads to STAT4 activation, which is important in driving Th1 cell-specific genes [e.g., interferon (IFN)-γ]

STAT3 itself is critical in Th17 cell development because deficiency of this factor greatly inhibits development of Th17 cells, whereas a hyperactive form enhances their polarization (Yang et al. 2007). Moreover, IL-21 can act similarly to IL-6, both cytokines being important activators of STAT3 (Yang et al. 2007; Zeng et al. 2007) and IL-21 can also induce Th17 polarization in combination with TGF-β (Korn et al. 2007). The importance of STAT3 is emphasized by its ability to directly bind and induce transcription of the *IL-23R* locus (Ghoreschi et al. 2010), which confers responsiveness to IL-23. Consequently, human data demonstrate that certain STAT3 mutations are associated with hyper-IgE syndrome with severe immunodeficiency, and with defective responsiveness to IL-6 (Minegishi et al. 2007) and impaired development of Th17 cells (Milner et al. 2008).

7.1.4 *Stabilization of Pathogenic Th17 Cells by IL-23*

Despite the ability of TG-β and IL-6 to induce IL-17-producing T cells, the development of these cells is incomplete in the absence of IL-23. Hence, IL-23R$^{-/-}$ cells have both reduced proliferation and IL-17 production and fail to accumulate in the central nervous system during autoimmune encephalomyelitis (McGeachy et al. 2009). In the absence of IL-23R there is impaired formation and maturation of Th17 cells as indicated by abnormal maintenance of the early T-cell differentiation factor IL-2, combined with defective upregulation of IL-7Rα (McGeachy et al. 2009). Through these studies, IL-23 has emerged as essential for the development of pathogenic inflammatory Th17 cells (McGeachy et al. 2009; Ghoreschi et al. 2010).

Because expression of IL-23R is known from this work to be essential for Th17 cells to fully adopt pathogenic effector functions, the factors promoting IL-23R itself are of immense interest. IL-6-induced STAT3 is important in induction of IL-23R expression, and once this receptor is expressed, IL-23 can then synergize with IL-6 by acting on STAT3 through the IL-23R. IL-23 thus promotes expression of its own receptor (Ghoreschi et al. 2010; Zúñiga et al. 2013). This positive feedback stabilizes the Th17 cell phenotype and confers ongoing responsiveness to IL-23. A key stabilization step for Th17 cell fate is obtained through induction of the lineage-determining transcription factor RORγt, analogous to T-bet for Th1 cells. This factor is essential for promotion of autoimmune pathology and differentiation of tissue-infiltrating Th17 cells in vivo, and its forced expression directly induces the production of IL-17 (Ivanov et al. 2006). Recent data demonstrate that IL-23 further reinforces Th17 cell pathogenicity by promoting expression of TGF-β3 during early Th17 cell development (Lee et al. 2012). TGF-β3 may contribute to pathogenic Th17 cell polarization via a feed-forward upregulation of IL-23R. Investigations have also revealed that GM-CSF is a key disease-promoting factor produced by pathogenic Th17 cells. This molecule, the expression of which is induced by RORγ, is sufficient to induce EAE in the genetic absence of IL-17 and IFN-γ, whereas GM-CSF-deficient T cells are unable to induce disease (Codarri et al. 2011). Moreover, IL-23 creates a positive feedback loop to promote maintenance of pathogenicity because IL-23 itself induces expression of GM-CSF from Th17 cells and this GM-CSF can then induce further production of IL-23 from antigen-presenting cells (El-Behi et al. 2011). The interplay of these factors allows continuance of pathological IL-23- and Th17-mediated autoimmune inflammation.

7.1.5 *TGF-β-Independent Induction of Th17 Cells*

TGF-β can participate in, but may not be sufficient for, Th17 cell generation. Although mice expressing a dominant-negative form of TGF-β receptor II in $CD4^+$ T cells display defective development of Th17 cells (Veldhoen et al. 2006b), both IL-17$^+$ and RORγt$^+$ cells are present in the intestinal lamina propria of these mice as well as in TGF-β-receptor I mutant mice (Ghoreschi et al. 2010). Furthermore, Th17 cells can be generated in vitro in the absence of TGF-β by using serum-free conditions and neutralizing anti-TGF-β antibodies (Ghoreschi et al. 2010), suggesting factors other than TGF-β can mediate Th17 cell development. In these TGF-β-deficient conditions, IL-6 and IL-23 in the presence of IL-1β are sufficient to drive Th17 cell development, with IL-6 acting through STAT3 to trigger expression of IL-23R and this expression being further enhanced by IL-23 itself. This novel, IL-23-dependent Th17 cell polarization is distinct from the previously described induction by TGF-β and IL-6, because under these conditions TGF-β inhibits Th17 polarization and IL-23R induction (Ghoreschi et al. 2010).

7.1.6 Responsiveness of Innate Immune Cells to IL-23

Following initial characterization of the critical role of IL-23 in autoimmune inflammation (Cua et al. 2003), investigations focused almost exclusively on the effects of this cytokine on adaptive, antigen-dependent T cells, perhaps partly because of the focus of studying the action of IL-23 in chronic, nonresolving autoimmune diseases driven by activated T cells, which respond to ubiquitously present self-antigens. However, the biology of the IL-23–IL-17 axis has also consistently demonstrated its physiological role in resolving acute infections as these cytokines are produced at very early stages (Lockhart et al. 2006) and act rapidly (Happel et al. 2003, 2005; Huang et al. 2004) to clear infection, after which the response is terminated.

The ability of IL-23 to drive innate inflammation in noninfectious situations was demonstrated using RAG-deficient mice in which lamina propria cells can respond directly to IL-23 in the absence of previous conditioning with other cytokines (Buonocore et al. 2010). Indeed, IL-23 is absolutely required for the intestinal pathology in RAG-deficient mice driven by an anti-CD40 antibody that stimulates antigen-presenting cells by mimicking CD40L on T lymphocytes, and is abrogated in IL-23p19-deficient mice (Uhlig et al. 2006). The cellular responders to IL-23 have recently been demonstrated to be c-Kit$^+$, SCA1$^+$, and RORγt$^+$ "innate lymphoid cells" (Buonocore et al. 2010) (Fig. 7.2).

A further IL-23-responsive innate lymphoid cell (ILC) was characterized in human mucosa-associated lymphoid tissue, and these cells express the NK cell receptor NKp44, RORγt, as well as CCR6, a chemokine receptor present on Th17 cells (Cella et al. 2009). Mucosal NKp44$^+$ cells do not function as classical NK cells, which release perforin and IFN-γ, but instead respond to IL-23 by producing IL-22. However, these mucosal NKp44$^+$ cells do not produce IL-17. Equivalent NK-like cell types have also been identified in mouse mucosal tissues, including the lamina propria and intraepithelial surfaces of the small intestine. These NK-like cells have low expression of the NK cell marker NK1.1 but express NKp46, CD127, and c-Kit (not NKp44, which is only expressed in humans), and are primed for rapid, immediate responsiveness to IL-23 (Cella et al. 2009; Luci et al. 2009; Sanos et al. 2009; Satoh-Takayama et al. 2008).

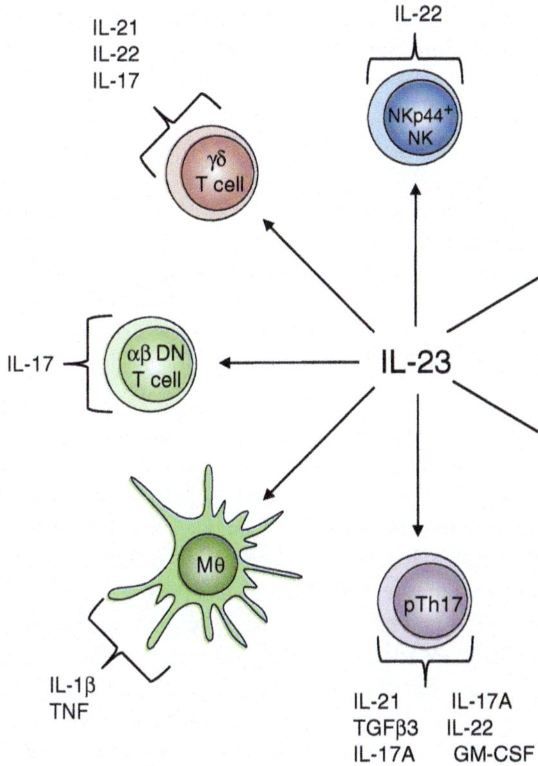

Fig. 7.2 IL-23-responsive cell subsets: various reported cell subsets that can respond to IL-23, and the cytokines associated with IL-23 signaling in these cells. It is important to note that only a subset of γδ T cells, ILC, macrophages, and NK cells are RORγt⁺ and respond to IL-23R-mediated stimulation. *ILC* innate lymphoid cell, *NK* natural killer, *iNKT cell* invariant NK T cell, *pTh17* pathogenic Th17 cells, *Mθ* macrophage, *DN* double negative, *GM-CSF* macrophage-granulocyte colony-stimulating factor

Cells that have historically been described as lymphoid tissue inducer (LTi)-like cells, because their surface phenotype is similar to those cells responsible for the formation of lymphoid tissues, have recently been shown to respond to IL-23. Through its action on these IL-23R$^+$ cells, IL-23 is able to stimulate the production of IL-17 and IL-22 in RAG-deficient splenocytes, and these cells, despite their extremely small numbers, allow RAG-deficient splenocytes to secrete as much as one third of the IL-17 produced by wild-type splenocytes (Takatori et al. 2009). The relevance of these cells was demonstrated in the innate response to in vivo challenge with a fungal molecule, zymosan, during which they provide an innate source of IL-17 and IL-22. Human LTi-like cells have similarly been shown to produce IL-17 and IL-22 (Cupedo et al. 2009). Similar to Th17 cells (Ivanov et al. 2006) and the intestinal "innate lymphocytes" discussed earlier (Buonocore et al. 2010), these LTi-like cells express RORγt, as well as aryl hydrocarbon receptor (AHR) and CCR6 (Takatori et al. 2009). However, in contrast to adaptive T lymphocytes, they have constitutive expression of IL-23R, which allows immediate IL-23 responsiveness. Given the similarities between the innate cell types responding to IL-23, it was speculated that these cells might represent different stages of a developmental lineage. Indeed, the LTi-like $CD127^+$ cells present in the human tonsil were demonstrated to directly develop into $CD56^+CD127^+$ NK-like cells and to remain distinct from conventional cytotoxic NK cells (Crellin et al. 2010). Together, these populations represent key early IL-23 responsive cells, which rapidly induce inflammation following exposure to IL-23.

7.1.7 IL-23-Responsive γδ T Cells

Other critically important acute cellular responders to IL-23 are naturally occurring, IL-17-producing γδ T cells. These cells are ideally situated to mediate rapid responses as they are tissue resident and can respond quickly to infections. They also mature earlier in the fetal thymus than αβ T cells and accumulate more rapidly at sites of infection (Hayday 2000). Recently, natural IL-17 producing γδ T cells were characterized and, similar to the other IL-23-responsive cells described, express RORγt, IL-23R, and AHR and produce the same set of signature cytokines including IL-17, IL-21, and IL-22 (Martin et al. 2009; Haas et al. 2009; Sutton et al. 2009). Indeed, γδ T cells account for the majority of $IL\text{-}23R^+$ cells in the lymph nodes of naive mice, with 40 % of nodal γδ T cells bearing the IL-23R, in contrast to the very small population of $IL\text{-}23R^+$ $CD4^+$ T lymphocytes (Awasthi et al. 2009). These IL-17-producing γδ T cells are thought to be programmed in the thymus for rapid production of this cytokine, and accumulate in the draining lymph node immediately in response to infection (Jensen et al. 2008; Ribot et al. 2009); this action is in contrast to the slow appearance of antigen-specific αβ T cells, which can take many days.

Consistent with their rapid kinetics, $IL\text{-}23R^+$ γδ T cells, such as LTi and NK-like cells, respond directly and immediately to IL-23 without the need for previous induction with TGF-β and IL-6 (Shibata et al. 2007; Umemura et al. 2007). Indeed, in contrast to Th17 cells, the latter two cytokines do not result in IL-17 production by γδ T cells (Shibata et al. 2007); rather, IL-17 production is enhanced by the addition of IL-1β (Duan et al. 2010). This cytokine, produced by antigen-presenting cells, confers greater responsiveness on $IL\text{-}1R1^+$ γδ T cells, which produce more IL-17 in response to IL-23 (Duan et al. 2010). Consistent with the necessity for a rapid, acute response to IL-23 following invasion of pathogens across mucosal surfaces, these highly IL-23-responsive $IL\text{-}1R1^+$ γδ T-cell subsets populate mucosal sites, including the peritoneum, intestinal lamina propria, and lung (Duan et al. 2010). IL-1R1 signalling is required for defense against mucosal infections including *Escherichia coli*, group B *Streptococcus*, and *Bacteriodes fragilis* (Shibata et al. 2007; Duan et al. 2010), possibly by enhancing local responsiveness to IL-23.

Although γδ T cells can respond to IL-23 produced by antigen-presenting cells activated by ligation of pattern-recognition receptors, IL-17-producing γδ T cells were found to express the receptors dectin-1, TLR1, and TLR2, but not TLR4 (Martin et al. 2009); this permits them to respond to fungal infections with rapid production of IL-17 and neutrophil recruitment. The significance of this in human disease is underscored by increased IL-17-producing γδ T cells observed in patients with active pulmonary infection with *Mycobacterium tuberculosis* (Peng et al. 2008). Furthermore, in mice, γδ T cells are the main producers of IL-17 in the lungs following the same bacterial infection (Lockhart et al. 2006). These cells also respond to *Candida albicans* infection in humans and, consistent with the higher infective load, expand in patients with human immunodeficiency virus (HIV) infection (Fenoglio et al. 2009).

7.1.8 A Novel Paradigm of IL-23 Action: Entheseal Resident IL-23R⁺ Cells

A novel mechanism of action of IL-23 has recently been described in which this cytokine induces rapid and severe inflammation with very specific anatomic distribution characteristic of spondyloarthropathy. The spondyloarthropathies (SpA) are rheumatic conditions characterized by inflammation localized to the insertional "entheseal" areas where tendons and ligaments attach to bone (Ball 1971, 1983; McGonagle et al. 1998). Numerous independent approaches from a range of diverse scientific disciplines have converged upon IL-23, suggesting that this is a pivotal cytokine in these conditions. Thus, the HLA-B27 molecule, which is tightly associated with SpA, being present in 90 % of patients with the prototypical subtype, ankylosing spondylitis (AS), misfolds and triggers a stress response in the endoplasmic reticulum, resulting in production of IL-23 (DeLay et al. 2009; Colbert et al. 2010). Bacteria associated with the reactive arthritis subtype of SpA also result in IL-23 production by this route, specifically through the induction of CHOP10, which binds to the IL-23p19 promoter to induce its expression (Goodall et al. 2010). Moreover, SpA is associated with an expansion of cells that can recognize abnormal surface forms of HLA-B27, and these cells are highly enriched for IL-23 responsiveness (Bowness et al. 2011). Indeed, polymorphisms in IL-23R suggestive of altered IL-23 responsiveness also correlate with the likelihood of developing AS (Wellcome Trust Case Control Consortium et al. 2007; Reveille et al. 2010). Clinical evidence has demonstrated that the subclinical inflammation present in almost 70 % of patients with SpA (Mielants et al. 1995) is associated with specific elaboration of IL-23 (Ciccia et al. 2009). IL-23 is also upregulated in the male genital tract in rodent models of spondyloarthropathy driven by the expression of HLA-B27 (Taurog et al. 2012).

The exact mechanistic reason why such specific cytokine dysregulations converging on IL-23 should result in the specific anatomic localization of pathology has remained unclear. A rational explanation for these findings is that entheses are uniquely sensitive to IL-23 because of the highly localized presence of IL-23R⁺

cells (Cua and Sherlock 2011). By using intravital multiphoton microscopy of naïve unmanipulated tissue, it was demonstrated that IL-23R$^+$ cells are highly localized to the entheses of the peripheral and axial skeleton. Furthermore, the expression of IL-23 alone, which parallels the IL-23 overactivity found in the human diseases, is sufficient to induce the hallmark pathogenic features of spondyloarthropathy (Sherlock et al. 2012). These cells nucleate pathology to the entheseal sites and induce the classical "dual" bone pathology seen in human SpA, which is characterized by both inflammatory bone erosion and new bone formation.

Similar to the γδ T cells and the other innate cellular responders to IL-23 already discussed, and in contrast to adaptive Th17 cells, the T cells resident in the canonical sites of inflammation in spondyloarthropathy constitutively express IL-23R: this allows them to respond immediately to IL-23 without a requirement for induction of responsiveness by other cytokines. Indeed, these cells express PLZF, a transcription regulator that confers a mature T-cell phenotype and the ability to immediately respond to cytokines without prior conditioning (Sherlock et al. 2012). Early entheseal pathology following experimental exposure to IL-23 can rapidly promote inflammation that is evident in 6 days. Such a time course is too fast for an adaptive antigen-mediated immune response to efficiently develop, and it also proceeds despite absence of CD4$^+$ T cells, including the Th17 subtype. Thus, the resident IL-23R$^+$ cells are likely to be able to mediate disease, and indeed, in vitro organ culture of entheseal tissue containing these cells can produce inflammatory cytokines within hours following IL-23 stimulation. These findings demonstrate a paradigm by which immune-mediated diseases can be localized to specific tissues and precise anatomic regions and demonstrate the role of IL-23 in this process by its action on a novel RORγt$^+$ PLZF$^+$ Thy1$^+$ SCA1$^+$ T-cell population.

7.2 Therapeutics Neutralizing IL-23

The central role of IL-23 in immune-mediated inflammation suggests that therapeutics targeting this cytokine will be highly effective in these disorders. Given the importance of this cytokine in inducing full pathogenicity of T cells in vivo (McGeachy et al. 2009), and its important role in numerous immune-mediated pathologies, IL-23 has emerged as an exciting therapeutic target. Moreover, the ability of IL-23 to reinforce its own production (El-Behi et al. 2011) suggests that neutralization of this cytokine will break a positive feedback loop that perpetuates inflammation, while at the same time preventing development of pathogenic potential in Th17 cells. Preclinical models confirmed that IL-23 is a key factor in inflammation of the central nervous system (Cua et al. 2003), bowel (Yen et al. 2006) and in models of rheumatoid arthritis (Murphy et al. 2003), psoriasis (Chan et al. 2006), and spondyloarthropathy (Sherlock et al. 2012). Many of these findings have begun to be translated to the setting of human diseases.

7.2.1 Psoriasis

The central role of IL-23 in psoriasis is suggested by the observation that polymorphisms in IL-23R are risk factors for its development (Capon et al. 2007; Nair et al. 2009; Cargill et al. 2007), and both IL-23 and IL-23R are directly upregulated in human psoriatic skin (Tonel et al. 2010), with IL-23p19 expression particularly in dendritic cells and epidermal Langerhans cells (Piskin et al. 2006; Lee et al. 2004). Confirmation of the direct role of IL-23 in mediating psoriasis came from experiments demonstrating that intradermal administration of IL-23 in mice results in the induction of psoriasis (Chan et al. 2006), and systemic exposure of mice to IL-23 similarly induces psoriatic skin changes (Sherlock et al. 2012). This effect is likely mediated, in part, by IL-17 and IL-22. IL-17 and the γδ T lymphocytes, which are a major source of this cytokine, provide a pivotal contribution to IL-23-driven psoriatic epidermal hyperplasia (Cai et al. 2011), and IL-22 mediates IL-23 induced psoriatic acanthosis (Zheng et al. 2007).

Neutralization of both IL-12 and IL-23 using an antibody against the common p40 subunit results in improvement of human psoriasis (Leonardi et al. 2008; Reich et al. 2011; Krueger et al. 2007). Therapeutics targeting IL-12p40 were designed to neutralize IL-12 before the discovery of IL-23; however, the subsequent research discussed earlier emphasizes that IL-23 is the key cytokine being targeted here. Consistently, a trial in which IL-23 was specifically inhibited using anti-IL-23p19 showed a dramatic improvement of psoriatic skin lesions (Jones et al. 2012). Moreover, administration of an antibody against IL-17A results in improvement of psoriasis (Leonardi et al. 2012), as does an antibody against IL-17RA (Papp et al. 2012), which by binding this receptor blocks the action not only of IL-17A but also IL-17F and IL-25. Therapeutics targeting IL-12p40 were designed to neutralize IL-12 before the discovery of IL-23, however; the subsequent research discussed earlier emphasizes that IL-23 is the key cytokine being targeted here. Consistently, a trial in which IL-23 was specifically inhibited using anti-IL-23p19 showed a dramatic improvement of psoriatic skin lesions at very low therapeutic doses (Jones et al. 2012). Moreover, administration of an antibody against IL-17A results in improvement of psoriasis (Leonardi et al. 2012), as does an antibody against IL-17RA (Papp et al. 2012), which by binding this receptor blocks the action not only of IL-17A but also IL-17F and IL-25.

7.2.2 Spondyloarthropathy Spectrum Disorders

Given the fundamental role of IL-23 in the pathogenesis of spondyloarthropathy, there is great interest in treatments targeting this molecule and related cytokines in these conditions. Recent work demonstrated that neutralization of IL-17 results in a reduction in spinal inflammation in ankylosing spondylitis, as visualized by magnetic resonance imaging (Baraliakos et al. 2011). Moreover, anti-IL-17 also results in improvements in noninfective uveitis (Hueber et al. 2010), which is of great importance because this condition is highly associated with ankylosing spondylitis. Additionally, antibody-mediated neutralization of IL-12/23p40 reduced signs and symptoms in psoriatic arthritis (Gottlieb et al. 2009). These findings suggest that neutralization of IL-23 is likely to be an effective treatment. As mentioned, IL-23 itself can induce both inflammation and bone changes associated with experimental ankylosing spondylitis. It is hoped that neutralization of IL-23 will therefore be able to not only terminate the inflammatory pathology but also inhibit the new bone growth that occurs in clinical spondyloarthropathy.

7.2.3 Inflammatory Bowel Disease

Inflammatory bowel disease shares predisposing genetic factors in the IL-23 pathway with psoriasis and ankylosing spondylitis, and agents targeting this pathway have been tested in therapeutic trials. However, neutralization of IL-17 in the context of Crohn's disease has been found to be ineffective and to increase the risks of adverse events, including local fungal infections (Hueber et al. 2012). As discussed next, IL-17 plays a crucial role in protection against fungi; hence, neutralization of this cytokine during mucosal inflammation at a key barrier surface is particularly challenging.

7.2.4 Consideration of the Potential Safety Profile of Targeting IL-23 and IL-17

A prime consideration in the use of any therapeutic agent that targets inflammatory cytokines is the risk of increasing predisposition to infection. Although no patients with specific deficiencies in IL-23p19 are known, patients lacking IL-12/IL-23p40 are predisposed to infections with *Mycobacteria* and *Salmonella*. The defects in the shared p40 subunit in these patients mean that their dendritic cells are unable to express IL-12p70, leading to defective IFN-γ expression in lymphocytes (Altare et al. 1998b). Mutations in the shared receptor chain for IL-12 and IL-23, IL-12Rβ1, have a similar infective predisposition (Altare et al. 1998a). Given the important role of IL-12, IFN-γ and Th1 cells in immunity to these specific organisms, it is highly likely that the infective predisposition results from deficiency of IL-12 rather than IL-23. Indeed, the same immunocompromised status is observed in patients with mutations in specific components of the IL-12/Th1/IFN-γ pathway. Thus, patients with defects in the IFN-γ receptor that prevent its cell-surface expression have macrophages that are unable to express tumor necrosis factor (TNF) in response to IFN-γ and a predisposition to mycobacterial infection (Newport et al. 1996; Jouanguy et al. 1996). Similarly, defects in STAT1, a molecule that mediates IFN-γR signaling, result in defective immunity to *Mycobacteria* (Dupuis et al. 2001).

Despite the absence of patients with specific mutations in IL-23p19 or IL-23R, animal studies demonstrate that IL-23 is necessary for defense against a restricted range of pathogenic species. Mice require IL-23 for defense against the rodent bacterium *Citrobacter rodentium* (Mangan et al. 2006) and the lung pathogen *Klebsiella pneumoniae*, which causes substantial mortality in IL-23-deficient mice as a result of defective IL-17 production (Happel et al. 2005). IL-17 itself is also necessary for defense against *Candida* in mice (Huang et al. 2004). Recent human studies show that Job's syndrome, an immunodeficiency associated with infections with *Candida* and *Staphylococcus aureus*, results from mutations in STAT3 (Minegishi et al. 2007; Holland et al. 2007), which is known to be a key signalling molecule in Th17 cell development. Indeed, these patients have impaired generation of Th17 cells because of defective induction of RORγt and impaired generation of responses when stimulated with *Candida* (Milner et al. 2008; Ma et al. 2008).

7.3 Conclusion

Investigation of the biology of IL-23 has lead to many unexpected insights associated with the characterization of a wide range of cellular responders in both the adaptive and innate immune system. Inconsistencies in prevailing paradigms to explain autoimmune inflammation based upon Th1 and Th2 cells alone were clarified and resolved by an understanding of the RORγt-dependent Th17 pathway. Most importantly, the fundamental nature of IL-23 and IL-23-responsive cells in a range of human pathologies means that treatments based on neutralization of these cytokines are likely to enable disease modification in illnesses for which the current therapies have limited efficacy. The past decade has witnessed a renaissance of immunology paradigms brought about by the discovery of IL-23 and Th17 cells. Our increased understanding of inflammatory disorders and the cellular and molecular biology underlying these conditions will help to practically address large areas of unmet medical need. We are indeed hopeful that the next decade will witness major therapeutic advances whereby patients will benefit from these innovations.

References

Aggarwal S, Ghilardi N, Xie M-H, de Sauvage FJ, Gurney AL (2003) Interleukin-23 promotes a distinct CD4 T cell activation state characterized by the production of interleukin-17. J Biol Chem 278:1910–1914. doi:10.1074/jbc.M207577200

Altare F, Durandy A, Lammas D, Emile JF, Lamhamedi S, Le Deist F, Drysdale P, Jouanguy E, Döffinger R, Bernaudin F, Jeppsson O, Gollob JA, Meinl E, Segal AW, Fischer A, Kumararatne D, Casanova JL (1998a) Impairment of mycobacterial immunity in human interleukin-12 receptor deficiency. Science 280:1432–1435

Altare F, Lammas D, Revy P, Jouanguy E, Döffinger R, Lamhamedi S, Drysdale P, Scheel-Toellner D, Girdlestone J, Darbyshire P, Wadhwa M, Dockrell H, Salmon M, Fischer A, Durandy A,

Casanova JL, Kumararatne DS (1998b) Inherited interleukin 12 deficiency in a child with bacille Calmette-Guérin and *Salmonella enteritidis* disseminated infection. J Clin Invest 102:2035–2040. doi:10.1172/JCI4950

Awasthi A, Riol-Blanco L, Jäger A, Korn T, Pot C, Galileos G, Bettelli E, Kuchroo VK, Oukka M (2009) Cutting edge: IL-23 receptor gfp reporter mice reveal distinct populations of IL-17-producing cells. J Immunol 182:5904–5908. doi:10.4049/jimmunol.0900732

Ball J (1971) Enthesopathy of rheumatoid and ankylosing spondylitis. Ann Rheum Dis 30: 213–223

Ball J (1983) The enthesopathy of ankylosing spondylitis. Br J Rheumatol 22:25–28

Baraliakos X, Braun J, Laurent DD, Baeten D, van der Heijde D, Sieper J, Emery P (2011) Interleukin-17A blockade with secukinumab reduces spinal inflammation in patients with ankylosing spondylitis as early as week 6, as detected by magnetic resonance imaging. Arthritis Rheum 63(Suppl 10):2486D. doi:10.1002/(ISSN)1529-0131

Becher B, Durell BG, Noelle RJ (2002) Experimental autoimmune encephalitis and inflammation in the absence of interleukin-12. J Clin Invest 110:493–497. doi:10.1172/JCI15751

Bettelli E, Carrier Y, Gao W, Korn T, Strom TB, Oukka M, Weiner HL, Kuchroo VK (2006) Reciprocal developmental pathways for the generation of pathogenic effector TH17 and regulatory T cells. Nature (Lond) 441:235–238. doi:10.1038/nature04753

Billiau A, Heremans H, Vandekerckhove F, Dijkmans R, Sobis H, Meulepas E, Carton H (1988) Enhancement of experimental allergic encephalomyelitis in mice by antibodies against IFN-gamma. J Immunol 140:1506–1510

Bowness P, Ridley A, Shaw J, Chan AT, Wong-Baeza I, Fleming M, Cummings F, McMichael A, Kollnberger S (2011) Th17 cells expressing KIR3DL2+ and responsive to HLA-B27 homodimers are increased in ankylosing spondylitis. J Immunol 186:2672–2680. doi:10.4049/jimmunol.1002653

Buonocore S, Ahern PP, Uhlig HH, Ivanov II, Littman DR, Maloy KJ, Powrie F (2010) Innate lymphoid cells drive interleukin-23-dependent innate intestinal pathology. Nature (Lond) 464:1371–1375. doi:10.1038/nature08949

Cai Y, Shen X, Ding C, Qi C, Li K, Li X, Jala VR, Zhang H-G, Wang T, Zheng J, Yan J (2011) Pivotal role of dermal IL-17-producing γδ T cells in skin inflammation. Immunity 35:596–610. doi:10.1016/j.immuni.2011.08.001

Capon F, Di Meglio P, Szaub J, Prescott NJ, Dunster C, Baumber L, Timms K, Gutin A, Abkevic V, Burden AD, Lanchbury J, Barker JN, Trembath RC, Nestle FO (2007) Sequence variants in the genes for the interleukin-23 receptor (IL23R) and its ligand (IL12B) confer protection against psoriasis. Hum Genet 122:201–206. doi:10.1007/s00439-007-0397-0

Cargill M, Schrodi SJ, Chang M, Garcia VE, Brandon R, Callis KP, Matsunami N, Ardlie KG, Civello D, Catanese JJ, Leong DU, Panko JM, McAllister LB, Hansen CB, Papenfuss J, Prescott SM, White TJ, Leppert MF, Krueger GG, Begovich AB (2007) A large-scale genetic association study confirms IL12B and leads to the identification of IL23R as psoriasis-risk genes. Am J Hum Genet 80:273–290. doi:10.1086/511051

Cella M, Fuchs A, Vermi W, Facchetti F, Otero K, Lennerz JKM, Doherty JM, Mills JC, Colonna M (2009) A human natural killer cell subset provides an innate source of IL-22 for mucosal immunity. Nature (Lond) 457:722–725. doi:10.1038/nature07537

Chan JR, Blumenschein W, Murphy E, Diveu C, Wiekowski M, Abbondanzo S, Lucian L, Geissler R, Brodie S, Kimball AB, Gorman DM, Smith K, de Waal-Malefyt R, Kastelein RA, McClanahan TK, Bowman EP (2006) IL-23 stimulates epidermal hyperplasia via TNF and IL-20R2-dependent mechanisms with implications for psoriasis pathogenesis. J Exp Med 203:2577–2587. doi:10.1084/jem.20060244

Ciccia F, Bombardieri M, Principato A, Giardina A, Tripodo C, Porcasi R, Peralta S, Franco V, Giardina E, Craxi A, Pitzalis C, Triolo G (2009) Overexpression of interleukin-23, but not interleukin-17, as an immunologic signature of subclinical intestinal inflammation in ankylosing spondylitis. Arthritis Rheum 60:955–965. doi:10.1002/art.24389

Codarri L, Gyülvészi G, Tosevski V, Hesske L, Fontana A, Magnenat L, Suter T, Becher B (2011) RORγt drives production of the cytokine GM-CSF in helper T cells, which is essential for the effector phase of autoimmune neuroinflammation. Nat Immunol 12:560–567. doi:10.1038/ni.2027

Colbert RA, DeLay ML, Klenk EI, Layh-Schmitt G (2010) From HLA-B27 to spondyloarthritis: a journey through the ER. Immunol Rev 233:181–202. doi:10.1111/j.0105-2896.2009.00865.x

Crellin NK, Trifari S, Kaplan CD, Cupedo T, Spits H (2010) Human NKp44+IL-22+ cells and LTi-like cells constitute a stable RORC+ lineage distinct from conventional natural killer cells. J Exp Med 207:281–290. doi:10.1084/jem.20091509

Cua DJ, Sherlock JP (2011) Autoimmunity's collateral damage: gut microbiota strikes 'back'. Nat Med 17:1055–1056. doi:10.1038/nm0911-1055

Cua DJ, Sherlock JP, Chen Y, Murphy CA, Joyce B, Seymour B, Lucian L, To W, Kwan S, Churakova T, Zurawski S, Wiekowski M, Lira SA, Gorman D, Kastelein RA, Sedgwick JD (2003) Interleukin-23 rather than interleukin-12 is the critical cytokine for autoimmune inflammation of the brain. Nature (Lond) 421:744–748. doi:10.1038/nature01355

Cui Y, Shao H, Lan C, Nian H, O'Brien RL, Born WK, Kaplan HJ, Sun D (2009) Major role of gamma delta T cells in the generation of IL-17+ uveitogenic T cells. J Immunol 183:560–567. doi:10.4049/jimmunol.0900241

Cupedo T, Crellin NK, Papazian N, Rombouts EJ, Weijer K, Grogan JL, Fibbe WE, Cornelissen JJ, Spits H (2009) Human fetal lymphoid tissue-inducer cells are interleukin 17-producing precursors to RORC+ CD127+ natural killer-like cells. Nat Immunol 10:66–74. doi:10.1038/ni.1668

DeLay ML, Turner MJ, Klenk EI, Smith JA, Sowders DP, Colbert RA (2009) HLA-B27 misfolding and the unfolded protein response augment interleukin-23 production and are associated with Th17 activation in transgenic rats. Arthritis Rheum 60:2633–2643. doi:10.1002/art.24763

Duan J, Chung H, Troy E, Kasper DL (2010) Microbial colonization drives expansion of IL-1 receptor 1-expressing and IL-17-producing gamma/delta T cells. Cell Host Microbe 7: 140–150. doi:10.1016/j.chom.2010.01.005

Duong TT, St. Louis J, Gilbert JJ, Finkelman FD, Strejan GH (1992) Effect of anti-interferon-gamma and anti-interleukin-2 monoclonal antibody treatment on the development of actively and passively induced experimental allergic encephalomyelitis in the SJL/J mouse. J Neuroimmunol 36:105–115

Dupuis S, Dargemont C, Fieschi C, Thomassin N, Rosenzweig S, Harris J, Holland SM, Schreiber RD, Casanova JL (2001) Impairment of mycobacterial but not viral immunity by a germline human STAT1 mutation. Science 293:300–303. doi:10.1126/science.1061154

El-Behi M, Ciric B, Dai H, Yan Y, Cullimore M, Safavi F, Zhang G-X, Dittel BN, Rostami A (2011) The encephalitogenicity of T(H)17 cells is dependent on IL-1- and IL-23-induced production of the cytokine GM-CSF. Nat Immunol 12:568–575. doi:10.1038/ni.2031

Fenoglio D, Poggi A, Catellani S, Battaglia F, Ferrera A, Setti M, Murdaca G, Zocchi MR (2009) Vdelta1 T lymphocytes producing IFN-gamma and IL-17 are expanded in HIV-1-infected patients and respond to *Candida albicans*. Blood 113:6611–6618. doi:10.1182/blood-2009-01-198028

Ferber IA, Brocke S, Taylor-Edwards C, Ridgway W, Dinisco C, Steinman L, Dalton D, Fathman CG (1996) Mice with a disrupted IFN-gamma gene are susceptible to the induction of experimental autoimmune encephalomyelitis (EAE). J Immunol 156:5–7

Ghoreschi K, Laurence A, Yang X-P, Tato CM, McGeachy MJ, Konkel JE, Ramos HL, Wei L, Davidson TS, Bouladoux N, Grainger JR, Chen Q, Kanno Y, Watford WT, Sun H-W, Eberl G, Shevach EM, Belkaid Y, Cua DJ, Chen W, O'Shea JJ (2010) Generation of pathogenic T(H)17 cells in the absence of TGF-β signalling. Nature (Lond) 467:967–971. doi:10.1038/nature09447

Goodall JC, Wu C, Zhang Y, McNeill L, Ellis L, Saudek V, Gaston JSH (2010) Endoplasmic reticulum stress-induced transcription factor, CHOP, is crucial for dendritic cell IL-23 expression. Proc Natl Acad Sci USA 107:17698–17703. doi:10.1073/pnas.1011736107

Gottlieb A, Menter A, Mendelsohn A, Shen Y-K, Li S, Guzzo C, Fretzin S, Kunynetz R, Kavanaugh A (2009) Ustekinumab, a human interleukin 12/23 monoclonal antibody, for psoriatic arthritis: randomised, double-blind, placebo-controlled, crossover trial. Lancet 373: 633–640. doi:10.1016/S0140-6736(09)60140-9

Haas JD, González FHM, Schmitz S, Chennupati V, Föhse L, Kremmer E, Förster R, Prinz I (2009) CCR6 and NK1.1 distinguish between IL-17A and IFN-gamma-producing gammadelta effector T cells. Eur J Immunol 39:3488–3497. doi:10.1002/eji.200939922

Happel KI, Zheng M, Young E, Quinton LJ, Lockhart E, Ramsay AJ, Shellito JE, Schurr JR, Bagby GJ, Nelson S, Kolls JK (2003) Cutting edge: roles of Toll-like receptor 4 and IL-23 in IL-17 expression in response to *Klebsiella pneumoniae* infection. J Immunol 170:4432–4436

Happel KI, Dubin PJ, Zheng M, Ghilardi N, Lockhart C, Quinton LJ, Odden AR, Shellito JE, Bagby GJ, Nelson S, Kolls JK (2005) Divergent roles of IL-23 and IL-12 in host defense against *Klebsiella pneumoniae*. J Exp Med 202:761–769. doi:10.1084/jem.20050193

Hayday AC (2000) γδ cells: a right time and a right place for a conserved third way of protection. Annu Rev Immunol 18:975–1026. doi:10.1146/annurev.immunol.18.1.975

Heremans H, Dillen C, Groenen M, Martens E, Billiau A (1996) Chronic relapsing experimental autoimmune encephalomyelitis (CREAE) in mice: enhancement by monoclonal antibodies against interferon-gamma. Eur J Immunol 26:2393–2398. doi:10.1002/eji.1830261019

Holland SM, DeLeo FR, Elloumi HZ, Hsu AP, Uzel G, Brodsky N, Freeman AF, Demidowich A, Davis J, Turner ML, Anderson VL, Darnell DN, Welch PA, Kuhns DB, Frucht DM, Malech HL, Gallin JI, Kobayashi SD, Whitney AR, Voyich JM, Musser JM, Woellner C, Schäffer AA, Puck JM, Grimbacher B (2007) STAT3 mutations in the hyper-IgE syndrome. N Engl J Med 357:1608–1619. doi:10.1056/NEJMoa073687

Huang W, Na L, Fidel PL, Schwarzenberger P (2004) Requirement of interleukin-17A for systemic anti-*Candida albicans* host defense in mice. J Infect Dis 190:624–631. doi:10.1086/422329

Hueber W, Patel DD, Dryja T, Wright AM, Koroleva I, Bruin G, Antoni C, Draelos Z, Gold MH, Psoriasis Study Group, Durez P, Tak PP, Gomez-Reino JJ, Rheumatoid Arthritis Study Group, Foster CS, Kim RY, Samson CM, Falk NS, Chu DS, Callanan D, Nguyen QD, Uveitis Study Group, Rose K, Haider A, Di Padova F (2010) Effects of AIN457, a fully human antibody to interleukin-17A, on psoriasis, rheumatoid arthritis, and uveitis. Sci Transl Med 2:52ra72. doi:10.1126/scitranslmed.3001107

Hueber W, Sands BE, Lewitzky S, Vandemeulebroecke M, Reinisch W, Higgins PDR, Wehkamp J, Feagan BG, Yao MD, Karczewski M, Karczewski J, Pezous N, Bek S, Bruin G, Mellgard B, Berger C, Londei M, Bertolino AP, Tougas G, Travis SPL, for the Secukinumab in Crohn's Disease Study Group (2012) Secukinumab, a human anti-IL-17A monoclonal antibody, for moderate to severe Crohn's disease: unexpected results of a randomised, double-blind placebo-controlled trial. Gut 61:1693–1700. doi:10.1136/gutjnl-2011-301668

Ivanov II, McKenzie BS, Zhou L, Tadokoro CE, Lepelley A, Lafaille JJ, Cua DJ, Littman DR (2006) The orphan nuclear receptor RORgammat directs the differentiation program of proinflammatory IL-17+ T helper cells. Cell 126:1121–1133. doi:10.1016/j.cell.2006.07.035

Jensen KDC, Su X, Shin S, Li L, Youssef S, Yamasaki S, Steinman L, Saito T, Locksley RM, Davis MM, Baumgarth N, Chien Y-H (2008) Thymic selection determines gammadelta T cell effector fate: antigen-naive cells make interleukin-17 and antigen-experienced cells make interferon gamma. Immunity 29:90–100. doi:10.1016/j.immuni.2008.04.022

Jones SA, Sutton CE, Cua DJ, Mills KHG (2012) Therapeutic potential of targeting IL-17. Nat Immunol 13:1022–1025. doi:10.1038/ni.2450

Jouanguy E, Altare F, Lamhamedi S, Revy P, Emile JF, Newport M, Levin M, Blanche S, Seboun E, Fischer A, Casanova JL (1996) Interferon-gamma-receptor deficiency in an infant with fatal bacille Calmette-Guérin infection. N Engl J Med 335:1956–1961. doi:10.1056/NEJM199612263352604

Korn T, Bettelli E, Gao W, Awasthi A, Jäger A, Strom TB, Oukka M, Kuchroo VK (2007) IL-21 initiates an alternative pathway to induce proinflammatory T(H)17 cells. Nature (Lond) 448:484–487. doi:10.1038/nature05970

Krueger GG, Langley RG, Leonardi C, Yeilding N, Guzzo C, Wang Y, Dooley LT, Lebwohl M, CNTO 1275 Psoriasis Study Group (2007) A human interleukin-12/23 monoclonal antibody for the treatment of psoriasis. N Engl J Med 356:580–592. doi:10.1056/NEJMoa062382

Lee E, Trepicchio WL, Oestreicher JL, Pittman D, Wang F, Chamian F, Dhodapkar M, Krueger JG (2004) Increased expression of interleukin 23 p19 and p40 in lesional skin of patients with psoriasis vulgaris. J Exp Med 199:125–130. doi:10.1084/jem.20030451

Lee Y, Awasthi A, Yosef N, Quintana FJ, Xiao S, Peters A, Wu C, Kleinewietfeld M, Kunder S, Hafler DA, Sobel RA, Regev A, Kuchroo VK (2012) Induction and molecular signature of pathogenic T(H)17 cells. Nat Immunol 13:991–999. doi:10.1038/ni.2416

Leonardi CL, Kimball AB, Papp KA, Yeilding N, Guzzo C, Wang Y, Li S, Dooley LT, Gordon KB, PHOENIX 1 study investigators (2008) Efficacy and safety of ustekinumab, a human interleukin-12/23 monoclonal antibody, in patients with psoriasis: 76-week results from a randomised, double-blind, placebo-controlled trial (PHOENIX 1). Lancet 371:1665–1674. doi:10.1016/S0140-6736(08)60725-4

Leonardi C, Matheson R, Zachariae C, Cameron G, Li L, Edson-Heredia E, Braun D, Banerjee S (2012) Anti-interleukin-17 monoclonal antibody ixekizumab in chronic plaque psoriasis. N Engl J Med 366:1190–1199. doi:10.1056/NEJMoa1109997

Lockhart E, Green AM, Flynn JL (2006) IL-17 production is dominated by gammadelta T cells rather than CD4 T cells during *Mycobacterium tuberculosis* infection. J Immunol 177:4662–4669

Lublin FD, Knobler RL, Kalman B, Goldhaber M, Marini J, Perrault M, D'Imperio C, Joseph J, Alkan SS, Korngold R (1993) Monoclonal anti-gamma interferon antibodies enhance experimental allergic encephalomyelitis. Autoimmunity 16:267–274

Luci C, Reynders A, Ivanov II, Cognet C, Chiche L, Chasson L, Hardwigsen J, Anguiano E, Banchereau J, Chaussabel D, Dalod M, Littman DR, Vivier E, Tomasello E (2009) Influence of the transcription factor RORgammat on the development of NKp46+ cell populations in gut and skin. Nat Immunol 10:75–82. doi:10.1038/ni.1681

Ma CS, Chew GYJ, Simpson N, Priyadarshi A, Wong M, Grimbacher B, Fulcher DA, Tangye SG, Cook MC (2008) Deficiency of Th17 cells in hyper IgE syndrome due to mutations in STAT3. J Exp Med 205:1551–1557. doi:10.1084/jem.20080218

Mangan PR, Harrington LE, O'Quinn DB, Helms WS, Bullard DC, Elson CO, Hatton RD, Wahl SM, Schoeb TR, Weaver CT (2006) Transforming growth factor-beta induces development of the T(H)17 lineage. Nature (Lond) 441:231–234. doi:10.1038/nature04754

Martin B, Hirota K, Cua DJ, Stockinger B, Veldhoen M (2009) Interleukin-17-producing gammadelta T cells selectively expand in response to pathogen products and environmental signals. Immunity 31:321–330. doi:10.1016/j.immuni.2009.06.020

McGeachy MJ, Chen Y, Tato CM, Laurence A, Joyce-Shaikh B, Blumenschein WM, McClanahan TK, O'Shea JJ, Cua DJ (2009) The interleukin 23 receptor is essential for the terminal differentiation of interleukin 17-producing effector T helper cells in vivo. Nat Immunol 10:314–324. doi:10.1038/ni.1698

McGonagle D, Gibbon W, Emery P (1998) Classification of inflammatory arthritis by enthesitis. Lancet 352:1137–1140. doi:10.1016/S0140-6736(97)12004-9

Mielants H, Veys EM, Cuvelier C, de Vos M, Goemaere S, de Clercq L, Schatteman L, Elewaut D (1995) The evolution of spondyloarthropathies in relation to gut histology. II. Histological aspects. J Rheumatol 22:2273–2278

Milner JD, Brenchley JM, Laurence A, Freeman AF, Hill BJ, Elias KM, Kanno Y, Spalding C, Elloumi HZ, Paulson ML, Davis J, Hsu A, Asher AI, O'Shea J, Holland SM, Paul WE, Douek DC (2008) Impaired T(H)17 cell differentiation in subjects with autosomal dominant hyper-IgE syndrome. Nature (Lond) 452:773–776. doi:10.1038/nature06764

Minegishi Y, Saito M, Tsuchiya S, Tsuge I, Takada H, Hara T, Kawamura N, Ariga T, Pasic S, Stojkovic O, Metin A, Karasuyama H (2007) Dominant-negative mutations in the DNA-binding domain of STAT3 cause hyper-IgE syndrome. Nature 448:1058–1062. doi:10.1038/nature06096

Mosmann TR, Cherwinski H, Bond MW, Giedlin MA, Coffman RL (1986) Two types of murine helper T cell clone. I. Definition according to profiles of lymphokine activities and secreted proteins. J Immunol 136:2348–2357

Murphy CA, Langrish CL, Chen Y, Blumenschein W, McClanahan T, Kastelein RA, Sedgwick JD, Cua DJ (2003) Divergent pro- and antiinflammatory roles for IL-23 and IL-12 in joint autoimmune inflammation. J Exp Med 198:1951–1957. doi:10.1084/jem.20030896

Nair RP, Duffin KC, Helms C, Ding J, Stuart PE, Goldgar D, Gudjonsson JE, Li Y, Tejasvi T, Feng B-J, Ruether A, Schreiber S, Weichenthal M, Gladman D, Rahman P, Schrodi SJ, Prahalad S, Guthery SL, Fischer J, Liao W, Kwok P-Y, Menter A, Lathrop GM, Wise CA, Begovich AB, Voorhees JJ, Elder JT, Krueger GG, Bowcock AM, Abecasis GR, Collaborative Association Study of Psoriasis (2009) Genome-wide scan reveals association of psoriasis with IL-23 and NF-kappaB pathways. Nat Genet 41:199–204. doi:10.1038/ng.311

Newport MJ, Huxley CM, Huston S, Hawrylowicz CM, Oostra BA, Williamson R, Levin M (1996) A mutation in the interferon-gamma-receptor gene and susceptibility to mycobacterial infection. N Engl J Med 335:1941–1949. doi:10.1056/NEJM199612263352602

Oppmann B, Lesley R, Blom B, Timans JC, Xu Y, Hunte B, Vega F, Yu N, Wang J, Singh K, Zonin F, Vaisberg E, Churakova T, Liu M, Gorman D, Wagner J, Zurawski S, Liu Y, Abrams JS, Moore KW, Rennick D, de Waal-Malefyt R, Hannum C, Bazan JF, Kastelein RA (2000) Novel p19 protein engages IL-12p40 to form a cytokine, IL-23, with biological activities similar as well as distinct from IL-12. Immunity 13:715–725

Papp KA, Leonardi C, Menter A, Ortonne J-P, Krueger JG, Kricorian G, Aras G, Li J, Russell CB, Thompson EHZ, Baumgartner S (2012) Brodalumab, an anti-interleukin-17-receptor antibody for psoriasis. N Engl J Med 366:1181–1189. doi:10.1056/NEJMoa1109017

Parham C, Chirica M, Timans J, Vaisberg E, Travis M, Cheung J, Pflanz S, Zhang R, Singh KP, Vega F, To W, Wagner J, O'Farrell A-M, McClanahan T, Zurawski S, Hannum C, Gorman D, Rennick DM, Kastelein RA, de Waal MR, Moore KW (2002) A receptor for the heterodimeric cytokine IL-23 is composed of IL-12Rbeta1 and a novel cytokine receptor subunit, IL-23R. J Immunol 168:5699–5708

Peng MY, Wang ZH, Yao CY, Jiang LN, Jin QL, Wang J, Li BQ (2008) Interleukin 17-producing gamma delta T cells increased in patients with active pulmonary tuberculosis. Cell Mol Immunol 5:203–208. doi:10.1038/cmi.2008.25

Piskin G, Sylva-Steenland RMR, Bos JD, Teunissen MBM (2006) In vitro and in situ expression of IL-23 by keratinocytes in healthy skin and psoriasis lesions: enhanced expression in psoriatic skin. J Immunol 176:1908–1915

Reich K, Langley RG, Papp KA, Ortonne J-P, Unnebrink K, Kaul M, Valdes JM (2011) A 52-week trial comparing briakinumab with methotrexate in patients with psoriasis. N Engl J Med 365:1586–1596. doi:10.1056/NEJMoa1010858

Reveille JD, Sims A-M, Danoy P, Evans DM, Leo P, Pointon JJ, Jin R, Zhou X, Bradbury LA, Appleton LH, Davis JC, Diekman L, Doan T, Dowling A, Duan R, Duncan EL, Farrar C, Hadler J, Harvey D, Karaderi T, Mogg R, Pomeroy E, Pryce K, Taylor J, Savage L, Deloukas P, Kumanduri V, Peltonen L, Ring SM, Whittaker P, Glazov E, Thomas GP, Maksymowych WP, Inman RD, Ward MM, Stone MA, Weisman MH, Wordsworth BP, Brown MA (2010) Genome-wide association study of ankylosing spondylitis identifies non-MHC susceptibility loci. Nat Genet 42:123–127. doi:10.1038/ng.513

Ribot JC, deBarros A, Pang DJ, Neves JF, Peperzak V, Roberts SJ, Girardi M, Borst J, Hayday AC, Pennington DJ, Silva-Santos B (2009) CD27 is a thymic determinant of the balance between interferon-gamma- and interleukin 17-producing gammadelta T cell subsets. Nat Immunol 10:427–436. doi:10.1038/ni.1717

Roark CL, French JD, Taylor MA, Bendele AM, Born WK, O'Brien RL (2007) Exacerbation of collagen-induced arthritis by oligoclonal, IL-17-producing gamma delta T cells. J Immunol 179:5576–5583

Sandborn WJ, Gasink C, Gao L-L, Blank MA, Johanns J, Guzzo C, Sands BE, Hanauer SB, Targan S, Rutgeerts P, Ghosh S, de Villiers WJS, Panaccione R, Greenberg G, Schreiber S, Lichtiger S, Feagan BG, CERTIFI Study Group (2012) Ustekinumab induction and maintenance therapy in refractory Crohn's disease. N Engl J Med 367:1519–1528. doi:10.1056/NEJMoa1203572

Sanos SL, Bui VL, Mortha A, Oberle K, Heners C, Johner C, Diefenbach A (2009) RORgammat and commensal microflora are required for the differentiation of mucosal interleukin 22-producing NKp46+ cells. Nat Immunol 10:83–91. doi:10.1038/ni.1684

Satoh-Takayama N, Vosshenrich CAJ, Lesjean-Pottier S, Sawa S, Lochner M, Rattis F, Mention J-J, Thiam K, Cerf-Bensussan N, Mandelboim O, Eberl G, Di Santo JP (2008) Microbial flora

drives interleukin 22 production in intestinal NKp46+ cells that provide innate mucosal immune defense. Immunity 29:958–970. doi:10.1016/j.immuni.2008.11.001
Segal BM, Dwyer BK, Shevach EM (1998) An interleukin (IL)-10/IL-12 immunoregulatory circuit controls susceptibility to autoimmune disease. J Exp Med 187:537–546
Sherlock JP, Joyce-Shaikh B, Turner SP, Chao C-C, Sathe M, Grein J, Gorman DM, Bowman EP, McClanahan TK, Yearley JH, Eberl G, Buckley CD, Kastelein RA, Pierce RH, LaFace DM, Cua DJ (2012) IL-23 induces spondyloarthropathy by acting on ROR-γt+ CD3+CD4-CD8- entheseal resident T cells. Nat Med 18:1069–1076. doi:10.1038/nm.2817
Shibata K, Yamada H, Hara H, Kishihara K, Yoshikai Y (2007) Resident Vdelta1+ gammadelta T cells control early infiltration of neutrophils after Escherichia coli infection via IL-17 production. J Immunol 178:4466–4472
Sutton CE, Lalor SJ, Sweeney CM, Brereton CF, Lavelle EC, Mills KHG (2009) Interleukin-1 and IL-23 induce innate IL-17 production from gammadelta T cells, amplifying Th17 responses and autoimmunity. Immunity 31:331–341. doi:10.1016/j.immuni.2009.08.001
Takatori H, Kanno Y, Watford WT, Tato CM, Weiss G, Ivanov II, Littman DR, O'Shea JJ (2009) Lymphoid tissue inducer-like cells are an innate source of IL-17 and IL-22. J Exp Med 206:35–41. doi:10.1084/jem.20072713
Taurog JD, Rival C, van Duivenvoorde LM, Satumtira N, Dorris ML, Sun M, Shelton JM, Richardson JA, Hamra FK, Hammer RE, Tung KSK (2012) Autoimmune epididymoorchitis is essential to the pathogenesis of male-specific spondylarthritis in HLA-B27-transgenic rats. Arthritis Rheum 64:2518–2528. doi:10.1002/art.34480
Tonel G, Conrad C, Laggner U, Di Meglio P, Grys K, McClanahan TK, Blumenschein WM, Qin J-Z, Xin H, Oldham E, Kastelein RA, Nickoloff BJ, Nestle FO (2010) Cutting edge: a critical functional role for IL-23 in psoriasis. J Immunol 185:5688–5691. doi:10.4049/jimmunol.1001538
Uhlig HH, McKenzie BS, Hue S, Thompson C, Joyce-Shaikh B, Stepankova R, Robinson N, Buonocore S, Tlaskalova-Hogenova H, Cua DJ, Powrie F (2006) Differential activity of IL-12 and IL-23 in mucosal and systemic innate immune pathology. Immunity 25:309–318. doi:10.1016/j.immuni.2006.05.017
Umemura M, Yahagi A, Hamada S, Begum MD, Watanabe H, Kawakami K, Suda T, Sudo K, Nakae S, Iwakura Y, Matsuzaki G (2007) IL-17-mediated regulation of innate and acquired immune response against pulmonary *Mycobacterium bovis* bacille Calmette-Guerin infection. J Immunol 178:3786–3796
Veldhoen M, Hocking RJ, Atkins CJ, Locksley RM, Stockinger B (2006a) TGFbeta in the context of an inflammatory cytokine milieu supports de novo differentiation of IL-17-producing T cells. Immunity 24:179–189. doi:10.1016/j.immuni.2006.01.001
Veldhoen M, Hocking RJ, Flavell RA, Stockinger B (2006b) Signals mediated by transforming growth factor-beta initiate autoimmune encephalomyelitis, but chronic inflammation is needed to sustain disease. Nat Immunol 7:1151–1156. doi:10.1038/ni1391
Voorthuis JA, Uitdehaag BM, De Groot CJ, Goede PH, van der Meide PH, Dijkstra CD (1990) Suppression of experimental allergic encephalomyelitis by intraventricular administration of interferon-gamma in Lewis rats. Clin Exp Immunol 81:183–188
Wellcome Trust Case Control Consortium, Australo-Anglo-American Spondylitis Consortium (TASC), Burton PR, Clayton DG, Cardon LR, Craddock N, Deloukas P, Duncanson A, Kwiatkowski DP, McCarthy MI, Ouwehand WH, Samani NJ, Todd JA, Donnelly P, Barrett JC, Davison D, Easton D, Evans DM, Leung H-T, Marchini JL, Morris AP, Spencer CCA, Tobin MD, Attwood AP, Boorman JP, Cant B, Everson U, Hussey JM, Jolley JD, Knight AS, Koch K, Meech E, Nutland S, Prowse CV, Stevens HE, Taylor NC, Walters GR, Walker NM, Watkins NA, Winzer T, Jones RW, McArdle WL, Ring SM, Strachan DP, Pembrey M, Breen G, St Clair D, Caesar S, Gordon-Smith K, Jones L, Fraser C, Green EK, Grozeva D, Hamshere ML, Holmans PA, Jones IR, Kirov G, Moskivina V, Nikolov I, O'Donovan MC, Owen MJ, Collier DA, Elkin A, Farmer A, Williamson R, McGuffin P, Young AH, Ferrier IN, Ball SG, Balmforth AJ, Barrett JH, Bishop TD, Iles MM, Maqbool A, Yuldasheva N, Hall AS, Braund PS, Dixon RJ, Mangino M, Stevens S, Thompson JR, Bredin F, Tremelling M, Parkes M, Drummond H,

Lees CW, Nimmo ER, Satsangi J, Fisher SA, Forbes A, Lewis CM, Onnie CM, Prescott NJ, Sanderson J, Matthew CG, Barbour J, Mohiuddin MK, Todhunter CE, Mansfield JC, Ahmad T, Cummings FR, Jewell DP, Webster J, Brown MJ, Lathrop MG, Connell J, Dominiczak A, Marcano CA, Burke B, Dobson R, Gungadoo J, Lee KL, Munroe PB, Newhouse SJ, Onipinla A, Wallace C, Xue M, Caulfield M, Farrall M, Barton A, Biologics in RA Genetics and Genomics Study Syndicate (BRAGGS) Steering Committee, Bruce IN, Donovan H, Eyre S, Gilbert PD, Hilder SL, Hinks AM, John SL, Potter C, Silman AJ, Symmons DP, Thomson W, Worthington J, Dunger DB, Widmer B, Frayling TM, Freathy RM, Lango H, Perry JR, Shields BM, Weedon MN, Hattersley AT, Hitman GA, Walker M, Elliott KS, Groves CJ, Lindgren CM, Rayner NW, Timpson NJ, Zeggini E, Newport M, Sirugo G, Lyons E, Vannberg F, Hill AV, Bradbury LA, Farrar C, Pointon JJ, Wordsworth P, Brown MA, Franklyn JA, Heward JM, Simmonds MJ, Gough SC, Seal S, Breast Cancer Susceptibility Collaboration (UK), Stratton MR, Rahman N, Ban M, Goris A, Sawcer SJ, Compston A, Conway D, Jallow M, Newport M, Sirugo G, Rockett KA, Bumpstead SJ, Chaney A, Downes K, Ghori MJ, Gwilliam R, Hunt SE, Inouye M, Keniry A, King E, McGinnis R, Potter S, Ravindrarajah R, Whittaker P, Widden C, Withers D, Cardin NJ, Davison D, Ferreira T, Pereira-Gale J, Hallgrimsdo'ttir IB, Howie BN, Su Z, Teo YY, Vukcevic D, Bentley D, Brown MA, Compston A, Farrall M, Hall AS, Hattersley AT, Hill AV, Parkes M, Pembrey M, Stratton MR, Mitchell SL, Newby PR, Brand OJ, Carr-Smith J, Pearce SH, McGinnis R, Keniry A, Deloukas P, Reveille JD, Zhou X, Sims AM, Dowling A, Taylor J, Doan T, Davis JC, Savage L, Ward MM, Learch TL, Weisman MH, Brown M (2007) Association scan of 14,500 nonsynonymous SNPs in four diseases identifies autoimmunity variants. Nat Genet 39:1329–1337. doi:10.1038/ng.2007.17

Willenborg DO, Fordham S, Bernard CC, Cowden WB, Ramshaw IA (1996) IFN-gamma plays a critical down-regulatory role in the induction and effector phase of myelin oligodendrocyte glycoprotein-induced autoimmune encephalomyelitis. J Immunol 157:3223–3227

Yang XO, Panopoulos AD, Nurieva R, Chang SH, Wang D, Watowich SS, Dong C (2007) STAT3 regulates cytokine-mediated generation of inflammatory helper T cells. J Biol Chem 282: 9358–9363. doi:10.1074/jbc.C600321200

Yen D, Cheung J, Scheerens H, Poulet F, McClanahan T, McKenzie B, Kleinschek MA, Owyang A, Mattson J, Blumenschein W, Murphy E, Sathe M, Cua DJ, Kastelein RA, Rennick D (2006) IL-23 is essential for T cell-mediated colitis and promotes inflammation via IL-17 and IL-6. J Clin Invest 116:1310–1316. doi:10.1172/JCI21404

Zeng R, Spolski R, Casas E, Zhu W, Levy DE, Leonard WJ (2007) The molecular basis of IL-21-mediated proliferation. Blood 109:4135–4142. doi:10.1182/blood-2006-10-054973

Zhang G-X, Gran B, Yu S, Li J, Siglienti I, Chen X, Kamoun M, Rostami A (2003) Induction of experimental autoimmune encephalomyelitis in IL-12 receptor-beta 2-deficient mice: IL-12 responsiveness is not required in the pathogenesis of inflammatory demyelination in the central nervous system. J Immunol 170:2153–2160

Zheng Y, Danilenko DM, Valdez P, Kasman I, Eastham-Anderson J, Wu J, Ouyang W (2007) Interleukin-22, a T(H)17 cytokine, mediates IL-23-induced dermal inflammation and acanthosis. Nature (Lond) 445:648–651. doi:10.1038/nature05505

Zúñiga LA, Jain R, Haines C, Cua DJ (2013) Th17 cell development: from the cradle to the grave. Immunol Rev 252:78–88. doi:10.1111/imr.12036

Chapter 8
IL-36: An Epithelial Cytokine Important in Psoriasis

John E. Sims, Solenne Vigne, Cem Gabay, and Jennifer E. Towne

Abstract Interleukin-36 (IL-36) refers to three cytokines that are members of the IL-1 family. These cytokines are expressed predominantly but not exclusively by epithelial tissues. Similar to other IL-1 family cytokines, they promote inflammatory responses by activating kinases such as JNK, p38, and the IKKs, and activating transcription factors such as NFκB. They only do this, however, after N-terminal processing, which increases their specific activity by three to five orders of magnitude. Extensive evidence links IL-36 to various forms of psoriasis. It is also very likely involved in lung physiology and pathology and may play a role in the biology of other tissues including kidney, joint, and adipose tissue. In part, these activities take place by direct action of IL-36 on tissue cells. In part, however, they are undoubtedly mediated by the action of IL-36 on cells of the immune system, which results in responses marked by production of IFN-γ. Although we are learning much about the role of IL-36 in different tissues, it is clear that we are at a very early stage in understanding the biology of this cytokine.

Keywords IL-1 family • IL-36 • Inflammation • Psoriasis

J.E. Sims (✉)
Department of Pathology and Immunology, 4207 43rd Avenue NE, Seattle, WA 98105, USA
e-mail: sims.johne@gmail.com

S. Vigne • C. Gabay
Department of Pathology and Immunology, CMU/Rue Michel-Servet 1, 1211 Geneva, Switzerland

J.E. Towne
Amgen, 1201 Amgen Place W, Seattle, WA 98119, USA

T. Yoshimoto and T. Yoshimoto (eds.), *Cytokine Frontiers: Regulation of Immune Responses in Health and Disease*, DOI 10.1007/978-4-431-54442-5_8, © Springer Japan 2014

8.1 Introduction

Interleukin-36 (IL-36) refers to three cytokines that are members of the IL-1 family. Despite being three separate protein molecules, they all use the same receptor and (so far as is known) deliver the same biological signals. They have been implicated in the biology of epithelial tissues, most notably skin, where there is strong evidence for a role in several types of psoriasis. They also influence the differentiation of naïve T cells in the immune system. This chapter reviews what is known about the biochemistry and biology of IL-36. The IL-36 family is currently a subject of active research, however, and no doubt much more will be learned in the coming years.

8.2 IL-36 Expression and Biochemistry

The cytokines comprising the IL-36 family were first discovered in sequence databases as expressed open reading frames with homology to IL-1α and IL-1β. Originally termed IL-1F5, IL-1F6, IL-1F8, and IL-1F9 (Sims et al. 2001), the names were subsequently changed to IL-36Ra, IL-36α, IL-36β, and IL-36γ, respectively, as information about their function became known (Dinarello et al. 2010).

The IL-36 receptor is a two-chain structure, similar to receptors for other IL-1 family members (Fig. 8.1: biochemistry). The primary ligand-binding subunit (IL-36R) was previously termed IL1RL2 or IL-1Rrp2 (Debets et al. 2001; Lovenberg et al. 1996; Towne et al. 2011). Once bound to IL-36α, IL-36β, or

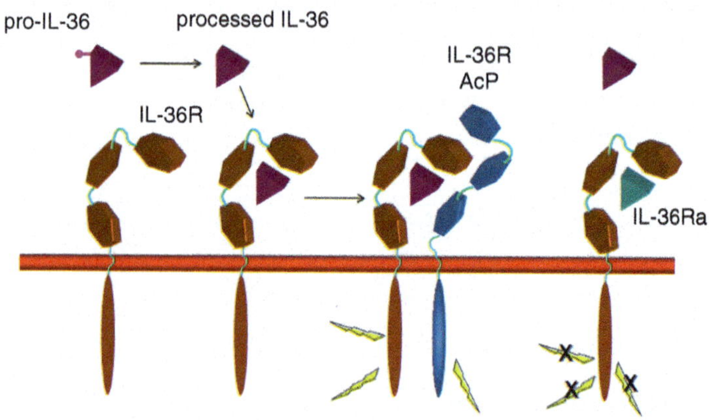

Fig. 8.1 Biochemistry. Interleukin (IL)-36 is originally made as a precursor (pro-IL36) that requires N-terminal processing before it can bind to its receptor with appreciable affinity. Once IL-36 has bound to the receptor, the IL-36–IL-36R complex recruits IL-1R AcP, and it is the heterotrimeric complex which is then capable of signaling. IL-36Ra antagonizes this process by binding to IL-36R, thereby blocking access to agonist forms of IL-36, but fails to recruit IL-1R AcP and therefore does not lead to signaling

IL-36γ, it recruits the IL-1 receptor accessory protein (AcP), which is also utilized by the receptors for IL-1 and IL-33 (Towne et al. 2004). The heterotrimeric complex is required for signal transduction and activates inflammatory signaling pathways similar to those activated by IL-1, including MAP kinases and the transcription factor NFκB (Sims and Smith 2010). IL-36Ra (IL-36-receptor antagonist) acts as a regulator of IL-36 signaling. It binds to IL-36R, competing for binding with the IL-36 agonists, but in contrast to IL-36α, IL-36β, and IL-36γ, receptor-bound IL-36Ra does not allow recruitment of AcP and thus does not initiate a signaling response (Towne et al. 2011). This is the same mechanism as is used by IL-1Ra to regulate the actions of IL-1.

The IL-36 cytokines are clustered on human chromosome 2, within a larger locus that also includes the genes for IL-1α, IL-1β, IL-1Ra, IL-37, and IL-1F10 (Nicklin et al. 2002; Taylor et al. 2002). The IL-36 receptor also lies on human chromosome 2, flanked on the centromeric side by IL-1R1 and IL-1R2, and on the telomeric side by IL-33R and both subunits of the IL-18 receptor (the order of genes is IL-1R2/IL-1R1/IL-36R/IL-33R/IL-18R/IL-18RAP). All members of the IL-1 family appear to have originated with a common ancestor, because there is almost exact conservation of intron placement within the genes. The same is true for the IL-1-receptor family (Sims 2002; Sims et al. 1995).

As is true for all IL-1 family members except IL-1Ra, IL-36 cytokines are synthesized without a signal peptide and thus are not secreted via the endoplasmic reticulum–Golgi pathway. How they are externalized from the cell is poorly understood (Martin et al. 2009). What is clear, however, is that N-terminal processing of the primary translation product is needed for activity at physiologically relevant concentrations (see Fig. 8.1). In the absence of processing, IL-36 biological activity requires concentrations of the order of 1–10 μg/ml. After appropriate processing, however, the specific activity increases by three to five orders of magnitude (Towne et al. 2011). N-terminal truncation must occur at precisely the right location to generate the more active form; removal of one more or one fewer amino acid does not increase specific activity over that of full-length cytokines. The processing enzyme or enzymes are not known. There is very little sequence similarity between the three isoforms of IL-36 in the region of the cleavage site. Notably, there are no aspartate residues or any other features that resemble a cleavage site for caspase-1.

IL-36 cytokines are widely expressed, although almost always at quite low levels. Expression of at least one family member has been reported in many different tissues (Sims and Smith 2010; Towne et al. 2004; Sims and Towne, unpublished data). Because there are no comprehensive studies in the literature, to avoid citing scores of references we simply give an unattributed summary here. Tissues with baseline expression of three or four family members, in either mouse or human, include skin, esophagus, uterus, and tonsil (see Table 8.1 for human data). IL-36 family members can be strongly upregulated in the skin by many agents [inflammatory cytokines, Toll-like receptor (TLR) ligands] or by pathological conditions (viral infection, contact hypersensitivity, psoriasis). All IL-36 family members are also upregulated from very low baseline levels in articular chondrocytes by IL-1. Airway, particularly tracheal and bronchial epithelium, expresses IL-36α and

Table 8.1 Expression of interleukin (IL)-36 forms in different tissues

	IL-36α		IL-36β		IL-36γ		IL-36Ra	
Tissue	Rank	Copies/cell	Rank	Copies/cell	Rank	Copies/cell	Rank	Copies/cell
Tonsil	1	70,590	2	1,729	1	32,925	1	43,090
Skin	4	216	1	3,794	2	7,601	2	33,417
Esophagus	2	20,355	3	392	3	1,450	3	8,186
Cervix	3	1,740			7	187	4	7,363
Umbilical cord					6	281	5	7,178
Placenta					9	134	6	4,182
Testis					5	583	7	552
Trachea	5	133			4	1,035		
Vas deferens					8	160	8	284
Rectum					10	128		
Thyroid							9	116

Determination of the number of copies per cell of transcripts derived from IL-36 family genes in resting human tissues was performed by Asterand (Xpressway Profiles; Asterand, 2 Orchard Road, Royston, Herts. SG8 5HD, UK; www.asterand.com) using quantitative RT-PCR. Probes and primers for IL-36α, IL-36β, IL-36γ, and IL-36Ra were designed using sequences NM_014440, NM_014438, NM_019618, and NM_012275, respectively

especially IL-36γ, particularly after stimulation with inflammatory cytokines, TLR ligands, bacteria, rhinovirus infection, or house dust mites. There is also significant expression of IL-36 family members in testis and large intestine. Of course, there are a number of tissues in which only one or two family members are expressed. In addition, most of the data (except as just noted) derive from resting tissues, so IL-36 expression in activated or diseased tissue is often unknown.

The IL-36 receptor, on the other hand, tends to be widely expressed throughout the body. There are a few tissues and cell types, most notably skin, in which expression is higher than elsewhere, but by and large expression is low and does not seem to be much regulated.

8.3 IL-36 in Psoriasis

Psoriasis is a chronic inflammatory disease of skin that is common in North American and European populations, although less so in Asians (Nestle et al. 2009). It is characterized visually by red patches on the skin that are often raised or appear covered by silver scales. Symptoms include intense itching and burning of the skin, as well as psychological distress (particularly when lesions occur on the face). Histologically, there is hyperplasia of the epidermis (acanthosis) including the stratum corneum (hyperkeratosis), altered keratinocyte differentiation, of which the most obvious features are retention of nuclei in the stratum corneum (parakeratosis), expansion of the capillary network in the superficial dermis, and a hematopoietic cell infiltrate, particularly in the dermis but also in other layers leading to neutrophilic pustules in both the epidermis and stratum corneum. Agents that inhibit tumor necrosis factor (TNF)-α or the shared p40 subunit of IL-12 and IL-23 are effective in treating psoriasis and have been approved by regulatory agencies in various countries. Preliminary clinical trial results suggest that inhibitors of IL-17 or of IL-23 alone are also highly effective (Johnson-Huang et al. 2012; Leonardi et al. 2012).

Gene expression studies have found that IL-36α and IL-36γ are among the most highly expressed and highly induced genes in psoriasis plaques (Blumberg et al. 2007; Debets et al. 2001; Zhou et al. 2003). These genes are not upregulated in the other common human skin disease, atopic dermatitis (Blumberg et al. 2007). In the skin, IL-36α and IL-36γ are expressed primarily by keratinocytes, and can be upregulated in human keratinocytes or skin cultures by TNF, IL-17, or IL-22 (Carrier et al. 2011; Johnston et al. 2011). Combinations of those molecules are even more effective. Similarly, in mouse, intradermal injection of TNF or IL-23 will upregulate IL-36 family members, and injection of combinations of TNF, IL-17, IL-22, IL-23, and IFN-γ work even better (Blumberg et al. 2010). Notably, imiquimod, a TLR7 agonist that can induce psoriatic-like changes in human skin and is often painted onto mouse skin to create a psoriatic-like condition, is an excellent inducer of IL-36 (Tortola et al. 2012).

IL-36 also acts on keratinocytes and other cells in skin. In human keratinocyte cultures, IL-36 induces inflammatory cytokines such as IL-6, IL-8, and TNF, as well as antimicrobial proteins that are abundant in psoriatic lesional skin including human beta-defensins 2 and -3, LL37, and S100A7 (sometimes called psoriasin) (Carrier et al. 2011; Johnston et al. 2011). It also strongly induces itself. In the mouse, intradermal injection of IL-36 induces many genes, including IL-17, IL-20, IL-22, IL-23, TNF, IFN-γ, and a host of other cytokines, chemokines, and antimicrobial proteins that are characteristic of the gene expression pattern in psoriasis (Blumberg et al. 2010).

Just as IL-36α and IL-36γ are strongly upregulated in human psoriasis lesions, forced overexpression of IL-36α in mouse skin leads to an inflammatory condition that is histologically very similar to psoriasis (Blumberg et al. 2010), including acanthosis, hyper- and parakeratosis, infiltration of various hematopoietic cells, including $CD11c^+$ dendritic cells, neutrophilic pustules, and a significant expansion of the capillary network in the superficial dermis. The expression of IL-17, IL-22, IL-23, TNF-α, IFN-γ, and antimicrobial peptides is highly elevated, as it is in psoriasis, and the inflammation and other features are greatly improved by treatment with the agents mentioned here that are successful in treating human psoriasis.

There is one major point on which this mouse model differs from the human disease, however, and that is that skin inflammation occurs in the IL-36α transgenic animals even when they lack T cells (i.e., on a $rag^{-/-}$ background). The authors note that IL-36 is capable of inducing IL-17, IL-22, IL-23, TNF-α, etc., from various types of skin cells, and that those same agents can in turn induce IL-36. They suggest that human psoriasis may be initiated by T cells expressing some of these genes which then upregulate IL-36, but that forced overexpression of IL-36 can substitute for the initial T-cell stimulus and promote a self-amplifying gene expression loop.

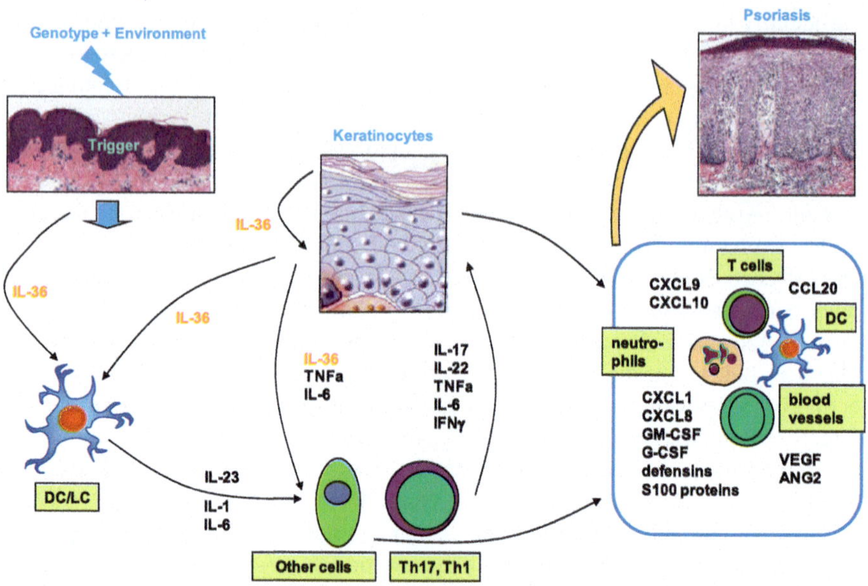

Fig. 8.2 Psoriasis. Upon encountering an appropriate environmental trigger (e.g., bacterial or viral infection), keratinocytes of individuals with an appropriate genetic background synthesize IL-36. IL-36 acts on skin-resident dendritic cells (DC) to induce chemokines and cytokines including IL-1, IL-6, and IL-23; these promote development and migration to skin of Th17 cells and perhaps Th1 cells. Cytokines made by T cells and skin-resident cells in response to DC cytokines include IL-17, IL-22, TNF-α, IL-6, and IFN-γ, which act on keratinocytes and other skin cells to induce more IL-36, resulting in a self-amplifying gene expression loop. Also induced are chemokines that attract neutrophils, monocytes, T cells, and inflammatory dendritic cells, as well as angiogenic factors and antimicrobial proteins, many of which also act as inflammatory stimuli. The net effect of all this cytokine and chemokine activity is the development of skin inflammation in the form of psoriasis

Determining whether this hypothesis is true will require further experimentation. It is worth noting, however, that there is a strong correlation between expression of IL-36α and IL-36γ in human psoriatic lesions and expression of other cytokines including IL-17, IL-23, TNF-α, and IFN-γ (Carrier et al. 2011), suggesting that such a self-reinforcing gene expression loop might occur in human disease.

Perhaps the most definitive experiment implicating IL-36 in human psoriasis has been done by transplanting human lesional skin onto immunodeficient mice (Blumberg et al. 2010). The skin retains its lesional phenotype, even several weeks after the transplant when it has fully engrafted. However, the psoriatic characteristics are substantially ameliorated by inhibition of IL-36 with an anti-human IL-36 receptor antibody. The implication is that the continued action of IL-36 produced by the human graft (mouse IL-36 has very little activity on human IL-36R) is required for the maintenance of the lesional phenotype. For an overview of the mechanism, see Fig. 8.2 (psoriasis).

The foregoing discussion has focused on the common form of psoriasis, plaque psoriasis or psoriasis vulgaris. There are many other less common forms of the disease. One, which is particularly severe, is termed generalized pustular psoriasis. The disease often flares quite suddenly, with large areas of the skin being covered with erythematous lesions and neutrophilic pustules, and can be life threatening without supportive care (Naldi and Gambini 2007). Smahi and colleagues studied a group of Tunisian families in which generalized pustular psoriasis (GPP) was inherited in Mendelian recessive fashion and discovered that the causative mutation led to a leucine-to-proline change at position 27 of IL-36Ra (Marrakchi et al. 2011). The mutant IL-36Ra protein was approximately tenfold less potent than the wild type and was also less stable. Onoufriadis and colleagues published a complementary study in which they determined that three unrelated cases of GPP all carried missense mutations in both copies of the IL-36Ra gene (Onoufriadis et al. 2011). Both groups demonstrated enhanced responses to IL-36 in patient cells, as would be expected if the natural regulator of IL-36 was poorly functional. More recently, Sugiura et al. have found a homozygous nonsense mutation in IL-36Ra in a Japanese patient with GPP (Sugiura et al. 2012) [although it should be noted that other reports, unsurprisingly, find that there are cases of GPP that are not associated with IL36Ra mutations (Farooq et al. 2013; Li et al. 2013)]. Even in the GPP cases with wild-type IL-36Ra, however, there is dramatic upregulation of IL-36α and IL-36γ (Farooq et al. 2013). Thus, overexpression of IL-36 seems to be associated with plaque psoriasis, whereas reduced expression of IL-36Ra leads to generalized pustular psoriasis. This finding is reminiscent of the findings with IL-1 in which mutations in the inflammasome component NLRP3 that result in overproduction of IL-1β lead to familial fever syndromes (Goldbach-Mansky 2012), whereas the absence of IL-1Ra results in a related but clinically distinct disease (Aksentijevich et al. 2009; Reddy et al. 2009). The latter has been called DIRA (deficiency of interleukin-1-receptor antagonist), and as a consequence, GPP (at least when known to be caused by mutation in IL-36Ra) has been renamed DITRA (deficiency of interleukin-thirty-six receptor antagonist) (Marrakchi et al. 2011; Park et al. 2012).

8.4 IL-36 Immune Effects

In addition to its role in skin, IL-36 plays significant roles in immune responses. Dendritic cells (DCs), the initiators of immune responses, respond strongly to IL-36. Mouse bone marrow-derived dendritic cells (BMDCs) express high levels of IL-36R and make IL-12 and IL-23 as well as IL-1β, TNF-α, IL-6, CCL1, CXCL1, and GM-CSF upon IL-36 stimulation (Vigne et al. 2011). These cytokines and chemokines enhance both Th1 and Th17 responses as well as innate responses. Dendritic cell stimulation is unique to IL-36; other IL-1 family cytokines, including IL-1 itself, have little effect on murine BMDCs. In addition, IL-36 has a modest maturing effect on DCs, upregulating MHC class II antigens as well as CD40, CD80 and CD86 to a level comparable to that achieved by LPS stimulation. Finally, BMDCs themselves constitutively express IL-36γ and upregulate IL-36α upon stimulation by various agents including IL-36 itself. The production of IL-36 by BMDC suggests the presence of positive autocrine and paracrine IL-36/IL-36R amplification loops.

Recent data in humans support the findings in mouse cells. IL-36 stimulation of human monocyte-derived DCs leads to induction of cytokines such as IL-23, IL-1β, TNF-α, and IL-6, and chemokines such as CXCL1 and CXCL8, similar to what is seen in mouse (Dinh, J.E.T., and J.E.S.; unpublished data). Mutamba demonstrated that IL-36R is present in monocyte-derived DCs but not in human peripheral blood monocytes and human monocyte-derived macrophages (Mutamba et al. 2012). Furthermore, IL-36 induced the maturation of monocyte-derived dendritic cells and stimulated the production of proinflammatory cytokines such as IL-18 and IL-12 (Mutamba et al. 2012). Taken together, these findings indicate that IL-36 exerts marked stimulatory effects on DC and may therefore play a critical role in early immune and inflammatory responses to tissue damage and pathogens.

In both mice and humans, members of the IL-1 family drive the development of $CD4^+$ T-cell adaptive responses. IL-1β, IL-18, and IL-33 influence $CD4^+$ T-cell responses and their polarization into Th17, Th1, and Th2 $CD4^+$ effector cells, respectively, although the selectivity of these responses may be modulated by the cytokine environment. More specifically, in addition to IL-1 cytokines, the presence of cytokines stimulating the janus kinase (JAK) and signal transducer and activator of transcription (STAT) pathways are required to fully stimulate the polarization of $CD4^+$ T cells into the different Th subsets (Ali et al. 2007; Chung et al. 2009; Dinarello 1999; Guo et al. 2009; Lasiglie et al. 2011; Schmitz et al. 2005; Sims and Smith 2010). IL-1 has been shown to promote the differentiation of naïve $CD4^+$ T cells into Th17 cells (Acosta-Rodriguez et al. 2007; Chung et al. 2009; Kryczek et al. 2007) and to promote expansion of IL-17-secreting memory $CD4^+$ T cells (Rao et al. 2007). IL-18 synergizes with IL-12 as an amplifying signal to induce Th1 responses (Akira 2000). Although IL-33 is also an IL-1 family member, its function on the adaptive immune response is distinct from IL-18 in that it predominantly drives Th2 responses (Trajkovic et al. 2004). The subset-specific effects of IL-1, IL-18, and IL-33 result in part from selective expression of their receptors [reviewed by Sims and Smith (2010)].

Recently, it was shown that IL-36 is also a potent regulator of T-cell responses. Similar to other members of the IL-1 family, IL-36α, IL-36β, and IL-36γ have been reported to have a direct effect on mouse $CD4^+$ T cells and splenocytes, inducing the production of several proinflammatory chemokines and cytokines including IFN-γ, IL-4, and IL-17 (Vigne et al. 2011). The critical difference between IL-36 and other IL-1 family members is that IL-36R is predominantly expressed by naïve $CD4^+$ T cells, whereas the receptors for other IL-1 family members instead are present on polarized $CD4^+$ T-cell subsets, as described here. Consequently, IL-36, but not other IL-1 family cytokines, potently induces IL-2 production, survival, and cell division in T-cell-receptor-activated naïve T cells (Th0 cells). This finding suggests a pivotal role for IL-36 in priming early immune responses (Vigne et al. 2012).

In addition, IL-36 has a role in the induction of T-cell polarization. IL-36β, as does IL-18, acts in combination with IL-12 to induce the in vitro differentiation of

Th0 cells into IFN-γ-producing Th1 cells (Vigne et al. 2012). The combinations IL-12+IL-36β and IL-12+IL-18 both induced T-bet and IL-12Rβ2 mRNA expression. Interestingly, IL-12Rβ2 mRNA levels were much higher when Th0 cells were primed in the presence of IL-12+IL-36β than with IL-12+IL-18, pointing toward a more efficient cooperation between IL-36 and IL-12 in Th0 cells. IL-36β was constitutively expressed in Th0 cells, and interestingly, Th1 polarization induced by IL-12+IL-18 was severely impaired in IL-36R$^{-/-}$ Th0 cells. Thus, endogenous IL-36 plays a prominent role in IL-12-mediated Th1 polarization, even when it is apparently stimulated by IL-18. The role of IL-36 in Th1 responses can also be seen in vivo. IL-36R$^{-/-}$ mice have decreased Th1 cell responses following *Mycobacterium bovis* BCG infection, suggesting that endogenous IL-36 signaling is required to obtain an efficacious response against mycobacteria (Vigne et al. 2012). In addition, the role of IL-36 in Th1 responses has also been reported in human cells. Mutamba et al demonstrated that IL-36β-matured monocyte-derived dendritic cells stimulated differentiation of Th1 cells (Mutamba et al. 2012).

The mechanism by which IL-36 induces the polarization of Th0 cells into the Th1 subset has recently been explored in vitro by using IL-2$^{-/-}$ Th0 cells. In the absence of IL-2, the IL-12+IL-36β combination failed to induce IL-12Rβ2 and stimulated production of very little IFN-γ. Thus, the influence of IL-36 on Th1 differentiation is completely dependent on IL-2 induction. In contrast, the effect of IL-36 on Th0 cell proliferation and survival is largely independently of IL-2 (Vigne et al. 2012).

The findings just described suggest the possibility of significant interplay between nonhematopoietic cells in tissues such as skin and lung with immune cells including T cells and resident and migratory dendritic cells. For example, IL-36 cytokines produced at high levels by epithelial cells, for example, keratinocytes, may have both autocrine and paracrine actions as a result of the presence of IL-36R on epithelial cells and on DC. Indeed, IL-36 produced by epithelial cells can stimulate the maturation of DC with the enhanced expression of MHC class II and of co-stimulatory molecules and also induce the production of proinflammatory chemokines and cytokines. Furthermore, DCs produce IL-36α after cell activation and IL-36γ in a constitutive manner, further enhancing the response. In lymph nodes, IL-36 produced by DC activates Th0 cells, resulting in cell proliferation, enhanced survival, and IL-2 secretion. The endogenous production of IL-36β by Th0 cells further contributes to their activation. Finally, IL-36 and IL-12 synergize to induce polarization into IFN-γ-secreting Th1 cells in an IL-2-dependent manner through the induction of IL-12Rβ2 expression. The levels of IFN-γ secretion are further enhanced by the proliferative effect of IL-36 on Th0 cells (Fig. 8.3: immune effects). Finally, although IL-36 provides an accelerating and amplifying signal for Th0 proliferation and IFN-γ production, it cannot be excluded that, as previously described for IL-18 and IL-33, the selectivity of the effect of IL-36 on Th1 cell polarization is dependent on the cytokine environment. Thus, in the light of the current findings, IL-36 has emerged as an important determinant in the stimulation of innate and adaptive immune responses.

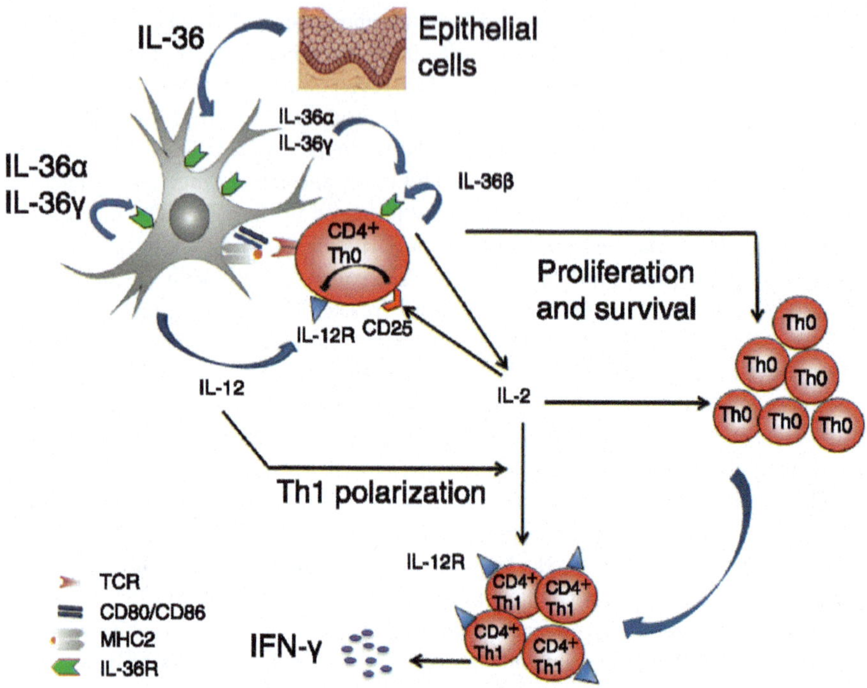

Fig. 8.3 Immune effects. Upon T-cell–dendritic cell (DC) interaction by TCR–MHC engagement, IL-36R, which is constitutively expressed on both DC and Th0 cells, mediates cell activation by IL-36. On the one hand, IL-36 produced by epithelial cells and DCs activates DCs to secrete IL-12, and on the other hand, IL-36 produced by Th0 cells activates Th0 cells, resulting in cell proliferation, survival of naïve T cells, and IL-2 secretion. By a synergistic effect, IL-36 and IL-12 induce Th1 polarization in an IL-2-dependent manner through the induction of IL-12Rβ2 expression, leading to IFN-γ secretion, which is further amplified by the proliferative effect of IL-36 on Th0 cells. The formation of a positive feedback loop created by IL-36/IL-36R leads to sustained IFN-γ-mediated immune responses. (This figure was originally published in Vigne et al. 2012)

8.5 IL-36 Actions on Other Tissues

Although the biology of IL-36 is best understood in the skin, IL-36 and IL-36R are expressed in multiple other tissues, most notably lung. IL-36γ is strongly induced in human primary bronchial epithelial cells (PBEC) by infectious agents including *Pseudomonas aeruginosa* and rhinovirus as well as by inflammatory cytokines such as TNF-α and IL-1β (Bochkov et al. 2010; Vos et al. 2005). Interestingly, rhinovirus induction of IL-36γ was stronger in PBEC from asthmatic donors than from controls (Bochkov et al. 2010). It is known that rhinovirus infections are frequently associated with asthma exacerbations, including more severe respiratory symptoms and reductions in lung function. Rhinoviral infection results in neutrophilic inflammation in the upper airways. Intranasal administration of IL-36α or IL-36γ in mice results in a rapid recruitment of neutrophils into the bronchoalveolar lavage fluid (BALF) (Ramadas et al. 2011, 2012). Thus, it is reasonable to postulate that IL-36 contributes to neutrophil-predominant lung inflammation and the associated asthma exacerbations that can accompany rhinoviral infection.

Published reports suggest a potential role for IL-36 in mouse models of experimental allergic asthma. IL-36γ is increased in ovalbumin-challenged A/J mice (Ramadas et al. 2006) as well as in the lungs of mice sensitized and challenged with house dust mites (Ramadas et al. 2011). Genetic mapping of loci associated with the asthma sensitivity of A/J mice compared to the asthma resistance of C3H/HeJ mice led to identification of a genomic region containing the genes for IL-1 receptor antagonist (IL-1Ra), IL-1 F10, IL-36Ra, IL-36β, IL-36α, and IL-36γ (Ewart et al. 2000). No functional sequence variants were identified in IL-1Ra between the two mouse strains. Several noncoding polymorphisms were found, however, in the *Il1f9* gene (coding for IL-36γ) between A/J and C3H/HeJ mice (Ramadas et al. 2006). There was also a clear difference in the expression level of IL-36γ in lung homogenates in response to ovalbumin between the two strains, with a substantially larger increase seen in the asthma-susceptible A/J mice. Notably, there was no difference in IL-1Ra levels. Therefore, polymorphisms in IL-36γ may be important in conferring increased allergic asthma disease susceptibility in A/J mice.

To explore the role of IL-36, recombinant IL-36γ was administered intranasally twice a day for 2 days to mice; this led to epithelial cell hypertrophy, cellular infiltration into alveolar spaces and around the airways, and mucus production (Ramadas et al. 2011). In addition, a single intratracheal challenge with IL-36γ was sufficient to increase lung resistance 24 h later. Intratracheal administration of IL-36γ resulted in increased cellularity of the BALF, predominantly neutrophils but also to a small extent lymphocytes. No increase in eosinophils was found in BALF after multiple intranasal challenges with IL-36γ. IL-36γ challenge induced neutrophil chemoattractants (CXCL1 and CXCL2) whereas expression of the eosinophil-specific chemokines CCL11 and CCL24 did not change. IL-1α levels were also increased following intranasal challenge with IL-36γ. Intranasal IL-1α administration to mice leads to many of the same responses in lung as IL-36γ, including neutrophil influx, but the effects of IL-36 (in this case, IL-36α) were unchanged in IL-1$\alpha^{-/-}\beta^{-/-}$ mice, thus demonstrating that the effect of IL-36 is independent of IL-1 (Ramadas et al. 2012).

Which cells in the lung respond to IL-36? More than 95 % of immune cells in a naïve mouse lung are CD11c$^+$ alveolar macrophages, which express both IL-36R and IL-1RAcP, suggesting these cells are poised to respond to IL-36 released into the lung airspaces (Ramadas et al. 2012). In support of this, the same authors demonstrated that IL-36 acts on splenic CD11c$^+$ antigen-presenting cells to induce the production of early response cytokines, neutrophil-specific chemokines, and T-cell costimulatory molecules (Ramadas et al. 2012). Further studies are necessary to determine whether alveolar macrophages can similarly respond to IL-36. In addition, lung fibroblasts also express IL-36R and respond to IL-36 with activation of NFκB and MAP kinases and upregulation of the neutrophil chemokines IL-8 and CXCL3 and the Th17 chemokine CCL20 (Chustz et al. 2011).

Bronchial epithelial cells produce IL-36 following stimulation with cytokines such as TNF, IL-1β, and IL-17, in response to infection with *Pseudomonas aeruginosa* or rhinovirus, and following challenge with cigarette smoke condensate (Bochkov et al. 2010; Chustz et al. 2011; Parsanejad et al. 2008; Vos et al. 2005). As already noted, the IL-36 produced by bronchial epithelial cells can act on the adjacent fibroblasts and alveolar macrophages to induce inflammatory cytokines and neutrophil chemokines, leading in turn to an increase in airway hyperresponsiveness and an influx of neutrophils into the lung. Neutrophil accumulation in the airways contributes to the pathogenesis of several lung diseases including asthma, chronic obstructive pulmonary disease (COPD), acute respiratory distress syndrome, and cystic fibrosis. These data suggest that IL-36 may play a significant role in pulmonary disease.

Unfortunately, there are only a few reports examining expression of IL-36 in pulmonary disease in humans. One study did find increased IL-36γ mRNA expression in biopsies of recurrent respiratory papillomas and expression levels correlated with disease severity (DeVoti et al. 2008). In another study, plasma levels of more than 500 inflammatory mediators were measured in patients with COPD and healthy controls (Chen et al. 2012). There was a general downregulation of systemic inflammatory responses in patients with COPD undergoing acute exacerbations, and IL-36α was among the proteins that were decreased (Chen et al. 2012). No information was reported on levels of IL-36 or other inflammatory cytokines in the lung.

IL-36 may also play a role in joint disease. IL-36β mRNA is expressed in human synovial tissues, and human synovial fibroblasts and articular chondrocytes express the IL-36R (Magne et al. 2006). IL-36β was induced in human synovial fibroblasts upon stimulation with IL-1β and TNF-α whereas human articular chondrocytes constitutively expressed IL-36β. Stimulation of these cell types with IL-36β led to induction of IL-6 and IL-8. However, there is no difference in the level of IL-36β in either synovial fluid or serum between healthy controls and rheumatoid arthritis and osteoarthritis patients (Magne et al. 2006).

There may be a role for IL-36 in kidney disease. Ichii and colleagues examined expression of 84 inflammatory genes in kidneys from control mice and mice with chronic glomerulonephritis (CGN) and found a strong and specific upregulation of IL-36α (Ichii et al. 2010). Expression of IL-36α was localized to the epithelium from the distal convoluted tubules to the cortical collecting ducts. The number of IL-36α-positive tubules correlated with proteinuria and fibrosis scores and with the presence of tubulointerstitial lesions. IL-36α was similarly shown to be induced in several other renal disease models, including the streptozotocin-induced diabetic model, where the presence of IL-36α-positive tubules correlated with severity of renal damage.

Other researchers have investigated IL-36 expression in human adipose tissue. IL-36α was found to be expressed whereas IL-36β was not (IL-36γ was not examined) (van Asseldonk et al. 2010). IL-36α was primarily expressed in the stromal vasculature but also in the adipocytes themselves. Expression of peroxisome proliferator-activated receptor (PPAR)-γ was significantly downregulated in human adipocytes upon stimulation with IL-36α, suggesting a reduction in adipocyte differentiation, and both IL-36α and IL-36β induced IL-6 and IL-8 in human adipocytes.

8.6 Concluding Remarks

Although the evidence implicating IL-36 in psoriasis is strong, it is clear that much remains to be discovered regarding the biology of this cytokine. For example, is its role in skin disease purely to stimulate production of other cytokines from DC and nonhematopoietic skin-resident cells? Or does it also influence adaptive immune responses (by either $T\alpha\beta$ or $T\gamma\delta$ cells) that occur in the skin? If so, this would imply an action on T cells other than naïve Th0 cells. Regarding Th0 cells, does IL-36 only promote IFN-γ responses? Or, with the right combination of co-stimuli, will it also lead to the induction of cytokines typical of other polarized T cells? In the lung, although IL-36 can be abundantly induced by multiple stimuli, there is as yet no strong linkage to any pathological (or, for that matter, physiological) condition. Finally, there are hints that IL-36 has a role to play in several other tissues (joint, kidney, adipose) but again, we have barely scratched the surface in determining exactly what that role is. And finally, there are many cell types and tissues in which the expression of IL-36 has not been examined, except in the resting state. Furthermore, in addition to the unknown biology just described, the enzymes involved in processing pro-IL-36 to generate its active form, and the regulation of this processing, remain a mystery. It is likely that IL-36 will turn out, after further study, to have as prominent a role in human health and disease as its fellow family members IL-1, IL-18, and IL-33.

References

Acosta-Rodriguez EV, Napolitani G, Lanzavecchia A, Sallusto F (2007) Interleukins 1beta and 6 but not transforming growth factor-beta are essential for the differentiation of interleukin 17-producing human T helper cells. Nat Immunol 8(9):942–949

Akira S (2000) The role of IL-18 in innate immunity. Curr Opin Immunol 12(1):59–63

Aksentijevich I, Masters SL, Ferguson PJ, Dancey P, Frenkel J, van Royen-Kerkhoff A, Laxer R, Tedgard U, Cowen EW, Pham TH, Booty M, Estes JD, Sandler NG, Plass N, Stone DL, Turner ML, Hill S, Butman JA, Schneider R, Babyn P, El-Shanti HI, Pope E, Barron K, Bing X, Laurence A, Lee CC, Chapelle D, Clarke GI, Ohson K, Nicholson M, Gadina M, Yang B, Korman BD, Gregersen PK, van Hagen PM, Hak AE, Huizing M, Rahman P, Douek DC, Remmers EF, Kastner DL, Goldbach-Mansky R (2009) An autoinflammatory disease with deficiency of the interleukin-1-receptor antagonist. N Engl J Med 360(23):2426–2437

Ali S, Huber M, Kollewe C, Bischoff SC, Falk W, Martin MU (2007) IL-1 receptor accessory protein is essential for IL-33-induced activation of T lymphocytes and mast cells. Proc Natl Acad Sci USA 104(47):18660–18665

Blumberg H, Dinh H, Trueblood ES, Pretorius J, Kugler D, Weng N, Kanaly ST, Towne JE, Willis CR, Kuechle MK, Sims JE, Peschon JJ (2007) Opposing activities of two novel members of the IL-1 ligand family regulate skin inflammation. J Exp Med 204(11):2603–2614

Blumberg H, Dinh H, Dean C Jr, Trueblood ES, Bailey K, Shows D, Bhagavathula N, Aslam MN, Varani J, Towne JE, Sims JE (2010) IL-1RL2 and its ligands contribute to the cytokine network in psoriasis. J Immunol 185(7):4354–4362

Bochkov YA, Hanson KM, Keles S, Brockman-Schneider RA, Jarjour NN, Gern JE (2010) Rhinovirus-induced modulation of gene expression in bronchial epithelial cells from subjects with asthma. Mucosal Immunol 3(1):69–80

Carrier Y, Ma HL, Ramon HE, Napierata L, Small C, O'Toole M, Young DA, Fouser LA, Nickerson-Nutter C, Collins M, Dunussi-Joannopoulos K, Medley QG (2011) Inter-regulation of Th17 cytokines and the IL-36 cytokines in vitro and in vivo: implications in psoriasis pathogenesis. J Invest Dermatol 131(12):2428–2437

Chen H, Wang Y, Bai C, Wang X (2012) Alterations of plasma inflammatory biomarkers in the healthy and chronic obstructive pulmonary disease patients with or without acute exacerbation. J Proteomics 75(10):2835–2843

Chung Y, Chang SH, Martinez GJ, Yang XO, Nurieva R, Kang HS, Ma L, Watowich SS, Jetten AM, Tian Q, Dong C (2009) Critical regulation of early Th17 cell differentiation by interleukin-1 signaling. Immunity 30(4):576–587

Chustz RT, Nagarkar DR, Poposki JA, Favoreto S Jr, Avila PC, Schleimer RP, Kato A (2011) Regulation and function of the IL-1 family cytokine IL-1F9 in human bronchial epithelial cells. Am J Respir Cell Mol Biol 45(1):145–153

Debets R, Timans JC, Homey B, Zurawski S, Sana TR, Lo S, Wagner J, Edwards G, Clifford T, Menon S, Bazan JF, Kastelein RA (2001) Two novel IL-1 family members, IL-1 delta and IL-1 epsilon, function as an antagonist and agonist of NF-kappa B activation through the orphan IL-1 receptor-related protein 2. J Immunol 167(3):1440–1446

DeVoti JA, Rosenthal DW, Wu R, Abramson AL, Steinberg BM, Bonagura VR (2008) Immune dysregulation and tumor-associated gene changes in recurrent respiratory papillomatosis: a paired microarray analysis. Mol Med 14(9–10):608–617

Dinarello CA (1999) IL-18: A TH1-inducing, proinflammatory cytokine and new member of the IL-1 family. J Allergy Clin Immunol 103(1 Pt 1):11–24

Dinarello C, Arend W, Sims J, Smith D, Blumberg H, O'Neill L, Goldbach-Mansky R, Pizarro T, Hoffman H, Bufler P, Nold M, Ghezzi P, Mantovani A, Garlanda C, Boraschi D, Rubartelli A, Netea M, van der Meer J, Joosten L, Mandrup-Poulsen T, Donath M, Lewis E, Pfeilschifter J, Martin M, Kracht M, Muehl H, Novick D, Lukic M, Conti B, Solinger A, Peyman K, van de Veerdonk F, Gabel C (2010) IL-1 family nomenclature. Nat Immunol 11(11):973

Ewart SL, Kuperman D, Schadt E, Tankersley C, Grupe A, Shubitowski DM, Peltz G, Wills-Karp M (2000) Quantitative trait loci controlling allergen-induced airway hyperresponsiveness in inbred mice. Am J Respir Cell Mol Biol 23(4):537–545

Farooq M, Nakai H, Fujimoto A, Fujikawa H, Matsuyama A, Kariya N, Aizawa A, Fujiwara H, Ito M, Shimomura Y (2013) Mutation analysis of the IL36RN gene in 14 Japanese patients with generalized pustular psoriasis. Hum Mutat 34(1):176–183

Goldbach-Mansky R (2012) Immunology in clinic review series; focus on autoinflammatory diseases: update on monogenic autoinflammatory diseases: the role of interleukin (IL)-1 and an emerging role for cytokines beyond IL-1. Clin Exp Immunol 167(3):391–404

Guo L, Wei G, Zhu J, Liao W, Leonard WJ, Zhao K, Paul W (2009) IL-1 family members and STAT activators induce cytokine production by Th2, Th17, and Th1 cells. Proc Natl Acad Sci USA 106(32):13463–13468

Ichii O, Otsuka S, Sasaki N, Yabuki A, Ohta H, Takiguchi M, Hashimoto Y, Endoh D, Kon Y (2010) Local overexpression of interleukin-1 family, member 6 relates to the development of tubulointerstitial lesions. Lab Invest 90(3):459–475

Johnson-Huang LM, Lowes MA, Krueger JG (2012) Putting together the psoriasis puzzle: an update on developing targeted therapies. Dis Model Mech 5(4):423–433

Johnston A, Xing X, Guzman AM, Riblett M, Loyd CM, Ward NL, Wohn C, Prens EP, Wang F, Maier LE, Kang S, Voorhees JJ, Elder JT, Gudjonsson JE (2011) IL-1F5, -F6, -F8, and -F9: a novel IL-1 family signaling system that is active in psoriasis and promotes keratinocyte antimicrobial peptide expression. J Immunol 186(4):2613–2622

Kryczek I, Wei S, Vatan L, Escara-Wilke J, Szeliga W, Keller ET, Zou W (2007) Cutting edge: opposite effects of IL-1 and IL-2 on the regulation of IL-17+ T cell pool IL-1 subverts IL-2-mediated suppression. J Immunol 179(3):1423–1426

Lasiglie D, Traggiai E, Federici S, Alessio M, Buoncompagni A, Accogli A, Chiesa S, Penco F, Martini A, Gattorno M (2011) Role of IL-1 beta in the development of human T(H)17 cells: lesson from NLPR3 mutated patients. PLoS One 6(5):e20014

Leonardi C, Matheson R, Zachariae C, Cameron G, Li L, Edson-Heredia E, Braun D, Banerjee S (2012) Anti-interleukin-17 monoclonal antibody ixekizumab in chronic plaque psoriasis. N Engl J Med 366(13):1190–1199

Li M, Lu Z, Cheng R, Li H, Guo Y, Yao Z (2013) IL36RN gene mutations are not associated with sporadic generalized pustular psoriasis in Chinese patients. Br J Dermatol 168(2):452–455

Lovenberg TW, Crowe PD, Liu C, Chalmers DT, Liu XJ, Liaw C, Clevenger W, Oltersdorf T, De Souza EB, Maki RA (1996) Cloning of a cDNA encoding a novel interleukin-1 receptor related protein (IL 1R-rp2). J Neuroimmunol 70(2):113–122

Magne D, Palmer G, Barton JL, Mezin F, Talabot-Ayer D, Bas S, Duffy T, Noger M, Guerne PA, Nicklin MJ, Gabay C (2006) The new IL-1 family member IL-1F8 stimulates production of inflammatory mediators by synovial fibroblasts and articular chondrocytes. Arthritis Res Ther 8(3):R80

Marrakchi S, Guigue P, Renshaw BR, Puel A, Pei XY, Fraitag S, Zribi J, Bal E, Cluzeau C, Chrabieh M, Towne JE, Douangpanya J, Pons C, Mansour S, Serre V, Makni H, Mahfoudh N, Fakhfakh F, Bodemer C, Feingold J, Hadj-Rabia S, Favre M, Genin E, Sahbatou M, Munnich A, Casanova JL, Sims JE, Turki H, Bachelez H, Smahi A (2011) Interleukin-36-receptor antagonist deficiency and generalized pustular psoriasis. N Engl J Med 365(7):620–628

Martin U, Scholler J, Gurgel J, Renshaw B, Sims JE, Gabel CA (2009) Externalization of the leaderless cytokine IL-1F6 occurs in response to lipopolysaccharide/ATP activation of transduced bone marrow macrophages. J Immunol 183(6):4021–4030

Mutamba S, Allison A, Mahida Y, Barrow P, Foster N (2012) Expression of IL-1Rrp2 by human myelomonocytic cells is unique to DCs and facilitates DC maturation by IL-1F8 and IL-1F9. Eur J Immunol 42(3):607–617

Naldi L, Gambini D (2007) The clinical spectrum of psoriasis. Clin Dermatol 25(6):510–518

Nestle FO, Kaplan DH, Barker J (2009) Psoriasis. N Engl J Med 361(5):496–509

Nicklin MJ, Barton JL, Nguyen M, FitzGerald MG, Duff GW, Kornman K (2002) A sequence-based map of the nine genes of the human interleukin-1 cluster. Genomics 79(5):718–725

Onoufriadis A, Simpson MA, Pink AE, Di Meglio P, Smith CH, Pullabhatla V, Knight J, Spain SL, Nestle FO, Burden AD, Capon F, Trembath RC, Barker JN (2011) Mutations in IL36RN/IL1F5 are associated with the severe episodic inflammatory skin disease known as generalized pustular psoriasis. Am J Hum Genet 89(3):432–437

Park H, Bourla AB, Kastner DL, Colbert RA, Siegel RM (2012) Lighting the fires within: the cell biology of autoinflammatory diseases. Nat Rev Immunol 12(8):570–580

Parsanejad R, Fields WR, Steichen TJ, Bombick BR, Doolittle DJ (2008) Distinct regulatory profiles of interleukins and chemokines in response to cigarette smoke condensate in normal human bronchial epithelial (NHBE) cells. J Interferon Cytokine Res 28(12):703–712

Ramadas RA, Li X, Shubitowski DM, Samineni S, Wills-Karp M, Ewart SL (2006) IL-1 Receptor antagonist as a positional candidate gene in a murine model of allergic asthma. Immunogenetics 58(10):851–855

Ramadas RA, Ewart SL, Medoff BD, Levine AM (2011) Interleukin-1 family member 9 stimulates chemokine production and neutrophil influx in mouse lungs. Am J Respir Cell Mol Biol 44(2):134–145

Ramadas RA, Ewart SL, Iwakura Y, Medoff BD, Levine AM (2012) IL-36alpha exerts pro-inflammatory effects in the lungs of mice. PLoS One 7(9):e45784

Rao DA, Tracey KJ, Pober JS (2007) IL-1alpha and IL-1beta are endogenous mediators linking cell injury to the adaptive alloimmune response. J Immunol 179(10):6536–6546

Reddy S, Jia S, Geoffrey R, Lorier R, Suchi M, Broeckel U, Hessner MJ, Verbsky J (2009) An autoinflammatory disease due to homozygous deletion of the IL1RN locus. N Engl J Med 360(23):2438–2444

Schmitz J, Owyang A, Oldham E, Song Y, Murphy E, McClanahan TK, Zurawski G, Moshrefi M, Qin J, Li X, Gorman DM, Bazan JF, Kastelein RA (2005) IL-33, an interleukin-1-like cytokine that signals via the IL-1 receptor-related protein ST2 and induces T helper type 2-associated cytokines. Immunity 23(5):479–490

Sims JE (2002) IL-1 and IL-18 receptors, and their extended family. Curr Opin Immunol 14(1):117–122

Sims JE, Nicklin MJ, Bazan JF, Barton JL, Barfield SJ, Ford JE, Kastelein RA, Kumar S, Liu H, Mulero JJ, Pan J, Pan Y, Smith DE, Young PR (2001) A new nomenclature for IL-1-famuly genes. Trends Immunol 22(10):536–537

Sims JE, Smith DE (2010) The IL-1 family: regulators of immunity. Nat Rev Immunol 10(2):89–102

Sims JE, Painter SL, Gow IR (1995) Genomic organization of the type I and type II IL-1 receptors. Cytokine 7(6):483–490

Sugiura K, Takeichi T, Kono M, Ogawa Y, Shimoyama Y, Muro Y, Akiyama M (2012) A novel IL36RN/IL1F5 homozygous nonsense mutation, p.Arg10X, in a Japanese patient with adult-onset generalized pustular psoriasis. Br J Dermatol 167(3):699–701

Taylor SL, Renshaw BR, Garka KE, Smith DE, Sims JE (2002) Genomic organization of the interleukin-1 locus. Genomics 79(5):726–733

Tortola L, Rosenwald E, Abel B, Blumberg H, Schafer M, Coyle AJ, Renauld JC, Werner S, Kisielow J, Kopf M (2012) Psoriasiform dermatitis is driven by IL-36-mediated DC-keratinocyte crosstalk. J Clin Invest 122(11):3965–3976

Towne JE, Garka KE, Renshaw BR, Virca GD, Sims JE (2004) Interleukin (IL)-1F6, IL-1F8, and IL-1F9 signal through IL-1Rrp2 and IL-1RAcP to activate the pathway leading to NF-kappaB and MAPKs. J Biol Chem 279(14):13677–13688

Towne JE, Renshaw BR, Douangpanya J, Lipsky BP, Shen M, Gabel CA, Sims JE (2011) Interleukin-36 (IL-36) ligands require processing for full agonist (IL-36alpha, IL-36beta, and IL-36gamma) or antagonist (IL-36Ra) activity. J Biol Chem 286(49):42594–42602

Trajkovic V, Sweet MJ, Xu D (2004) T1/ST2–an IL-1 receptor-like modulator of immune responses. Cytokine Growth Factor Rev 15(2–3):87–95

van Asseldonk EJ, Stienstra R, Koenen TB, van Tits LJ, Joosten LA, Tack CJ, Netea MG (2010) The effect of the interleukin-1 cytokine family members IL-1F6 and IL-1F8 on adipocyte differentiation. Obesity (Silver Spring) 18(11):2234–2236

Vigne S, Palmer G, Lamacchia C, Martin P, Talabot-Ayer D, Rodriguez E, Ronchi F, Sallusto F, Dinh H, Sims JE, Gabay C (2011) IL-36R ligands are potent regulators of dendritic and T cells. Blood 118(22):5813–5823

Vigne S, Palmer G, Martin P, Lamacchia C, Strebel D, Rodriguez E, Olleros ML, Vesin D, Garcia I, Ronchi F, Sallusto F, Sims JE, Gabay C (2012) IL-36 signaling amplifies Th1 responses by enhancing proliferation and Th1 polarization of naive CD4+ T cells. Blood 120(17): 3478–3487

Vos JB, van Sterkenburg MA, Rabe KF, Schalkwijk J, Hiemstra PS, Datson NA (2005) Transcriptional response of bronchial epithelial cells to *Pseudomonas aeruginosa*: identification of early mediators of host defense. Physiol Genomics 21(3):324–336

Zhou X, Krueger JG, Kao MC, Lee E, Du F, Menter A, Wong WH, Bowcock AM (2003) Novel mechanisms of T-cell and dendritic cell activation revealed by profiling of psoriasis on the 63,100-element oligonucleotide array. Physiol Genomics 13(1):69–78

Chapter 9
Anti-TNF Therapy: 20 Years from Our First Therapeutic Adventure

Jagdeep Nanchahal, Peter Taylor, Richard Williams, and Marc Feldmann

Abstract Anti-tumor necrosis factor (TNF) therapy was the first successful anti-cytokine therapy, and its discovery was based on detailed examination of human disease tissue in rheumatoid arthritis and on exploring the dysregulation of cytokine expression within that tissue.

From those first successful clinical trials we conducted in 1992 with infliximab, which were widely publicized more than a year before publication, a competition emerged between a set of companies that had made anti-TNF monoclonals and receptor fusion proteins to treat sepsis to get their products on the market for rheumatoid arthritis (RA). Subsequent clinical trials based on RA success took place in other diseases: Crohn's disease, ankylosing spondylitis, psoriasis, psoriatic arthritis, and ulcerative colitis. By 1999 etanercept, a tumor necrosis factor-receptor (TNF-R) p75: IgG Fc fusion protein, was approved for RA and infliximab, a chimeric anti-TNF monoclonal antibody, was approved for RA and Crohn's disease. The foregoing diseases subsequently all became approved indications for anti-TNF therapy that has emerged as the first successful anti-cytokine therapy. By 2013 anti-TNF had become a huge commercial success with global sales of US$27 billion, replacing statins as the most valuable drug class.

But there are many challenges remaining. There are many other diseases that could benefit from anti-TNF therapy, and we discuss many but not all of them in this review. And then, it is clear that anti-TNF therapy is not a cure. We need to get closer to a cure and find complementary therapeutics that when added to

J. Nanchahal • R. Williams • M. Feldmann (✉)
Nuffield Department of Orthopaedics, Rheumatology and Musculoskeletal Sciences,
Kennedy Institute of Rheumatology, University of Oxford, Roosevelt Drive, Headington,
Oxford, OX3 7FY, UK
e-mail: marc.feldmann@kennedy.ox.ac.uk

P. Taylor
Nuffield Department of Orthopaedics, Rheumatology and Musculoskeletal Sciences,
Botnar Research Centre, University of Oxford, Old Road, Headington, Oxford, OX3 7LD, UK

T. Yoshimoto and T. Yoshimoto (eds.), *Cytokine Frontiers: Regulation of Immune Responses in Health and Disease*, DOI 10.1007/978-4-431-54442-5_9, © Springer Japan 2014

anti-TNF would take us closer to a cure. Doing so safely and reproducibly is a challenge, but great understanding of the pathogenesis of RA and how anti-TNF works will be helpful.

Keywords Anti-TNF • Monoclonal antibody therapy • New indications

9.1 Overview of Cytokines and TNF

Tumor necrosis factor (TNF) is one of a family of more than 100 protein messenger molecules generically termed "cytokines" that regulate immunity, inflammation, tissue destruction, repair, fibrosis, and angiogenesis: in fact, all interesting biological processes. TNF is unusual in many ways, which possibly have bearing on the effectiveness of its blockade in diverse clinical situations.

TNF is a homotrimer, which exists in two molecular forms: membrane TNF, with a long cytoplasmic tail of 86 amino acids (AA), and a soluble form, which is shed from the cell surface by "TNF-α converting enzyme," abbreviated TACE. Adam 17 is the major but not only TACE, because Adam 17 "knockouts" still produce reduced soluble TNF but at reduced levels (Black et al. 1997; Peschon et al. 1998).

TNF resides in normal individuals or animals on the surface of many blood cells, particularly on neutrophils and macrophages. Cleavage of preformed membrane TNF is one of the mechanisms by which TNF can be released rapidly after almost all types of stress. Release of TNF from granules in mast cells, basophils, also helps to make TNF the most rapidly released pro-inflammatory cytokine preceding IL-1β and IL-6, in mice or humans in a variety of circumstances where it has been evaluated. After the initial burst of preformed TNF, the cytokine is synthesized rapidly as the TNF gene chromatin is "open," not needing to be unwound.

TNF was reported to induce a "cascade" of other pro-inflammatory cytokines in mice in response to lipopolysaccharide (LPS) or bacteria (Fong et al. 1989). The evidence is that blocking TNF reduced IL-1β and IL-6 markedly in the serum. A similar "TNF-dependent cytokine cascade" was reported in cultures of human rheumatoid synovial tissue, which spontaneously produce the cytokines, enzymes, and eicosanoids expected to be present in inflamed joints. Blockade of TNF with rabbit anti-serum or monoclonal antibodies to TNF reduced the production of IL-1β and IL-6: this was the first clue that TNF might be a good therapeutic target for RA (Brennan et al. 1989).

There are other interesting aspects of the biology of TNF. Very little is detected in blood; what little there is in blood has limited bioactivity and is rapidly cleared from the circulation. There are in biological fluids soluble TNF receptors, both TNF-R1 and, more abundantly, TNF-R2, generated by proteolytic cleavage, in considerable molar excess to TNF itself and thus able to limit the diffusion of TNF and partly neutralize it (Cope et al. 1992).

TNF is one of the few cytokines for which there are two distinct receptors, which have been conserved over millions of years of evolution and signal in a different manner. The expression of the two receptors is discrete on different cells, with TNF-R1 on all cells but TNF-R2 much more limited in distribution, chiefly on cells of the lymphoid system. This conservation and selective distribution raise the question of the respective function of the receptors. Multiple studies have documented that TNF-R1 (p55) is the receptor that drives the majority of inflammation, raising the hypothesis that perhaps TNF-R1 might be a better therapeutic target than TNF itself, as it would leave part of the diverse functions and potentially beneficial functions of TNF intact (Bluml et al. 2012; Kollias and Kontoyiannis 2002). TNF-R2 has been implicated in immune function and repair, which appear to be desirable in the context of inflammatory disease.

The first observations that had suggested a role of TNF in autoimmune disease were the upregulation of both TNF and its receptor in active RA synovial tissue. However, many other pro-inflammatory cytokines were also upregulated, and so the dilemma was which, if any, of these might be an effective therapeutic target (Feldmann et al. 1996). The inflammatory functions of IL-1β, IL-6, GM-CSF, TNF, and interferon (IFN)-γ are very similar, so the question remained as to why would blocking just one of these make any biological or clinical difference? By culturing rheumatoid synovial cells in the absence of extrinsic stimulation, the fact that TNF blockade downregulated all the other pro-inflammatory cytokines was discovered, suggesting it might be a therapeutic target (Brennan et al. 1989).

The work leading up to the discovery of anti-TNF therapy has been reviewed in many other publications (e.g., Feldmann 2009; Feldmann et al. 1996; Feldmann and Maini 2003; Feldmann and Maini 2010). Hence, to avoid "multiple publications" we are selecting here certain aspects to highlight and are not repeating many aspects of these prior reviews.

9.2 Lessons from Animal Models

9.2.1 Introduction

Analysis of the physiological role played by TNF in animal models has in general confirmed its pro-inflammatory properties, as predicted on the basis of in vitro studies. However, a number of recent findings have appeared that reveal a more complex role for TNF in immune-mediated pathologies than was hitherto anticipated. This idea is not surprising in an evolutionary context because TNF is a key host defense mediator. For example, not all models of inflammation respond to TNF blockade in the same way, and a number of findings suggest a role for TNFR signaling in promoting regulatory T-cell responses and in suppressing Th1 and Th17 responses.

9.2.2 TNF as a Driver of Inflammation

Early evidence of a pro-inflammatory role for TNF in vivo came from TNF injection experiments and from the group of Kollias et al. who generated a strain of mice bearing a human TNF (hTNF) transgene (including its endogenous promoter region). The 3′-AU rich region of the promoter region of the transgene had been substituted by the 3′-untranslated region of the human β-globin gene, resulting in overexpression. hTNF transgenic mice develop a spontaneous form of arthritis that could be prevented by continuous administration of anti-human TNF monoclonal antibody (mAb) (Keffer et al. 1991). Histopathological analysis of the joints of hTNF transgenic mice revealed similarities to human RA and demonstrated that the disease was highly erosive in nature, with subchondral bone, rather than cartilage, being mostly affected. It was subsequently observed that treatment of hTNF transgenic mice with a blocking anti-IL-1R antibody prevented arthritis, demonstrating that IL-1β is an important downstream pathological mediator in this model (Probert et al. 1995). This result is consistent with findings in human RA synovial cell cultures in which TNF inhibition was found to block IL-1β production (Brennan et al. 1989), indicating the dependence of IL-1β production on TNF.

TNF expression in hTNF transgenic mice was not confined to the joint but was expressed in various tissues, including lung, spleen, and the joint, and it is curious that the joint should be the main target of the disease while other tissues are spared. In fact, another strain of TNF transgenic mice in which the TNF transgene lacked an AU-rich region were found to develop not only arthritis but also inflammatory bowel disease (IBD) (Kontoyiannis et al. 1999). TNF-overproducing mice can be back-crossed to $RAG^{-/-}$ mice without altering the arthritis phenotype, showing that TNF can drive joint inflammation in the absence of the adaptive immune system. In contrast, the severity of IBD was reduced in TNF-transgenic $RAG^{-/-}$ mice, indicating the heterogeneity of the role of TNF-α and the involvement of lymphocytes in this model (Kontoyiannis et al. 1999).

Subsequent to the demonstration of spontaneous arthritis in hTNF transgenic mice, a number of studies have assessed the effect of TNF blockade in collagen-induced arthritis (CIA), an animal model of RA. These studies showed that treatment of mice with anti-TNF antibodies, or soluble TNF receptors, reduced the severity of arthritis when administered before the onset of clinical disease (Piguet et al. 1992; Thorbecke et al. 1992; Williams et al. 1992). Subsequently, we demonstrated that TNF blockade using anti-TNF mAb or p75 TNFR-Fc fusion protein was effective in established arthritis (Williams et al. 1992; Williams et al. 1995). These findings confirmed the importance of TNF in CIA and provided an important part of the rationale for the testing of anti-TNF antibody therapy in human RA. In addition to arthritis, TNF has been shown to play a major pro-inflammatory role in a number of animal models of autoimmune disease, including IBD (Kollias 2005).

9.2.3 The Relationship Between TNF and Regulatory T Cells

Some controversy surrounds the relationship between TNF and the regulatory T-cell responses. On the one hand, many reports have highlighted an inhibitory effect of inflammation in general, and TNF in particular, on the size and level of activity of the regulatory T-cell population. A seminal study by Ehrenstein et al. demonstrated that regulatory T cells from RA patients had reduced capacity to modulate pro-inflammatory cytokine production by effector T cells stimulated mitogenically with anti-CD3 (Ehrenstein et al. 2004). Thus, treatment with anti-TNF mAb led to increased numbers of regulatory T cells in peripheral blood and restored their suppressive capacity. Subsequent studies demonstrated that TNF inhibition in RA results in the emergence of a $FoxP3^+$ regulatory T population that mediates suppression via transforming growth factor (TGF)-β and IL-10 (Nadkarni et al. 2007). In another study of patients with active RA, increased spontaneous apoptosis of regulatory T cells was shown, which was downregulated by treatment with TNF-blocking mAb (Toubi et al. 2005). Consistent with these findings, regulatory T-cell suppressor activity in RA was reported to be downregulated by myeloid cell-derived inflammatory mediators (van Amelsfort et al. 2007), and in another study it was shown that compromised regulatory T-cell activity in RA was associated with reduced FoxP3 expression, which could reversed by TNF blockade (Valencia et al. 2006). In CIA, regulatory T-cell activity has also been shown to be diminished (Kelchtermans et al. 2005), and similar results were obtained in experimental autoimmune encephalomyelitis (EAE) (Korn et al. 2007), multiple sclerosis (Viglietta et al. 2004), and type 1 diabetes (Chee et al. 2011; Lindley et al. 2005; Ryba et al. 2011).

On the other hand, there is evidence to suggest that TNF promotes regulatory T-cell responses. It was shown, for example, that TNF-α may increase regulatory T-cell responses in a TNFR2-dependent manner, resulting in proliferation, upregulation of FoxP3 expression, and increased suppressive capacity (Chen et al. 2007). TNF blockade was also shown to abrogate lipopolysaccharide (LPS)-induced expansion of splenic regulatory T cells in vivo (Hamano et al. 2011). Furthermore, anti-TNF prevented the expansion of regulatory T cells in murine psoriasis (Ma et al. 2010), and the activity of natural, although not inducible, regulatory T cells was shown to be dependent on TNF (Housley et al. 2011). It is of interest that TNFR2 is predominantly expressed by regulatory T cells and that TNFR2-expressing $CD4^+FoxP3^+$ regulatory T cells represent the maximally suppressive subset of regulatory T cells (Chen et al. 2010; Chen et al. 2008). A critical role for TNFR2 signaling has also been demonstrated in the induction of human antigen-specific regulatory T cells by tolerogenic dendritic cells (Kleijwegt et al. 2010). Much remains to be unraveled about TNF interaction with T cells.

9.2.4 Paradoxical Effects of TNF on Effector T-Cell Responses

Recent findings from our (Williams') laboratory have shed new light on the complex relationship between TNF and the immune system by the discovery of an endogenous regulatory pathway, triggered by TNFR1, that specifically targets Th1 and Th17 cells (Notley et al. 2008). Thus, blockade of TNF, or deletion of TNFR1, was found to cause an expansion of collagen-specific Th1 and Th17 cells in the spleen and lymph nodes of mice with CIA; this was caused not only by a diversion of Th1/Th17 cells away from the site of inflammation and into the lymphoid organs, because the same phenomenon was observed in collagen-immunized TNFR1$^{-/-}$ (but not TNFR2$^{-/-}$) mice in the absence of inflammation (Notley et al. 2008). Furthermore, a more recent study that reproduces our findings in a model of reactive arthritis documents a global expansion of Th1/Th17 cells in the lymph nodes, spleen, and joints of TNFR1$^{-/-}$ mice (Elicabe et al. 2010). Importantly, we confirmed using an adoptive transfer system that this expanded population of Th1/Th17 cells is highly pathogenic when the anti-TNF "brake" is removed (Notley et al. 2008), which may help to explain why disease flares following withdrawal of TNF-blocking drugs. This phenomenon may also explain some of the adverse effects of TNF inhibitors, such as new onset or exacerbation of multiple sclerosis or cutaneous psoriasis. This result is supported by the finding that anti-TNF exacerbates murine psoriasis-like disease by enhancing the activity of Th17 cells (Ma et al. 2010).

The differentiation and survival of Th1 and Th17 cells are largely controlled by IL-12 and IL-23, respectively, and these two cytokines share a common p40 subunit. We therefore hypothesized that the expansion of Th1 and Th17 cells following blockade of TNF was caused by an increased IL-12/IL-23 p40 expression; this was subsequently confirmed in vitro and in vivo at the level of mRNA and protein. In addition, we were able to show that the expansion of Th1 and Th17 cells could be reduced in TNFR1$^{-/-}$ mice by blockade of IL-12/IL-23 p40 (Notley et al. 2008). Hence, our data show that at least part of the expansion of Th1 and Th17 cells by TNF inhibition is by augmented p40 expression. It was also reported that TNF selectively inhibits IL-12/IL-23 p40 expression in human and mouse myeloid cells in vitro, which is in agreement with this concept (Ma et al. 2000; Zakharova and Ziegler 2005). In addition to these findings, Cope et al. have shown that prolonged exposure of lymphocytes to TNF in RA leads to a state of T-cell hyporesponsiveness from perturbation of TCR signal transduction pathways, which is also consistent with an immunoregulatory role for TNF-α (Cope 2003).

Further support for an immunoregulatory and protective role for TNF is provided by the results of a very early study which showed that administration of rTNF to young lupus-prone mice led to protection against disease and reduced levels of anti-nuclear antibodies (Jacob and McDevitt 1988). More recently it was reported that TNF deficiency was associated with increased production of anti-nuclear antibodies and accelerated onset of murine lupus (Kontoyiannis and Kollias 2000). Similar findings were reported in murine type 1 diabetes (Satoh et al. 1989), and in another study in autoimmune diabetes TNF was also shown to have anti-inflammatory properties depending on the timing of TNF expression (Christen et al. 2001). TNF was also reported to play a protective role in adjuvant arthritis (Kim et al. 2008) and in EAE, TNF$^{-/-}$ mice developed enhanced inflammation and demyelination, whereas treatment of susceptible mice with rTNF reduced the severity of disease (Liu et al. 1998a). In a different study it was shown that EAE failed to resolve in TNFα$^{-/-}$ or

$TNFR^{-/-}$ mice, indicating a role for TNF in the resolution of inflammation in this model (Kassiotis and Kollias 2001). But in other models of EAD (chronic, relapsing), anti-TNF was protective especially if administered locally, intrathecally (Triantaphyllopoulos et al. 1998).

In view of the increased Th1/Th17 responses in anti-TNF-α-treated mice with CIA (Notley et al. 2008), the question arises as to whether similar changes occur in patients treated with TNF inhibitors. One recent study in RA has provided evidence of increased numbers of Th17 cells in peripheral blood following anti-TNF therapy (Aerts et al. 2010). In addition, we carried out a longitudinal study of patients with RA treated with TNF-α inhibitors to assess their Th17/IL-17 levels before and after the start of therapy. Significant increases in circulating Th17 cells, but not Th1 cells, were detected in patients after anti-TNF therapy, accompanied by increased production of IL-12/IL-23 p40 (Alzabin et al. 2012). There was also suggestive evidence of an inverse relationship between baseline Th17 levels and the subsequent response of patients with RA to anti-TNF therapy. These findings, if confirmed in larger studies, show that anti-TNF therapy results in an expansion of Th17 responses, and a Th17-targeted therapeutic approach may be useful for anti-TNF non-responder patients or in combination with anti-TNF therapy. How and when to combine Th17-targeted therapy safely presents a real challenge.

9.3 Role of TNF in Other Diseases

The key role played by TNF in inflammatory arthritis and IBD is well established, together with the therapeutic benefit of systemic inhibition (currently TNF inhibition is approved for rheumatoid arthritis, juvenile rheumatoid arthritis, Crohn's disease, ulcerative colitis, psoriasis, psoriatic arthritis, and ankylosing spondylitis). However, despite ample evidence that TNF also acts in a variety of other disorders and preclinical data that blockade may be of therapeutic benefit, the few clinical trials in these conditions have been disappointing. This finding has led us to look beyond the pathological effects of TNF in chronic inflammation as emerging data show that TNF is also involved in physiological regeneration. There is now evidence from a diverse range of tissues that TNF enhances repair and the effects are very context dependent. For example, in fracture healing low doses of TNF are key in recruiting mesenchymal stromal cells (Glass et al. 2011), with short exposure enhancing osteogenesis while longer exposure is inhibitory (Lu et al. 2012). It is not surprising that this pleiotropic cytokine has vastly different effects, but there are some general trends emerging whereby short exposure within a critical low-dose range may be beneficial when present at the appropriate time, whereas sustained high levels are detrimental. Furthermore, TNF appears to have varying effects on different cell types, perhaps related to their expression of TNF receptors. The situation is complicated as the relative expression of TNFR1 and R2 also appears to determine the response. Hence, it is not surprising that global inhibition has resulted in very mixed results.

Here we review the role of TNF in three diverse areas—the brain, metabolic syndrome, and fibrosis. Other concerns are not included for lack of space, for example, cancer. The safety profile of TNF inhibition has been well established in clinical studies, and the available data would suggest that we cannot simply extrapolate our success with TNF inhibition in chronic inflammatory disorders such as rheumatoid arthritis and Crohn's disease to other conditions. Perhaps the area that may be most amenable to target next is localized musculoskeletal fibrosis, where the anti-TNF is delivered at the site of the pathology (Verjee et al. 2013).

9.3.1 TNF and the Brain

TNF signaling is involved in several important central nervous system (CNS) functions, including activation of microglia and astrocytes by injury, regulation of blood–brain barrier permeability, neurogenesis, glutaminergic transmission, synaptic plasticity, and scaling. Synaptic scaling, whereby the strength of all synapses on a cell are adjusted in response to prolonged changes in the cell's electrical activity, has been shown to be mediated by TNF (Stellwagen and Malenka 2006). Microglia, astrocytes, and neurons are all sources of TNF in the central nervous system (CNS) (Montgomery and Bowers 2012). Astrocytes, which represent the predominant glial cell of the CNS, have a major role in scaling. A single astrocyte contacts thousands of synapses, and astrocytes have been shown to release glutamate on stimulation by TNF (Domercq et al. 2006).

9.3.2 TNF in Memory and Learning

Cytokines play an important role in hippocampal-dependent learning and memory, that is, spatial memory, object recognition, and contextual fear conditioning (McAfoose and Baune 2009). Mechanisms involved in memory consolidation include synaptic scaling and adult neurogenesis. Overexpression and absence of cytokines has been shown to directly influence hippocampal-dependent memory. For example, basal levels of IL-1β are required for normal memory function, and any deviation from the physiological range results in impaired memory (Goshen et al. 2007). Endogenous glial-derived TNF is critical for homeostatic synaptic scaling, with transgenic mice that overexpress TNF demonstrating cognitive impairment (Fiore et al. 2000). In a murine model, orthopedic surgery was associated with elevated IL-1β and upstream TNF levels, which led to impairment of hippocampal-dependent memory. These effects were reversed by cytokine inhibition (Terrando et al. 2011; Terrando et al. 2010). In the brain, the main source of TNF is the microglia. Microglia are the primary immune cell in the CNS and share many phenotypic features with macrophages. Although TNFR2 is predominantly expressed by macrophages, both TNFR1 and TNFR2 are present in microglia (Dziewulska and Mossakowski 2003), and mice deficient in both TNFR1 and R2 were more susceptible to hippocampal ischemic injury (Bruce et al. 1996). The density of microglia varies across different brain regions, being especially high in the hippocampus and substantia nigra (Lawson et al. 1990). There are also marked regional differences in basal cytokine expression, with hippocampal microglia expressing higher levels of TNF (Ren et al. 1999). The cognitive decline that occurred following orthopedic surgery in the murine model was associated with activation of macrophage-like cells in the hippocampus (Terrando et al. 2011) and the recruitment of monocytes into the CNS.

It has been suggested that microglial responses are tailored in regional and insult-specific manner (Carson et al. 2007). The effect of TNF also depends on the expression of TNF receptors. TNFR1 is constitutively expressed on most cell types whereas TNFR2 expression is induced on upon inflammatory stimulation and is limited to cells of the lymphoid lineage, microglia, and endothelial cells. Soluble TNF (sTNF) binds with high affinity to TNFR1 but readily dissociates from TNFR2. Close spatial association between the two receptors means that the ligand could be rapidly passed from TNFR2 to TNFR1, and it has been hypothesized that this may lead to increased efficacy of signaling (Tartaglia et al. 1993). Indeed, it has been suggested that sTNFR2 may prolong the inflammatory response by intermittent release of sTNF to TNFR1. There is evidence that TNFR2 may contribute to the chronicity of inflammation in Crohn's disease (Holtmann et al. 2002). TNFR2 effectively binds transmembrane TNF (tmTNF), which can act both as a ligand and as a receptor (Grell et al. 1995; Zhang et al. 2008). In this context, it is of interest that the intracytoplasmic part of transmembrane TNF is extensive, with 76 amino acids in humans or 79 in mice, and multiple signaling motifs (Eissner et al. 2004; Pennica et al. 1985). The latter allows for reverse signaling and bidirectional communication between cells. The balance between sTNF and tmTNF is influenced by cell type, activation status of the cell, the stimulus eliciting TNF production, activity of ADAM 17 TNF converting enzyme (TACE) (Gearing et al. 1994), and endogenous inhibitors such as TIMP3 (Smookler et al. 2006). Both TNFR1 and -R2 lack enzymatic activity within their intracellular domains and signal via intracellular adaptor molecules, resulting in activation of a number of pathways, including NF-κB, p38, and c-jun N-terminal kinase (JNK). In turn, this results in responses ranging from inflammation to proliferation, cell migration, apoptosis, and necrosis. TNFR2 activation can activate both inflammatory and neuronal survival pathways. Activation of NF-κB can be either neuroprotective or neurotoxic, the effect being dependent on the duration of NF-κB activation and differential activity of regulatory proteins.

Soluble TNF has been found to be elevated in a number of pathological conditions including ischemic stroke, Alzheimer's disease, Parkinson's disease, and multiple sclerosis.

9.3.3 TNF in Multiple Sclerosis

Patients with multiple sclerosis (MS) have elevated levels of TNF in the serum, cerebrospinal fluid (CSF), and active lesions. Based on preclinical evidence, including neutralization studies (Triantaphyllopoulos et al. 1998; Baker et al. 1994) a phase II trial of the TNF inhibitor lenercept was commenced in multiple sclerosis but it showed a dose-dependent increase in attack severity and trend of increase in attack duration (1999). Further evidence for the potential detrimental effect of TNF inhibition on the nervous system was presented as a series of three cases where patients with RA treated with TNF inhibitors developed a demyelinating disorder, which resolved in two individuals after discontinuation of anti-TNF but persisted in the third patient (Fromont et al. 2009). Subsequent studies in murine models showed that although TNFR1 signaling was associated with demyelination, TNFR2-mediated pathways were crucial for remyelination (Arnett et al. 2001). Transgenic murine models that maintained tmTNF in the absence of sTNF showed suppression of disease onset and progression of EAE (Alexopoulou et al. 2006). The dual neurotoxic and neuroprotective role of TNF means that there is a potential for negative consequences for TNF inhibition for pathological conditions, and caution needs to be exercised in extrapolating preclinical data to drive clinical trials. Animal models that develop degenerative neurological conditions are relatively short term and hence may not truly emulate the pathological processes encountered in the clinical setting.

9.3.4 TNF in Acute Stroke

TNF also appears to have a dual role in ischemic brain injury, again depending on levels, timing, duration, relative expression of TNF receptors, and anatomic site. Mild cerebral ischemia, which is seen clinically following transient ischemic attacks, is associated with raised serum levels of TNF, which remain elevated for 72 h after the event (Castillo et al. 2003). Interestingly, this conferred a neuroprotective effect within this time window (Castillo et al. 2003). TNF inhibition following middle cerebral artery occlusion in a murine model resulted in reduction of the size of cortical infarct (Nawashiro et al. 1997). However, hippocampal and striatal neurogenesis, most likely mediated by TNFR2, was markedly reduced in a rat model following middle cerebral artery occlusion (Heldmann et al. 2005). Furthermore, tmTNF and TNFR2 signaling was found to be protective against epileptic seizures induced by injection of kainic acid into the hippocampus (Balosso et al. 2005).

9.3.5 TNF in Parkinson's Disease

Inflammatory pathways appear to play a key role in Parkinson's disease (PD), which appears to arise through a complex interplay between genetic susceptibility and environmental factors (Gorell et al. 2004). As in other neurodegenerative diseases, elevated levels of sTNF and sTNFR1 have been detected in the CSF and tissues of patients with PD, and TNF and TNFR1 polymorphisms have been associated with the risk of developing PD (McCoy and Tansey 2008). The role of TNF is supported by in vitro and in vivo studies showing that TNF is highly toxic to dopaminergic neurones. Conversely, in other animal models TNF appears to be beneficial. Again the role of TNF appears to be contextual, and low levels have been shown to be protective whereas chronic high levels were detrimental (Chertoff et al. 2011).

9.3.6 TNF in Brain Injury

TNF can be potently induced following brain injury and promotes neuroinflammation and neurodegeneration (Sriram and O'Callaghan 2007). Elevated levels are found following traumatic brain injury, and TNF neutralization was protective against cerebral ischemia if administered *before* the insult (Lavine et al. 1998). TNF adversely affects the brain by promoting immune cell infiltration by altering the integrity of the blood–brain barrier, stimulating apoptosis of microvascular endothelium, and activation of microglial cells that in turn perpetuate the inflammatory cycle. Paradoxically, mice lacking both TNF-R1 and TNF-R2 were more susceptible to hippocampal ischemic injury (Bruce et al. 1996). The neuroprotective effect of TNF appears to be region specific; TNF promoted neurodegeneration in the striatum nigra but was protective in the hippocampus (Sriram et al. 2006). The mechanisms involved include activation of astroglia to release neurotrophic factors, activation of repair of peripheral nerves and cerebral microvasculature, induction of anti-oxidant pathways and anti-apoptotic factors, and stimulation of neuronal plasticity (Sriram and O'Callaghan 2007).

9.3.7 Role of TNF in Metabolic Syndrome

Metabolic syndrome is characterized by obesity, impaired glucose metabolism, atherogenic dyslipidemia, and elevated blood pressure, which all independently increase the risk of atherosclerotic disease and type 2 diabetes mellitus (Grundy et al. 2005). Excessive visceral (intraperitoneal) fat represents one of the key underlying abnormalities together with low-grade chronic inflammation. The latter is commonly defined by C-reactive protein (CRP) levels above 3 mg/l.

Adipocyte hypertrophy as a result of excessive oral intake results in their secreting the chemokine CCL2, which in turn recruits macrophages. Interestingly, the majority of macrophages are located around dead adipocytes, whose numbers increase dramatically in obese individuals (Cinti et al. 2005). It is also possible that the non-viable adipocytes release chemoattractants for macrophages. The macrophages produce TNF, which induces the release of saturated fatty acids from adipocytes via lipolysis, that in turn induce further inflammatory changes in the macrophages via TLR-4. This paracrine signaling loop sets up a vicious cycle, leading to more inflammation as the macrophages polarize from the alternatively activated M2 phenotype to the classically activated M1 (Suganami and Ogawa 2010).

There is also evidence for the role of increased numbers of TNF-expressing mast cells in the visceral fat of obese animals (Altintas et al. 2011).

TNF affects lipid metabolism by adipocytes at multiple levels. It decreases free fatty acid uptake from the circulation, reduces lipogenesis, and also enhances lipolysis. TNF also prevents the normal development of pre-adipocytes to fully differentiated adipose cells by activating the Wnt pathway (Cawthorn et al. 2007; Gustafson and Smith 2006).

There is clear evidence that TNF mediates insulin resistance both in vitro and in rodent models (Sethi and Hotamisligil 1999). TNF influences the sensitivity of fat cells to insulin by several mechanisms. It attenuates insulin receptor signaling, downregulates genes involved in insulin action, and counteracts transcription factors that regulate insulin sensitivity (Ryden and Arner 2007).

Insulin resistance is an important component of metabolic syndrome that can be observed in some individuals with obesity or chronic inflammatory disease such as rheumatoid arthritis. Individuals with rheumatoid arthritis have a 50 % higher risk of mortality related to cardiovascular disease than the general population (Avina-Zubieta et al. 2008), and the accelerated atherogenesis and cardiovascular complications are independent of other risk factors (Gonzalez-Gay et al. 2005). Inhibition of TNF in patients with RA or ankylosing spondylitis resulted in an improvement in insulin sensitivity in those with the greatest insulin resistance (Yazdani-Biuki et al. 2004) (Kiortsis et al. 2005). TNF blockade in patients with inflammatory arthritis also leads to changes in lipid profiles, including increases in HDL cholesterol as well as total and LDL cholesterol and perhaps also triglycerides. Data from the biological registries of patients with rheumatic disorders treated with anti-TNF indicate reduced risk of cardio- and cerebrovascular events compared to patients not treated with TNF inhibitors (McKellar et al. 2009). A randomized trial that studied the effects of TNF inhibition for 6 months resulted in improvement in insulin resistance only in normal weight but not in obese patients with RA (Stavropoulos-Kalinoglou et al. 2012). Furthermore, TNF-neutralizing antibodies failed to improve insulin sensitivity in obese patients with type 2 diabetes mellitus (Ofei et al. 1996). Therefore, although there appears to be a clear link between TNF and cardiovascular disease in patients with inflammatory arthritis, it does not necessarily follow that it also has a pivotal role in patients who do not suffer from rheumatic diseases and effects are limited in obese individuals with inflammatory arthropathy. However, TNF upregulates fibrinogen levels and platelet levels in RA, as judged by their rapid reduction following anti-TNF therapy (Feldmann and Maini 2001).

The lack of effect in clinical studies may in part relate to selective targeting as the TNF is active only in certain tissues such as adipose tissue and systemic inhibition may fail to achieve the desired effect (Nieto-Vazquez et al. 2008). Furthermore, although TNF expression is increased in adipose tissue of males with reduced insulin sensitivity, females seemed to lack a similar correlation with insulin sensitivity (Pfeiffer et al. 1997), perhaps because of differential expression of the TNF receptors. The local effect of TNF on the adipose tissue is emphasised by the finding that, in contrast to TNF, plasma (rather than adipose) IL-6 demonstrated the strongest relationship with obesity and insulin resistance (Kern et al. 2001). The conflicting levels of circulating levels of TNF and IL-6 in obese children of different racial origins may also in part be explained by the fact that serum levels may not accurately reflect the situation in adipose tissue (Tam et al. 2010).

9.3.8 TNF in Fibrosis

Fibrosis is a common pathological endpoint of many inflammatory disorders affecting critical visceral organs and contributes to 45 % of deaths in the Western world (Wynn 2008). Although inflammation often precedes fibrosis, the role of TNF remains unclear. In an uncontrolled case series of patients with inflammatory arthritis treated with TNF inhibitors, there was an improvement in the cutaneous fibrosis element of systemic sclerosis (Bosello et al. 2005; Lam et al. 2007; Ramos-Casals et al. 2011).

The results from animal models are also not conclusive. In a murine model of silica particle-induced pulmonary fibrosis there was a marked increase in levels of mRNA for TNFR but not of transforming growth factor (TGF)-β. Infusion of anti-TNF resulted in reduced fibrosis and infusion of exogenous TNF increased the fibrosis (Piguet et al. 1990). TNF production was found to be essential for the development of bleomycin-induced pulmonary fibrosis in a murine model, in part through upregulation of TGF-β1 expression (Ortiz et al. 1998), and systemic administration of soluble TNF receptor led to reduction in fibrosis (Piguet and Vesin 1994). The utility of these models may, however, be limited as not all strains of mice develop pulmonary fibrosis in response to bleomycin, with BALB/c animals being resistant, in contrast to C57Bl/6 mice (Ortiz et al. 1998), despite both strains demonstrating elevated TNF levels in response to the bleomycin. In support of the pro-fibrotic role of TNF, transgenic mice overexpressing TNF under the control of human surfactant protein developed pulmonary changes resembling those seen in human idiopathic pulmonary fibrosis (Miyazaki et al. 1995). However, others have shown that TNF-overexpressing transgenic mice do not develop pulmonary fibrosis but instead a disorder similar to emphysema. In another study, TNF transgenics tolerated bleomycin, whereas littermate controls developed severe pulmonary fibrosis. One potential explanation for the protection against bleomycin may be that the transgenics were less susceptible because of downregulation of TNFR1 and altered inflammatory response (Fujita et al. 2003).

The development of pulmonary fibrosis has also been considered in respect of TNF-receptor expression. Intratracheal instillation of bleomycin or silica resulted in upregulation of TNFR2 but not TNFR1 in C56Bl/6 mice (Ortiz et al. 1999); this was accompanied by pulmonary fibrosis in wild-type animals but not in dual TNFR1/R2 knockouts. Furthermore, TNFR1/R2 combined knockout mice failed to develop pulmonary fibrosis on exposure to asbestos fibers (Liu et al. 1998b). Here it appears that TNFR2 has a pathological role, in contrast to some data from some animal models of neurological disease.

The role of TNF in fibrosis of other organs has also been investigated. A hepatotoxic agent administered intraperitoneally led to hepatic fibrosis in wild-type mice but this response was attenuated in TNFR1 knockout mice (Kitamura et al. 2002). In the latter group there was reduced TNF expression and also reduced production of CCL2. CCL2 is an important chemokine for macrophages so the effects of TNFR1 deletion may be indirect because of attenuation of the inflammatory response. Carbon tetrachloride led to prominent hepatic fibrosis in wild-type and TNFR2 knockout but not in TNFR1 knockout mice (Sudo et al. 2005).

TNF blockade reduced interstitial inflammation and fibrosis in a rat model of crescentic glomerulonephritis (Khan et al. 2005). In a murine model of obstructive nephropathy leading to tubulointerstitial fibrosis, absence of either TNFR1 or TNFR2 resulted in attenuated fibrosis (Guo et al. 1999).

Heart failure is characterized by cardiac hypertrophy, ventricular dilatation, and fibrosis (Gullestad et al. 2012), and transgenic mice with cardiac-specific overexpression of TNF developed congestive cardiac failure (Kubota et al. 1997). Systemic infusion of TNF intraperitoneally into rats at the levels comparable to those seen in patients with heart failure was associated with left ventricular dysfunction and dilatation (Bozkurt et al. 1998). However, a randomized trial of infliximab in patients with heart failure and reduced left ventricular ejection fraction was stopped early because of increased mortality in the group receiving high-dose anti-TNF (Chung et al. 2003). This finding emphasizes the fact that elevated levels of a cytokine are not necessarily indicative of it having a key role in pathogenesis and, indeed, TNF has been shown to play a role in myocardial repair by enhancing the engraftment into a myocardial infarct of mesenchymal stromal cells (Kim et al. 2009). Also, TNF may have different roles in subgroups of patients. Approximately half of patients with heart failure have preserved left ventricular ejection fraction, and this group are characterized by cardiac hypertrophy and fibrosis. It has been suggested that TNF inhibition may have a beneficial effect in this subgroup (Gullestad et al. 2012). In patients with rheumatoid arthritis, anti-TNF appeared to reduce the risk of heart failure and did not worsen preexisting cardiac failure (Listing et al. 2008).

The mechanism by which TNF may have a pro-fibrotic role remains unclear. TGF-β1 is known to be involved in a variety of fibrotic disorders (Wynn 2008). There is some evidence that TNF may act indirectly via TGF-β1. Upregulation of TNF in pulmonary tissue of mice by adenoviral transduction resulted in increased inflammatory cellular infiltrate, elevation of TGF-β1, accumulation of myofibroblasts, and pulmonary fibrosis (Sime et al. 1998). An in vitro study using three T3 and murine pulmonary fibroblasts appeared to confirm this (Sullivan et al. 2005), whereas in human dermal fibroblasts from neonatal foreskins TNF inhibited TGF-β1 Smad signaling, again via AP-1 activation (Verrecchia et al. 2000). Our recent data show that in Dupuytren's disease TNF acts via the Wnt pathway to promote fibrosis through the generation of myofibroblasts and matrix production (Verjee et al. 2013). Importantly, we also showed in vitro that TNF blockade effectively inhibits myofibroblast activity without affecting cell viability (Verjee et al. 2013).

In vivo there is a delicate balance between deposition of extracellular matrix components and degradation. There are some data to suggest that TNF may promote fibrosis by downregulating matrix breakdown. Human fibroblasts treated with TNF showed increased expression of matrix metalloproteinases and reduced expression of TIMP (Meikle et al. 1989), thereby on balance enhancing degradation of the matrix (Distler et al. 2008).

In summary, the role of TNF in fibrosis remains unclear based on limited data in murine models and clinical and in vitro studies, partly because the role of TNF is context dependent, as well as depending on the precise levels and, importantly, the target cell population. Our findings based on studying early-passage cells from patients with Dupuytren's disease, a condition of localized fibrosis in the palm, suggest a key role for TNF in its pathogenesis. We found increased expression of TNF receptors, especially TNFR2 in the myofibroblasts (Verjee et al. 2013). TNFR2 may effectively increase the efficacy of soluble TNF by "passing the ligand" to TNFR1, an effect potentiated by the fact that TNFR2 has a higher affinity for soluble TNF (Tartaglia et al. 1993) (Barbara et al. 1994). TNFR2 appears to be crucial in signalling a pro-fibrotic response as intestinal myofibroblasts from wild-type or TNFR1-deficient mice showed increased proliferation and collagen production on treatment with TNF whereas myofibroblasts from TNFR2$^{-/-}$ or combined TNFR1$^{-/-}$/TNFR2$^{-/-}$ were unresponsive (Theiss et al. 2005; Verrecchia et al. 2000). Furthermore, we found that palmar dermal fibroblasts from patients with Dupuytren's disease exhibited a dose response, with maximal induction of myofibroblast phenotype on exposure to TNF at 0.1 ng/ml, with higher levels being inhibitory. In contrast, TNF resulted in inhibition of contraction of non-palmar dermal fibroblasts from Dupuytren's patients and palmar fibroblasts from normal individuals (Verjee et al. 2013). Results of clinical studies would be of great interest.

From the foregoing examples it is clear that TNF has divergent roles that in part depend on dose, duration of exposure, tissue type, and the local expression of TNF receptors. The relative proportion of TNFR1 and TNFR2 appears to be particularly important, and this makes it challenging to replicate the success of systemic inhibition of TNF in rheumatoid arthritis for other diseases. Future success may depend on the development of a new generation of TNF inhibitors, for example, those that can be targeted to specific tissues or selectively neutralize ligation of TNF to TNFR1 while preserving TNFR2 signaling. Perhaps the most likely next success will be for localized musculoskeletal fibrotic disorders where anti-TNF is delivered locally and remains restricted to the site of delivery.

9.4 Clinical Aspects of TNF Blockade

The introduction of biological agents targeting TNF has significantly modified the treatment paradigm for RA and unequivocally validated TNF as a molecular target for therapy. The clinical success and impact of TNF inhibitors has fostered the development of other biological and small molecule modulators of pro-inflammatory cytokines, expanded our understanding of rheumatoid pathogenesis, the role of TNF in health and inflammation, and the proper use of parenterally administered biological agents.

Clinical studies have shown that TNF blockade in RA consistently improves disease activity and quality of life. Five biological agents that inhibit TNF have been approved for RA (Taylor 2010): etanercept, a soluble dimeric TNFR2 IgG1-Fc fusion construct (Enbrel); infliximab, a chimeric anti-TNF-α IgG1 monoclonal antibody (mAb) (Remicade); adalimumab, a human IgG1 anti-TNFα mAb (Humira); golimumab, a human IgG1 anti-TNFα mAb (Simponi), and certolizumab pegol, a TNF-specific, pegylated Fab′ antibody fragment (Cimzia). Monoclonal antibodies or Fab′ antibody fragments have specificity for TNF-α. In contrast, the fusion protein etanercept comprises two TNFR2 receptor extracellular domains and the Fc portion of human IgG-binding members of the lymphotoxin (LT) family, namely, soluble LTα3 and cell-surface LTα2β1. Etanercept neutralizes LTα3 and sTNF with similar potency. Although all biological TNF inhibitors share many biological, clinical, and adverse effects, there are differences among them beyond their pharmacokinetics (Taylor 2010). Infliximab, adalimumab, and golimumab are IgG1 antibodies and etanercept contains an IgG1 Fc portion; certolizumab lacks an Fc portion. In addition to sharing the ability to inhibit TNF, all five agents bind to TNF with high affinity, and all three are virtually the same in their ability to neutralize soluble and transmembrane TNF.

Numerous advances in the therapeutic possibilities for RA have been realized with the development of this class of therapy, including more complete therapeutic responses and higher remission rates, protection against radiographic deterioration, improved mortality, and possibly a reduced rate of cardiovascular events in RA patients (Carmona et al. 2007; Jacobsson et al. 2007).

9.4.1 *Mechanism of Action*

TNF is an attractive therapeutic target owing to its abundant expression in the rheumatoid joint and plethora of pro-inflammatory effects that include regulation of other pro-inflammatory mediators. Definitive verification of a primary role for TNF in RA came from clinical trials demonstrating the impressive amelioration of disease with TNF blockade.

TNF-α primarily mediates inflammation by promoting cellular activation and trafficking of leukocytes to inflammatory sites. TNF-α may further contribute to the pathogenesis of RA by induction of pro-inflammatory cytokines such as IL-1β and IL-6, increasing endothelial layer permeability, activation of neutrophils and eosinophils, induction of the synthesis of acute-phase reactants, and the induction of tissue-degrading enzymes (matrix metalloproteinase enzymes) produced by synoviocytes or chondrocytes.

The first formal proof that TNF regulates other pro-inflammatory cytokines in vivo was the observation that there is a rapid reduction in serum IL-6 concentrations, closely followed by falling serum CRP, following administration of infliximab (Charles et al. 1999; Elliott et al. 1994; Lorenz et al. 1996). Although IL-1β concentrations are often below the limit of detection in the peripheral blood of RA patients, where it is detectable, downregulation has been reported in a proportion of patients (Elliott et al. 1994). Similarly, in a small study of repeat synovial biopsies obtained before and 2 weeks after a single infusion of 10 mg/kg infliximab, computerized image analysis of sections stained for cytokine-producing cells demonstrated a reduction in synovial IL-1α and IL-1α in a subgroup (Ulfgren et al. 2000). It is clear that the benefits of anti-TNF therapy are not mediated by upregulation of endogenous pro-inflammatory cytokine inhibitors because circulating IL-1ra and soluble TNF receptor levels fall after infliximab infusion (Charles et al. 1999).

It is thought that a major mechanism of action of TNF inhibitors is likely to be mediated by modulation of inflammatory cell traffic. A dose-dependent rise in peripheral blood lymphocyte counts is observed following infliximab infusion, with a maximum rise within 24 h of treatment (Paleolog et al. 1996); this is mediated by modulation of both arms of the inflammatory cell recruitment cascade. Thus, there is reduced histological expression of synovial cytokine-induced vascular adhesion molecules, such as E-selectin and VCAM-1, following anti-TNF treatment (Tak et al. 1996), and a significant dose-dependent reduction in soluble serum E-selectin and ICAM-1 concentrations (Paleolog et al. 1996), as well as significantly diminished immunohistological expression of the chemokines IL-8 and CCL2, with a trend toward reduction in a number of other chemokines (Taylor et al. 2000).

Further indirect evidence to suggest TNF blockade reduces inflammatory cell recruitment to the joint is based on the observation that infliximab therapy is associated with a reduction in numbers of synovial tissue macrophages and lymphocytes (Tak et al. 1996; Taylor et al. 2000). However, the most definitive confirmation that TNF blockade reduces leukocyte traffic to inflamed joints was obtained in an open-label clinical trial demonstrating a 40–50 % decrease in retention of autologous indium-111-labeled granulocytes in the hands, wrists, and knees 2 weeks after infliximab treatment (Taylor et al. 2000). There is a reduction in the marginating granulocyte pool after infliximab treatment, an observation that would normally be associated with a rise in peripheral blood granulocyte counts (Taylor et al. 1999). However, in contrast to peripheral blood lymphocyte counts, numbers of peripheral blood granulocytes decrease after infliximab with maximal changes within 24h. The reason for this is that myeloid cell production is reduced secondary to downregulation of granulocyte-macrophage colony-stimulating factor (GM-CSF) as a consequence of TNF blockade. Because of the short circulating half-life of the granulocyte, approximately 8h, a diminished rate of cell production dominates the peripheral blood picture.

One factor contributing to the rapid reduction in joint swelling observed by both patients and physicians after anti-TNF therapy is likely to be a reduction in tissue edema and capillary leak, mediated by vascular endothelial growth factor (VEGF), a cytokine implicated in new blood vessel formation and found to be elevated in the serum of RA patients (Paleolog et al. 1998). Serum concentrations of VEGF show a dose-dependent reduction following infliximab infusions, but without normalization. There is also reduction in synovial vascular density and in particular a reduction in angiogenesis, as assessed by diminished number of vessels expressing the αVβ3 integrin (Taylor 2005). Further evidence for a reduction in synovial vascularity following TNF blockade is the observation that the vascular signal on quantitative power Doppler imaging is significantly reduced following infliximab therapy (Taylor et al. 2004; Taylor et al. 2006).

Relatively early in the history of clinical trials of TNF blockade, marked reduction in circulating concentrations of the precursors of the matrix metalloproteinase enzymes MMP-1 and MMP-3 was reported (Brennan et al. 1997), as well as a significant reduction in synovial tissue expression of matrix metalloproteinases (Catrina et al. 2002). Similarly, serum levels of osteoprotegerin (OPG) and soluble receptor activator of nuclear factor κB ligands (sRANKL), both of which are elevated in RA compared with normal sera, are normalized following infliximab therapy without influencing the OPG to sRANKL ratio (Ziolkowska et al. 2002). These observations predicted the disease-modifying capability of anti-TNF inhibitors, which is now established.

One hypothesis for the failure of etanercept to provide clinical benefits in trials of Crohn's disease, in contrast to the marked benefits demonstrated with monoclonal antibodies to TNF-α, is that the antibodies may cause an increase in apoptosis of lamina propria and peripheral T cells through caspase-dependent mechanisms. However, this topic remains controversial, and the relevance of modulation of apoptosis as the mechanism of action of TNF inhibitors in RA is unclear. In one study, decreased synovial cellularity was reported as early as 48h after infliximab administration, but with no corresponding evidence of apoptosis (Smeets et al. 2003). In another study, however, apoptosis in synovial macrophages has been reported to be induced by both etanercept and infliximab, with a corresponding increase in active caspase-3 expression. No increase in lymphocyte apoptosis was observed, however (Catrina et al. 2005). The relevance of these interesting observations to the mode of action of TNF inhibitors in RA is not clear.

9.4.2 Clinical Use in Rheumatoid Arthritis

TNF blockers were initially approved in rheumatoid RA for patients with moderately to severely active disease who had failed an adequate trial of methotrexate (MTX) or other disease-modifying anti-rheumatic drugs (DMARDs). TNF blockers were also indicated for the control of signs and symptoms and the prevention of functional or radiographic deterioration. Following the findings of large controlled trials in early rheumatoid arthritis patients, the FDA recommendations for infliximab, adalimumab, and etanercept were modified to allow their use as the initial DMARD in RA, and thus the provision to fail MTX (or other DMARD) was removed. Thus, although MTX and TNF inhibitors appear to be equally effective in clinical trials, TNF inhibitor therapy has been established as yielding better radiographic outcomes (alone or in combination with MTX). Despite the clinical and radiographic potency of TNF blockade, particularly with concomitant MTX therapy, the circumstances in which this combination should be initiated as a first-line therapy have not been firmly established because a variety of other factors determine these choices, a major consideration being the cost-effectiveness of biological therapies within the context of particular healthcare economies.

The European League Against Rheumatism (EULAR) recommendations for the pharmacological management of RA were published in 2010 (Smolen et al. 2010). These guidelines recommend that for patients responding insufficiently to MTX and/or other synthetic DMARDs with or without glucocorticoids, a TNF inhibitor should be initiated and, unless contraindicated, in combination with concomitant MTX. EULAR guidelines recommend that treatment should be aimed at reaching a target of remission or low disease activity as soon as possible in every patient; so long as the target has not been reached, treatment should be adjusted by frequent (every 1–3 months) and strict monitoring. However, although desirable, it is recognized that these targets remain aspirational for many patients and particularly so for those with longer disease duration.

There are situations in which TNF inhibitor use is not advisable. Relative contraindications include overlap syndromes, a history of demyelinating disorder (MS, optic neuritis), congestive heart failure, and pregnancy. Absolute contraindications include untreated active or latent tuberculosis.

Following initiation of therapy with MTX, dosed appropriately for no more than 3–6 months, if low disease activity is not reached, it remains unclear whether a switch to combination DMARDs is as effective as addition of a biological anti-TNF agent. There are data from the BeST (a Dutch acronym for strategy study) trial suggesting that MTX plus TNF inhibitor is superior to add-on or sequential combination DMARD regimens (Goekoop-Ruiteman et al. 2005). In the SWEFOT (short for Swedish Pharmacotherapy) trial, patients with early RA had initial MTX treatment, and only after failing this agent were they randomized to one of two therapeutic strategies: the addition of SSZ plus hydroxychloroquine, or the addition of infliximab (van Vollenhoven et al. 2009). After 1 year, 25 % of patients with the former combination and 39 % of patients with the latter had achieved a EULAR good response, the difference being statistically significant. Recently published data from The Randomized Comparative Effectiveness Study of Oral Triple Therapy versus Etanercept plus MTX in the Early, Aggressive Rheumatoid Arthritis (TEAR) trial suggested that treatment with the combination of MTX plus etanercept provides superior radiographic benefit compared with oral triple therapy (Moreland et al. 2012).

DMARD-naïve patients with poor prognostic markers might be considered for combination therapy of MTX plus a biological in very exceptional situations. However, the EULAR recommendations for treatment of RA consider that, in general, this approach is not cost effective (Smolen et al. 2010). Such patients must manifest a weighty combination of risk factors that portends a severe outcome. Such poor prognostic factors include active polysynovitis, radiographic erosions, seropositivity for RF and/or anti-CCP antibodies, functional impairment, and sustained elevations of acute-phase markers. The identification of such high-risk patients mandates the combined use of MTX and TNF inhibition because studies have consistently shown the best clinical and radiographic outcomes when this combination is used instead of TNF inhibitor or MTX alone.

Another compelling reason for the use of TNF inhibitors is the rapid onset of effects and enhanced well-being reported by patients receiving TNF blockade. The brisk mitigation of stiffness and malaise and subsequent resumption of physical or pleasurable activities previously forfeited commonly follow TNF inhibitor initiation. Whether this represents a central nervous system or systemic effect of TNF inhibition is unknown. However, infliximab is known to decrease cerebrospinal fluid (CSF) prostaglandin levels, while having no effect on CSF TNF, interferon-γ, IL-10, or nitric oxide.

The popularity of these agents is evident from their growth worldwide. By 2013 TNF inhibitors have displaced statins as the world's best-selling pharmaceuticals with global sales exceeding $25 billion (the 2011 sales). Their high cost contributes to this figure. Nonetheless, several factors have limited their more widespread use, including patient or physician reluctance, limited or variable coverage by third-party insurance carriers, toxicity concerns, and, in particular, high costs.

9.5 Conclusions

Anti-TNF therapy has produced considerable clinical benefit, with a good safety profile, not only for the disease for which it was first rationally clinically evaluated on the basis of preclinical research, RA, but also a collection of other chronic diseases for which the rationale was chiefly the success of anti-TNF in RA. These other diseases include both forms of IBD, Crohn's disease and ulcerative colitis, psoriasis, psoriatic arthritis, ankylosing spondylitis, and juvenile RA. There is also off-label use, for example, for Behçet's syndrome. However, only 60–70 % of patients respond, and a minority go into remission. Thus, efficacy needs to improve in the future.

Because anti-TNF was the first "biological" (monoclonal antibody or related protein) therapeutic to be widely used for a chronic disease, the sales were high and durable, and thus it triggered an appreciation in the pharmaceutical industry that monoclonal antibodies were no longer a niche product, but could be "mainstream" therapeutics, even "blockbusters." Big Pharma rapidly acquired the major antibody-producing companies, e.g., Centocor.

Sales have grown, driven by long-term use (often 5 years or more), ready patient acceptance, the fear of injections proving relatively groundless, and also by high compliance: in contrast to MTX therapy, which may improve RA joints but make you feel unwell, anti-TNF promotes a "feel-good" factor, possibly by a central neural mechanism, although of course the relief from pain and tiredness is helpful. By 2011, the global sales of all five anti-TNFs reached $25 billion, and the estimates are that by end of 2012 anti-TNF will be the biggest selling drug class—a surprising result for our project, starting in a small research institute in London in the 1980s, based on a hypothesis published in 1983. But there no time for complacency: there is still much to do to reach our current goal, to move much closer to a cure.

Acknowledgments The work described here involved very many colleagues. Chief among them are Professor Sir Ravinder Maini, who started working with Marc Feldmann in 1984. A colleague since 1987 who contributed hugely was Professor Fionula Brennan. Sadly, she died in June 2012. This was the topic to which she devoted most of her career, making invaluable contributions, and therefore this article is dedicated to her memory.

References

The Lenercept Multiple Sclerosis Study Group and The University of British Columbia MS/MRI Analysis Group (1999) TNF neutralization in MS: results of a randomized, placebo-controlled multicenter study. Neurology 53:457–465

Aerts NE, De Knop KJ, Leysen J, Ebo DG, Bridts CH, Weyler JJ, Stevens WJ, De Clerck LS (2010) Increased IL-17 production by peripheral T helper cells after tumour necrosis factor blockade in rheumatoid arthritis is accompanied by inhibition of migration-associated chemokine receptor expression. Rheumatology 49:2264–2272

Alexopoulou L, Kranidioti K, Xanthoulea S, Denis M, Kotanidou A, Douni E, Blackshear PJ, Kontoyiannis DL, Kollias G (2006) Transmembrane TNF protects mutant mice against intracellular bacterial infections, chronic inflammation and autoimmunity. Eur J Immunol 36: 2768–2780

Altintas MM, Azad A, Nayer B, Contreras G, Zaias J, Faul C, Reiser J, Nayer A (2011) Mast cells, macrophages, and crown-like structures distinguish subcutaneous from visceral fat in mice. J Lipid Res 52:480–488

Alzabin S, Abraham SM, Taher TE, Palfreeman A, Hull D, McNamee KE, Jawad A, Pathan E, Kinderlerer A, Taylor PC, Williams RO, Mageed R (2012) Incomplete response of inflammatory arthritis to TNFα blockade is associated with the Th17 pathway. Ann Rheum Dis 71: 1471–1478

Arnett HA, Mason J, Marino M, Suzuki K, Matsushima GK, Ting JP (2001) TNF alpha promotes proliferation of oligodendrocyte progenitors and remyelination. Nat Neurosci 4:1116–1122

Avina-Zubieta JA, Choi HK, Sadatsafavi M, Etminan M, Esdaile JM, Lacaille D (2008) Risk of cardiovascular mortality in patients with rheumatoid arthritis: a meta-analysis of observational studies. Arthritis Rheum 59:1690–1697

Baker D, Butler D, Scallon BJ, O'Neill JK, Turk JL, Feldmann M (1994) Control of established experimental allergic encephalomyelitis by inhibitionof tumor necrosis factor activity within the central nervous system using monoclonal antibodies and TNF receptor–immunoglobulin fusion proteins. Eur J Immunol 24:2040–2048

Balosso S, Ravizza T, Perego C, Peschon J, Campbell IL, De Simoni MG, Vezzani A (2005) Tumor necrosis factor-alpha inhibits seizures in mice via p75 receptors. Ann Neurol 57:804–812

Barbara JA, Smith WB, Gamble JR, Van Ostade X, Vandenabeele P, Tavernier J, Fiers W, Vadas MA, Lopez AF (1994) Dissociation of TNF-alpha cytotoxic and proinflammatory activities by p55 receptor- and p75 receptor-selective TNF-alpha mutants. EMBO J 13:843–850

Black RA, Rauch CT, Kozlosky CJ, Peschon JJ, Slack JL, Wolfson MF, Castner BJ, Stocking KL, Reddy P, Srinivasan S, Nelson N, Boiani N, Schooley KA, Gerhart M, Davis R, Fitzner JN, Johnson RS, Paxton RJ, March CJ, Cerretti DP (1997) A metalloproteinase disintegrin that releases tumour-necrosis factor-alpha from cells. Nature (Lond) 385:729–733

Bluml S, Scheinecker C, Smolen JS, Redlich K (2012) Targeting TNF receptors in rheumatoid arthritis. Int Immunol 24:275–281

Bosello S, De Santis M, Tolusso B, Zoli A, Ferraccioli G (2005) Tumor necrosis factor-alpha inhibitor therapy in erosive polyarthritis secondary to systemic sclerosis. Ann Intern Med 143:918–920

Bozkurt B, Kribbs SB, Clubb FJ Jr, Michael LH, Didenko VV, Hornsby PJ, Seta Y, Oral H, Spinale FG, Mann DL (1998) Pathophysiologically relevant concentrations of tumor necrosis

factor-alpha promote progressive left ventricular dysfunction and remodeling in rats. Circulation 97:1382–1391
Brennan FM, Chantry D, Jackson A, Maini R, Feldmann M (1989) Inhibitory effect of TNFα antibodies on synovial cell interleukin-1 production in rheumatoid arthritis. Lancet 2:244–247
Brennan FM, Browne KA, Green PA, Jaspar JM, Maini RN, Feldmann M (1997) Reduction of serum matrix metalloproteinase 1 and matrix metalloproteinase 3 in rheumatoid arthritis patients following anti-tumour necrosis factor-alpha (cA2) therapy. Br J Rheumatol 36: 643–650
Bruce AJ, Boling W, Kindy MS, Peschon J, Kraemer PJ, Carpenter MK, Holtsberg FW, Mattson MP (1996) Altered neuronal and microglial responses to excitotoxic and ischemic brain injury in mice lacking TNF receptors. Nat Med 2:788–794
Carmona L, Descalzo MA, Perez-Pampin E, Ruiz-Montesinos D, Erra A, Cobo T, Gomez-Reino JJ (2007) All-cause and cause-specific mortality in rheumatoid arthritis are not greater than expected when treated with tumour necrosis factor antagonists. Ann Rheum Dis 66:880–885
Carson MJ, Bilousova TV, Puntambekar SS, Melchior B, Doose JM, Ethell IM (2007) A rose by any other name? The potential consequences of microglial heterogeneity during CNS health and disease. Neurotherapeutics 4:571–579
Castillo J, Moro MA, Blanco M, Leira R, Serena J, Lizasoain I, Davalos A (2003) The release of tumor necrosis factor-alpha is associated with ischemic tolerance in human stroke. Ann Neurol 54:811–819
Catrina AI, Lampa J, Ernestam S, af Klint E, Bratt J, Klareskog L, Ulfgren AK (2002) Anti-tumour necrosis factor (TNF)-alpha therapy (etanercept) down-regulates serum matrix metalloproteinase (MMP)-3 and MMP-1 in rheumatoid arthritis. Rheumatology (Oxf) 41:484–489
Catrina AI, Trollmo C, af Klint E, Engstrom M, Lampa J, Hermansson Y, Klareskog L, Ulfgren AK (2005) Evidence that anti-tumor necrosis factor therapy with both etanercept and infliximab induces apoptosis in macrophages, but not lymphocytes, in rheumatoid arthritis joints: extended report. Arthritis Rheum 52:61–72
Cawthorn WP, Heyd F, Hegyi K, Sethi JK (2007) Tumour necrosis factor-alpha inhibits adipogenesis via a beta-catenin/TCF4(TCF7L2)-dependent pathway. Cell Death Differ 14:1361–1373
Charles P, Elliott MJ, Davis D, Potter A, Kalden JR, Antoni C, Breedveld FC, Smolen JS, Eberl G, deWoody K, Feldmann M, Maini RN (1999) Regulation of cytokines, cytokine inhibitors, and acute-phase proteins following anti-TNF-alpha therapy in rheumatoid arthritis. J Immunol 163:1521–1528
Chee J, Angstetra E, Mariana L, Graham KL, Carrington EM, Bluethmann H, Santamaria P, Allison J, Kay TW, Krishnamurthy B, Thomas HE (2011) TNF receptor 1 deficiency increases regulatory T cell function in nonobese diabetic mice. J Immunol 187:1702–1712
Chen X, Baumel M, Mannel DN, Howard OM, Oppenheim JJ (2007) Interaction of TNF with TNF receptor type 2 promotes expansion and function of mouse CD4+CD25+ T regulatory cells. J Immunol 179:154–161
Chen X, Subleski JJ, Kopf H, Howard OM, Mannel DN, Oppenheim JJ (2008) Cutting edge: expression of TNFR2 defines a maximally suppressive subset of mouse CD4+CD25+FoxP3+ T regulatory cells: applicability to tumor-infiltrating T regulatory cells. J Immunol 180: 6467–6471
Chen X, Subleski JJ, Hamano R, Howard OM, Wiltrout RH, Oppenheim JJ (2010) Co-expression of TNFR2 and CD25 identifies more of the functional CD4+FOXP3+ regulatory T cells in human peripheral blood. Eur J Immunol 40:1099–1106
Chertoff M, Di Paolo N, Schoeneberg A, Depino A, Ferrari C, Wurst W, Pfizenmaier K, Eisel U, Pitossi F (2011) Neuroprotective and neurodegenerative effects of the chronic expression of tumor necrosis factor alpha in the nigrostriatal dopaminergic circuit of adult mice. Exp Neurol 227:237–251
Christen U, Wolfe T, Mohrle U, Hughes AC, Rodrigo E, Green EA, Flavell RA, von Herrath MG (2001) A dual role for TNF-alpha in type 1 diabetes: islet-specific expression abrogates the

ongoing autoimmune process when induced late but not early during pathogenesis. J Immunol 166:7023–7032

Chung ES, Packer M, Lo KH, Fasanmade AA, Willerson JT, Anti TNFTACHFI (2003) Randomized, double-blind, placebo-controlled, pilot trial of infliximab, a chimeric monoclonal antibody to tumor necrosis factor-alpha, in patients with moderate-to-severe heart failure: results of the anti-TNF Therapy Against Congestive Heart Failure (ATTACH) trial. Circulation 107:3133–3140

Cinti S, Mitchell G, Barbatelli G, Murano I, Ceresi E, Faloia E, Wang S, Fortier M, Greenberg AS, Obin MS (2005) Adipocyte death defines macrophage localization and function in adipose tissue of obese mice and humans. J Lipid Res 46:2347–2355

Cope AP (2003) Exploring the reciprocal relationship between immunity and inflammation in chronic inflammatory arthritis. Rheumatology (Oxf) 42:716–731

Cope AP, Aderka D, Doherty M, Engelmann H, Gibbons D, Jones AC, Brennan FM, Maini RN, Wallach D, Feldmann M (1992) Increased levels of soluble tumor necrosis factor receptors in the sera and synovial fluid of patients with rheumatic diseases. Arthritis Rheum 35:1160–1169

Distler JH, Schett G, Gay S, Distler O (2008) The controversial role of tumor necrosis factor alpha in fibrotic diseases. Arthritis Rheum 58:2228–2235

Domercq M, Brambilla L, Pilati E, Marchaland J, Volterra A, Bezzi P (2006) P2Y1 receptor-evoked glutamate exocytosis from astrocytes: control by tumor necrosis factor-alpha and prostaglandins. J Biol Chem 281:30684–30696

Dziewulska D, Mossakowski MJ (2003) Cellular expression of tumor necrosis factor α and its receptors in human ischemic stroke. Clin Neuropathol 22:35–40

Ehrenstein MR, Evans JG, Singh A, Moore S, Warnes G, Isenberg DA, Mauri C (2004) Compromised function of regulatory T cells in rheumatoid arthritis and reversal by anti-TNFalpha therapy. J Exp Med 200:277–285

Eissner G, Kolch W, Scheurich P (2004) Ligands working as receptors: reverse signaling by members of the TNF superfamily enhance the plasticity of the immune system. Cytokine Growth Factor Rev 15:353–366

Elicabe RJ, Cargnelutti E, Serer MI, Stege PW, Valdez SR, Toscano MA, Rabinovich GA, Di Genaro MS (2010) Lack of TNFR p55 results in heightened expression of IFN-gamma and IL-17 during the development of reactive arthritis. J Immunol 185:4485–4495

Elliott MJ, Maini RN, Feldmann M, Long-Fox A, Charles P, Bijl H, Woody JN (1994) Repeated therapy with monoclonal antibody to tumour necrosis factor alpha (cA2) in patients with rheumatoid arthritis. Lancet 344:1125–1127

Feldmann M (2009) Translating molecular insights in autoimmunity into effective therapy. Annu Rev Immunol 27:1–27

Feldmann M, Maini RN (2001) Anti-TNF alpha therapy of rheumatoid arthritis: what have we learned? Annu Rev Immunol 19:163–196

Feldmann M, Maini RN (2003) Lasker Clinical Medical Research Award. TNF defined as a therapeutic target for rheumatoid arthritis and other autoimmune diseases. Nat Med 9:1245–1250

Feldmann M, Maini RN (2010) Anti-TNF therapy, from rationale to standard of care: what lessons has it taught us? J Immunol 185:791–794

Feldmann M, Brennan FM, Maini RN (1996) Role of cytokines in rheumatoid arthritis. Annu Rev Immunol 14:397–440

Fiore M, Angelucci F, Alleva E, Branchi I, Probert L, Aloe L (2000) Learning performances, brain NGF distribution and NPY levels in transgenic mice expressing TNF-alpha. Behav Brain Res 112:165–175

Fong Y, Tracey KJ, Moldawer LL, Hesse DG, Manogue KB, Kenney JS, Lee AT, Kuo GC, Allison AC, Lowry SF, Cerami A (1989) Antibodies to cachectin/tumor necrosis factor reduce interleukin 1 beta and interleukin 6 appearance during lethal bacteremia. J Exp Med 170:1627–1633

Fromont A, De Seze J, Fleury MC, Maillefert JF, Moreau T (2009) Inflammatory demyelinating events following treatment with anti-tumor necrosis factor. Cytokine 45:55–57

Fujita M, Shannon JM, Morikawa O, Gauldie J, Hara N, Mason RJ (2003) Overexpression of tumor necrosis factor-alpha diminishes pulmonary fibrosis induced by bleomycin or transforming growth factor-beta. Am J Respir Cell Mol Biol 29:669–676

Gearing AJ, Beckett P, Christodoulou M, Churchill M, Clements J, Davidson AH, Drummond AH, Galloway WA, Gilbert R, Gordon JL et al (1994) Processing of tumour necrosis factor-alpha precursor by metalloproteinases. Nature (Lond) 370:555–557

Glass GE, Chan JK, Freidin A, Feldmann M, Horwood NJ, Nanchahal J (2011) TNF-alpha promotes fracture repair by augmenting the recruitment and differentiation of muscle-derived stromal cells. Proc Natl Acad Sci USA 108:1585–1590

Goekoop-Ruiteman YP, de Vries-Bouwstra JK, Allaart CF et al (2005) Clinical and radiographic outcomes of four different treatment strategies in patients with early rheumatoid arthritis (the BeSt study): a randomized, controlled trial. Arthritis Rheum 11:3381–3390

Gonzalez-Gay MA, Gonzalez-Juanatey C, Martin J (2005) Rheumatoid arthritis: a disease associated with accelerated atherogenesis. Semin Arthritis Rheum 35:8–17

Gorell JM, Peterson EL, Rybicki BA, Johnson CC (2004) Multiple risk factors for Parkinson's disease. J Neurol Sci 217:169–174

Goshen I, Kreisel T, Ounallah-Saad H, Renbaum P, Zalzstein Y, Ben-Hur T, Levy-Lahad E, Yirmiya R (2007) A dual role for interleukin-1 in hippocampal-dependent memory processes. Psychoneuroendocrinology 32:1106–1115

Grell M, Douni E, Wajant H, Lohden M, Clauss M, Maxeiner B, Georgopolous S, Lesslauer W, Kollias G, Pfizenmaier K, Scheurich P (1995) The transmembrane form of tumor necrosis factor is the prime activating ligand of the 80 kDa tumor necrosis factor receptor. Cell 83(5):793–802

Grundy SM, Cleeman JI, Daniels SR, Donato KA, Eckel RH, Franklin BA, Gordon DJ, Krauss RM, Savage PJ, Smith SC, Spertus JA, Costa F, American Heart Association, National Heart, Lung, and Blood Institute (2005) Diagnosis and management of the metabolic syndrome: an American Heart Association/National Heart, Lung, and Blood Institute scientific statement. Circulation 112:2735–2752

Gullestad L, Ueland T, Vinge LE, Finsen A, Yndestad A, Aukrust P (2012) Inflammatory cytokines in heart failure: mediators and markers. Cardiology 122:23–35

Guo G, Morrissey J, McCracken R, Tolley T, Klahr S (1999) Role of TNFR1 and TNFR2 receptors in tubulointerstitial fibrosis of obstructive nephropathy. Am J Physiol 277:F766–F772

Gustafson B, Smith U (2006) Cytokines promote Wnt signaling and inflammation and impair the normal differentiation and lipid accumulation in 3T3-L1 preadipocytes. J Biol Chem 281:9507–9516

Hamano R, Huang J, Yoshimura T, Oppenheim JJ, Chen X (2011) TNF optimally activatives regulatory T cells by inducing TNF receptor superfamily members TNFR2, 4-1BB and OX40. Eur J Immunol 41:2010–2020

Heldmann U, Thored P, Claasen JH, Arvidsson A, Kokaia Z, Lindvall O (2005) TNF-alpha antibody infusion impairs survival of stroke-generated neuroblasts in adult rat brain. Exp Neurol 196:204–208

Holtmann MH, Schuchmann M, Zeller G, Galle PR, Neurath MF (2002) The emerging distinct role of TNF-receptor 2 (p80) signaling in chronic inflammatory disorders. Arch Immunol Ther Exp (Warsz) 50:279–288

Housley WJ, Adams CO, Nichols FC, Puddington L, Lingenheld EG, Zhu L, Rajan TV, Clark RB (2011) Natural but not inducible regulatory T cells require TNF-alpha signaling for in vivo function. J Immunol 186:6779–6787

Jacob CO, McDevitt HO (1988) Tumour necrosis factor-alpha in murine autoimmune "lupus" nephritis. Nature (Lond) 331:356–358

Jacobsson LT, Turesson C, Nilsson JA, Petersson IF, Lindqvist E, Saxne T, Geborek P (2007) Treatment with TNF blockers and mortality risk in patients with rheumatoid arthritis. Ann Rheum Dis 66:670–675

Kassiotis G, Kollias G (2001) Uncoupling the proinflammatory from the immunosuppressive properties of tumor necrosis factor (TNF) at the p55 TNF receptor level: implications for pathogenesis and therapy of autoimmune demyelination. J Exp Med 193:427–434

Keffer J, Probert L, Cazlaris H, Georgopoulos S, Kaslaris E, Kioussis D, Kollias G (1991) Transgenic mice expressing human tumour necrosis factor: a predictive genetic model of arthritis. EMBO J 10:4025–4031

Kelchtermans H, De Klerck B, Mitera T, Van Balen M, Bullens D, Billiau A, Leclercq G, Matthys P (2005) Defective CD4+CD25+ regulatory T cell functioning in collagen-induced arthritis: an important factor in pathogenesis, counter-regulated by endogenous IFN-gamma. Arthritis Res Ther 7:R402–R415

Kern PA, Ranganathan S, Li C, Wood L, Ranganathan G (2001) Adipose tissue tumor necrosis factor and interleukin-6 expression in human obesity and insulin resistance. Am J Physiol Endocrinol Metab 280:E745–E751

Khan SB, Cook HT, Bhangal G, Smith J, Tam FW, Pusey CD (2005) Antibody blockade of TNF-alpha reduces inflammation and scarring in experimental crescentic glomerulonephritis. Kidney Int 67:1812–1820

Kim EY, Chi HH, Rajaiah R, Moudgil KD (2008) Exogenous tumor necrosis factor-alpha induces suppression of autoimmune arthritis. Arthritis Res Ther 10:R38

Kim YS, Park HJ, Hong MH, Kang PM, Morgan JP, Jeong MH, Cho JG, Park JC, Ahn Y (2009) TNF-alpha enhances engraftment of mesenchymal stem cells into infarcted myocardium. Front Biosci 14:2845–2856

Kiortsis DN, Mavridis AK, Vasakos S, Nikas SN, Drosos AA (2005) Effects of infliximab treatment on insulin resistance in patients with rheumatoid arthritis and ankylosing spondylitis. Ann Rheum Dis 64:765–766

Kitamura K, Nakamoto Y, Akiyama M, Fujii C, Kondo T, Kobayashi K, Kaneko S, Mukaida N (2002) Pathogenic roles of tumor necrosis factor receptor p55-mediated signals in dimethylnitrosamine-induced murine liver fibrosis. Lab Invest 82:571–583

Kleijwegt FS, Laban S, Duinkerken G, Joosten AM, Zaldumbide A, Nikolic T, Roep BO (2010) Critical role for TNF in the induction of human antigen-specific regulatory T cells by tolerogenic dendritic cells. J Immunol 185:1412–1418

Kollias G (2005) TNF pathophysiology in murine models of chronic inflammation and autoimmunity. Semin Arthritis Rheum 34:3–6

Kollias G, Kontoyiannis D (2002) Role of TNF/TNFR in autoimmunity: specific TNF receptor blockade may be advantageous to anti-TNF treatments. Cytokine Growth Factor Rev 13: 315–321

Kontoyiannis D, Kollias G (2000) Accelerated autoimmunity and lupus nephritis in NZB mice with an engineered heterozygous deficiency in tumor necrosis factor. Eur J Immunol 30: 2038–2047

Kontoyiannis D, Pasparakis M, Pizarro TT, Cominelli F, Kollias G (1999) Impaired on/off regulation of TNF biosynthesis in mice lacking TNF AU-rich elements: implications for joint and gut-associated immunopathologies. Immunity 10:387–398

Korn T, Reddy J, Gao W, Bettelli E, Awasthi A, Petersen TR, Backstrom BT, Sobel RA, Wucherpfennig KW, Strom TB, Oukka M, Kuchroo VK (2007) Myelin-specific regulatory T cells accumulate in the CNS but fail to control autoimmune inflammation. Nat Med 13: 423–431

Kubota T, McTiernan CF, Frye CS, Slawson SE, Lemster BH, Koretsky AP, Demetris AJ, Feldman AM (1997) Dilated cardiomyopathy in transgenic mice with cardiac-specific overexpression of tumor necrosis factor-alpha. Circ Res 81:627–635

Lam GK, Hummers LK, Woods A, Wigley FM (2007) Efficacy and safety of etanercept in the treatment of scleroderma-associated joint disease. J Rheumatol 34:1636–1637

Lavine SD, Hofman FM, Zlokovic BV (1998) Circulating antibody against tumor necrosis factor-alpha protects rat brain from reperfusion injury. J Cereb Blood Flow Metab 18:52–58

Lawson LJ, Perry VH, Dri P, Gordon S (1990) Heterogeneity in the distribution and morphology of microglia in the normal adult mouse brain. Neuroscience 39:151–170

Lindley S, Dayan CM, Bishop A, Roep BO, Peakman M, Tree TI (2005) Defective suppressor function in CD4(+)CD25(+) T-cells from patients with type 1 diabetes. Diabetes 54:92–99

Listing J, Strangfeld A, Kekow J, Schneider M, Kapelle A, Wassenberg S, Zink A (2008) Does tumor necrosis factor alpha inhibition promote or prevent heart failure in patients with rheumatoid arthritis? Arthritis Rheum 58:667–677

Liu J, Marino MW, Wong G, Grail D, Dunn A, Bettadapura J, Slavin AJ, Old L, Bernard CC (1998a) TNF is a potent anti-inflammatory cytokine in autoimmune-mediated demyelination. Nat Med 4:78–83

Liu JY, Brass DM, Hoyle GW, Brody AR (1998b) TNF-alpha receptor knockout mice are protected from the fibroproliferative effects of inhaled asbestos fibers. Am J Pathol 153: 1839–1847

Lorenz HM, Antoni C, Valerius T, Repp R, Grunke M, Schwerdtner N, Nusslein H, Woody J, Kalden JR, Manger B (1996) In vivo blockade of TNF-alpha by intravenous infusion of a chimeric monoclonal TNF-alpha antibody in patients with rheumatoid arthritis—short term cellular and molecular effects. J Immunol 156:1646–1653

Lu Z, Wang G, Dunstan CR, Zreiqat H (2012) Short-term exposure to tumor necrosis factor-alpha enables human osteoblasts to direct adipose tissue-derived mesenchymal stem cells into osteogenic differentiation. Stem Cells Dev 21:2420–2429

Ma X, Sun J, Papasavvas E, Riemann H, Robertson S, Marshall J, Bailer RT, Moore A, Donnelly RP, Trinchieri G, Montaner LJ (2000) Inhibition of IL-12 production in human monocyte-derived macrophages by TNF. J Immunol 164:1722–1729

Ma HL, Napierata L, Stedman N, Benoit S, Collins M, Nickerson-Nutter C, Young DA (2010) Tumor necrosis factor alpha blockade exacerbates murine psoriasis-like disease by enhancing Th17 function and decreasing expansion of Treg cells. Arthritis Rheum 62:430–440

McAfoose J, Baune BT (2009) Evidence for a cytokine model of cognitive function. Neurosci Biobehav Rev 33:355–366

McCoy MK, Tansey MG (2008) TNF signaling inhibition in the CNS: implications for normal brain function and neurodegenerative disease. J Neuroinflammation 5:45

McKellar GE, McCarey DW, Sattar N, McInnes IB (2009) Role for TNF in atherosclerosis? Lessons from autoimmune disease. Nat Rev Cardiol 6:410–417

Meikle MC, Atkinson SJ, Ward RV, Murphy G, Reynolds JJ (1989) Gingival fibroblasts degrade type I collagen films when stimulated with tumor necrosis factor and interleukin 1: evidence that breakdown is mediated by metalloproteinases. J Periodont Res 24:207–213

Miyazaki Y, Araki K, Vesin C, Garcia I, Kapanci Y, Whitsett JA, Piguet PF, Vassalli P (1995) Expression of a tumor necrosis factor-alpha transgene in murine lung causes lymphocytic and fibrosing alveolitis. A mouse model of progressive pulmonary fibrosis. J Clin Invest 96: 250–259

Montgomery SL, Bowers WJ (2012) Tumor necrosis factor-alpha and the roles it plays in homeostatic and degenerative processes within the central nervous system. J Neuroimmune Pharmacol 7:42–59

Moreland LW, O'Dell JR, Paulus HE, Curtis JR, Bathon JM, St. Clair EW, Bridges SL Jr, Zhang J, McVie T, Howard G, van der Heijde D, Cofield SS (2012) A randomized comparative effectiveness study of oral triple therapy versus etanercept plus methotrexate in early aggressive rheumatoid arthritis: the treatment of early aggressive rheumatoid arthritis trial. Arthritis Rheum 64:2824–2835

Nadkarni S, Mauri C, Ehrenstein MR (2007) Anti-TNF-alpha therapy induces a distinct regulatory T cell population in patients with rheumatoid arthritis via TGF-beta. J Exp Med 204:33–39

Nawashiro H, Martin D, Hallenbeck JM (1997) Neuroprotective effects of TNF binding protein in focal cerebral ischemia. Brain Res 778:265–271

Nieto-Vazquez I, Fernandez-Veledo S, Kramer DK, Vila-Bedmar R, Garcia-Guerra L, Lorenzo M (2008) Insulin resistance associated to obesity: the link TNF-alpha. Arch Physiol Biochem 114:183–194

Notley CA, Inglis JJ, Alzabin S, McCann FE, McNamee KE, Williams RO (2008) Blockade of tumor necrosis factor in collagen-induced arthritis reveals a novel immunoregulatory pathway for Th1 and Th17 cells. J Exp Med 205:2491–2497

Ofei F, Hurel S, Newkirk J, Sopwith M, Taylor R (1996) Effects of an engineered human anti-TNF-alpha antibody (CDP571) on insulin sensitivity and glycemic control in patients with NIDDM. Diabetes 45:881–885

Ortiz LA, Lasky J, Hamilton RF Jr, Holian A, Hoyle GW, Banks W, Peschon JJ, Brody AR, Lungarella G, Friedman M (1998) Expression of TNF and the necessity of TNF receptors in bleomycin-induced lung injury in mice. Exp Lung Res 24:721–743

Ortiz LA, Lasky J, Lungarella G, Cavarra E, Martorana P, Banks WA, Peschon JJ, Schmidts HL, Brody AR, Friedman M (1999) Upregulation of the p75 but not the p55 TNF-alpha receptor mRNA after silica and bleomycin exposure and protection from lung injury in double receptor knockout mice. Am J Respir Cell Mol Biol 20:825–833

Paleolog EM, Hunt M, Elliott MJ, Feldmann M, Maini RN, Woody JN (1996) Deactivation of vascular endothelium by monoclonal anti-tumor necrosis factor α antibody in rheumatoid arthritis. Arthritis Rheum 39:1082–1091

Paleolog EM, Young S, Stark AC, McCloskey RV, Feldmann M, Maini RN (1998) Modulation of angiogenic vascular endothelial growth factor (VEGF) by TNFα and IL-1 in rheumatoid arthritis. Arthritis Rheum 41:1258–1265

Pennica D, Hayflick JS, Bringman TS, Palladino MA, Goeddel DV (1985) Cloning and expression in *Escherichia coli* of the cDNA for murine tumour necrosis factor. Proc Natl Acad Sci USA 82(18):6060–6064

Peschon JJ, Slack JL, Reddy P, Stocking KL, Sunnarborg SW, Lee DC, Russell WE, Castner BJ, Johnson RS, Fitzner JN, Boyce RW, Nelson N, Kozlosky CJ, Wolfson MF, Rauch CT, Cerretti DP, Paxton RJ, March CJ, Black RA (1998) An essential role for ectodomain shedding in mammalian development. Science 282:1281–1284

Pfeiffer A, Janott J, Mohlig M, Ristow M, Rochlitz H, Busch K, Schatz H, Schifferdecker E (1997) Circulating tumor necrosis factor alpha is elevated in male but not in female patients with type II diabetes mellitus. Horm Metab Res 29:111–114

Piguet PF, Vesin C (1994) Treatment by human recombinant soluble TNF receptor of pulmonary fibrosis induced by bleomycin or silica in mice. Eur Respir J 7:515–518

Piguet PF, Collart MA, Grau GE, Sappino AP, Vassalli P (1990) Requirement of tumour necrosis factor for development of silica-induced pulmonary fibrosis. Nature 344:245–247

Piguet PF, Grau GE, Vesin C, Loetscher H, Gentz R, Lesslauer W (1992) Evolution of collagen arthritis in mice is arrested by treatment with anti-tumour necrosis factor (TNF) antibody or a recombinant soluble TNF receptor. Immunology 77:510–514

Probert L, Plows D, Kontogeorgos G, Kollias G (1995) The type I interleukin-1 receptor acts in series with tumor necrosis factor (TNF) to induce arthritis in TNF-transgenic mice. Eur J Immunol 25:1794–1797

Ramos-Casals M, Perez-Alvarez R, Perez-de-Lis M, Xaubet A, Bosch X, Group BS (2011) Pulmonary disorders induced by monoclonal antibodies in patients with rheumatologic autoimmune diseases. Am J Med 124:386–394

Ren L, Lubrich B, Biber K, Gebicke-Haerter PJ (1999) Differential expression of inflammatory mediators in rat microglia cultured from different brain regions. Brain Res Mol Brain Res 65:198–205

Ryba M, Marek N, Hak L, Rybarczyk-Kapturska K, Mysliwiec M, Trzonkowski P, Mysliwska J (2011) Anti-TNF rescue CD4+Foxp3+ regulatory T cells in patients with type 1 diabetes from effects mediated by TNF. Cytokine 55:353–361

Ryden M, Arner P (2007) Tumour necrosis factor-alpha in human adipose tissue—from signalling mechanisms to clinical implications. J Intern Med 262:431–438

Satoh J, Seino H, Abo T, Tanaka S, Shintani S, Ohta S, Tamura K, Sawai T, Nobunaga T, Oteki T et al (1989) Recombinant human tumor necrosis factor alpha suppresses autoimmune diabetes in nonobese diabetic mice. J Clin Invest 84:1345–1348

Sethi JK, Hotamisligil GS (1999) The role of TNF alpha in adipocyte metabolism. Semin Cell Dev Biol 10:19–29

Sime PJ, Marr RA, Gauldie D, Xing Z, Hewlett BR, Graham FL, Gauldie J (1998) Transfer of tumor necrosis factor-alpha to rat lung induces severe pulmonary inflammation and patchy interstitial fibrogenesis with induction of transforming growth factor-beta1 and myofibroblasts. Am J Pathol 153:825–832

Smeets TJ, Kraan MC, van Loon ME, Tak PP (2003) Tumor necrosis factor alpha blockade reduces the synovial cell infiltrate early after initiation of treatment, but apparently not by induction of apoptosis in synovial tissue. Arthritis Rheum 48:2155–2162

Smolen JS, Landewe R, Breedveld FC, Dougados M, Emery P, Gaujoux-Viala C, Gorter S, Knevel R, Nam J, Schoels M, Aletaha D, Buch M, Gossec L, Huizinga T, Bijlsma JW, Burmester G, Combe B, Cutolo M, Gabay C, Gomez-Reino J, Kouloumas M, Kvien TK, Martin-Mola E, McInnes I, Pavelka K, van Riel P, Scholte M, Scott DL, Sokka T, Valesini G, van Vollenhoven R, Winthrop KL, Wong J, Zink A, van der Heijde D (2010) EULAR recommendations for the management of rheumatoid arthritis with synthetic and biological disease-modifying antirheumatic drugs. Ann Rheum Dis 69:964–975

Smookler DS, Mohammed FF, Kassiri Z, Duncan GS, Mak TW, Khokha R (2006) Tissue inhibitor of metalloproteinase 3 regulates TNF-dependent systemic inflammation. J Immunol 176: 721–725

Sriram K, O'Callaghan JP (2007) Divergent roles for tumor necrosis factor-alpha in the brain. J Neuroimmune Pharmacol 2:140–153

Sriram K, Matheson JM, Benkovic SA, Miller DB, Luster MI, O'Callaghan JP (2006) Deficiency of TNF receptors suppresses microglial activation and alters the susceptibility of brain regions to MPTP-induced neurotoxicity: role of TNF-alpha. FASEB J 20:670–682

Stavropoulos-Kalinoglou A, Metsios GS, Panoulas VF, Nightingale P, Koutedakis Y, Kitas GD (2012) Anti-tumour necrosis factor alpha therapy improves insulin sensitivity in normal-weight but not in obese patients with rheumatoid arthritis. Arthritis Res Ther 14:R160

Stellwagen D, Malenka RC (2006) Synaptic scaling mediated by glial TNF-alpha. Nature (Lond) 440:1054–1059

Sudo K, Yamada Y, Moriwaki H, Saito K, Seishima M (2005) Lack of tumor necrosis factor receptor type 1 inhibits liver fibrosis induced by carbon tetrachloride in mice. Cytokine 29: 236–244

Suganami T, Ogawa Y (2010) Adipose tissue macrophages: their role in adipose tissue remodeling. J Leukoc Biol 88:33–39

Sullivan DE, Ferris M, Pociask D, Brody AR (2005) Tumor necrosis factor-alpha induces transforming growth factor-beta1 expression in lung fibroblasts through the extracellular signal-regulated kinase pathway. Am J Respir Cell Mol Biol 32:342–349

Tak PP, Taylor PC, Breedveld FC, Smeets TJM, Daha MR, Kluin PM, Meinders AE, Maini RN (1996) Decrease in cellularity and expression of adhesion molecules by anti-tumor necrosis factor α monoclonal antibody treatment in patients with rheumatoid arthritis. Arthritis Rheum 39:1077–1081

Tam CS, Clement K, Baur LA, Tordjman J (2010) Obesity and low-grade inflammation: a paediatric perspective. Obes Rev 11:118–126

Tartaglia LA, Pennica D, Goeddel DV (1993) Ligand passing: the 75-kDa tumor necrosis factor (TNF) receptor recruits TNF for signaling by the 55-kDa TNF receptor. J Biol Chem 268: 18542–18548

Taylor PC (2005) Serum vascular markers and vascular imaging in assessment of rheumatoid arthritis disease activity and response to therapy. Rheumatology (Oxf) 44:721–728

Taylor PC (2010) Pharmacology of TNF blockade in rheumatoid arthritis and other chronic inflammatory diseases. Curr Opin Pharmacol 10:308–315

Taylor PC, Peters AM, Glass DM, Maini RN (1999) Effects of treatment of rheumatoid arthritis patients with an antibody against tumour necrosis factor alpha on reticuloendothelial and intrapulmonary granulocyte traffic. Clin Sci (Lond) 97:85–89

Taylor PC, Peters AM, Paleolog E, Chapman PT, Elliott MJ, McCloskey R, Feldmann M, Maini RN (2000) Reduction of chemokine levels and leukocyte traffic to joints by tumor necrosis factor α blockade in patients with rheumatoid arthritis. Arthritis Rheum 43:38–47

Taylor PC, Steuer A, Gruber J, Cosgrove DO, Blomley MJ, Marsters PA, Wagner CL, McClinton C, Maini RN (2004) Comparison of ultrasonographic assessment of synovitis and joint vascularity with radiographic evaluation in a randomized, placebo-controlled study of infliximab therapy in early rheumatoid arthritis. Arthritis Rheum 50:1107–1116

Taylor PC, Steuer A, Gruber J, McClinton C, Cosgrove DO, Blomley MJ, Marsters PA, Wagner CL, Maini RN (2006) Ultrasonographic and radiographic results from a two-year controlled trial of immediate or one-year-delayed addition of infliximab to ongoing methotrexate therapy in patients with erosive early rheumatoid arthritis. Arthritis Rheum 54:47–53

Terrando N, Monaco C, Ma D, Foxwell BM, Feldmann M, Maze M (2010) Tumor necrosis factor-alpha triggers a cytokine cascade yielding postoperative cognitive decline. Proc Natl Acad Sci USA 107:20518–20522

Terrando N, Eriksson LI, Ryu JK, Yang T, Monaco C, Feldmann M, Jonsson Fagerlund M, Charo IF, Akassoglou K, Maze M (2011) Resolving postoperative neuroinflammation and cognitive decline. Ann Neurol 70:986–995

Theiss AL, Simmons JG, Jobin C, Lund PK (2005) Tumor necrosis factor (TNF) alpha increases collagen accumulation and proliferation in intestinal myofibroblasts via TNF receptor 2. J Biol Chem 280:36099–36109

Thorbecke GJ, Shah R, Leu CH, Kuruvilla AP, Hardison AM, Palladino MA (1992) Involvement of endogenous tumor necrosis factor α and transforming growth factor β during induction of collagen type II arthritis in mice. Proc Natl Acad Sci USA 89:7375–7379

Toubi E, Kessel A, Mahmudov Z, Hallas K, Rozenbaum M, Rosner I (2005) Increased spontaneous apoptosis of CD4+CD25+ T cells in patients with active rheumatoid arthritis is reduced by infliximab. Ann N Y Acad Sci 1051:506–514

Triantaphyllopoulos K, Croxford J, Baker D, Chernajovsky Y (1998) Cloning and expression of murine IFN beta and a TNF antagonist for gene therapy of experimental allergic encephalomyelitis. Gene Ther 5:253–263

Ulfgren AK, Andersson U, Engstrom M, Klareskog L, Maini RN, Taylor PC (2000) Systemic anti-tumor necrosis factor alpha therapy in rheumatoid arthritis down-regulates synovial tumor necrosis factor alpha synthesis. Arthritis Rheum 43:2391–2396

Valencia X, Stephens G, Goldbach-Mansky R, Wilson M, Shevach EM, Lipsky PE (2006) TNF downmodulates the function of human CD4+CD25hi T-regulatory cells. Blood 108:253–261

van Amelsfort JM, van Roon JA, Noordegraaf M, Jacobs KM, Bijlsma JW, Lafeber FP, Taams LS (2007) Proinflammatory mediator-induced reversal of CD4+, CD25+ regulatory T cell-mediated suppression in rheumatoid arthritis. Arthritis Rheum 56:732–742

van Vollenhoven RF, Ernestam S, Geborek P, Petersson IF, Coster L, Waltbrand E, Zickert A, Theander J, Thorner A, Hellstrom H, Teleman A, Dackhammar C, Akre F, Forslind K, Ljung L, Oding R, Chatzidionysiou A, Wornert M, Bratt J (2009) Addition of infliximab compared with addition of sulfasalazine and hydroxychloroquine to methotrexate in patients with early rheumatoid arthritis (Swefot trial): 1-year results of a randomised trial. Lancet 374:459–466

Verjee LS, Verhoekx JSN, Chan JK, Krausgruber T, Nicolaidou V, Izadi D, Davidson D, Feldmann M, Midwood KS, Nanchahal J (2013) Unravelling the signaling pathways promoting fibrosis in Dupuytren's disease reveals TNF as a therapeutic target. Proc Natl Acad Sci USA 110: E928–E937

Verrecchia F, Pessah M, Atfi A, Mauviel A (2000) Tumor necrosis factor-alpha inhibits transforming growth factor-beta/smad signaling in human dermal fibroblasts via AP-1 activation. J Biol Chem 275:30226–30231

Viglietta V, Baecher-Allan C, Weiner HL, Hafler DA (2004) Loss of functional suppression by CD4+CD25+ regulatory T cells in patients with multiple sclerosis. J Exp Med 199:971–979

Williams RO, Feldmann M, Maini RN (1992) Anti-tumor necrosis factor ameliorates joint disease in murine collagen-induced arthritis. Proc Natl Acad Sci USA 89:9784–9788

Williams RO, Ghrayeb J, Feldmann M, Maini RN (1995) Successful therapy of collagen-induced arthritis with TNF receptor-IgG fusion protein and combination with anti-CD4. Immunology 84:433–439

Wynn TA (2008) Cellular and molecular mechanisms of fibrosis. J Pathol 214:199–210

Yazdani-Biuki B, Stelzl H, Brezinschek HP, Hermann J, Mueller T, Krippl P, Graninger W, Wascher TC (2004) Improvement of insulin sensitivity in insulin resistant subjects during prolonged treatment with the anti-TNF-alpha antibody infliximab. Eur J Clin Invest 34:641–642

Zakharova M, Ziegler HK (2005) Paradoxical anti-inflammatory actions of TNF-alpha: inhibition of IL-12 and IL-23 via TNF receptor 1 in macrophages and dendritic cells. J Immunol 175:5024–5033

Zhang H, Yan D, Shi X, Liang H, Pang Y, Qin N, Chen H, Wang J, Yin B, Jiang X, Feng W, Zhang W, Zhou M, Li Z (2008) Transmembrane TNF-alpha mediates "forward" and "reverse" signaling, inducing cell death or survival via the NF-kappaB pathway in Raji Burkitt lymphoma cells. J Leukoc Biol 84:789–797

Ziolkowska M, Kurowska M, Radzikowska A, Luszczykiewicz G, Wiland P, Dziewczopolski W, Filipowicz-Sosnowska A, Pazdur J, Szechinski J, Kowalczewski J, Rell-Bakalarska M, Maslinski W (2002) High levels of osteoprotegerin and soluble receptor activator of nuclear factor kappa B ligand in serum of rheumatoid arthritis patients and their normalization after anti-tumor necrosis factor alpha treatment. Arthritis Rheum 46:1744–1753

Part II
Cytokines in Allergic Disorders

Chapter 10
Interleukin-25: A Key Initiator of Type 2 Immune Responses

Hiroshi Nakajima, Tomohiro Tamachi, and Koichi Hirose

Abstract T helper 2 (Th2) cells induce allergic inflammation through the production of Th2 cytokines such as interleukin (IL)-4, IL-5, and IL-13. In addition, it has been demonstrated that IL-25 (IL-17E) is involved in the initiation and amplification of type 2 immune responses. Recent studies have shown that a number of cell types including activated Th2 cells, epithelial cells, mast cells, eosinophils, and macrophages produce IL-25 in response to a variety of stimuli. Regarding IL-25-responding cells, accumulating evidence has revealed that IL-25 induces type 2 immune responses through the activation of Th2 cells, natural killer T (NKT) cells, eosinophils, and innate immune cells including multipotent progenitor type 2 (MPP^{type2}) cells, natural helper cells (NH cells), nuocytes, innate type 2 helper (Ih2) cells, and type 2 myeloid (T2M) cells. In vivo, we and others have shown that IL-25 is expressed in the airways in murine asthma models and is involved in the induction of antigen-induced airway inflammation and remodeling. In humans, IL-25 is suggested to be involved in the pathogenesis of allergic diseases such as asthma and Churg–Strauss syndrome. In this chapter, we summarize recent advances regarding the biology of IL-25.

Keywords Allergy • Host defense • Innate immune cells • Interleukin-25 • T cells

Abbreviations

αGalCer	α-Galactosylceramide
AHR	Airway hyperreactivity
CSS	Churg–Strauss syndrome

H. Nakajima (✉) • T. Tamachi • K. Hirose
Department of Allergy and Clinical Immunology, Graduate School of Medicine, Chiba University, 1-8-1 Inohana, Chiba City, Chiba, 260-8670, Japan
e-mail: nakajimh@faculty.chiba-u.jp

T. Yoshimoto and T. Yoshimoto (eds.), *Cytokine Frontiers: Regulation of Immune Responses in Health and Disease*, DOI 10.1007/978-4-431-54442-5_10, © Springer Japan 2014

EAE	Experimental autoimmune encephalomyelitis
Ih2 cells	Innate type 2 helper cells
MPP^{type2}	Multipotent progenitor type 2
NBNT cells	Non-B/non-T cells
NH cells	Natural helper cells
NKT cells	Natural killer T cells
T2M cells	Type 2 myeloid cells
TSLP	Thymic stromal lymphopoietin

10.1 Interleukin-25 (IL-25) and IL-25R

10.1.1 Structure of IL-25

Interleukin (IL)-17 (also known as IL-17A) has been identified as cytotoxic T lymphocyte-associated antigen-8 (CTLA-8) in the mouse and shown to exhibit multiple biological activities including the induction of various pro-inflammatory cytokines, chemokines, and cell adhesion molecules in many cell types (Reynolds et al. 2010; Iwakura et al. 2011). Subsequently, IL-17B, IL-17C, IL-17D, IL-17E (IL-25), and IL-17F have been identified by homology search of IL-17A (Li et al. 2000; Shi et al. 2000; Fort et al. 2001; Hymowitz et al. 2001; Lee et al. 2001; Starnes et al. 2001, 2002) (Table 10.1). As the IL-17F gene is located adjacent to IL-17A, it is suggested that IL-17A and IL-17F may share the same regulatory elements and may have evolved from gene duplication. IL-17A is composed of 177 amino acids, containing the N-terminal signal peptide and cysteine residues conserved among the IL-17 family cytokines (Lee et al. 2001). Sequence alignment of IL-17 family cytokines shows that IL-17A and IL-17F are closely related, with an approximately 50 % sequence identity (Hymowitz et al. 2001; Starnes et al. 2001), that IL-17B and IL-17D are less homologous, and that IL-25 is the most divergent member, showing less than 20 % homology with IL-17A (Lee et al. 2001), which may explain different characteristics of IL-25 from IL-17A (Pan et al. 2001; Angkasekwinai et al. 2007). Although there is limited homology in the sequence, the conformational state is conserved among the IL-17 family cytokines, especially for the four β-strands in the C-terminal region and four cysteine and two serine residues, which are critical to form a cysteine knot fold of the IL-17 family cytokines (Hymowitz et al. 2001; Gerhardt et al. 2009).

10.1.2 IL-25 Receptors

IL-17RB (also known as EV127) has been identified by homology searches for sequences related to IL-17RA (Lee et al. 2001; Shi et al. 2000; Tian et al. 2000). IL-17RB is a 502-amino-acid single transmembrane protein that shows 26 % amino

Table 10.1 Members of interleukin (IL)-17 family cytokines

Ligand	Receptor	Homology with IL-17A (%)	Producers	Target cells
IL-17A (IL-17)	IL-17RA/ IL-17RC/ IL-17RD	100	$CD4^+$ T cell (Th17 cell), $CD8^+$ T cell, γδ T cell, NKT cell, LTi-like cell, neutrophil	Epithelial cell, fibroblast, keratinocyte, synoviocyte, endothelial cell, T cell, B cell, macrophage
IL-17B	IL-17RB	24	Chondrocyte, neuron	Monocyte, endothelial cell
IL-17C	IL-17RE/ IL-17RA	26	$CD4^+$ T cell, DC, macrophage, keratinocyte	Monocyte
IL-17D	Unknown	30	$CD4^+$ T cell, B cell	Endothelial cell myeloid progenitor
IL-17E (IL-25)	IL-17RB/ IL-17RA	16	$CD4^+$ T cell (Th2 cell), mast cell, eosinophil, basophil, macrophage, epithelial cell	Th2 cell, Th9 cell, NKT cell, NH cell, MPP^{type2} cell, nuocyte, Ih2 cell, T2M cell, epithelial cell, endothelial cell
IL-17F	IL-17RA/ IL-17RC	50	$CD4^+$ T cell (Th17 cell), $CD8^+$ T cell, γδ T cell, NKT cell, LTi-like cell	Epithelial cell, fibroblast, keratinocyte, synoviocyte, endothelial cell

LTi-like cell lymphoid tissue inducer-like cell

acid identity to IL-17RA. IL-17RB has been identified as a receptor for IL-17B (Shi et al. 2000); however, subsequent study has revealed that IL-25 shows higher affinity to IL-17RB than does IL-17B (Lee et al. 2001; Hymowitz et al. 2001). Further studies using mice deficient in either IL-17RA or IL-17RB have revealed that the functional IL-25 receptor requires IL-17RA as well as IL-17RB (Rickel et al. 2008). Thus, analogous to the IL-17A-receptor complex, which is composed of IL-17RA and IL-17RC, the IL-25 receptor is composed of IL-17RA and IL-17RB (Table 10.1).

10.2 IL-25-Producing Cells

IL-25, whose expression was initially reported to be restricted to highly polarized Th2 cells (Fort et al. 2001), has been found to be expressed by several cell types in hematopoietic and nonhematopoietic lineages. In hematopoietic lineages, IL-25 mRNA is expressed in mast cells (Ikeda et al. 2003), basophils (Wang et al. 2007), eosinophils (Wang et al. 2007; Dolgachev et al. 2009), alveolar macrophages (Kang et al. 2005), and microglia (Kleinschek et al. 2007). We have shown that bone marrow-derived mast cells express IL-25 mRNA upon stimulation with cross-linking of IgE receptors in a calcineurin-dependent manner, and that the expression levels of IL-25 mRNA were comparable to those of activated Th2 cells (Ikeda et al. 2003). Interestingly, activated basophils from allergic subjects produce more IL-25 than those from nonallergic subjects upon stimulation with IgE receptor cross-linking (Wang et al. 2007). Eosinophils also express IL-25 mRNA upon stimulation with IL-5 (Wang et al. 2007) or stem cell factor (Dolgachev et al. 2009). Importantly, Terrier et al. have shown that the levels of IL-25 in sera are increased in active Churg–Strauss syndrome (CSS) patients and that eosinophils are the main IL-25-producing cells in CSS patients (Terrier et al. 2010). Kang et al. have shown that alveolar macrophages also express IL-25 mRNA upon stimulation with small particles (Kang et al. 2005). In mouse brain, Kleinschek et al. have shown that IL-25 mRNA is constitutively expressed by microglia and that its expression is involved in the protection from experimental autoimmune encephalomyelitis (EAE) (Kleinschek et al. 2007).

IL-25 is also expressed in nonhematopoietic cells including epithelial cells (Angkasekwinai et al. 2007; Zaph et al. 2008), endothelial cells (Sonobe et al. 2009), and keratinocytes (Wang et al. 2007). Lung epithelial cells express IL-25 mRNA upon stimulation with allergens such as ragweed or *Aspergillus oryzae* protease (Angkasekwinai et al. 2007). The induction of IL-25 by protease allergens in fibroblasts is suppressed by Erk, JNK, and p38 inhibitors (Yu et al. 2010), suggesting that protease allergens induce IL-25 expression through the activation of MAPK pathways. In the gut, IL-25 is detected in the intestinal epithelial cells of the large intestine in conventionally reared mice but not in germfree mice (Zaph et al. 2008), suggesting that commensal bacteria or their products trigger IL-25 expression. The epithelial cell-derived IL-25 is reported to be involved in the maintenance of intestinal homeostasis, presumably by inhibiting TLR-mediated IL-23 expression in macrophages (Zaph et al. 2008).

10.3 IL-25-Responding Cells

It has been shown that in vivo and in vitro biological activities of IL-25 are markedly different from those described for other IL-17 family cytokines (Reynolds et al. 2010; Iwakura et al. 2011). The biological function of IL-25 was first revealed by the administration of exogenous IL-25 or transgenic mice expressing IL-25 (Fort et al. 2001; Pan et al. 2001; Hurst et al. 2002; Kim et al. 2002). The systemic administration of IL-25 results in eosinophilia through the production of IL-5 (Fort et al. 2001; Hurst et al. 2002; Pan et al. 2001), whereas the systemic administration of other IL-17 family cytokines induces neutrophilia (Schwarzenberger et al. 1998; Linden et al. 2000; Shi et al. 2000). IL-25 induces the gene expression of IL-4 and IL-13 in multiple tissues and the resultant type 2 immune responses including increased serum IgE levels and pathological changes in multiple tissues (Fort et al. 2001; Hurst et al. 2002; Pan et al. 2001). Importantly, previous studies have demonstrated that IL-25 is capable of inducing Th2 cytokine production even in RAG-deficient mice (Fort et al. 2001; Pan et al. 2001), indicating that T cells and B cells are not required for the IL-25-induced Th2 cytokine production, and that once IL-25 is produced, IL-25 enhances allergic responses even in the absence of Th2 cells.

Although IL-17A mainly functions on nonhematopoietic cells such as epithelial cells and fibroblasts to induce pro-inflammatory cytokines and chemokines, IL-25 mostly targets hematopoietic cells including innate and adaptive immune cells and promotes type 2 immune responses. Earlier studies showed that the major IL-25-responding cell in vivo is a novel population of non-B/non-T cells (NBNT cells), which is characterized as Lin^- MHC class II^{high} $CD11c^{dull}$ (Fort et al. 2001) or $c\text{-}Kit^+$ $Fc\varepsilon R1^-$ (Fallon et al. 2006). NBNT cells with the $c\text{-}Kit^+$ $Fc\varepsilon R1^-$ phenotype are shown to be important for the initiation of *Nippostrongylus brasiliensis* worm expulsion in vivo (Fallon et al. 2006).

Regarding the IL-25-responding NBNT cells in the myeloid lineage, it has been shown that IL-25 regulates the surface expression of ICAM-1, ICAM-3, and L-selectin on eosinophils through the activation of p38 MAPK, JNK and NF-κB pathways (Wong et al. 2005; Cheung et al. 2006). Dolgachev et al. have shown that IL-17RB is induced in a myeloid cell population that is identified as $IL\text{-}4^+$ $CD11b^+$ $Gr\text{-}1^+$ cells in the lung and bone marrow during chronic allergen responses (Dolgachev et al. 2009). Recently, the same group has shown that the majority of IL-4- and IL-13-producing cells in the lungs upon repeated allergen exposure is $IL\text{-}17RB^+$ $CD11b^+$ $Gr\text{-}1^+$ type 2 myeloid (T2M) cells, which represent a novel distinct granulocytic population closely related to eosinophils (Petersen et al. 2012). Intratracheal administration of IL-25 strongly induces IL-4 and IL-13 production from T2M cells, and the Th2 cytokine production by T2M cells is resistant to dexamethasone (Petersen et al. 2012). In addition, human $CD14^+$ monocytes from peripheral blood mononuclear cells are reported to express functional IL-17RB and respond to IL-25 (Caruso et al. 2009b).

On the other hand, recent studies have shown that IL-17RB is preferentially expressed on a fraction of α-galactosylceramide (αGalCer)/CD1d $dimer^+$ NKT cells and that $IL\text{-}17RB^+$ NKT cells produce Th2 cytokines and induce airway hyperreactivity in response to IL-25 (Terashima et al. 2008; Stock et al. 2009). In addition to Th2 cytokines, $CD4^+$ $IL\text{-}17RB^+$ iNKT cells are shown to be capable of producing Th9 cytokines (IL-9 and IL-10) and Th17 cytokines (IL-17A and IL-22) in response to IL-25 (Watarai et al. 2012).

Recently, several groups identified a novel population of lineage-negative innate effector cells that respond to IL-25, including multipotent progenitor type 2 (MPP^{type2}) cells (Saenz et al. 2010a, b), adipose tissue-associated $c\text{-}Kit^+$ $Sca\text{-}1^+$ lymphoid cells [termed natural helper cells (NH cells)] (Moro et al. 2010), nuocytes (Neill et al. 2010), or innate type 2 helper (Ih2) cells (Price et al. 2010). Although the lineage relationships among these cells remain unclear, these innate cell populations seem to be responsible for the production of Th2 cytokines, in particular, IL-13, during helminthic parasite infection and serve to mediate worm expulsion (Saenz et al. 2010a; Oliphant et al. 2011; Spits and Cupedo 2012) (Table 10.2).

In addition to innate immune cells, T cells have also been shown to respond to IL-25. IL-25 directs naive T helper cells toward Th2 commitment as well as augmenting the cytokine production of effector/memory Th2 cells (Wang et al. 2007;

Table 10.2 Interleukin (IL)-25-responding innate immune cells

		NH cell	Nuocyte	Ih2 cell	MPP^{type2} cell	T2M cell
Anatomic location		FALC	LN, spleen	LN, spleen, liver	LN, GALT	Lung, bone marrow
Surface phenotype	Lineage marker	–	–	–	–	Gr-1$^+$ CD11b$^+$
	c-Kit	+	+	+	+	–/+
	Sca1	+	+	–	+	–
	ST2	+	+	Unknown	–	Unknown
	IL-7Rα	+	+	Unknown	–	Unknown
	CD90.2	+	+	+	–	Unknown
	CD44	+	+	+	–	Unknown
Progenitor capacity	–	–	–	–	+	Unknown

FALC fat-associated lymphoid cluster, *LN* lymph node, *GALT* gut-associated lymphoid tissue

Angkasekwinai et al. 2007). Angkasekwinai et al. have shown that IL-17RB mRNA is detected in freshly isolated naive CD4$^+$ T cells and is highly upregulated in Th2 cells but not in Th1 or Th17 cells (Angkasekwinai et al. 2007, 2010). Addition of IL-25 to the culture of naive T cells promotes the differentiation of Th2 cells through the production of IL-4 and subsequent activation of STAT6, although IL-25 has no effect on the polarization of Th1 cells and Th17 cells (Angkasekwinai et al. 2007). IL-25 seems to regulate early IL-4 expression by potentiating the induction of NFATc1 and JunB, which leads to IL-4-dependent upregulation of GATA3 expression (Angkasekwinai et al. 2007; Wang et al. 2007). In humans, the expression of IL-17RB is detected in circulating memory Th2 cells and upregulated in CD3/CD28-activated T cells (Wang et al. 2006, 2007).

Thymic stromal lymphopoietin (TSLP) is shown to activate dendritic cells to directly promote Th2 cell differentiation (Liu 2006). In the presence of TSLP-activated dendritic cells, IL-17RB is strongly induced on Th2 memory cells, and thus IL-25 enhances cytokine production and proliferation of Th2 cells (Wang et al. 2007). Moreover, Angkasekwinai et al. have shown that Th9 cells, a recently identified T helper lineage that is induced in the presence of IL-4 and TGF-β and produces IL-9 and IL-10 (Veldhoen et al. 2008; Dardalhon et al. 2008), express IL-17RB and that IL-25 enhances IL-9 and IL-10 production in differentiating Th9 cells (Angkasekwinai et al. 2010). Taken together, these findings indicate that IL-25 enhances type 2 immune responses through the activation of a variety of immune cells in both innate and acquired immunity (Fig. 10.1).

10.4 IL-25 Receptor Signaling

Recently, Rickel et al. have shown that IL-25-induced production of IL-5 and IL-13 in splenocytes is defective not only in IL-17RB-deficient mice but also in IL-17RA-deficient mice (Rickel et al. 2008). In addition, Angkasekwinai et al. have reported

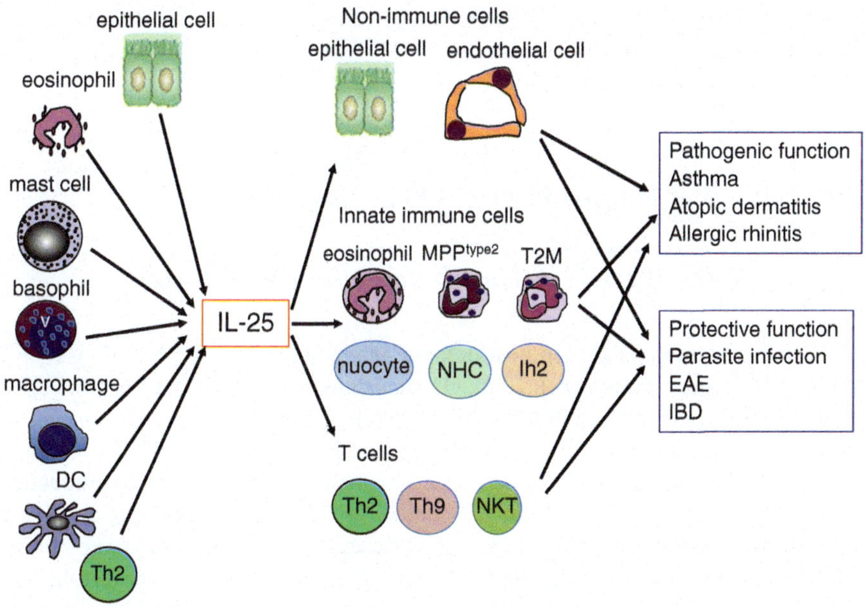

Fig. 10.1 Interleukin (IL)-25-producing and IL-25-responding cells that have been reported and their possible roles

that IL-25-mediated IL-9 production under Th9-inducing conditions depends on both IL-17RB and IL-17RA (Angkasekwinai et al. 2010). Although IL-25 does not bind to IL-17RA in vitro (Hymowitz et al. 2001), it has been shown that when IL-17RB and IL-17RA are expressed in 293T cells, they interact with each other (Angkasekwinai et al. 2010). Moreover, Ely et al. have shown that IL-17RA binds to the IL-17RB–IL-25 complex with high affinity (Ely et al. 2009). Furthermore, similar to IL-17RB-deficient mice, IL-17RA-deficient mice fail to exhibit IL-25-mediated expansion of innate c-Kit^{int} cells (Saenz et al. 2010b). These findings suggest that the functional IL-25 receptor is composed of IL-17RA and IL-17RB in both innate immune cells and T cells.

Regarding IL-25R-mediated signaling, we have shown that IL-25R-cross-linking induces NF-κB activation in a TRAF6-dependent manner (Maezawa et al. 2006). We also found that TRAF6 is bound to the cytoplasmic region of IL-25R through the conserved TRAF6-binding motif (Maezawa et al. 2006). Act1, an adaptor of IL-17RA signaling, is also shown to be involved in IL-17RB signaling (Claudio et al. 2009). The interaction between Act1 and IL-17RB is abolished when SEFIR domain is deleted in either Act1 or IL-17RB, suggesting that Act1 interacts with IL-17RB by the dimerization of the SEFIR domain (Swaidani et al. 2009). Recently, Swaidani et al. have reported that IL-25-induced Th2 polarization and cytokine production are impaired in Act1-deficient T cells (Swaidani et al. 2011). In addition, they have shown that the in vivo generation of antigen-specific Th2 cells is defective in Act1-deficient T cells and that IL-25-mediated induction of GATA3 is attenuated in Act1-deficient T cells (Swaidani et al. 2011). Taken together, these results suggest that TRAF6 and Act1 play a critical role in IL-25R-mediated signaling.

10.5 Roles of Interleukin (IL)-25 in Host Defense

10.5.1 Roles of IL-25 in Host Defense Against Parasite

IL-25 has an important role in host defense against parasites by promoting type 2 immune responses (Fallon et al. 2006; Owyang et al. 2006; Zhao et al. 2010). Fallon et al. have shown that IL-25-deficient mice infected with *Nippostrongylus brasiliensis* (*N. brasiliensis*) show delayed worm expulsion and induction of Th2 cytokines, whereas administration of IL-25 accelerates worm expulsion through the induction of Th2 cytokine production by c-Kit$^+$ FcεR1$^-$ NBNT cells (Fallon et al. 2006). Interestingly, administration of IL-25 induces worm expulsion even in *N. brasiliensis-infected* RAG-deficient mice (Fallon et al. 2006), suggesting that T cells are not essential for IL-25-mediated worm expulsion. Similarly, Owyang et al. have shown that IL-25-deficient mice on a C57BL/6 background fail to expel *Trichuris muris* (Owyang et al. 2006). Furthermore, Zhao et al. have shown that epithelial cells but not immune cells are the major source of IL-25 in the small intestine, and that *N. brasiliensis* infection upregulates IL-25 expression in an IL-13- and STAT6-dependent manner (Zhao et al. 2010). These findings suggest that IL-25 produced by epithelial cells may initiate type 2 immune responses by activating innate immune cells.

Recent studies have identified innate immune cell populations such as NH cells (Moro et al. 2010), MPP^{type2} cells (Saenz et al. 2010b), nuocytes (Neill et al. 2010), and Ih2 cells (Price et al. 2010), which promote type 2 immune responses in response to IL-25. Interestingly, Price et al. have shown that Ih2 cells are the major IL-13-expressing innate cells in mice infected with *N. brasiliensis* (Price et al. 2010). Although the expression of some cell-surface molecules and anatomic location are different among these innate lymphoid cell populations, they share many characteristics including their ability in Th2 cytokine production upon IL-25 or IL-33 stimulation and in the host defense against parasites (Saenz et al. 2010a; Oliphant et al. 2011; Spits and Cupedo 2012) (Table 10.2). The lineage relationships among NH cells, MPP^{type2} cells, nuocytes, and Ih2 cells, and the precise roles of these innate populations in type 2 immune responses should be determined in future.

More recently, it has been demonstrated that Act1 deficiency in epithelial cells results in a marked delay in worm expulsion and in reduced expansion of the Lin$^-$ c-Kit$^+$ innate cell population in the mesenteric lymph node, lung, and liver (Kang et al. 2012). It has also been demonstrated that the expression of IL-25 and IL-33 is reduced in the intestinal epithelial cells in epithelial-specific Act1-deficient mice and that adoptive transfer of Lin$^-$ c-Kit$^+$ cells or combined injection of IL-25 and

IL-33 restores the type 2 immune responses in epithelial cell-specific Act1-deficient mice (Kang et al. 2012). These results may suggest that Act1 expressed in epithelial cells initiate type 2 immunity against helminth infection by positive-feedback regulation of IL-25.

10.5.2 IL-25 Functions as a Negative Regulator of Inflammatory Responses

Several studies have shown that IL-25 suppresses Th1 cell- or Th17 cell-mediated inflammatory immune responses. Kleinschek et al. have shown that IL-25-deficient mice are highly susceptible to EAE, a Th17 cell-driven inflammatory disease that serves as a disease model for human multiple sclerosis, and that the accelerated EAE in IL-25-deficient mice is associated with the acceleration of the IL-23–Th17 cell axis (Kleinschek et al. 2007). They have also demonstrated that IL-25 treatment induces the production of IL-13, which is crucial for the suppression of Th17 responses through the inhibition of IL-23, IL-1β and IL-6 expression in activated dendritic cells (Kleinschek et al. 2007). IL-25 treatment also attenuates another type of EAE that is induced by the transfer of auto-reactive $CD4^+$ T cells (Sonobe et al. 2009). Importantly, Sonobe et al. have shown that IL-25 is produced by endothelial cells in the lumbar spinal cord in normal mice, and that IL-25 production at this site is reduced after the development of EAE (Sonobe et al. 2009). Commensal-dependent expression of IL-25 in intestinal epithelial cells is also shown to limit the expansion of Th17 cells by inhibiting IL-23 production from macrophages (Zaph et al. 2008). A study on NOD mice, a model of type 1 diabetes, has revealed that IL-25 treatment during the active phase of autoimmunity reduces the frequency of auto-reactive Th17 cells and inhibits Th17 cell-mediated development of insulitis (Emamaullee et al. 2009).

IL-25 also reduces inflammatory cytokine response driven by Toll-like receptor ligands in human blood monocytes and intestinal $CD14^+$ cells (Caruso et al. 2009a, b). In vivo administration of IL-25 in mice has confirmed the protective effect of IL-25 on monocyte-derived cytokine production in a lipopolysaccharide (LPS)-induced lethal endotoxin shock model (Caruso et al. 2009b). In addition, IL-25 inhibits the production of Th1 cell-related cytokines and prevents experimental murine colitis (Mchenga et al. 2008; Caruso et al. 2009a). The inhibition of colitis by IL-25 is also shown to associate with the induction of alternatively activated macrophages (Rizzo et al. 2012), a subset of macrophages with anti-inflammatory properties. Furthermore, it has been shown that IL-25 is constitutively produced by subepithelial macrophages in human intestine, and that the production is markedly reduced in the inflamed gut in patients with Crohn's disease or ulcerative colitis, the two major forms of inflammatory bowel diseases in humans (Caruso et al. 2009a). Taken together, these findings indicate that IL-25 may be an important counter-regulator of non-Th2 cell-mediated inflammatory processes in various organs.

10.6 Roles of IL-25 in Allergic Responses

10.6.1 Roles of IL-25 in Allergic Airway Inflammation

Many studies have shown that allergic airway inflammation is mainly induced by Th2 cells through the production of Th2 cytokines such as IL-4, IL-5, and IL-13 (Cohn et al. 2004; Nakajima and Takatsu 2007). By using murine asthma models, we and others have provided direct evidence that IL-5-producing $CD4^+$ T cells mediate antigen-induced eosinophil recruitment into the airways in sensitized mice (Nakajima et al. 1992; Foster et al. 1996). In addition, it has been shown that IL-13 is a key cytokine that induces goblet cell hyperplasia and airway hyperreactivity (AHR) (Wills-Karp et al. 1998; Grunig et al. 1998).

Regarding the roles of IL-25 in allergic airway inflammation, we have shown that IL-25 mRNA is expressed in the airways of antigen-sensitized, antigen-inhaled mice (Tamachi et al. 2006). Neutralization of IL-25 by IL-17RB-Fc fusion protein significantly inhibits antigen-induced eosinophil and $CD4^+$ T cell recruitment into the airways (Tamachi et al. 2006). Studies using anti-IL-25 antibody confirmed that the blockade of IL-25 reduces eosinophil recruitment and Th2 cytokine production in murine asthma models, indicating that IL-25 blockade attenuates allergic airway inflammation through the reduction of Th2 responses (Ballantyne et al. 2007; Angkasekwinai et al. 2007). Conversely, transgenic overexpression of IL-25 in epithelial cells leads to the enhancement of antigen-induced eosinophil recruitment and mucus production (Tamachi et al. 2006; Angkasekwinai et al. 2007). Importantly, we found that IL-25-induced airway inflammation completely depends on STAT6 (Tamachi et al. 2006). Sharkhuu et al. have also shown that IL-25-induced AHR and mucous hypersecretion depend on IL-13-STAT6 signaling (Sharkhuu et al. 2006). These studies suggest that IL-25 plays an important role in the enhancement of antigen-induced allergic airway inflammation.

10.6.2 IL-25-Producing Cells at the Site of Allergic Airway Inflammation

As mentioned earlier, accumulating evidence has shown that not only hematopoietic cells such as Th2 cells, mast cells, basophils, eosinophils, and alveolar macrophages but also nonhematopoietic cells such as lung and intestinal epithelial cells, fibroblasts, endothelial cells, and keratinocytes express IL-25 mRNA.

Recent studies have revealed critical immunogenic and immunomodulatory functions of airway epithelial cells. In particular, the importance of a group of epithelial cell-derived cytokines that promotes type 2 immune responses such as IL-25, IL-33, and TSLP has gained attention (Paul and Zhu 2010; Barnes 2011; Lambrecht and Hammad 2012). IL-25, IL-33, and TSLP are produced and released by airway epithelial cells in response to various environmental and microbial stimuli or cellular damage, and induce type 2 immune responses and cause tissue damage and remodeling in the airways, suggesting their pivotal roles in the pathophysiology of asthma (Paul and Zhu 2010; Barnes 2011; Lambrecht and Hammad 2012). Indeed, by using IL-25-deficient mice, it has been shown that airway structural cell-derived IL-25 is responsible for induction of eosinophilic airway inflammation in murine asthma models (Suzukawa et al. 2012). However, at present, the details of IL-25-producing cells and the stimuli for IL-25 production at the site of antigen-induced airway inflammation are not fully understood.

In addition, it has been demonstrated that IL-25 is cleaved at multiple sites by matrix metalloproteinase (MMP)-7 in the asthmatic airways and that the cleaved IL-25 strongly induces Th2 cytokine production as compared to native IL-25 (Goswami et al. 2009). Therefore, it is suggested that the activity of IL-25 is regulated not only by the levels of transcription but also by posttranslational modifications. In other aspects of the regulation of IL-25 production, we have shown that IL-22, one of the Th17 cell-derived cytokines (Akdis et al. 2012), attenuates antigen-induced airway inflammation, possibly by inhibiting IL-25 production by lung epithelial cells (Takahashi et al. 2011).

10.6.3 IL-25-Responding Cells at the Site of Allergic Airway Inflammation

The functional IL-25 receptor complex is constituted of IL-17RB and IL-17RA (Reynolds et al. 2010; Iwakura et al. 2011). Earlier studies using RAG-deficient mice have demonstrated that IL-25 acts on NBNT cells expressing high levels of MHC class II and low levels of CD11c (Fort et al. 2001). In addition, T cells, especially memory Th2 cells (Wang et al. 2007), Th9 cells (Angkasekwinai et al. 2010), a subpopulation of NKT cells (Terashima et al. 2008; Stock et al. 2009), and $CD14^+$ cells (Caruso et al. 2009b) have been shown to respond to IL-25. Moreover, recent studies have shown that innate immune cells including MPP^{type2} cells (Saenz et al. 2010a, b), NH cells (Moro et al. 2010), nuocytes (Neill et al. 2010), Ih2 cells (Price et al. 2010), and T2M cells (Petersen et al. 2012) respond to IL-25 and produce Th2 cytokines.

Regarding IL-25-responding cells at the site of allergic airway inflammation, we have shown that the depletion of CD4-expressing cells by in vivo administration of anti-CD4 antibody attenuates IL-25-mediated enhancement of Th2 cytokine production and airway inflammation upon antigen inhalation (Tamachi et al. 2006), suggesting that cells expressing CD4 are key players of IL-25-mediated airway inflammation. Subsequently, it has been shown that a subpopulation of NKT cells expressing CD4 and IL-17RB ($CD4^+$ IL-$17RB^+$ iNKT cells) is preferentially located in the lung and produces IL-13 upon stimulation with IL-25 (Terashima et al. 2008; Stock et al. 2009). It has also been shown that the transfer of $CD4^+$ IL-$17RB^+$ iNKT cells but not of $CD4^+$ IL-$17RB^-$ iNKT cells into NKT cell-deficient mice reconstitutes IL-25-mediated AHR (Terashima et al. 2008), suggesting that $CD4^+$

IL-17RB⁻ iNKT cells play a nonredundant role in IL-25-mediated AHR. In addition, a recent report has shown that Th9 cells, a new subset of helper T cells producing IL-9 (Veldhoen et al. 2008), express IL-17RB and are able to produce IL-9 in response to IL-25 in murine asthma models (Angkasekwinai et al. 2010). These findings suggest that T cells including Th2 cells, Th9 cells, and NKT cells could be involved in antigen-induced, IL-25-mediated asthmatic responses in mice.

On the other hand, previous studies have demonstrated that intranasal or intratracheal administration of large amounts of recombinant IL-25 induces Th2 cytokine production and eosinophil infiltration into the lungs, even in RAG-deficient mice, which lack both helper T cells and NKT cells (Fort et al. 2001; Pan et al. 2001). Regarding NBNT cell populations, Claudio et al. have identified $CD11c^+$ macrophage-like lung cells as physiological targets of IL-25 in vivo (Claudio et al. 2009). In addition, the involvement of lineage-marker-negative type 2 innate lymphoid cells expressing IL-7Rα, CD25, Sca-1, and T1/ST2 in the production of IL-5 and IL-13 is shown in house dust mite- or ovalbumin-induced asthma models (Wolterink et al. 2012). Moreover, Barlow et al. have shown that allergen challenge induces lung infiltration of nuocytes, which are the major innate source of IL-13, and that nuocytes directly induce AHR even in the absence of IL-13-producing $CD4^+$ T cells (Barlow et al. 2012). These findings indicate the presence of innate immune cells that could produce Th2 cytokines in the asthma lungs. Taken together, we suppose that IL-25 could induce allergic inflammation by two distinct mechanisms. In a situation in which IL-25 is abundant, innate immune cells themselves are sufficient for causing allergic inflammation through the production of IL-5 and IL-13 (Fig. 10.2a). In contrast, in another situation where the amounts of IL-25 are limited, collaboration between innate immune cells and acquired immune cells such as Th2 cells, Th9 cells, and NKT cells is required for inducing IL-25-mediated allergic inflammation (Fig. 10.2b). Because the amounts of endogenously produced IL-25 in the lungs seem to be limited, it is suggested that in a physiological setting, IL-25 requires both innate and acquired immune cells to exert its function. Future studies identifying IL-25-responding cells at the site of allergic airway inflammation could help the understanding of the precise mechanism by which IL-25 induces type 2 immune responses in the lung.

10.6.4 Roles of IL-25 in Tissue Remodeling

Recently, Daley et al. have demonstrated that repeated antigen exposure to the lungs in antigen-sensitized mice induces not only Th2 cell-mediated eosinophilic airway inflammation but also pulmonary arterial remodeling (Daley et al. 2008), suggesting that Th2 cell-mediated immune responses cause arterial remodeling at the site of allergic inflammation. Interestingly, Fort et al. have shown that continuous exposure to IL-25 results in prominent histological changes in the lung, including perivascular eosinophilic infiltration and medial thickening of small- to medium-size pulmonary arteries (Fort et al. 2001). In addition, Angkasekwinai et al. have shown that a substantial number of $CD4^+$ T cells and $CD11C^+$ dendritic cells infiltrates

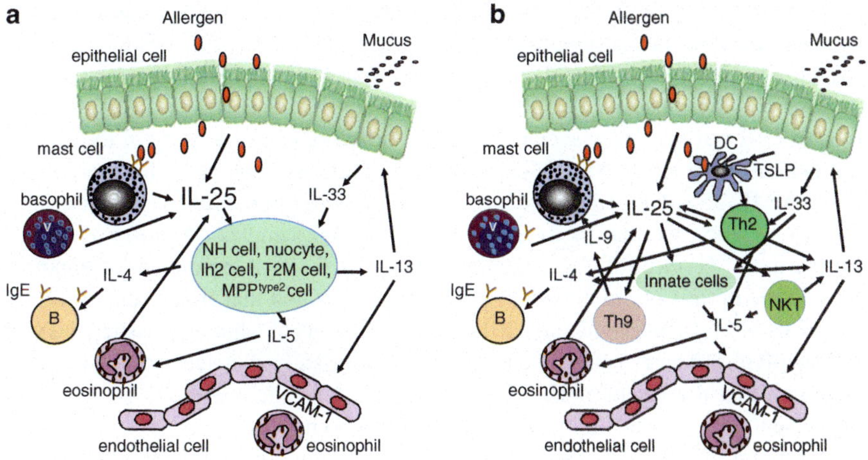

Fig. 10.2 Possible mechanisms for IL-25-mediated enhancement of type 2 immune responses in the airways. Th2 cells, mast cells, basophils, eosinophils, and epithelial cells produce IL-25 upon activation. In a situation in which IL-25 is abundant (**a**), IL-25 directly functions on innate immune cells such as NH cells, nuocytes, Ih2 cells, T2M cells, or MPP^{type2} cells to induce Th2 cytokine production and then causes allergic inflammation. In contrast, if the amounts of IL-25 are limited (**b**), collaboration between innate immune cells and acquired immune cells such as Th2 cells, Th9 cells, and NKT cells is required for inducing IL-25-mediated allergic inflammation. Because the amounts of endogenously produced IL-25 in the lungs seem to be limited, it is suggested that in a pathophysiological setting such as bronchial asthma, IL-25 needs antigen-activated Th2 cells, Th9 cells, or NKT cells to exert its function

around blood vessels in aged lung-specific IL-25 transgenic mice (Angkasekwinai et al. 2007). Moreover, we have recently shown that prolonged expression of IL-25 in the lungs induces the development of pulmonary arterial remodeling in NKT cell-dependent mechanisms (Kawashima et al. 2013). IL-25 has also been shown to contribute to angiogenesis partly by increasing endothelial cell VEGF/VEGF receptor expression through PI3K/Akt and Erk/MAPK pathways in asthma (Corrigan et al. 2011b). Moreover, Gregory et al. have shown that the recruitment of endothelial progenitor cells to the lungs and subsequent neovascularization depends on IL-25 (Gregory et al. 2013). Gregory et al. have also shown that the blocking of IL-25 in allergen-exposed mice abrogates peribronchial collagen deposition and airway smooth muscle hyperplasia (Gregory et al. 2013). Taken together, these results suggest that IL-25 is involved not only in the development of type 2 inflammation in the airways but also in the orchestration of airway remodeling.

10.7 Roles of IL-25 in Human Diseases

Regarding the roles of IL-25 in human inflammatory diseases, Létuvé et al. have reported that IL-25-expressing cells are present in the bronchial submucosa of asthmatic subjects, that some of these cells have an eosinophil-like morphology, and that primary lung fibroblasts express IL-17RB and produce several mediators that could induce tissue eosinophilia, namely, CCL-11, CCL-5, and CXCL-8, in response to IL-25 (Létuvé et al. 2006). In addition, Jung et al. have reported that the intronic polymorphism of the IL-17RB gene is correlated with the expression levels of IL-17RB mRNA and associated with the development of asthma in the Korean population (Jung et al. 2009). Moreover, Wang et al. have demonstrated that mRNA levels for IL-25 and IL-17RB are elevated not only in asthma lungs but also in skin lesions of atopic dermatitis as compared to normal samples (Wang et al. 2007). Furthermore, Hvid et al. have demonstrated that IL-25 is produced by dendritic cells in the dermis of patients with atopic dermatitis and that IL-25 suppresses the synthesis of filaggrin (Hvid et al. 2011), reduced levels of which are associated with an impaired skin barrier in atopic dermatitis. Importantly, Corrigan et al. have demonstrated that the levels of IL-25 are correlated with the magnitudes of the late-phase antigen-induced clinical responses in skin (Corrigan et al. 2011a). Furthermore, IL-25 production by phytohemagglutinin-stimulated blood samples is linked to the levels of allergen-specific IgE antibodies (Herberth et al. 2010). These findings suggest the possible involvement of IL-25 in allergic diseases in humans. The outgrowth of Th2 cells in response to respiratory syncytial virus (RSV) has also been shown to be dependent on airway epithelial cell-derived IL-25 (Kaiko et al. 2010), suggesting that IL-25 expression may participate in virus-induced exacerbation of allergic asthma.

As for IL-25-producing cells in humans, Corrigan et al. have demonstrated that the expression of IL-25 is increased in asthmatic bronchial mucosa and skin dermis of atopic subjects after allergen challenge and that IL-25 expression colocalizes with eosinophils, mast cells, and endothelial cells (Corrigan et al. 2011a). On the other hand, Hvid et al. have demonstrated that DCs is likely to produce IL-25 in atopic dermatitis patients (Hvid et al. 2011). With respect to IL-25-responding cells in humans, it has been demonstrated that the expression of IL-17RB colocalizes with eosinophils, mast cells, endothelial cells, and T lymphocytes (Corrigan et al. 2011a). In addition, Mjösberg et al. have shown that lineage-negative $CD127^+$ $CD161^+$ $CRTH2^+$ innate lymphoid cells produce IL-13 in response to IL-25 (Mjösberg et al. 2011). IL-4- and IL-13-producing T2M-like cells are also increased in peripheral blood in asthmatic patients (Petersen et al. 2012). These results suggest that several cell types including innate immune cells are involved in IL-25-mediated type 2 responses in allergic disorders in humans.

In addition, a recent study has shown that IL-25 levels are increased in sera of patients with active CSS, an inflammatory disease that is characterized by systemic vasculitis and blood and tissue eosinophilia, and that IL-25 levels are correlated with disease activity and blood eosinophil counts (Terrier et al. 2010). The same group has also shown that eosinophils are the main IL-25-producing cells, whereas memory $CD4^+$ T cells are the main IL-25-responding cells in patients with CSS (Terrier et al. 2010). These findings suggest that IL-25 is involved in the pathogenesis of CSS.

10.8 Closing Remarks

Accumulating evidence, including ours, suggests the importance of nonhematopoietic cell-derived cytokines such as IL-25, IL-33, and TSLP in the initiation or enhancement of type 2 immune responses. These cytokines are potent inducers of a panel of recently described innate immune cells, which produce large amounts of IL-5 and IL-13 independent of adaptive immunity. Although further studies are needed to evaluate the roles of these cytokines in allergic diseases in humans, these cytokines could be possible therapeutic targets of allergic diseases.

References

Akdis M, Palomares O, van de Veen W, van Splunter M, Akdis CA (2012) TH17 and TH22 cells: a confusion of antimicrobial response with tissue inflammation versus protection. J Allergy Clin Immunol 129:1438–1449

Angkasekwinai P, Park H, Wang YH, Wang YH, Chang SH, Corry DB, Liu YJ, Zhu Z, Dong C (2007) Interleukin 25 promotes the initiation of proallergic type 2 responses. J Exp Med 204:1509–1517

Angkasekwinai P, Chang SH, Thapa M, Watarai H, Dong C (2010) Regulation of IL-9 expression by IL-25 signaling. Nat Immunol 11:250–256

Ballantyne SJ, Barlow JL, Jolin HE, Nath P, Williams AS, Chung KF, Sturton G, Wong SH, McKenzie AN (2007) Blocking IL-25 prevents airway hyperresponsiveness in allergic asthma. J Allergy Clin Immunol 120:1324–1331

Barlow JL, Bellosi A, Hardman CS, Drynan LF, Wong SH, Cruickshank JP, McKenzie AN (2012) Innate IL-13-producing nuocytes arise during allergic lung inflammation and contribute to airways hyperreactivity. J Allergy Clin Immunol 129:191–198

Barnes PJ (2011) Pathophysiology of allergic inflammation. Immunol Rev 242:31–50

Caruso R, Sarra M, Stolfi C, Rizzo A, Fina D, Fantini MC, Pallone F, MacDonald TT, Monteleone G (2009a) Interleukin-25 inhibits interleukin-12 production and Th1 cell-driven inflammation in the gut. Gastroenterology 136:2270–2279

Caruso R, Stolfi C, Sarra M, Rizzo A, Fantini MC, Pallone F, MacDonald TT, Monteleone G (2009b) Inhibition of monocyte-derived inflammatory cytokines by IL-25 occurs via p38 Map kinase-dependent induction of Socs-3. Blood 113:3512–3519

Cheung PF, Wong CK, Ip WK, Lam CW (2006) IL-25 regulates the expression of adhesion molecules on eosinophils: mechanism of eosinophilia in allergic inflammation. Allergy 61:878–885

Claudio E, Sønder SU, Saret S, Carvalho G, Ramalingam TR, Wynn TA, Chariot A, Garcia-Perganeda A, Leonardi A, Paun A, Chen A, Ren NY, Wang H, Siebenlist U (2009) The adaptor protein CIKS/Act1 is essential for IL-25-mediated allergic airway inflammation. J Immunol 182:1617–1630

Cohn L, Elias JA, Chupp GL (2004) Asthma: mechanisms of disease persistence and progression. Annu Rev Immunol 22:789–815

Corrigan CJ, Wang W, Meng Q, Fang C, Eid G, Caballero MR, Lv Z, An Y, Wang YH, Liu YJ, Kay AB, Lee TH, Ying S (2011a) Allergen-induced expression of IL-25 and IL-25 receptor in atopic asthmatic airways and late-phase cutaneous responses. J Allergy Clin Immunol 128: 116–124

Corrigan CJ, Wang W, Meng Q, Fang C, Wu H, Reay V, Lv Z, Fan Y, An Y, Wang YH, Liu YJ, Lee TH, Ying S (2011b) T-helper cell type 2 (Th2) memory T cell-potentiating cytokine IL-25 has the potential to promote angiogenesis in asthma. Proc Natl Acad Sci USA 108:1579–1584

Daley E, Emson C, Guignabert C, de Waal MR, Louten J, Kurup VP, Hogaboam C, Taraseviciene-Stewart L, Voelkel NF, Rabinovitch M, Grunig E, Grunig G (2008) Pulmonary arterial remodeling induced by a TH2 immune response. J Exp Med 205:361–372

Dardalhon V, Awasthi A, Kwon H, Galileos G, Gao W, Sobel RA, Mitsdoerffer M, Strom TB, Elyaman W, Ho IC, Khoury S, Oukka M, Kuchroo VK (2008) IL-4 inhibits TGF-β-induced Foxp3$^+$ T cells and, together with TGF-β, generates IL-9$^+$ IL-10$^+$ Foxp3$^-$ effector T cells. Nat Immunol 9:1347–1355

Dolgachev V, Petersen BC, Budelsky AL, Berlin AA, Lukacs NW (2009) Pulmonary IL-17E (IL-25) production and IL-17RB$^+$ myeloid cell-derived Th2 cytokine production are dependent upon stem cell factor-induced responses during chronic allergic pulmonary disease. J Immunol 183:5705–5715

Ely LK, Fischer S, Garcia KC (2009) Structural basis of receptor sharing by interleukin 17 cytokines. Nat Immunol 10:1245–1251

Emamaullee JA, Davis J, Merani S, Toso C, Elliott JF, Thiesen A, Shapiro AM (2009) Inhibition of Th17 cells regulates autoimmune diabetes in NOD mice. Diabetes 58:1302–1311

Fallon PG, Ballantyne SJ, Mangan NE, Barlow JL, Dasvarma A, Hewett DR, McIlgorm A, Jolin HE, McKenzie AN (2006) Identification of an interleukin(IL)-25dependent cell population that provides IL-4, IL-5, and IL-13 at the onset of helminth expulsion. J Exp Med 203:1105–1116

Fort MM, Cheung J, Yen D, Li J, Zurawski SM, Lo S, Menon S, Clifford T, Hunte B, Lesley R, Muchamuel T, Hurst SD, Zurawski G, Leach MW, Gorman DM, Rennick DM (2001) IL-25, a novel molecule that induces IL-4, IL-5, and IL-13 and Th2-associated pathologies in vivo. Immunity 15:985–995

Foster PS, Hogan SP, Ramsay AJ, Matthaei KI, Young IG (1996) Interleukin 5 deficiency abolishes eosinophilia, airways hyperreactivity, and lung damage in a mouse asthma model. J Exp Med 183:195–201

Gerhardt S, Abbott WM, Hargreaves D, Pauptit RA, Davies RA, Needham MR, Langham C, Barker W, Aziz A, Snow MJ, Dawson S, Welsh F, Wilkinson T, Vaugan T, Beste G, Bishop S, Popovic B, Rees G, Sleeman M, Tuske SJ, Coales SJ, Hamuro Y, Russell C (2009) Structure of IL-17A in complex with a potent, fully human neutralizing antibody. J Mol Biol 394:905–921

Goswami S, Angkasekwinai P, Shan M, Greenlee KJ, Barranco WT, Polikepahad S, Seryshev A, Song LZ, Redding D, Singh B, Sur S, Woodruff P, Dong C, Corry DB, Kheradmand F (2009) Divergent functions for airway epithelial matrix metalloproteinase 7 and retinoic acid in experimental asthma. Nat Immunol 10:496–503

Gregory LG, Jones CP, Walker SA, Sawant D, Gowers KH, Campbell GA, McKenzie AN, Lloyd CM (2013) IL-25 drives remodelling in allergic airways disease induced by house dust mite. Thorax 68:82–90

Grunig G, Warnock M, Wakil AE, Venkayya R, Brombacher F, Rennick DM, Sheppard D, Mohrs M, Donaldson DD, Locksley RM, Corry DB (1998) Requirement for IL-13 independently of IL-4 in experimental asthma. Science 282:2261–2263

Herberth G, Daegelmann C, Röder S, Behrendt H, Krämer U, Borte M, Heinrich J, Herbarth O, Lehmann I; LISA plus study group (2010) IL-17E but not IL-17A is associated with allergic sensitization: results from the LISA study. Pediatr Allergy Immunol 21:1086–1090

Hurst SD, Muchamuel T, Gorman DM, Gilbert JM, Clifford T, Kwan S, Menon S, Seymour B, Jackson C, Kung TT, Brieland JK, Zurawski SM, Chapman RW, Zurawski G, Coffman RL (2002) New IL-17 family members promote Th1 or Th2 responses in the lung: in vivo function of the novel cytokine IL-25. J Immunol 169:443–453

Hvid M, Vestergaard C, Kemp K, Christensen GB, Deleuran B, Deleuran M (2011) IL-25 in atopic dermatitis: a possible link between inflammation and skin barrier dysfunction? J Invest Dermatol 131:150–157

Hymowitz SG, Filvaroff EH, Yin JP, Lee J, Cai L, Risser P, Maruoka M, Mao W, Foster J, Kelley RF, Pan G, Gurney AL, de Vos AM, Starovasnik MA (2001) IL-17s adopt a cystine knot fold: structure and activity of a novel cytokine, IL-17F, and implications for receptor binding. EMBO J 20:5332–5341

Ikeda K, Nakajima H, Suzuki K, Kagami S-I, Hirose K, Suto A, Saito Y, Iwamoto I (2003) Mast cells produce interleukin-25 upon FcεRI-mediated activation. Blood 101:3594–3596

Iwakura Y, Ishigame H, Saijo S, Nakae S (2011) Functional specialization of interleukin-17 family members. Immunity 34:149–162

Jung JS, Park BL, Cheong HS, Bae JS, Kim JH, Chang HS, Rhim T, Park JS, Jang AS, Lee YM, Kim KU, Uh ST, Na JO, Kim YH, Park CS, Shin HD (2009) Association of IL-17RB gene polymorphism with asthma. Chest 135:1173–1180

Kaiko GE, Phipps S, Angkasekwinai P, Dong C, Foster PS (2010) NK cell deficiency predisposes to viral-induced Th2-type allergic inflammation via epithelial-derived IL-25. J Immunol 185:4681–4690

Kang CM, Jang AS, Ahn MH, Shin JA, Kim JH, Choi YS, Rhim TY, Park CS (2005) Interleukin-25 and interleukin-13 production by alveolar macrophages in response to particles. Am J Respir Cell Mol Biol 33:290–296

Kang Z, Swaidani S, Yin W, Wang C, Barlow JL, Gulen MF, Bulek K, Do JS, Aronica M, McKenzie AN, Min B, Li X (2012) Epithelial cell-specific Act1 adaptor mediates interleukin-25-dependent helminth expulsion through expansion of Lin(−) c-Kit(+) innate cell population. Immunity 36:821–833

Kawashima S, Hirose K, Takahashi T, Tamachi T, Ikeda K, Tokoyoda K, Nakayama T, Nakajima H (2013) IL-25 induces pulmonary arterial remodeling by natural killer T cell-dependent mechanisms. Int Arch Allergy Immunol 161 Suppl 2:118–124

Kim MR, Manoukian R, Yeh R, Silbiger SM, Danilenko DM, Scully S, Sun J, DeRose ML, Stolina M, Chang D, Van GY, Clarkin K, Nguyen HQ, Yu YB, Jing S, Senaldi G, Elliott G, Medlock ES (2002) Transgenic overexpression of human IL-17E results in eosinophilia, B-lymphocyte hyperplasia, and altered antibody production. Blood 100:2330–2340

Kleinschek MA, Owyang AM, Joyce-Shaikh B, Langrish CL, Chen Y, Gorman DM, Blumenschein WM, McClanahan T, Brombacher F, Hurst SD, Kastelein RA, Cua DJ (2007) IL-25 regulates Th17 function in autoimmune inflammation. J Exp Med 204:161–170

Lambrecht BN, Hammad H (2012) The airway epithelium in asthma. Nat Med 18:684–692

Lee J, Ho WH, Maruoka M, Corpuz RT, Baldwin DT, Foster JS, Goddard AD, Yansura DG, Vandlen RL, Wood WI, Gurney AL (2001) IL-17E, a novel proinflammatory ligand for the IL-17 receptor homolog IL-17Rh1. J Biol Chem 276:1660–1664

Létuvé S, Lajoie-Kadoch S, Audusseau S, Rothenberg ME, Fiset PO, Ludwig MS, Hamid Q (2006) IL-17E upregulates the expression of proinflammatory cytokines in lung fibroblasts. J Allergy Clin Immunol 117:590–596

Li H, Chen J, Huang A, Stinson J, Heldens S, Foster J, Dowd P, Gurney AL, Wood WI (2000) Cloning and characterization of IL-17B and IL-17C, two new members of the IL-17 cytokine family. Proc Natl Acad Sci USA 97:773–778

Linden A, Hoshino H, Laan M (2000) Airway neutrophils and interleukin-17. Eur Respir J 15:973–977

Liu YJ (2006) Thymic stromal lymphopoietin: master switch for allergic inflammation. J Exp Med 203:269–273

Maezawa Y, Nakajima H, Suzuki K, Tamachi T, Ikeda K, Inoue J, Saito Y, Iwamoto I (2006) Involvement of TNF receptor-associated factor 6 in IL-25 receptor signaling. J Immunol 176:1013–1018

Mchenga SS, Wang D, Li C, Shan F, Lu C (2008) Inhibitory effect of recombinant IL-25 on the development of dextran sulfate sodium-induced experimental colitis in mice. Cell Mol Immunol 5:425–431

Mjösberg JM, Trifari S, Crellin NK, Peters CP, van Drunen CM, Piet B, Fokkens WJ, Cupedo T, Spits H (2011) Human IL-25- and IL-33-responsive type 2 innate lymphoid cells are defined by expression of CRTH2 and CD161. Nat Immunol 12:1055–1062

Moro K, Yamada T, Tanabe M, Takeuchi T, Ikawa T, Kawamoto H, Furusawa J, Ohtani M, Fujii H, Koyasu S (2010) Innate production of T(H)2 cytokines by adipose tissue-associated c-Kit(+) Sca-1(+) lymphoid cells. Nature (Lond) 463:540–544

Nakajima H, Takatsu K (2007) Role of cytokines in allergic airway inflammation. Int Arch Allergy Immunol 142:265–273

Nakajima H, Iwamoto I, Tomoe S, Matsumura R, Tomioka H, Takatsu K, Yoshida S (1992) CD4+ T-lymphocytes and interleukin-5 mediate antigen-induced eosinophil infiltration into the mouse trachea. Am Rev Respir Dis 146:374–377

Neill DR, Wong SH, Bellosi A, Flynn RJ, Daly M, Langford TK, Bucks C, Kane CM, Fallon PG, Pannell R, Jolin HE, McKenzie AN (2010) Nuocytes represent a new innate effector leukocyte that mediates type-2 immunity. Nature (Lond) 464:1367–1370

Oliphant CJ, Barlow JL, McKenzie AN (2011) Insights into the initiation of type 2 immune responses. Immunology 134:378–385

Owyang AM, Zaph C, Wilson EH, Guild KJ, McClanahan T, Miller HR, Cua DJ, Goldschmidt M, Hunter CA, Kastelein RA, Artis D (2006) Interleukin 25 regulates type 2 cytokine-dependent immunity and limits chronic inflammation in the gastrointestinal tract. J Exp Med 203: 843–849

Pan G, French D, Mao W, Maruoka M, Risser P, Lee J, Foster J, Aggarwal S, Nicholes K, Guillet S, Schow P, Gurney AL (2001) Forced expression of murine IL-17E induces growth retardation, jaundice, a Th2-biased response, and multiorgan inflammation in mice. J Immunol 167:6559–6567

Paul WE, Zhu J (2010) How are T(H)2-type immune responses initiated and amplified? Nat Rev Immunol 10:225–235

Petersen BC, Budelsky AL, Baptist AP, Schaller MA, Lukacs NW (2012) Interleukin-25 induces type 2 cytokine production in a steroid-resistant interleukin-17RB + myeloid population that exacerbates asthmatic pathology. Nat Med 18:751–758

Price AE, Liang HE, Sullivan BM, Reinhardt RL, Eisley CJ, Erle DJ, Locksley RM (2010) Systemically dispersed innate IL-13-expressing cells in type 2 immunity. Proc Natl Acad Sci USA 107:11489–11494

Reynolds JM, Angkasekwinai P, Dong C (2010) IL-17 family member cytokines: regulation and function in innate immunity. Cytokine Growth Factor Rev 21:413–423

Rickel EA, Siegel LA, Yoon BR, Rottman JB, Kugler DG, Swart DA, Anders PM, Tocker JE, Comeau MR, Budelsky AL (2008) Identification of functional roles for both IL-17RB and IL-17RA in mediating IL-25-induced activities. J Immunol 181:4299–4310

Rizzo A, Monteleone I, Fina D, Stolfi C, Caruso R, Fantini MC, Franzè E, Schwendener R, Pallone F, Monteleone G (2012) Inhibition of colitis by IL-25 associates with induction of alternatively activated macrophages. Inflamm Bowel Dis 18:449–459

Saenz SA, Noti M, Artis D (2010a) Innate immune cell populations function as initiators and effectors in Th2 cytokine responses. Trends Immunol 31:407–413

Saenz SA, Siracusa MC, Perrigoue JG, Spencer SP, Urban JF Jr, Tocker JE, Budelsky AL, Kleinschek MA, Kastelein RA, Kambayashi T, Bhandoola A, Artis D (2010b) IL-25 elicits a multipotent progenitor cell population that promotes Th2 cytokine responses. Nature (Lond) 464:1362–1366

Schwarzenberger P, La Russa V, Miller A, Ye P, Huang W, Zieske A, Nelson S, Bagby GJ, Stoltz D, Mynatt RL, Spriggs M, Kolls JK (1998) IL-17 stimulates granulopoiesis in mice: use of an alternate, novel gene therapy-derived method for in vivo evaluation of cytokines. J Immunol 161:6383–6389

Sharkhuu T, Matthaei KI, Forbes E, Mahalingam S, Hogan SP, Hansbro PM, Foster PS (2006) Mechanism of interleukin-25 (IL-17E)-induced pulmonary inflammation and airways hyperreactivity. Clin Exp Allergy 36:1575–1583

Shi Y, Ullrich SJ, Zhang J, Connolly K, Grzegorzewski KJ, Barber MC, Wang W, Wathen K, Hodge V, Fisher CL, Olsen H, Ruben SM, Knyazev I, Cho YH, Kao V, Wilkinson KA, Carrell JA, Ebner R (2000) A novel cytokine receptor-ligand pair: identification, molecular characterization, and in vivo immunomodulatory activity. J Biol Chem 275:19167–19176

Sonobe Y, Takeuchi H, Kataoka K, Li H, Jin S, Mimuro M, Hashizume Y, Sano Y, Kanda T, Mizuno T, Suzumura A (2009) Interleukin-25 expressed by brain capillary endothelial cells maintains blood–brain barrier function in a protein kinase Cε-dependent manner. J Biol Chem 284:31834–31842

Spits H, Cupedo T (2012) Innate lymphoid cells: emerging insights in development, lineage relationships, and function. Annu Rev Immunol 30:647–675

Starnes T, Robertson MJ, Sledge G, Kelich S, Nakshatri H, Broxmeyer HE, Hromas R (2001) IL-17F, a novel cytokine selectively expressed in activated T cells and monocytes, regulates angiogenesis and endothelial cell cytokine production. J Immunol 167:4137–4140

Starnes T, Broxmeyer HE, Robertson MJ, Hromas R (2002) IL-17D, a novel member of the IL-17 family, stimulates cytokine production and inhibits hemopoiesis. J Immunol 169:642–646

Stock P, Lombardi V, Kohlrautz V, Akbari O (2009) Induction of airway hyperreactivity by IL-25 is dependent on a subset of invariant NKT cells expressing IL-17RB. J Immunol 182: 5116–5122

Suzukawa M, Morita H, Nambu A, Arae K, Shimura E, Shibui A, Yamaguchi S, Suzukawa K, Nakanishi W, Oboki K, Kajiwara N, Ohno T, Ishii A, Körner H, Cua DJ, Suto H, Yoshimoto T, Iwakura Y, Yamasoba T, Ohta K, Sudo K, Saito H, Okumura K, Broide DH, Matsumoto K, Nakae S (2012) Epithelial cell-derived IL-25, but not Th17 cell-derived IL-17 or IL-17F, is crucial for murine asthma. J Immunol 189:3641–3652

Swaidani S, Bulek K, Kang Z, Liu C, Lu Y, Yin W, Aronica M, Li X (2009) The critical role of epithelial-derived Act1 in IL-17- and IL-25-mediated pulmonary inflammation. J Immunol 182:1631–1640

Swaidani S, Bulek K, Kang Z, Gulen MF, Liu C, Yin W, Abbadi A, Aronica M, Li X (2011) T cell-derived Act1 is necessary for IL-25-mediated Th2 responses and allergic airway inflammation. J Immunol 187:3155–3164

Takahashi K, Hirose K, Kawashima S, Niwa Y, Wakashin H, Iwata A, Tokoyoda K, Renauld JC, Iwamoto I, Nakayama T, Nakajima H (2011) IL-22 attenuates IL-25 production by lung epithelial cells and inhibits antigen-induced eosinophilic airway inflammation. J Allergy Clin Immunol 128:1067–1076

Tamachi T, Maezawa Y, Ikeda K, Kagami S, Hatano M, Seto Y, Suto A, Suzuki K, Watanabe N, Saito Y, Tokuhisa T, Iwamoto I, Nakajima H (2006) IL-25 enhances allergic airway inflammation by amplifying a TH2 cell-dependent pathway in mice. J Allergy Clin Immunol 118: 606–614

Terashima A, Watarai H, Inoue S, Sekine E, Nakagawa R, Hase K, Iwamura C, Nakajima H, Nakayama T, Taniguchi M (2008) A novel subset of mouse NKT cells bearing the IL-17 receptor B responds to IL-25 and contributes to airway hyperreactivity. J Exp Med 205:2727–2733

Terrier B, Bièche I, Maisonobe T, Laurendeau I, Rosenzwajg M, Kahn JE, Diemert MC, Musset L, Vidaud M, Sène D, Costedoat-Chalumeau N, Le Thi-Huong D, Amoura Z, Klatzmann D, Cacoub P, Saadoun D (2010) Interleukin-25: a cytokine linking eosinophils and adaptive immunity in Churg–Strauss syndrome. Blood 116:4523–4531

Tian E, Sawyer JR, Largaespada DA, Jenkins NA, Copeland NG, Shaughnessy JD Jr (2000) Evi27 encodes a novel membrane protein with homology to the IL17 receptor. Oncogene 19: 2098–2109

Veldhoen M, Uyttenhove C, van Snick J, Helmby H, Westendorf A, Buer J, Martin B, Wilhelm C, Stockinger B (2008) Transforming growth factor-β "reprograms" the differentiation of T helper 2 cells and promotes an interleukin 9-producing subset. Nat Immunol 9:1341–1346

Wang YH, Ito T, Wang YH, Homey B, Watanabe N, Martin R, Barnes CJ, McIntyre BW, Gilliet M, Kumar R, Yao Z, Liu YJ (2006) Maintenance and polarization of human TH2 central memory T cells by thymic stromal lymphopoietin-activated dendritic cells. Immunity 24:827–838

Wang YH, Angkasekwinai P, Lu N, Voo KS, Arima K, Hanabuchi S, Hippe A, Corrigan CJ, Dong C, Homey B, Yao Z, Ying S, Huston DP, Liu YJ (2007) IL-25 augments type 2 immune responses by enhancing the expansion and functions of TSLP-DC-activated Th2 memory cells. J Exp Med 204:1837–1847

Watarai H, Sekine-Kondo E, Shigeura T, Motomura Y, Yasuda T, Satoh R, Yoshida H, Kubo M, Kawamoto H, Koseki H, Taniguchi M (2012) Development and function of invariant natural killer T cells producing T(h)2- and T(h)17-cytokines. PLoS Biol 10:e1001255

Wills-Karp M, Luyimbazi J, Xu X, Schofield B, Neben TY, Karp CL, Donaldson DD (1998) Interleukin-13: central mediator of allergic asthma. Science 282:2258–2261

Wolterink RG, Kleinjan A, van Nimwegen M, Bergen I, de Bruijn M, Levani Y, Hendriks RW (2012) Pulmonary innate lymphoid cells are major producers of IL-5 and IL-13 in murine models of allergic asthma. Eur J Immunol 42:1106–1116

Wong CK, Cheung PF, Ip WK, Lam CW (2005) Interleukin-25-induced chemokines and interleukin-6 release from eosinophils is mediated by p38 mitogen-activated protein kinase, c-Jun N-terminal kinase, and nuclear factor-κB. Am J Respir Cell Mol Biol 33:186–194

Yu HS, Angkasekwinai P, Chang SH, Chung Y, Dong C (2010) Protease allergens induce the expression of IL-25 via Erk and p38 MAPK pathway. J Korean Med Sci 25:829–834

Zaph C, Du Y, Saenz SA, Nair MG, Perrigoue JG, Taylor BC, Troy AE, Kobuley DE, Kastelein RA, Cua DJ, Yu Y, Artis D (2008) Commensal-dependent expression of IL-25 regulates the IL-23-IL-17 axis in the intestine. J Exp Med 205:2191–2198

Zhao A, Urban JF Jr, Sun R, Stiltz J, Morimoto M, Notari L, Madden KB, Yang Z, Grinchuk V, Ramalingam TR, Wynn TA, Shea-Donohue T (2010) Critical role of IL-25 in nematode infection-induced alterations in intestinal function. J Immunol 185:6921–6929

Chapter 11
Interleukin-33: A Multifunctional Alarmin that Promotes Both Health and Disease

Kazufumi Matsushita and Tomohiro Yoshimoto

Abstract Interleukin-33 (IL-33) is a member of the IL-1 family of cytokines. IL-33 is present in the cell nucleus and is released into extracellular spaces, where it acts as an alarmin by binding to the receptor ST2. The receptor ST2 is widely expressed in the human body, and IL-33 exerts a wide range of biological activities depending on the target cells and cellular context. IL-33 is essential for host protection from helminth infections by promoting type 2 immune responses; IL-33 thus has an adverse effect on allergic inflammation. Furthermore, it is becoming clear that in addition to type 2 immunity-mediated diseases, IL-33 is involved in a wide variety of human diseases. IL-33 may be a predisposing factor for some inflammatory diseases, such as rheumatoid arthritis and inflammatory bowel disease, in which type 2 immunity is not involved. By contrast, IL-33 also has cardioprotective and neuroprotective properties in such conditions as ischemic heart disease, atherosclerosis, Alzheimer's disease, and multiple sclerosis. Therefore, thorough understanding of IL-33 is essential for achieving full therapeutic application of this molecule.

Keywords Alarmin • Allergy • IL-33 • Inflammation • ST2

11.1 Introduction

Interleukin-33 (IL-33), also known as IL-1F11, is a member of the IL-1 family of cytokines. The molecule was first described in 1999 as the gene *Dvs27*, which encodes a nuclear protein that is highly upregulated in vasospastic arteries in canines (Onda et al. 1999). In 2003, another group identified the human orthologue of *Dvs27* as a nuclear protein that is preferentially expressed in human high endothelial

K. Matsushita • T. Yoshimoto (✉)
Laboratory of Allergic Diseases, Institute for Advanced Medical Sciences,
Hyogo College of Medicine, 1-1 Mukogawa-cho, Nishinomiya, Hyogo 663-8501, Japan
e-mail: tomo@hyo-med.ac.jp

T. Yoshimoto and T. Yoshimoto (eds.), *Cytokine Frontiers: Regulation of Immune Responses in Health and Disease*, DOI 10.1007/978-4-431-54442-5_11, © Springer Japan 2014

venule (HEV) endothelial cells; the group named it nuclear factor from HEV (NF-HEV) (Baekkevold et al. 2003). In 2005, however, a computational database search of the IL-1R family identified a ligand of the orphan receptor T1/ST2 (ST2), also known as IL-1R-like 1; it became designated IL-33, and this opened research efforts into the role of IL-33 as a cytokine (Schmitz et al. 2005). Later studies revealed that IL-33 is identical to DVS27 and NF-HEV (Carriere et al. 2007; Liew et al. 2010). Because of its constitutive nuclear expression and IL-1-like cytokine domain, IL-33 is now considered a dual-functional protein that acts as both a cytokine and a nuclear factor. The first discovery of the receptor ST2 dates back to 1989 (Klemenz et al. 1989; Tominaga 1989). It was not until 1998, however, that ST2 was first shown to be involved in Th2 immune responses (Lohning et al. 1998). After identification of IL-33/ST2 as a ligand–receptor pair, many efforts have been made to reveal the roles of the IL-33/ST2 pathway in human diseases. Although initial studies suggested the specific involvement of IL-33 and ST2 in Th2 cell-mediated allergic diseases, this cytokine is now recognized as being involved in protection or exacerbation in a wide variety of diseases (Liew et al. 2010; Palmer and Gabay 2011). Here, we discuss IL-33 and ST2 from basic biology to disease-associated functions.

11.2 Biochemical Properties of IL-33

11.2.1 IL-33 Protein and mRNA

In full length, IL-33 is a 30-kDa protein (Fig. 11.1a). Human IL-33 possesses 270 amino acids, and it contains a short chromatin-binding domain at its N terminus (amino acids 40–58); this mediates the nuclear localization of IL-33 by binding to the acidic groove of the histone H2A–H2B complex (Roussel et al. 2008). Human IL-33 also contains a C-terminal IL-1-like cytokine domain (amino acids 112–270), which provides IL-33 with cytokine activity (Schmitz et al. 2005). The C-terminal cytokine domain is sufficient to bind and activate the receptor ST2. In fact, the commercially available recombinant human IL-33 widely used in studies is the 18-kDa IL-$33_{112–270}$ C-terminal portion. One study using multidimensional heteronuclear nuclear magnetic resonance spectroscopy determined the solution structure of the cytokine domain of human IL-33 (Lingel et al. 2009). The structure is characterized by a β-trefoil fold similar to that previously described for IL-1α, IL-1β, IL-1R antagonist, and IL-18. Mouse IL-33 is a 266-amino-acid protein. Comparing human and murine IL-33, the N-terminal part (amino acids 1–65 of human IL-33) and the C-terminal cytokine domain are particularly well conserved between the species, with overall amino acid identity of 55 %; this suggests the functional importance of both the N- and C-terminal domains of this protein.

Splice variants of IL-33 mRNA, which lack single or multiple exons (exons 3, 4, and 5), have been reported in human cells (Tsuda et al. 2012). Among them, a splice

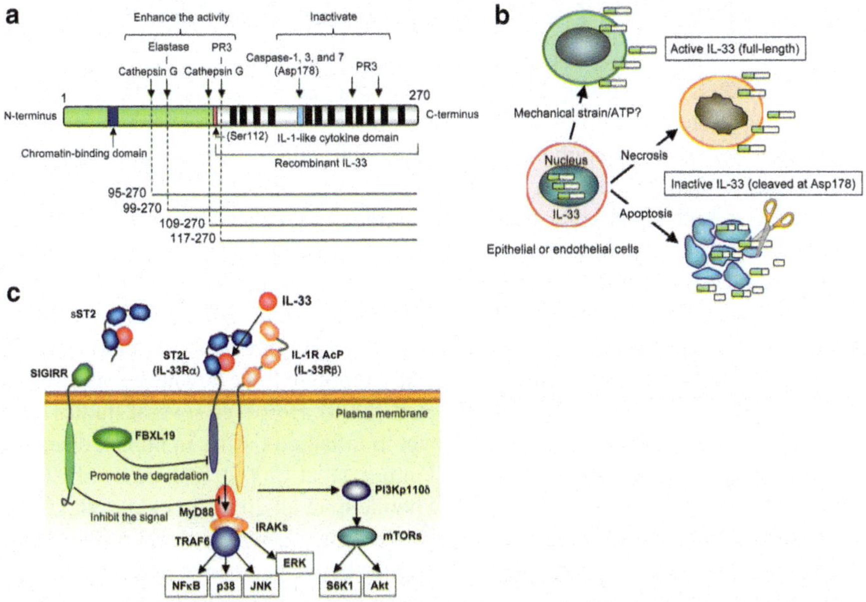

Fig. 11.1 Structure, production, and signaling pathway of interleukin (IL)-33. (**a**) Human IL-33 is a 270-amino-acid, 30-kDa protein and contains an N-terminal chromatin-binding domain and a C-terminal IL-1-like cytokine domain. The IL-1-like cytokine domain mediates the cytokine activity of IL-33. Commercially available recombinant IL-33, widely used, especially in initial studies of IL-33, is the 18-kDa IL-$33_{112\text{–}270}$. IL-33 undergoes intra- and extracellular processing by several proteases. Processing in the N-terminal part enhances the activity, whereas cleavage in the cytokine domain inactivates it. (**b**) Full-length IL-33 present in the cell nucleus is released by necrotic cell death. However, IL-33 undergoes caspase-mediated cleavage to be inactivated during apoptotic cell death. Recent studies have proposed IL-33 release through mechanical strain and/or ATP-mediated, cell death-independent processes. (**c**) Ligation of IL-33 to the receptor ST2L (IL-33Rα) induces the recruitment of co-receptor IL-1RAcP (IL-33Rβ) and initiates the intracellular signaling cascade. The signaling is dependent on the adaptor molecule MyD88. IRAK1/4 and TRAF6 are recruited to the signaling complex and activate NF-κB and MAP kinases. mTOR signaling is activated in a PI3Kp110δ-dependent manner. sST2, SIGIRR, and FBXL19 work as negative regulators for IL-33/ST2 signaling. sST2 acts as a decoy receptor for IL-33. SIGIRR interferes with the signal transduction from ST2L. FBXL19 facilitates the ubiquitin-proteasome-mediated degradation of signaling receptor ST2L

variant that lacks a single exon 3 has been functionally studied. The lack of 126-bp exon 3 results in the deletion of 42 amino acids (Arg74–Gly115), including the initially assumed caspase-1 cleavage site (Ser111–Ser112). In addition to the proposed caspase-1 cleavage site, the exon 3 contains other protease target sites (discussed in Sect. 11.2.2; Fig. 11.1a). Although the exact importance of IL-33 cleavage is still uncertain, it has been shown that this short variant is a "constitutive active" form of IL-33 (Hong et al. 2011). Two mRNAs encoding the same IL-33 coding sequence with different 5′-untranslated regions are transcribed from different promoters in mice. The two promoters are used in a cell type- and stimulation-specific manner for IL-33 expression (Polumuri et al. 2012; Talabot-Ayer et al. 2012).

11.2.2 Expression, Localization, and Secretion of IL-33

IL-33 mRNA is ubiquitously expressed in human and murine cells (Schmitz et al. 2005). The protein is also constitutively expressed in several cell types, and it is especially abundant in epithelial and endothelial cells (Baekkevold et al. 2003; Carriere et al. 2007; Kuchler et al. 2008; Moussion et al. 2008). The level of IL-33 expression in both mRNA and protein is dynamically regulated by cell proliferation, activation, or inflammation. A body of evidence indicates that pro-inflammatory stimuli induce IL-33 mRNA upregulation both in vitro and in vivo (Hudson et al. 2008; Ohno et al. 2009; Palmer et al. 2009; Kobori et al. 2010; Nile et al. 2010; Sponheim et al. 2010). The changes in tissue IL-33 protein levels are more complicated. IL-33 protein level in endothelial cells is reported to be downregulated during angiogenesis and inflammation (Kuchler et al. 2008). However, several studies have identified an upregulated IL-33 protein level in inflamed tissues in human diseases, as well as in murine disease models (Palmer et al. 2009; Kobori et al. 2010; Matsuyama et al. 2010; Nile et al. 2010; Sponheim et al. 2010; Yasuda et al. 2012). This heterogeneity can be explained by the dynamic change in IL-33 protein level during inflammation because it is on the balance between de novo protein synthesis and secretion (Haenuki et al. 2012). Because IL-33 was initially described as a nuclear protein, it is exclusively detected in the cell nucleus with a few exceptions (Onda et al. 1999; Baekkevold et al. 2003; Carriere et al. 2007; Hudson et al. 2008; Kuchler et al. 2008; Moussion et al. 2008; Ohno et al. 2009; Palmer et al. 2009; Kobori et al. 2010; Matsuyama et al. 2010; Nile et al. 2010; Sponheim et al. 2010; Haenuki et al. 2012; Yasuda et al. 2012). Thus, it is reasonable to consider that IL-33 possesses a function or functions as a nuclear factor. However, even though emerging evidence suggests that IL-33 transcriptionally regulates gene expression by activating (Choi et al. 2012) or repressing (Ali et al. 2011) nuclear factor kappa B (NF-κB), or by interacting with chromatins (Roussel et al. 2008), very little is known at present about the role of IL-33 as a nuclear factor.

Although it is clear that IL-33 is an important cytokine in various diseases, the mechanisms of IL-33 secretion remain enigmatic. Because it does not have a secretory leader sequence, IL-33 cannot be secreted through the classic endoplasmic reticulum-Golgi pathway. The initial study of human IL-33 suggested that full-length pro-form IL-33 is cleaved by caspase-1 (Ser111–Ser112) within the cells to secrete 18-kDa mature-form IL-33 (the C-terminal cytokine domain) in similar manner to structurally related family members, IL-1β and IL-18 (Schmitz et al. 2005). However the proposed caspase-1 cleavage site in human IL-33 is not conserved among species, and it is also different from the typical consensus caspase-1 target motif. Subsequent studies have clarified that the proposed site in human IL-33 is not cleaved by caspase-1 under physiological conditions. Rather, caspase-1—and also caspase-3 and -7—cleave IL-33 at Asp178–Gly179, but not at Ser111–Ser112, to inactivate it. Furthermore, full-length IL-33, initially referred to as pro-form IL-33, has been shown to be biologically active (Cayrol and Girard 2009; Luthi et al. 2009; Ohno et al. 2009; Talabot-Ayer et al. 2009).

An important—and the most recognized—mode of IL-33 secretion is necrotic cell death-mediated release (Fig. 11.1b). It has been shown that necrosis-inducing, but not apoptosis-inducing, stimuli cause secretion of full-length IL-33 into the supernatant from in vitro cultured cells. During apoptosis, however, IL-33 undergoes inactivation through caspase-1-, 3-, or -7-mediated cleavage (Cayrol and Girard 2009; Luthi et al. 2009; Ohno et al. 2009). Thus, the secretion of IL-33 is not like a secretory signal sequence containing conventional cytokines or inflammasome-regulated cytokines, IL-1β and IL-18, whose secretion is mediated through intracellular cleavage by caspase-1. IL-33 is more akin to an alarmin or damage-associated molecular patterns (DAMPs), such as IL-1α and high mobility group box 1 (HMGB1), endogenous pro-inflammatory factors released from dying or damaged cells (Bianchi 2007; Moussion et al. 2008; Liew et al. 2010; Palmer and Gabay 2011).

However, necrotic cell death may not be the only way of releasing IL-33. An aeroallergen, the fungus *Alternaria alternata*, has been identified as an inducer of IL-33 release without causing cell death. Airway exposure of mice to *Alternaria* extract induces the rapid secretion of IL-33 into the bronchoalveolar lavage fluid (BALF). *Alternaria* extract stimulation also induces IL-33 secretion from in vitro cultured epithelial cells. This allergen-induced IL-33 secretion is mediated by acute accumulation of extracellular ATP and subsequent elevation of intracellular Ca^{2+} level (Kouzaki et al. 2011). More recently, other allergens have also been identified as in vivo IL-33 inducers. Nasal exposure to ragweed pollen (a major aeroallergen in central Europe and North America) induces IL-33 secretion into nasal lavage fluid (Haenuki et al. 2012). Furthermore, intratracheal administration of chitin (a widespread environmental biopolymer of *N*-acetylglucosamine found in insects, crustaceans, and helminths) induces IL-33 secretion into the BALF (Yasuda et al. 2012). However, the mechanisms involved in pollen- or chitin-mediated IL-33 secretion have yet to be determined. From the perspective of therapeutic application, it would be very intriguing if all allergens used similar ways of inducing IL-33 secretion.

One interesting study has revealed that a biomechanical strain, cyclic biaxial stretching, induced IL-33 secretion from a cultured fibroblast cell line as well as from primary human skin fibroblasts without cellular necrosis (Kakkar et al. 2012). This study also indicated that mechanical stress caused by transaortic construction of the left ventricle induced IL-33 secretion from cardiac cells in mice. These results demonstrate that mechanical strains can induce cellular IL-33 release both in vitro and in vivo. Although the precise mechanisms of IL-33 secretion caused by the strain were not determined in that study, it is widely accepted that mechanical strain can induce ATP release and intracellular Ca^{2+} influx (Cockayne et al. 2000; Yamamoto et al. 2000); these are essential in the aforementioned *Alternaria*-induced IL-33 release. Because ragweed pollen and chitin are particulate matter, they can cause physical stress to the epithelial cells they contact. Furthermore, ragweed pollen possesses protease activities, which are another known ATP inducer (Gunawan et al. 2008; Kreda et al. 2010). Considering these facts, the physical stress and/or ATP-Ca^{2+} response may explain the "alternative" IL-33 secretion pathway induced by allergens or mechanical stimuli (Fig. 11.1b). This active IL-33 release pathway expands the alarmin IL-33 response to threats that do not induce cellular necrosis.

Although stimulation by Toll-like receptors (TLR) induces the upregulation of IL-33 mRNA expression in several cell types, many studies have failed to detect IL-33 protein secretion from cells stimulated with TLR ligands in vitro. Surprisingly, however, one study has shown that a TLR7 ligand, R848, induces IL-33 secretion in in vitro cultured alveolar macrophages, which suggests another pathway for IL-33 secretion (Chang et al. 2011). However, that result does not exclude the possibility of IL-33 release by TLR stimulation being attributable to cell death.

11.2.3 Posttranslational Processing of IL-33

As noted, it was once considered that IL-33 requires caspase-1-mediated cleavage for maturation. It is now widely accepted that IL-33 is biologically active in its full-length form. However, some studies have shown that IL-33 becomes processed and that the biological activity is modified by some proteases, which indicates the presence of "fine-tuning" mechanisms related to IL-33 function in a local milieu (Fig. 11.1a).

Caspase-1, -3, and -7, as already mentioned, cleave IL-33 at Asp178 (Asp175 in the mouse) for inactivation when cells are undergoing apoptotic cell death (Cayrol and Girard 2009; Luthi et al. 2009). The neutrophil serine proteases elastase and cathepsin G, however, have been reported to enhance the biological activity of IL-33 up to about tenfold by extracellular processing. Cathepsin G produces IL-33_{95-270} and IL-$33_{109-270}$, and elastase produces IL-33_{99-270}. All three truncated IL-33s show both in vitro and in vivo pro-inflammatory activity in mice. One study has also shown that, in oleic acid-induced acute lung injury model mice, the cleaved form of IL-33 is detected in the BALF; this suggests the potential roles of extracellular processing of IL-33 in disease exacerbation (Lefrancais et al. 2012). Another neutrophil protease, proteinase 3 (PR3), has also been shown to be an IL-33-processing enzyme. PR3 generates IL-$33_{117-270}$ by in vitro incubation of IL-33. Truncated IL-33 is more active than full-length IL-33. Longer incubation of IL-33 with PR3, however, further digests IL-$33_{117-270}$ at two other sites in the cytokine domain (Val219–Leu220 and Ile240–Gly241) and inactivates it (Bae et al. 2012). Although the cleavage sites and resulting products of IL-33 have been clearly indicated in in vitro studies, the roles of posttranslational processing of IL-33 in physiological and pathological conditions are largely unknown.

11.3 IL-33 Receptors and Biological Function

11.3.1 IL-33 Receptors and Intracellular Signaling

IL-33 directly binds the receptor ST2 (Schmitz et al. 2005). In humans and mice, there are at least two splicing forms of ST2, ST2L and sST2 (Yanagisawa et al. 1992, 1993; Li et al. 2000) (Fig. 11.1c). The "L" in ST2L denotes the "long" form, and the product is the functional activating receptor for IL-33. ST2L is a type 1 transmembrane receptor composed of an ectodomain with three-linked immunoglobulin-like motifs, a transmembrane domain, and a cytoplasmic Toll-IL-1R (TIR) domain. sST2, however, signifies "soluble" ST2, and the protein lacks transmembrane and intracellular domains. sST2 also directly binds to IL-33 as a "decoy" receptor, which prevents the association of IL-33 and ST2L, thereby blocking the signaling (Hayakawa et al. 2007).

Another TIR domain containing a receptor, IL-1R accessory protein (IL-1RAcP), is known to function as a co-receptor for IL-33 signaling. IL-1RAcP is shared by other IL-1 family members, IL-1α and β. After stimulation with IL-33, IL-1RAcP associates with ST2L in a ligand-dependent manner, increasing the binding affinity of the receptor and ligand (Ali et al. 2007; Chackerian et al. 2007; Palmer et al. 2008). Bone marrow-derived mast cells obtained from IL-1RAcP-deficient (IL-1RAcP$^{-/-}$) mice do not respond to IL-33 stimulation, which clearly indicates the importance of this co-receptor in IL-33 signaling (Chackerian et al. 2007). Interestingly, sIL-1RAcP, a short-splice variant of IL-1RAcP, also increases the binding affinity of sST2 to IL-33, which enhances the blocking capacity (Palmer et al. 2008).

Because both ST2L and IL-1RAcP are TIR domain-containing receptors, IL-33-mediated signaling resembles that of IL-1R and IL-18R (Fig. 11.1c). Stimulation of ST2L with IL-33 induces the recruitment of the TIR domain-containing adaptor molecule, the myeloid differentiation primary-response gene 88 (MyD88). MyD88 then recruits IL-1R-associated kinase 1 (IRAK1), IRAK4, and tumor necrosis factor (TNF)-receptor-associated factor 6 (TRAF6) to the signaling complex; this leads to the activation of NF-κB and mitogen-activated protein (MAP) kinases (Schmitz et al. 2005; Ali et al. 2007; Bae et al. 2012). TRAF6 activates NF-κB, MAP kinase p38, and c-Jun N-terminal kinase, whereas the activation of extracellular signal-regulated kinase (ERK) does not require TRAF6 (Funakoshi-Tago et al. 2008). Janus kinase 2 (JAK2) is specifically involved in NF-κB activation in IL-33 signaling. Fibroblasts derived from JAK2$^{-/-}$ mice fail to activate NF-κB in response to IL-33, although the MAP kinase activation is intact (Funakoshi-Tago et al. 2011). IL-33 signaling also involves PI3K activation (Choi et al. 2009; Salmond et al. 2012; Wong et al. 2012). IL-33 stimulation induces rapid activation of mammalian target of rapamycin (mTOR) in a PI3Kp110δ-dependent manner, which in turn activates S6K1 and protein kinase B (PKB)/Akt. In response to IL-33, this mTOR-signaling pathway has been reported as being required for optimal IL-5 and IL-13 production from Th2 cells and recently identified group 2 innate lymphoid cells (ILCs): natural helper cells, nuocytes, and I_H2 cells, hereafter called ILC2s, according to the proposal of Spits et al. (2013). An mTOR inhibitor, rapamycin, has been shown to block IL-33-induced airway inflammation as well as IL-5 and IL-13 production into the BALF in vivo (Salmond et al. 2012).

As IL-1R and IL-18R signaling is negatively regulated by single Ig IL-1-related receptor (SIGIRR) by attenuating the recruitment of IRAKs and TRAF6 to the MyD88-associated signaling complex, SIGIRR also interferes with the activation of NF-κB and MAP kinases in IL-33-mediated signaling (Garlanda et al. 2009).

SIGIRR$^{-/-}$ mice show an enhanced Th2 response induced by intraperitoneal injection of IL-33, which points to the in vivo regulatory function of SIGIRR in IL-33-mediated allergic inflammation (Bulek et al. 2009). After stimulation with cognate cytokines, the cell-surface expression of cytokine receptors is downregulated as a negative feedback loop through receptor endocytosis and subsequent degradation by the ubiquitin-proteasome pathway (Rocca et al. 2001; Martinez-Moczygemba et al. 2007). Recently, the precise negative feedback pathway that regulates ST2L expression has been determined (Zhao et al. 2012). Ligation of IL-33 induces a rapid decrease in cell-surface expression of ST2L. Activated ST2L undergoes glycogen synthase kinase 3β (GSK3B)-mediated phosphorylation at Ser442, which provides the binding site for FBXL19, an Skp-Cullin-F-box family E3 ubiquitin ligase. FBXL19 catalyzes polyubiquitination at Lys326 of ST2L, thereby leading to proteasomal degradation. That study also demonstrated that overexpression of FBXL19 in the lung prevents IL-33-induced epithelial cell death and endotoxin- or *Pseudomonas aeruginosa*-induced acute lung injury. Because FBXL19 targets the cytoplasmic tail of ST2L for polyubiquitination, the protein does not facilitate degradation of sST2, which intriguingly suggests that FBXL19 could be a therapeutic target for IL-33-mediated inflammatory diseases (Zhao et al. 2012).

11.3.2 Crosstalk of IL-33 with Other Signaling Pathways

Recently, many studies have focused not only on single receptor signaling pathways but also on the crosstalk among multiple signaling pathways. Indeed, several signals must be simultaneously activated in the body under physiological conditions. Thus, it is unsurprising that IL-33-mediated signaling is also orchestrated with other receptors. One unique study has shown that IL-33 signaling coordinates with c-Kit signaling in mast cells (Drube et al. 2010). IL-33 stimulation induces the physical assembly of ST2L, IL-1RAcP, and c-Kit in mast cells; phosphorylation of c-Kit is then facilitated together with subsequent activation of the c-Kit-dependent signaling pathway, including STAT3 in addition to the original IL-33 signaling. c-Kit$^{-/-}$ mast cells show decreased IL-33 responses. Conversely, co-stimulation with stem cell factor, a ligand for c-Kit receptor, amplifies IL-33-induced signaling, which results in augmentation of IL-6 production (Drube et al. 2010). In mast cells, phospholipase D and sphingosine kinase are also activated by IL-33 stimulation, which in turn induces Ca^{2+} influx (Pushparaj et al. 2009). Because Ca^{2+} signaling is critical for mast cell degranulation, several studies have shown that IL-33 enhances cross-linkage of FcεRI-induced mast cell degranulation (Pushparaj et al. 2009; Xu et al. 2010; Haenuki et al. 2012). Furthermore, IgE sensitization of mast cells increases ST2L expression, which potentiates IL-33-induced cytokine, chemokine, and eicosanoid production by cells (Ho et al. 2007; Pushparaj et al. 2009). These unique functions of IL-33 in mast cells seem to highlight the essential role of IL-33 in mast cell-related pathology, such as allergy, helminthic infection, and arthritis, as described in Sects. 11.4.1.1, 11.4.1.2, and 11.4.2.3.

TLR and dectin-1 have also been reported to orchestrate with IL-33 signaling. IL-33 priming of neutrophils potentiates the response to *Candida albicans*, a fungus that activates mammalian cells via TLRs and dectin-1. IL-33 induces TLRs- or dectin-1-mediated upregulation of complement receptor CR3, for which amplification of p38 MAP kinase activation has been suggested as being responsible (Le et al. 2012). In addition, IL-33 reverses the TLR-mediated, but not dectin-1-mediated, reduction of CXCR2 expression by regulating the expression of G-protein-coupled receptor kinase 2 (GRK2), which downregulates CXCR2 cell-surface expression. Thus, IL-33 amplifies neutrophil bactericidal activity and is critical for bacterial clearance in vivo (described in detail in Sect. 11.4.2.3) (Alves-Filho et al. 2010; Le et al. 2012). Studies into the crosstalk of IL-33 signaling with other receptors have just recently started. This signaling crosstalk is very complex and should be strongly affected by the cellular context. Future studies will unveil the full picture of the comprehensive signaling network involving IL-33.

11.3.3 IL-33 Target Cells and Responses

ST2L is expressed on a variety of cell types, epithelia, and endothelia in various tissues, fibroblasts, T cells, B-1 cells, basophils, mast cells, eosinophils, neutrophils, natural killer cells, invariant natural killer T cells, monocytes, macrophages, dendritic cells, and ILC2s (Ohno et al. 2012) (Fig. 11.2). Therefore, these cells are responsible for the cytokine activity of IL-33. As ST2L was initially shown to be abundant on Th2, but not Th1 cells (Lohning et al. 1998), IL-33 is a potent inducer of type 2 cytokines. IL-33 treatment of Th2 cells augments their IL-5 and IL-13 production (Schmitz et al. 2005). Similarly, the culture of naive T cells with IL-33 induces IL-5- and IL-13-, but not IL-4-, producing T cells (Kurowska-Stolarska et al. 2008). IL-33 also potently activates innate type 2 cytokine producers: basophils (Kondo et al. 2008; Smithgall et al. 2008; Suzukawa et al. 2008), mast cells (Ho et al. 2007; Iikura et al. 2007; Pushparaj et al. 2009), eosinophils (Matsuba-Kitamura et al. 2010), and ILC2s (Moro et al. 2010; Neill et al. 2010; Wilhelm et al. 2011; Salmond et al. 2012). It is striking that IL-33 induces IL-5 and IL-13 production from basophils and mast cells even in the absence of their cross-linkage with FcεRI (Ho et al. 2007; Iikura et al. 2007; Kondo et al. 2008; Smithgall et al. 2008; Suzukawa et al. 2008; Pushparaj et al. 2009). Eosinophils stimulated with IL-5 and/or GM-CSF upregulate ST2L, and they are responsible for IL-33 inducing IL-4 and chemokine production (Matsuba-Kitamura et al. 2010). Furthermore, IL-33 stimulates ILC2s to produce large amounts of IL-5, IL-9, and IL-13 (Wilhelm et al. 2011; Salmond et al. 2012). These type 2 cytokines produced from innate-type cells play critical roles in the development of IgE-independent "innate-type" allergy as well as in exacerbating IgE-dependent classical "acquired-type" allergy (described next in detail) (Yoshimoto 2010). Further, considering type 2 inflammation, IL-33 enhances differentiation of alternatively activated or M2 macrophages, which is characterized by the surface expression of mannose receptor and IL-4Rα and the production of

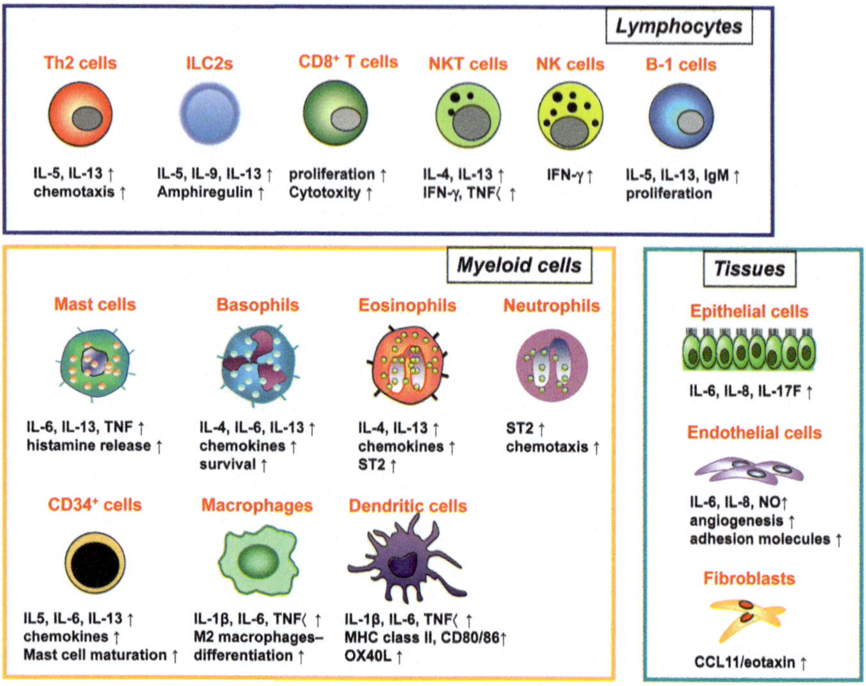

Fig. 11.2 Pleiotropic bioactivity of IL-33. Various lymphocytes and myeloid cells express ST2L and can respond to IL-33 to promote the expression or production of various cytokines and functions. In particular, IL-33 is a potent inducer of type 2 cytokines from allergy-related cells (Th2 cells, basophils, mast cells, eosinophils, and ILC2s). IL-33 also mediates inflammatory responses in IL-33-producing tissues (epithelial cells, endothelial cells, and fibroblasts). This broad array of effector functions of IL-33 on various cells indicates the importance and complex involvement of IL-33 in a wide range of diseases

CCL17 and CCL24. IL-33 directly induces M2 macrophage differentiation from bone marrow cells or alveolar macrophages in concert with IL-13 (Kurowska-Stolarska et al. 2009). In addition, IL-33-activated eosinophils enhance M2 macrophage polarization, which is mediated in part by IL-13 production (Stolarski et al. 2010). IL-33 also induces pro-inflammatory cytokines, such as IL-1β, IL-6, and TNF-α, from mast cells (Pushparaj et al. 2009; Drube et al. 2010; Xu et al. 2010) and basophils (Kondo et al. 2008; Smithgall et al. 2008; Suzukawa et al. 2008), as well as the Th1 cytokine interferon (IFN)-γ from NK cells and NKT cells (Smithgall et al. 2008; Bourgeois et al. 2009). IL-33 has been reported to be a chemoattractant for human Th2 cells (Komai-Koma et al. 2007). Furthermore, IL-33 recruits eosinophils, basophils, and neutrophils to the site of inflammation by facilitating the production of chemokines (Schmitz et al. 2005; Kondo et al. 2008; Kurowska-Stolarska et al. 2009; Haenuki et al. 2012; Yasuda et al. 2012). This broad array of effector functions of IL-33 on various cell types indicates the importance and complex involvement of IL-33 in a wide range of diseases.

Table 11.1 The role of interleukin (IL)-33 in diseases

Diseases exacerbated by IL-33
Asthma
Administration of IL-33 induces experimental asthma (mouse)
IL-33-deficient mice have less severe OVA-specific experimental asthma (mouse)
IL-33 mRNA/protein levels are increased in asthma (mouse/human)
IL-33 gene polymorphism is correlated to incidence of asthma susceptibility (human)
IL-33-ST2 signaling pathway is required in influenza-induced airway hyperactivity (mouse)
Allergic rhinitis
IL-33 protein levels are increased in serum and nasal fluid in Japanese cedar pollinosis (human)
IL-33 gene polymorphism is associated with Japanese cedar pollinosis (human)
IL-33-deficient mice show diminished pollen-specific experimental allergic rhinitis (mouse)
Allergic conjunctivitis
IL-33 protein levels are increased in conjunctival epithelium of allergic conjunctivitis (human)
Exogenous IL-33 augments pollen-driven conjunctival eosinophilic infiltration (mouse)
Atopic dermatitis
IL-33 mRNA/protein levels are increased in lesional skin of atopic dermatitis (human)
ST2 gene polymorphism is associated with atopic dermatitis (human)
Anaphylactic shock
IL-33 protein levels are elevated in serum of atopic patients during anaphylactic shock (human)
IL-33 triggers and enhances IgE-mediated systemic anaphylaxis (mouse)
Arthritis
IL-33 protein levels are increased in serum and synovial fluid in RA (human)
IL-33 mRNA/protein levels are increased in inflamed synovium in RA (human)
IL-33 exacerbates collagen-induced experimental arthritis (mouse)
Blocking ST2 attenuates collagen-induced experimental arthritis (mouse)
Inflammatory bowel disease
IL-33 mRNA/protein levels are increased in active lesions of ulcerative colitis (human)
IL-33/ST2 protein levels are increased in serum in IBD (human)
Anti-TNF treatment of ulcerative colitis decreases serum levels of IL-33 (human)
Diseases protected by IL-33
Ischemic heart disease
IL-33 improves outcome in acute myocardial infarction and heart failure (mouse)
ST2-deficient mice develop more severe experimental heart failure (mouse)
sST2 protein levels are increased in serum in acute myocardial infarction (human)
Atherosclerosis
IL-33 reduces development of atherosclerosis by inducing protective ox-LDL antibodies (mouse)
IL-33 reduces macrophage-derived foam cell formation (mouse)
Cardiac allograft
IL-33 prolongs cardiac allograft survival through induction of $ST2^{+}Foxp3^{+}$Treg (mouse)
Alzheimer's disease
IL-33 gene polymorphism is associated with reduced risk of Alzheimer's disease (human)
IL-33 expression is decreased in brain of patients with Alzheimer's disease (human)
Multiple sclerosis
IL-33 protein levels are increased in periphery and CNS of patients with MS (human)
IL-33 improves outcome with EAE (mouse)
Sepsis
IL-33 reduces mortality in sepsis by enhancing neutrophil function (mouse)
IL-33 prevents TLR4-induced CXCR2 inhibition in neutrophils (mouse/human)
IL-33 and sST2 protein levels are increased in serum in sepsis (human)

(continued)

Table 11.1 (continued)

Parasite infection
IL-33 accelerates expulsion of intestinal nematode *Trichuris muris* (mouse)
ST2-deficient mice have increased susceptibility to infection with *Toxoplasma gondii* (mouse)
IL-33 induces IL-5/IL-13-producing ILC2s to expel *Nippostrongylus brasiliensis* (mouse)
IL-33 is induced in lungs after *Strongyloides venezuelensis* infection (mouse)
IL-33-deficient mice show reduced *S. venezuelensis* expulsion (mouse)
Viral infection
IL-33 is induced in lungs after influenza virus infection (mouse)
IL-33 promotes wound healing of lungs after influenza virus infection (mouse)
IL-33 drives protective cytotoxic CD8^{+} T-cell responses to RNA or DNA viruses (mouse)
Liver disease
IL-33/sST2 protein levels are increased in serum of patients with liver diseases (human)
IL-33 improves outcomes of Con A-induced liver injury (mouse)

RA rheumatoid arthritis, *IBD* inflammatory bowel disease, *CNS* central nervous system, *MS* multiple sclerosis, *EAE* experimental autoimmune encephalomyelitis, *ILC2s* group 2 innate lymphoid cells

11.4 IL-33 in Various Diseases

Accumulating evidence indicates that IL-33 is associated with various diseases in mouse models and humans (Table 11.1). IL-33 acts as an exacerbating factor in various inflammatory diseases. Conversely, there are reports of IL-33 having produced improvement with several diseases. In this section, we describe multiple pathological roles of IL-33 and particularly focus on its double-edged nature for diseases in mouse models and humans.

11.4.1 Diseases Exacerbated by IL-33

11.4.1.1 Allergic Inflammation

Judging from its physiological functions in target cells (Fig. 11.2), IL-33 plays an important role in the pathogenesis of allergic inflammation. It has been speculated that aeroallergen (such as pollen grain or fungus)-induced IL-33 from the nuclei of airway epithelial cells induces the recruitment of allergic cells (Th2 cells, eosinophils, basophils, mast cells, or ILC2s) in the airways and stimulates them to induce cytokines or chemokines, which results in exacerbated allergic inflammation.

Asthma

Intranasal administration of IL-33 into mice induces airway hyperresponsiveness (AHR), goblet cell hyperplasia, and airway eosinophilic infiltration in an ST2/MyD88-dependent fashion by induction of IL-5 and IL-13 in the lungs (Kondo et al.

2008). Administration of IL-33 induces goblet cell hyperplasia in the airways of wild-type mice but not in those of IL-13$^{-/-}$ mice, which clearly indicates that IL-33 produces these responses by induction of endogenous IL-13. Furthermore, similar treatment to RAG2$^{-/-}$ mice, lacking T and B cells, more strikingly induces AHR with marked goblet cell hyperplasia and eosinophilic infiltration in the lungs. Thus, IL-33 induces asthma-like symptoms entirely independently of the acquired immune system. We have designated this type of allergen-independent- and IL-33-dependent asthma "innate-type asthma" (Kondo et al. 2008). In contrast to RAG2$^{-/-}$ mice, $\gamma c^{-/-}$RAG2$^{-/-}$ mice fail to induce innate-type asthma, which suggests that some bone marrow-derived cells are the target cells for IL-33. Indeed, recently identified innate lymphoid cells (ILC2s; NH cells, nuocytes, and I_H2 cells) (Spits et al. 2013), which are negative for lineage markers, express ST2$^+$; these lymphoid cells produce large amounts of IL-5 and IL-13 in response to IL-33 and have been identified as the predominant IL-5-/IL-13-producing population induced by IL-33 in murine lungs (Salmond et al. 2012).

It is possible to develop experimental allergic bronchial asthma by the inhalation of ovalbumin (OVA) in mice sensitized with OVA. Two groups have examined the effect of endogenous IL-33 in OVA-induced experimental asthma using IL-33$^{-/-}$ mice, which showed significantly diminished AHR and eosinophilic infiltration. These findings led to endogenous IL-33 induced by antigen exposure being implicated in the pathogenesis of asthma (Oboki et al. 2010; Louten et al. 2011). Indeed, OVA exposure increases the expression of IL-33 protein in the nuclei of large cells in the alveolar epithelium (Louten et al. 2011). Recently, Yasuda et al. demonstrated that alveolar epithelial type II cells (ATII) express IL-33 protein in their nuclei and that infection with gastrointestinal nematodes or intranasal administration of chitin increases the number of ATII cells and the level of IL-33 (Yasuda et al. 2012). ATII cells are well known as playing a protective role for the alveolus because, in response to lung injury, they synthesize and secrete components of the surfactant that regulate alveolar surface tension (Fehrenbach 2001). Judging from the cell structure (large cells in the alveolar epithelium), OVA-induced IL-33-producing cells could be ATII cells.

In asthma patients, serum levels of sST2, a secreted form of ST2, are higher than in healthy controls. Furthermore, in patients that have suffered asthma attacks, sST2 serum levels are acutely increased in proportion to the severity [decreased percentage of peak expiratory flow rate (% PEF), increased $PaCO_2$], which indicates that the IL-33 axis may be activated during such attacks. However, less is known about the relationship between serum levels of IL-33 and asthma severity. We have analyzed serum levels of IL-33 in 55 patients with asthma (45 atopic, 10 non-atopic) using a human IL-33 enzyme-linked immunosorbent assay. The average amount of IL-33 in healthy control sera is approximately 360 pg/ml compared with approximately 513 pg/ml in asthma patients. In addition, circulating IL-33 is increased in proportion to asthma severity according to the 2002 Global Initiative for Asthma Guideline (unpublished data). Another report has shown that IL-33 expression levels are significantly increased in bronchial biopsies from asthma patients compared with controls, especially in patients with severe asthma (Prefontaine et al. 2009). Furthermore, large-scale genome-wide association studies show that the single nucleotide polymorphism (SNP) in genes for IL-33 (*IL33*) and ST2 (*IL1RL1*) is associated with asthma onset and elevated eosinophil numbers (Gudbjartsson et al. 2009). From these results, IL-33 could be an induction or exacerbation factor for asthma. However, as mentioned in Sect. 11.2.2, the precise mechanisms involved in the allergen-mediated alarmin IL-33 induction from bronchial epithelia cells have yet to be elucidated.

Allergic Rhinitis

Allergic rhinitis (AR) is one of the most common allergic inflammatory diseases; globally, more than 600 million people suffer from it (Bousquet et al. 2007). AR is divided into two categories—seasonal and perennial AR. A previous study of ours showed that the serum level of IL-33 is significantly elevated in Japanese patients with seasonal AR, and that there is a significant association between susceptibility to AR and polymorphisms in the *IL33* gene (Sakashita et al. 2008). However, very little is known about the role of IL-33 in the development of AR. Recently, we established a novel murine model of ragweed pollen-specific AR and examined the pathological role of ragweed-induced IL-33 in the development of AR using IL-33$^{-/-}$ mice (Haenuki et al. 2012). After nasal challenge with ragweed pollen, ragweed-immunized wild-type mice manifest early-phase (sneezing) and late-phase (eosinophilic and basophilic accumulation) responses. In contrast, IL-33$^{-/-}$ and FcεRI$^{-/-}$ mice fail to develop either phase of AR response. IL-33 protein is constitutively expressed in the nucleus of nasal epithelial cells and is promptly released into nasal fluids in response to nasal exposure to ragweed pollen. This ragweed pollen-driven IL-33 is essential for sneezing and the nasal accumulation of eosinophils and basophils by increasing histamine release and inducing the production of chemoattractants from FcεRI^{+} mast cells and basophils, respectively. Together with the contribution of IL-33 in stimulating eosinophils, basophils, and mast cells to produce allergic inflammatory mediators, this process may lead to the recurrent seizures and irreversible mucosal hypertrophy in AR (Fig. 11.3a).

In humans, constitutive expression of IL-33 protein in the nasal epithelial cells of healthy controls has been demonstrated in addition to downregulated expression of IL-33 protein in the inflamed nasal epithelial cells of AR patients. However, IL-33 mRNA expression in the nasal epithelial cells of AR patients is significantly upregulated during the pollen season (Haenuki et al. 2012), which indicates that enhanced extracellular IL-33 release is associated with the reduced IL-33 protein expression in inflamed nasal epithelial cells. Indeed, especially during the pollen season, the IL-33 protein level in nasal secretions is significantly elevated, and it correlates with nasal symptoms in Japanese patients with seasonal AR (Asaka et al. 2012). Although the mechanism of ragweed pollen-driven IL-33 release from nasal epithelial cells remains unclear, taken together these results indicate that alarmin IL-33 may present an important therapeutic target for the prevention of AR.

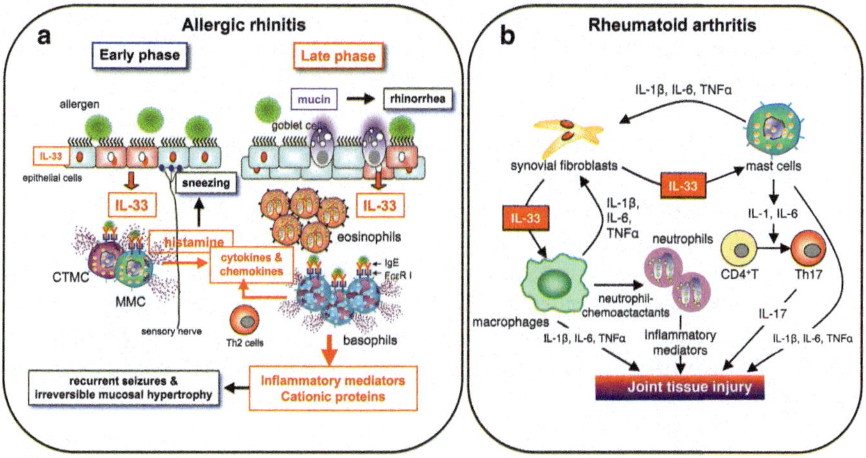

Fig. 11.3 Pathogenic role of IL-33 in allergic rhinitis (AR) (**a**) and rheumatoid arthritis (RA) (**b**). **a** Ragweed pollen-driven endogenous IL-33 from nasal epithelial cells contributes to both early-phase (sneezing) and late-phase (nasal accumulation of eosinophils and basophils) responses in AR by increasing histamine release and inducing the production of chemoattractants from, respectively, mast cells and basophils. Together with the contribution of IL-33 to stimulate Th2 cells, eosinophils, basophils, and mast cells in producing allergic inflammatory mediators, this process may lead to recurrent seizures and irreversible mucosal hypertrophy in AR. **b** IL-33 produced by activated synovial fibroblasts stimulates macrophages and/or mast cells of synovial tissue to produce IL-1**β**, IL-6, and TNF-**α**, which further upregulates IL-33 production from synovial fibroblasts. IL-33 stimulates macrophages to produce neutrophil chemoattractants, leading to the recruitment of neutrophils into the joints. IL-1**β** and IL-6 produced by mast cells and/or macrophages can promote antigen-specific Th17 cell development. All these inflammatory cytokines and mediators may combine to induce RA

Atopic Dermatitis

Genetic links have been made between atopic dermatitis (AD) and polymorphisms in the gene encoding ST2. The SNP in the distal promoter of the *ST2* gene (chromosome 2q12) is significantly associated with AD, and there is a strong correlation with high sST2 and total IgE levels in the sera of AD patients (Shimizu et al. 2005). These results indicate that ST2 genetic polymorphism is a candidate for conferring susceptibility to AD. Less is known about circulating IL-33; however, mRNA levels for IL-33 and ST2 are substantially increased in the lesional skin of patients with AD compared with non-lesional skin and healthy controls. Immunohistochemical staining has revealed that the numbers of IL-33^{+} cells were increased in keratinocytes and in endothelial cells, whereas those of ST2^{+} cells were increased in infiltrating inflammatory cells in the dermis of lesional skin compared with non-lesional skin (Savinko et al. 2012). It is well known that the common allergen house dust mite (HDM) is an external triggering factor in AD, and patients with AD are susceptible to *Staphylococcus aureus* infections. mRNA levels for IL-33 and ST2 are significantly and immediately increased in the skin of patients with AD after HDM or staphylococcal enterotoxin B (SEB) exposure (Savinko et al. 2012). Furthermore, an unexpected function of IL-33 with respect to AD has been reported: in in vitro experiments with human primary keratinocytes, IL-33 downregulated the secretion of human beta defensin 2 (hBD2), which is an inducible antimicrobial peptide produced by epithelial cells and is linked to the increased susceptibility of AD skin to *S. aureus* infection (Alase et al. 2012).

Taken together, these findings suggest that allergen- or SEB-induced alarmin IL-33 from keratinocytes and endothelial cells stimulates infiltrated $ST2^{+}$ cells (e.g., mast cells and eosinophils) or keratinocytes to produce pro-inflammatory cytokines or downregulate hBD2, respectively, resulting in induction and exacerbation of the allergic inflammation in AD skin. Consequently, IL-33 may be a danger molecule in AD. However the precise pathological role of IL-33 in AD demands further investigation.

Other Allergic Diseases

IL-33 activates mast cells and directly induces histamine following IgE sensitization (Haenuki et al. 2012), which suggests that IL-33 plays an important role in anaphylactic shock. Indeed, the serum levels of IL-33 are significantly elevated in atopic patients during anaphylactic shock (Pushparaj et al. 2009). In addition to nasal epithelial cells, IL-33 protein is constitutively expressed in the nucleus of epithelial cells in the conjunctiva, and IL-33 contributes to induction and exacerbation of allergic conjunctivitis in mice and humans (Matsuda et al. 2009; Matsuba-Kitamura et al. 2010).

11.4.1.2 Rheumatoid Arthritis

By means of immunostaining, IL-33 protein has been found to be expressed in the synovial tissue of patients with rheumatoid arthritis (RA) (Xu et al. 2008; Palmer et al. 2009). IL-33 mRNA and protein expression in synovial fibroblasts are strongly increased by stimulation with TNF-α and IL-1β (Xu et al. 2008; Palmer et al. 2009). In cell lysates, predominantly unprocessed 30-kDa full-length IL-33 has been detected using Western blotting (Palmer et al. 2009). In addition, the protein levels of IL-33 in the synovial fluid and serum of patients with RA are significantly higher than in controls, and the increase is proportionate to the severity (Matsuyama et al. 2010). Thus, the contribution of alarmin IL-33 in the pathogenesis of RA has attracted attention.

TNF inhibitors (anti-human TNF-α antibody, infliximab; TNF receptor inhibitor, etanercept) have been established as effective treatments of RA. It has been reported that IL-33 levels in the sera of patients who respond to TNF inhibitors decrease markedly, whereas IL-33 levels remain high in patients that do not show improvement. Interestingly, the levels of IL-33 correlate highly with those of IL-1β in the synovial fluid of RA patients who have not undergone any treatment. In addition, compared with TNF-α, stimulation with IL-1β more strongly induces IL-33 mRNA expression in synovial fibroblasts (Matsuyama et al. 2012). TNF inhibitors are successfully used in RA therapy; however, an inadequate response occurs in 30–40 % of treated patients (Weinblatt et al. 1999; Lipsky et al. 2000). In these low responders, IL-1β may contribute to the pathogenesis of RA by induction of IL-33 from synovial fibroblasts.

From studies using arthritis mouse models, it is apparent that IL-33 contributes to exacerbation of arthritis (Xu et al. 2008; Verri et al. 2010). Compared with normal mice, $ST2^{-/-}$ mice show significantly attenuated clinical symptoms of collagen-induced arthritis (CIA); they also display a diminished degree of collagen-specific pro-inflammatory cytokines (IL-17, IFN-γ, TNF-α) and anti-collagen antibody production (Xu et al. 2008). Administration of blocking antibody against ST2 at the onset of disease decreases the severity of CIA associated with markedly reduced IFN-γ and IL-17 production. This treatment further decreases mRNA expression of receptor activator nuclear factor-κB ligand (RANKL), which is an inducer of osteoclast differentiation and is linked with bone pathology in RA (Palmer et al. 2009). In addition, the administration of IL-33 exacerbates CIA in wild-type mice, but not in $ST2^{-/-}$ mice. The administration of $ST2^{+/+}$ mast cells into $ST2^{-/-}$ mice, however, restores the ability to increase the exacerbated responses to IL-33 seen in wild-type mice; this indicates the contribution of mast cells in IL-33-mediated CIA (Xu et al. 2008).

In another arthritis mouse model, methylated bovine serum albumin (BSA)-induced arthritis, IL-33 mRNA expression and infiltrated $ST2^{+}$-activated neutrophils are increased in intraarticular lesions. For the pathogenesis in this arthritis mouse model, Verri et al. have revealed that antigen-induced IL-33 from synovial fibroblasts stimulates macrophages of the synovial tissue to produce TNF-α, IL-1β, and chemokines; this leads to upregulation of IL-33 production from synovial fibroblasts and the recruitment of neutrophils into the joints (Fig. 11.3b) (Verri et al. 2010). Conversely, though methylated BSA-induced arthritis is completely suppressed in $IL\text{-}1\alpha/\beta^{-/-}$ mice, $IL\text{-}33^{-/-}$ mice develop arthritis similar to control mice (Oboki et al. 2010), which clearly indicates that IL-33 acts as an exacerbating factor, rather than an inducing factor, for arthritis. Nevertheless, from these results of human and murine studies, IL-33 may be a new biomarker for assessing disease activity in patients with RA and be a therapeutic target for RA.

11.4.1.3 Inflammatory Bowel Disease

Because IL-1 family members, such as IL-1β and IL-18, have been recognized as potent inflammatory mediators in the development of inflammatory bowel disease (IBD) (Dinarello 2002), the contribution of IL-33 in IBD has recently been examined.

In 2010, four independent groups reported the association of IL-33 with IBD, elevated expression being specifically observed in patients with active ulcerative colitis (UC) (Beltran et al. 2010; Kobori et al. 2010; Pastorelli et al. 2010; Seidelin et al. 2010). The expression of IL-33 mRNA is significantly elevated in biopsy specimens from inflamed colon lesions in patients with active UC compared with those with Crohn's disease and healthy controls (Pastorelli et al. 2010; Sponheim et al. 2010). In addition, the expression of IL-33 mRNA is increased in proportion to UC severity (Sponheim et al. 2010). IL-33 protein is highly expressed on epithelial cells in the inflamed colon (Pastorelli et al. 2010). Other reports have shown that IL-33 protein is increased primarily in ulceration-associated myofibroblasts in UC lesions (Kobori et al. 2010; Sponheim et al. 2010). ST2 is abundantly expressed in the epithelium in control specimens, whereas this expression is significantly decreased or lost in active UC at both the mRNA and protein level. In contrast, ST2 is upregulated on lamina propria macrophages and $CD4^+$ T cells in active UC (Beltran et al. 2010; Pastorelli et al. 2010). These observations of IL-33 and ST2 being dominantly expressed, respectively, in epithelial cells/myofibroblasts and lamina propria immune cells, may indicate the functional crosstalk between IL-33 and $ST2^+$ cells in active UC.

Regarding the mechanism underlying IL-33 production in UC, it has been reported that isolated human colonic subepithelial myofibroblasts and fibroblasts increase the expression of IL-33 mRNA and IL-33 protein when stimulated with TNF-α and IL-1β, both of which are recognized as important cytokines in the pathogenesis of UC (Sponheim et al. 2010). Furthermore, a TNF-α-stimulated intestinal epithelial cell line was found to increase both IL-33 mRNA and protein (Pastorelli et al. 2010). These results indicate the contribution of inflammatory cytokines, such as TNF-α and IL-1β, to IL-33 production in UC. Indeed, serum levels of IL-33 are elevated in patients with UC compared with healthy controls, whereas treatment with anti-human TNF-α antibody (infliximab) decreases IL-33 in sera (Pastorelli et al. 2010). Thus, IL-33 may play an important role in patients with UC. However, in a murine model of UC, dextran sulfate sodium-induced colitis, $IL\text{-}33^{-/-}$ mice displayed only partially decreased severity of colitis (Oboki et al. 2010). The precise pathophysiological role of alarmin IL-33 in UC therefore requires further investigation.

11.4.2 Diseases Improved by IL-33

11.4.2.1 Cardiovascular Disease

Despite its roles as an exacerbating factor in inflammatory diseases, such as allergy, RA, and IBD, IL-33 has been shown to participate as a protective factor in cardiovascular pathophysiology. Myocardial production of IL-33 can safeguard cardiac function in response to cardiac pressure overload. Furthermore, IL-33 plays a protective role in the development of atherosclerosis.

Ischemic Heart Disease

Serum levels of sST2 are increased in patients 1 day after myocardial infarction (Weinberg et al. 2002), and elevations of sST2 predict mortality in patients with acute myocardial infarction (Shimpo et al. 2004); this indicates ST2 as being a potentially useful biomarker in heart failure. Furthermore, serum levels of IL-33, especially the full-length form, are elevated in patients with advanced systolic chronic heart failure (Zhang et al. 2012). One study has shown that IL-33 antagonizes angiotensin II-induced hypertrophy in cardiomyocytes in vitro (Sanada et al. 2007). Furthermore, following pressure overload by transverse aortic constriction, $ST2^{-/-}$ mice show enhanced cardiac hypertrophy and fibrosis as well as impaired survival compared with control mice. By contrast, administration of IL-33 improves pathological changes and survival in wild-type mice but not in $ST2^{-/-}$ mice. These data clearly demonstrate that IL-33/ST2 signaling protects the myocardium under mechanical overload. As a mechanism for this protective activity, IL-33 reduces cardiomyocyte apoptosis by suppressing caspase-3 activity and increasing anti-apoptotic proteins. Furthermore, in vivo treatment with IL-33 can prevent cardiomyocyte apoptosis and improve cardiac function and survival after myocardial infarction in wild-type mice, but not in $ST2^{-/-}$ mice (Seki et al. 2009).

Serum sST2 may be synthesized and released from cardiac fibroblasts and cardiomyocytes remote from the site of injury or mechanical strain. IL-33 expression in cardiac fibroblasts is upregulated upon mechanical strain (Sanada et al. 2007). Furthermore, as noted before (Sect. 11.2.2), full-length IL-33 can be secreted directly by fibroblasts upon biomechanical strain, and a mechanical stress caused by transaortic construction of the left ventricle induces IL-33 secretion from cardiac cells in mice (Kakkar et al. 2012). Thus, it is possible that increased sST2 reduces IL-33 signaling, thereby resulting in heart failure; by contrast, mechanical strain-induced endogenous alarmin IL-33 from cardiac fibroblasts or exogenous IL-33 could increase cardioprotection and reduce cardiac hypertrophy and fibrosis by preventing cardiomyocyte apoptosis. Thus, IL-33 may have the therapeutic potential to regulate the myocardium under mechanical overload.

Atherosclerosis

Atherosclerosis is a chronic inflammatory disease characterized by specific initial lesions, in which lipids, macrophages, T cells, and smooth muscle cells accumulate in the arterial wall over time. Infiltrated activated immune cells produce cytokines and pro-inflammatory mediators in response to the presence of oxidized low-density lipoprotein (ox-LDL) (Hansson and Libby 2006). Macrophage-derived foam cells form the fatty streaks of the plaques of atherosclerosis. The disease appears to be driven by a Th1 response with an IFN-γ-inducing pathogenesis (McLaren and Ramji 2009). Therefore, it has been hypothesized that IL-33 exerts a protective effect during atherosclerosis by inducing a Th1 to Th2 switch of immune responses.

Mice lacking the gene for apolipoprotein E (ApoE$^{-/-}$ mice) fed a high-fat diet have high serum cholesterol levels and develop atherosclerosis. Treatment of ApoE$^{-/-}$ mice with IL-33 significantly reduces the atherosclerotic lesion size in the aortic sinus and the accumulation of macrophages and CD3$^+$ T cells in the plaque area. Furthermore, IL-33 administration into ApoE$^{-/-}$ mice increases Th2 cytokines and protective ox-LDL antibodies, whereas it decreases Th1 cytokine IFN-γ in the serum and lymph node cells (Miller et al. 2008). In addition, the accumulation of macrophage-derived foam cells in atherosclerotic plaques is significantly reduced in IL-33-treated ApoE$^{-/-}$ mice (McLaren et al. 2010). Conversely, administration of decoy IL-33R, sST2, results in much larger atherosclerotic plaques than in control mice. Interestingly, anti-IL-5 monoclonal antibody treatment prevents the protective effect of IL-33 and reduces the induction of ox-LDL antibodies. One study has shown that IL-5 induces ox-LDL antibodies (Binder et al. 2004). Thus, these results suggest that IL-33 plays a protective role in the development of atherosclerosis by inducing IL-5-stimulated protective ox-LDL antibodies and by reducing foam cell formation.

11.4.2.2 Central Nervous System Diseases

Recently, there has been increasing focus on the production and function of IL-33 in the central nervous system (CNS). Schmitz et al. initially reported that IL-33 mRNA levels are extremely high in the brain and spinal cord (Schmitz et al. 2005). Recent studies have shown that mouse astrocytes, the nonhematopoietic epithelial cell-like cells in the CNS, express IL-33 mRNA and protein, which increase in response to pathogen-associated molecular patterns (PAMPs) (Hudson et al. 2008; Yasuoka et al. 2011). These results indicate that IL-33 may have CNS-specific functions in addition to its role in immune regulation. Indeed, several studies have examined IL-33-associated neurodegenerative diseases, such as Alzheimer's disease and multiple sclerosis.

Alzheimer's Disease

Alzheimer's disease is a neurodegenerative disease typically found in people over the age of 65 years. In 2010, 36 million people worldwide were suffering from Alzheimer's disease and other forms of dementia (World Alzheimer Report 2010; www.alz.co.uk/research/world-report). The definitive cause of Alzheimer's disease is unknown, although genetic susceptibility at multiple genes affects the risk of the disease. The ε4 allele of the apolipoprotein E (APOE) gene is the strongest genetic risk factor for the common late-onset forms of Alzheimer's disease (Corder et al. 1993). However, patients without APOE-ε4 allele also develop Alzheimer's disease, which suggests that additional risk genes for the disease remain to be identified. IL-33 is located on chromosome 9p24, a chromosomal region of interest in Alzheimer's disease that has been defined in genome-wide studies (Li et al. 2008). A recent report has revealed a gene polymorphism that is associated with reduced risk of Alzheimer's disease within the *IL-33* gene region in addition to decreased IL-33 expression in the brains of patients with non-ε4 Alzheimer's disease compared with controls in Caucasian populations (Chapuis et al. 2009).

One of the causes of Alzheimer's disease is considered to be neuronal cell death induced by the deposition of beta-amyloid (Aβ40) in the brain (Hardy and Selkoe 2002). It is notable that the protective genetic variants in the *IL-33* gene are associated with a lesser degree of cerebral amyloid angiopathy, which contains mainly Aβ40 (Chapuis et al. 2009). Furthermore, overexpression of IL-33 in a neuroblastoma cell line transfected with amyloid precursor protein gene led to a significant decrease in the secretion of Aβ40 peptides. In contrast, exogenous IL-33 failed to modify the level of Aβ40 secretion, although this cell line expresses ST2. These results indicate that IL-33 functions as an intracellular nuclear factor with a transcriptional regulator. Thus, in the pathogenesis of Alzheimer's disease, it can be speculated that reduced IL-33 expression in the brain induces increased Aβ40, leading to neuronal cell death. The precise mechanism of reduced IL-33 expression and transcriptional activity by IL-33 in the brain remain unclear. Nonetheless, expectations are high that IL-33 could modulate Aβ40 formation as a new, revolutionary, next-generation treatment of Alzheimer's disease.

Multiple Sclerosis

Multiple sclerosis (MS) is an autoimmune disease of the CNS. MS involves autoreactive T cells that recognize myelin as epitopes, and this leads to demyelination in the brain and spinal cord tissue. IL-33 levels are increased in the peripheral and central nervous system of patients with active relapsing remitting MS. Treatment with IFN-β, a commonly used therapy for relapsing remitting MS, significantly suppresses IL-33 levels in MS patients. Furthermore, compared with normal subjects, IL-33 expression is increased in the astrocyte nuclei of both normal-appearing white matter and plaque tissue from the MS brain (Christophi et al. 2012). These results suggest the contribution of IL-33 to the pathogenesis of MS. However, the requirements for IL-33 release and the exact role of increased IL-33 in MS remain unclear. Recently, there has been increasing evidence that viruses contribute to MS pathogenesis by acting as these environmental triggers. IL-33 can be increased and subsequently released from astrocytes by neurotropic virus infection (Hudson et al. 2008). Thus, one possibility is that virus-induced alarmin IL-33 release in the CNS may be linked to the augmented IL-33 levels in MS.

In MS, the functional roles of IL-33 in the CNS have recently been examined by experimental autoimmune encephalomyelitis (EAE) with a mouse model of MS. Autoreactive Th17 and Th1 cells as well as M1 macrophages are associated with pathological CNS inflammation in EAE. In contrast, Th2 cytokines, IL-27, $CD4^{+}Foxp3^{+}$ T regulatory (Treg) cells (Pierson et al. 2012), and M2 macrophages (Tierney et al. 2009) are protective against EAE. Regarding the contribution of IL-33/ST2 in EAE, It has been reported that $ST2^{-/-}$ mice develop more severe EAE than wild-type (WT) mice (Jiang et al. 2012). Furthermore, IL-33 treatment after onset significantly attenuates EAE in WT mice, but not in $ST2^{-/-}$ mice; it does so by switching from pathogenic Th17- and Th1-type immunity to protective Th2-type immunity and by inducing $CD4^{+}Foxp3^{+}$ Treg cells and M2 macrophages. In normal mice, according to this report, IL-33 and ST2 are respectively expressed as follows: in both astrocytes and neurons, and in neurons in spinal cord tissues. In mice with EAE, the expression of ST2 is markedly increased, whereas the level of IL-33 change is not clear (Jiang et al. 2012). Another report shows that compared with control mice, most astrocytes express IL-33 significantly in EAE (Yasuoka et al. 2011); this is similar to the situation in human MS, as described earlier (Christophi et al. 2012). Nevertheless, these results taken together suggest that the IL-33/ST2 pathway may contribute to the pathogenesis of MS and that increased expression of alarmin IL-33 in the CNS can attenuate the development of MS by inducing protective immunity for MS.

11.4.2.3 Infectious Diseases

Sepsis

Sepsis is a potentially lethal medical condition; it is characterized by a whole-body inflammatory state that is triggered by a bacterial infection. Successful clearance of bacterial infection depends on efficient neutrophil migration to the infectious site. The chemokine receptor CXCR2 plays a central role in the recruitment of neutrophils (Olson and Ley 2002). In sepsis, pathogenic microorganism-derived substances induce the expression of G-protein-coupled receptor kinase 2 (GRK2) in neutrophils through TLRs (Alves-Filho et al. 2009). This GRK2 suppresses the expression of CXCR2 and the migratory capacity for chemokine (CXCL2) of neutrophils, which results in impaired neutrophil migration to the infection site for bacterial clearance and serious infections (Arraes et al. 2006). It has been reported that IL-33 restores the function of neutrophils (CXCR2 expression and migratory capacity for CXCL2) by preventing the induction of GRK2 (Alves-Filho et al. 2010). IL-33 can reduce mortality in mice with experimental sepsis from cecal ligation and puncture by blocking GRK2 expression and reversely maintaining the expression of CXCR2, which allows neutrophils to migrate to the site of infection for bacterial clearance (Alves-Filho et al. 2010) (Fig. 11.4a).

In clinical experiments, the sera of patients with sepsis show significantly increased IL-33 and sST2 levels, whereas healthy controls display no or hardly detectable IL-33 and sST2. Importantly, patients who do not recover from sepsis have significantly higher serum sST2 levels than do survivors (Alves-Filho et al. 2010). Because sST2 is a decoy receptor of IL-33, this result means that the function of IL-33 is suppressed in lethal septic patients. Indeed, neutrophils from lethal septic patients have significantly lower CXCR2 expression and chemotaxis than those from survivors. As described in the posttranslational processing of IL-33 (Sect. 11.2.3), the neutrophil serine proteases elastase and cathepsin G, significantly enhance the biological activity of IL-33 by extracellular processing in a mouse model of sepsis (Lefrancais et al. 2012). Thus, activated alarmin IL-33 by

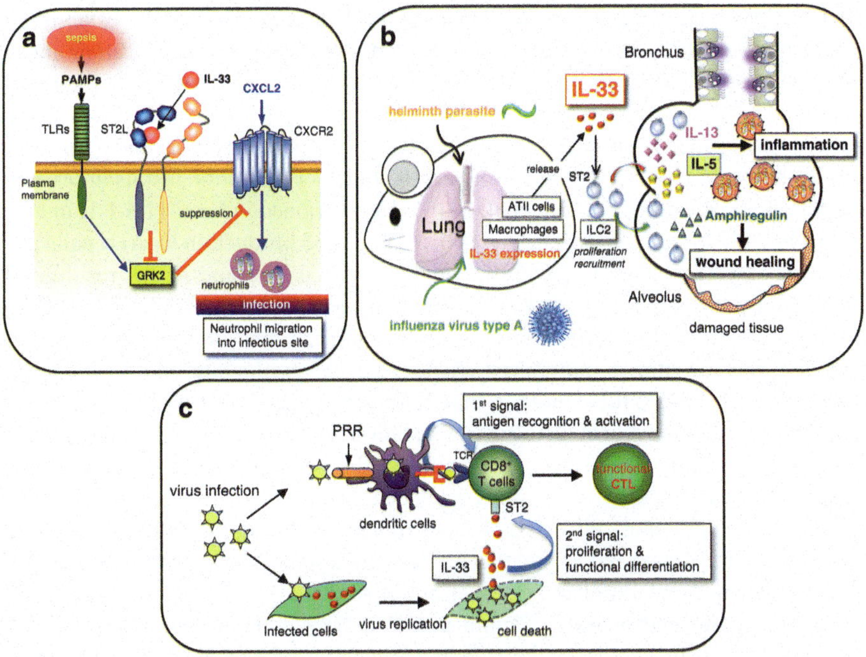

Fig. 11.4 Protective role of IL-33 in infectious diseases. (**a**) In sepsis, pathogen products such as lipopolysaccharide (LPS) induce the expression of GRK2 in neutrophils through TLRs; this suppresses the expression of CXCR2 and the migratory capacity for CXCL2 of neutrophils, and it results in impaired neutrophil recruitment to the site of infection for bacterial clearance and in serious infections. IL-33 can restore the function of neutrophils by preventing the induction of GRK2, which allows neutrophils to migrate to the site of infection for bacterial clearance. (**b**) In mice infected with a helminth parasite, rapid IL-33 production by alveolar epithelial type II (ATII) cells stimulates ILC2s in the lung to produce IL-5 and IL-13, which results in severe lung eosinophilia and worm expulsion. In mice infected with influenza virus, virus-induced IL-33 from the lung increases a number of ILC2s and stimulates ILC2s to produce amphiregulin in the lung, which accelerates tissue recovery from influenza virus-induced pulmonary damage. (**c**) There are two steps for generating protective antiviral $CD8^+$ T lymphocytes (CTLs). First, viruses (PAMPs) act on dendritic cells via pattern recognition receptors (PRR) and thereby are crucial for efficient priming of CTLs. Second, IL-33, an alarmin from virus-infected, damaged cells, has complementary and nonredundant functions, and it acts by generating potent CTL responses

neutrophils restores the function of neutrophils themselves by preventing the induction of GRK2 as a host defense against sepsis. These results reveal the new function of IL-33 and may provide a therapeutic benefit of IL-33 in sepsis.

Infections

Parasite Infection

Protective immunity to helminth parasite infection is critically dependent upon Th2 cytokines orchestrating goblet cell hyperplasia, mucous production, eosinophilia, and IgE production. However, before adaptive Th2 responses can be established, helminth expulsion is ongoing or completed, which suggests protective innate immune responses to helminths. Recently, ILC2s have been revealed to be important for innate protective immunity to the gastrointestinal nematode *Nippostrongylus brasiliensis* (Moro et al. 2010; Nile et al. 2010). ILC2s represent an early source of IL-13 during *N. brasiliensis* infection, and their lack of activation results in severely impaired worm clearance (Nile et al. 2010). As mentioned in Sect. 11.4.1.1, in mice infected with *Strongyloides venezuelensis*, early IL-33 production by ATII cells proliferates and stimulates ILC2s in the lung to produce IL-5 and IL-13, resulting in severe lung eosinophilia and worm expulsion (Yasuda et al. 2012) (Fig. 11.4b). Intranasal administration of chitin (a polysaccharide found on the surface of helminths) induces similar pulmonary changes (IL-33 production and eosinophilia) (Yasuda et al. 2012); this suggests that *S. venezuelensis* infection provokes these alterations principally through the action of chitin. However, the precise mechanism whereby IL-33 is released from ATII cells by the nematode remains unclear. One report has shown that IL-33 release is dependent on prior production of trefoil factor 2 (TFF2), an epithelial cell-derived repair molecule, during migration through the lung by *N. brasiliensis* larvae and that $TFF2^{-/-}$ mice accordingly show impaired Th2 responses and worm expulsion (Wills-Karp et al. 2012). Both ATII cells and TFF2 contribute to wound healing (Taupin and Podolsky 2003). Thus, alarmin IL-33 released from damaged epithelial cells by nematodes plays a host defense role via ILC2s driven by IL-5 and IL-13 production.

Virus Infection

In 2012, two previously unrecognized, very important protective functions of IL-33 against viral infection were revealed (Fig. 11.4b, c). One was IL-33- and ILC2s-mediated lung tissue repair after infection with influenza virus. As the name indicates, pandemic 2009 H1N1 caused a global pandemic virus in 2009: according to the World Health Organization, the deaths of more than 15,000 people were reported. Many influenza-related deaths were not a direct result of the invading virus but were linked to the body's failure to efficiently repair lung tissue damaged by the virus (Bramley et al. 2012). However, the molecular mechanisms involved in the wound healing and lung damage caused by the influenza virus have remained elusive. Monticelli et al. revealed that both ILC2s and IL-33 play an important role in "wound healing" of the lungs caused by influenza virus type A (H1N1) (Monticelli et al. 2011). They demonstrated that influenza virus-induced alarmin IL-33 from the lung increases a number of ILC2s in the lung; they also showed that IL-33-stimulated ILC2s produce amphiregulin, a ligand to the epidermal growth factor receptor involved in tissue remodeling and repair (Enomoto et al. 2009) (Fig. 11.4b). Depletion of ILC2s or blocking IL-33/ST2 signaling impairs lung function and tissue repair in mice infected with influenza virus. In addition, infusion of ILC2s or amphiregulin to the lungs of infected mice normalizes lung function, which suggests that the activation of ILC2s by IL-33 is essential to tissue repair at the lung surface. Indeed, stimulation of lung ILC2s with IL-2 and IL-7 in combination with

IL-33 results in tenfold-greater production of amphiregulin than in cultures stimulated with IL-2 and IL-7 alone. Monticelli et al. also revealed similar ILC2s in healthy human lung tissue. These findings raise the possibility that ILC2s may also coordinate lung tissue repair in humans and that targeting activation of ILC2s by alarmin IL-33 to increase amphiregulin production may accelerate tissue recovery in patients suffering from influenza virus-induced pulmonary damage.

The other previously unrecognized function of IL-33 is that it is required to generate protective cytotoxic $CD8^{+}$ T lymphocyte (CTL) responses to RNA or DNA viruses. In microorganisms, PAMPs are molecules that trigger the adaptive immune responses to infection. However, it remains unclear whether alarmins also help enhance antiviral defense. Bonilla et al. reported that both IL-$33^{-/-}$ and $ST2^{-/-}$ mice have a significantly impaired ability to induce CTLs during infection with various wild-type RNA or DNA viruses, but that these mice had similar levels of CTL induction to wild-type mice in response to replication-deficient virus strains (Bonilla et al. 2012). These authors proposed that viral replication is required for IL-33 production by damaged cells. Indeed, co-injection of IL-33 significantly increases CTL induction in mice vaccinated with attenuated-virus vectors or virus-like particles. In addition, radioresistant, nonhematopoietic cells in the spleen are identified as the main source of IL-33 during virus infections. It has been concluded that there are two steps for generating protective antiviral CTLs. First, PAMPs act primarily on dendritic cells via pattern recognition receptors and thereby are crucial for efficient priming of CTLs. Second, IL-33, an alarmin from virus-infected damaged cells, has complementary and nonredundant functions and acts to generate potent CTLs responses (Fig. 11.4c). These results reveal an important protective role of alarmin IL-33 in virus infection and establish a critical foundation for the development of effective vaccines against infectious diseases and cancer.

11.5 Concluding Remarks

IL-33 is constitutively expressed mainly in epithelial and endothelial cells of barrier tissues, and it is released by inflammation or tissue damage to act as an alarmin; it rapidly activates the cells that express the receptor ST2. Because ST2 is expressed on various types of cells, IL-33 has a wide range of biological activities. As initial studies demonstrated, the IL-33/ST2 pathway plays an essential role in regulating type 2 immune responses; thus, IL-33 is an attractive therapeutic target for allergic diseases. Furthermore, IL-33 is also upregulated in patients with RA and IBD, which suggests the possibility of using IL-33 as a therapeutic target or biomarker in disease diagnosis. However, it is necessary to keep in mind the adverse effect of IL-33 blocking. It is now becoming clear that IL-33 exerts a protective effect on cardiovascular and neurodegenerative diseases as well as in infectious diseases. Exogenous administration or endogenous upregulation of IL-33 may offer new therapeutic possibilities with the diseases for which this molecule can offer some protection.

To accomplish therapeutic application of IL-33, a number of issues need to be addressed. First, the exact manner of IL-33 release in vivo has to be clarified. As yet, there is far from a full understanding of the pathways for IL-33 release, especially in pathological conditions in vivo. Studies suggest that IL-33 is not released by a single mechanism, so it is conceivable that specific suppression of IL-33 release operates under certain disease conditions. Second, the role of IL-33 processing in vivo requires elucidation. Although full-length IL-33 is recognized as biologically active, that bioactivity can be modified by posttranslational processing through several proteases. However, there is a total lack of understanding of the role of IL-33-processing events in human diseases. Furthermore, it is important point to note that most experiments for IL-33 were conducted using 18-kDa truncated IL-33. Full-length and truncated IL-33 have similar activities in vitro and in vivo, but studies have reported their specific bioactivities. It is necessary to exercise caution in interpreting the data obtained from past examinations. Third, the role of IL-33 as a nuclear factor demands investigation. We have mainly discussed here the cytokine activity of IL-33, although the molecule could play another role in the cell nucleus. Little is known about the role of IL-33 as a nuclear factor. The heterogeneity between the phenotypes of IL-$33^{-/-}$ and $ST2^{-/-}$ mice may be partly attributable to a role of IL-33 other than as a cytokine. Fourth, there is a lack of understanding of the mechanisms of IL-33 in regulating human diseases. The exact role of IL-33 in regulating type 2 immune responses is becoming clear. Studies have revealed that IL-33 is more widely involved in human diseases other than Th2-related diseases. However, the exact mechanisms whereby IL-33 regulates such diseases are largely unexplained.

The complex involvement and double-edged role of alarmin IL-33 in human health and disease present challenges to the development of IL-33-targeting therapies. Nevertheless, IL-33 is still an attractive therapeutic target: by neutralizing it in the case of allergic or inflammatory diseases, or by administrating it in the case of cardiovascular or neurodegenerative diseases. The resolution of unanswered questions may offer opportunities for clinical interventions.

Acknowledgments This work was supported by a Strategic Program Grant for Research Institute Development in Private Institute from the Ministry of Education, Culture, Sports, Science and Technology (MEXT) in Japan.

References

Alase A, Seltmann J, Werfel T, Wittmann M (2012) Interleukin-33 modulates the expression of human beta-defensin 2 in human primary keratinocytes and may influence the susceptibility to bacterial superinfection in acute atopic dermatitis. Br J Dermatol 167:1386–1389

Ali S, Huber M, Kollewe C, Bischoff SC, Falk W, Martin MU (2007) IL-1 receptor accessory protein is essential for IL-33-induced activation of T lymphocytes and mast cells. Proc Natl Acad Sci USA 104:18660–18665

Ali S, Mohs A, Thomas M, Klare J, Ross R, Schmitz ML, Martin MU (2011) The dual function cytokine IL-33 interacts with the transcription factor NF-kappaB to dampen NF-kappaB-stimulated gene transcription. J Immunol 187:1609–1616

Alves-Filho JC, Freitas A, Souto FO, Spiller F, Paula-Neto H, Silva JS, Gazzinelli RT, Teixeira MM, Ferreira SH, Cunha FQ (2009) Regulation of chemokine receptor by Toll-like receptor 2 is critical to neutrophil migration and resistance to polymicrobial sepsis. Proc Natl Acad Sci USA 106:4018–4023

Alves-Filho JC, Sonego F, Souto FO, Freitas A, Verri WA Jr, Auxiliadora-Martins M, Basile-Filho A, McKenzie AN, Xu D, Cunha FQ, Liew FY (2010) Interleukin-33 attenuates sepsis by enhancing neutrophil influx to the site of infection. Nat Med 16:708–712

Arraes SM, Freitas MS, da Silva SV, de Paula Neto HA, Alves-Filho JC, Auxiliadora Martins M, Basile-Filho A, Tavares-Murta BM, Barja-Fidalgo C, Cunha FQ (2006) Impaired neutrophil chemotaxis in sepsis associates with GRK expression and inhibition of actin assembly and tyrosine phosphorylation. Blood 108:2906–2913

Asaka D, Yoshikawa M, Nakayama T, Yoshimura T, Moriyama H, Otori N (2012) Elevated levels of interleukin-33 in the nasal secretions of patients with allergic rhinitis. Int Arch Allergy Immunol 158(suppl 1):47–50

Bae S, Kang T, Hong J, Lee S, Choi J, Jhun H, Kwak A, Hong K, Kim E, Jo S, Kim S (2012) Contradictory functions (activation/termination) of neutrophil proteinase 3 enzyme (PR3) in interleukin-33 biological activity. J Biol Chem 287:8205–8213

Baekkevold ES, Roussigne M, Yamanaka T, Johansen FE, Jahnsen FL, Amalric F, Brandtzaeg P, Erard M, Haraldsen G, Girard JP (2003) Molecular characterization of NF-HEV, a nuclear factor preferentially expressed in human high endothelial venules. Am J Pathol 163:69–79

Beltran CJ, Nunez LE, Diaz-Jimenez D, Farfan N, Candia E, Heine C, Lopez F, Gonzalez MJ, Quera R, Hermoso MA (2010) Characterization of the novel ST2/IL-33 system in patients with inflammatory bowel disease. Inflamm Bowel Dis 16:1097–1107

Bianchi ME (2007) DAMPs, PAMPs and alarmins: all we need to know about danger. J Leukoc Biol 81:1–5

Binder CJ, Hartvigsen K, Chang MK, Miller M, Broide D, Palinski W, Curtiss LK, Corr M, Witztum JL (2004) IL-5 links adaptive and natural immunity specific for epitopes of oxidized LDL and protects from atherosclerosis. J Clin Invest 114:427–437

Bonilla WV, Frohlich A, Senn K, Kallert S, Fernandez M, Johnson S, Kreutzfeldt M, Hegazy AN, Schrick C, Fallon PG, Klemenz R, Nakae S, Adler H, Merkler D, Lohning M, Pinschewer DD (2012) The alarmin interleukin-33 drives protective antiviral CD8(+) T cell responses. Science 335:984–989

Bourgeois E, Van LP, Samson M, Diem S, Barra A, Roga S, Gombert JM, Schneider E, Dy M, Gourdy P, Girard JP, Herbelin A (2009) The pro-Th2 cytokine IL-33 directly interacts with invariant NKT and NK cells to induce IFN-gamma production. Eur J Immunol 39:1046–1055

Bousquet J, Dahl R, Khaltaev N (2007) Global alliance against chronic respiratory diseases. Allergy 62:216–223

Bramley AM, Dasgupta S, Skarbinski J, Kamimoto L, Fry AM, Finelli L, Jain S (2012) Intensive care unit patients with 2009 pandemic influenza A (H1N1pdm09) virus infection: United States, 2009. Influenza Other Respir Viruses 6:e134–e142

Bulek K, Swaidani S, Qin J, Lu Y, Gulen MF, Herjan T, Min B, Kastelein RA, Aronica M, Kosz-Vnenchak M, Li X (2009) The essential role of single Ig IL-1 receptor-related molecule/Toll IL-1R8 in regulation of Th2 immune response. J Immunol 182:2601–2609

Carriere V, Roussel L, Ortega N, Lacorre DA, Americh L, Aguilar L, Bouche G, Girard JP (2007) IL-33, the IL-1-like cytokine ligand for ST2 receptor, is a chromatin-associated nuclear factor in vivo. Proc Natl Acad Sci USA 104:282–287

Cayrol C, Girard JP (2009) The IL-1-like cytokine IL-33 is inactivated after maturation by caspase-1. Proc Natl Acad Sci USA 106:9021–9026

Chackerian AA, Oldham ER, Murphy EE, Schmitz J, Pflanz S, Kastelein RA (2007) IL-1 receptor accessory protein and ST2 comprise the IL-33 receptor complex. J Immunol 179:2551–2555

Chang YJ, Kim HY, Albacker LA, Baumgarth N, McKenzie AN, Smith DE, Dekruyff RH, Umetsu DT (2011) Innate lymphoid cells mediate influenza-induced airway hyper-reactivity independently of adaptive immunity. Nat Immunol 12:631–638

Chapuis J, Hot D, Hansmannel F, Kerdraon O, Ferreira S, Hubans C, Maurage CA, Huot L, Bensemain F, Laumet G, Ayral AM, Fievet N, Hauw JJ, DeKosky ST, Lemoine Y, Iwatsubo T, Wavrant-Devrieze F, Dartigues JF, Tzourio C, Buee L, Pasquier F, Berr C, Mann D, Lendon C, Alperovitch A, Kamboh MI, Amouyel P, Lambert JC (2009) Transcriptomic and genetic studies identify IL-33 as a candidate gene for Alzheimer's disease. Mol Psychiatry 14:1004–1016

Choi YS, Choi HJ, Min JK, Pyun BJ, Maeng YS, Park H, Kim J, Kim YM, Kwon YG (2009) Interleukin-33 induces angiogenesis and vascular permeability through ST2/TRAF6-mediated endothelial nitric oxide production. Blood 114:3117–3126

Choi YS, Park JA, Kim J, Rho SS, Park H, Kim YM, Kwon YG (2012) Nuclear IL-33 is a transcriptional regulator of NF-kappaB p65 and induces endothelial cell activation. Biochem Biophys Res Commun 421:305–311

Christophi GP, Gruber RC, Panos M, Christophi RL, Jubelt B, Massa PT (2012) Interleukin-33 upregulation in peripheral leukocytes and CNS of multiple sclerosis patients. Clin Immunol 142:308–319

Cockayne DA, Hamilton SG, Zhu QM, Dunn PM, Zhong Y, Novakovic S, Malmberg AB, Cain G, Berson A, Kassotakis L, Hedley L, Lachnit WG, Burnstock G, McMahon SB, Ford AP (2000) Urinary bladder hyporeflexia and reduced pain-related behaviour in P2X3-deficient mice. Nature (Lond) 407:1011–1015

Corder EH, Saunders AM, Strittmatter WJ, Schmechel DE, Gaskell PC, Small GW, Roses AD, Haines JL, Pericak-Vance MA (1993) Gene dose of apolipoprotein E type 4 allele and the risk of Alzheimer's disease in late onset families. Science 261:921–923

Dinarello CA (2002) The IL-1 family and inflammatory diseases. Clin Exp Rheumatol 20:S1–S13

Drube S, Heink S, Walter S, Lohn T, Grusser M, Gerbaulet A, Berod L, Schons J, Dudeck A, Freitag J, Grotha S, Reich D, Rudeschko O, Norgauer J, Hartmann K, Roers A, Kamradt T (2010) The receptor tyrosine kinase c-Kit controls IL-33 receptor signaling in mast cells. Blood 115:3899–3906

Enomoto Y, Orihara K, Takamasu T, Matsuda A, Gon Y, Saito H, Ra C, Okayama Y (2009) Tissue remodeling induced by hypersecreted epidermal growth factor and amphiregulin in the airway after an acute asthma attack. J Allergy Clin Immunol 124(913–920):e911–e917

Fehrenbach H (2001) Alveolar epithelial type II cell: defender of the alveolus revisited. Respir Res 2:33–46

Funakoshi-Tago M, Tago K, Hayakawa M, Tominaga S, Ohshio T, Sonoda Y, Kasahara T (2008) TRAF6 is a critical signal transducer in IL-33 signaling pathway. Cell Signal 20:1679–1686

Funakoshi-Tago M, Tago K, Sato Y, Tominaga S, Kasahara T (2011) JAK2 is an important signal transducer in IL-33-induced NF-kappaB activation. Cell Signal 23:363–370

Garlanda C, Anders HJ, Mantovani A (2009) TIR8/SIGIRR: an IL-1R/TLR family member with regulatory functions in inflammation and T cell polarization. Trends Immunol 30:439–446

Gudbjartsson DF, Bjornsdottir US, Halapi E, Helgadottir A, Sulem P, Jonsdottir GM, Thorleifsson G, Helgadottir H, Steinthorsdottir V, Stefansson H, Williams C, Hui J, Beilby J, Warrington NM, James A, Palmer LJ, Koppelman GH, Heinzmann A, Krueger M, Boezen HM, Wheatley A, Altmuller J, Shin HD, Uh ST, Cheong HS, Jonsdottir B, Gislason D, Park CS, Rasmussen LM, Porsbjerg C, Hansen JW, Backer V, Werge T, Janson C, Jonsson UB, Ng MC, Chan J, So WY, Ma R, Shah SH, Granger CB, Quyyumi AA, Levey AI, Vaccarino V, Reilly MP, Rader DJ, Williams MJ, van Rij AM, Jones GT, Trabetti E, Malerba G, Pignatti PF, Boner A, Pescollderungg L, Girelli D, Olivieri O, Martinelli N, Ludviksson BR, Ludviksdottir D, Eyjolfsson GI, Arnar D, Thorgeirsson G, Deichmann K, Thompson PJ, Wjst M, Hall IP, Postma DS, Gislason T, Gulcher J, Kong A, Jonsdottir I, Thorsteinsdottir U, Stefansson K (2009) Sequence variants affecting eosinophil numbers associate with asthma and myocardial infarction. Nat Genet 41:342–347

Gunawan H, Takai T, Ikeda S, Okumura K, Ogawa H (2008) Protease activity of allergenic pollen of cedar, cypress, juniper, birch and ragweed. Allergol Int 57:83–91

Haenuki Y, Matsushita K, Futatsugi-Yumikura S, Ishii KJ, Kawagoe T, Imoto Y, Fujieda S, Yasuda M, Hisa Y, Akira S, Nakanishi K, Yoshimoto T (2012) A critical role of IL-33 in experimental allergic rhinitis. J Allergy Clin Immunol 130(184–194):e111

Hansson GK, Libby P (2006) The immune response in atherosclerosis: a double-edged sword. Nat Rev Immunol 6:508–519

Hardy J, Selkoe DJ (2002) The amyloid hypothesis of Alzheimer's disease: progress and problems on the road to therapeutics. Science 297:353–356

Hayakawa H, Hayakawa M, Kume A, Tominaga S (2007) Soluble ST2 blocks interleukin-33 signaling in allergic airway inflammation. J Biol Chem 282:26369–26380

Ho LH, Ohno T, Oboki K, Kajiwara N, Suto H, Iikura M, Okayama Y, Akira S, Saito H, Galli SJ, Nakae S (2007) IL-33 induces IL-13 production by mouse mast cells independently of IgE-Fc epsilon RI signals. J Leukoc Biol 82:1481–1490

Hong J, Bae S, Jhun H, Lee S, Choi J, Kang T, Kwak A, Hong K, Kim E, Jo S, Kim S (2011) Identification of constitutively active interleukin 33 (IL-33) splice variant. J Biol Chem 286:20078–20086

Hudson CA, Christophi GP, Gruber RC, Wilmore JR, Lawrence DA, Massa PT (2008) Induction of IL-33 expression and activity in central nervous system glia. J Leukoc Biol 84:631–643

Iikura M, Suto H, Kajiwara N, Oboki K, Ohno T, Okayama Y, Saito H, Galli SJ, Nakae S (2007) IL-33 can promote survival, adhesion and cytokine production in human mast cells. Lab Invest 87:971–978

Jiang HR, Milovanovic M, Allan D, Niedbala W, Besnard AG, Fukada SY, Alves-Filho JC, Togbe D, Goodyear CS, Linington C, Xu D, Lukic ML, Liew FY (2012) IL-33 attenuates EAE by suppressing IL-17 and IFN-gamma production and inducing alternatively activated macrophages. Eur J Immunol 42:1804–1814

Kakkar R, Hei H, Dobner S, Lee RT (2012) Interleukin 33 as a mechanically responsive cytokine secreted by living cells. J Biol Chem 287:6941–6948

Klemenz R, Hoffmann S, Werenskiold AK (1989) Serum- and oncoprotein-mediated induction of a gene with sequence similarity to the gene encoding carcinoembryonic antigen. Proc Natl Acad Sci USA 86:5708–5712

Kobori A, Yagi Y, Imaeda H, Ban H, Bamba S, Tsujikawa T, Saito Y, Fujiyama Y, Andoh A (2010) Interleukin-33 expression is specifically enhanced in inflamed mucosa of ulcerative colitis. J Gastroenterol 45:999–1007

Komai-Koma M, Xu D, Li Y, McKenzie AN, McInnes IB, Liew FY (2007) IL-33 is a chemoattractant for human Th2 cells. Eur J Immunol 37:2779–2786

Kondo Y, Yoshimoto T, Yasuda K, Futatsugi-Yumikura S, Morimoto M, Hayashi N, Hoshino T, Fujimoto J, Nakanishi K (2008) Administration of IL-33 induces airway hyperresponsiveness and goblet cell hyperplasia in the lungs in the absence of adaptive immune system. Int Immunol 20:791–800

Kouzaki H, Iijima K, Kobayashi T, O'Grady SM, Kita H (2011) The danger signal, extracellular ATP, is a sensor for an airborne allergen and triggers IL-33 release and innate Th2-type responses. J Immunol 186:4375–4387

Kreda SM, Seminario-Vidal L, van Heusden CA, O'Neal W, Jones L, Boucher RC, Lazarowski ER (2010) Receptor-promoted exocytosis of airway epithelial mucin granules containing a spectrum of adenine nucleotides. J Physiol 588:2255–2267

Kuchler AM, Pollheimer J, Balogh J, Sponheim J, Manley L, Sorensen DR, De Angelis PM, Scott H, Haraldsen G (2008) Nuclear interleukin-33 is generally expressed in resting endothelium but rapidly lost upon angiogenic or proinflammatory activation. Am J Pathol 173:1229–1242

Kurowska-Stolarska M, Kewin P, Murphy G, Russo RC, Stolarski B, Garcia CC, Komai-Koma M, Pitman N, Li Y, Niedbala W, McKenzie AN, Teixeira MM, Liew FY, Xu D (2008) IL-33 induces antigen-specific IL-5+ T cells and promotes allergic-induced airway inflammation independent of IL-4. J Immunol 181:4780–4790

Kurowska-Stolarska M, Stolarski B, Kewin P, Murphy G, Corrigan CJ, Ying S, Pitman N, Mirchandani A, Rana B, van Rooijen N, Shepherd M, McSharry C, McInnes IB, Xu D, Liew FY (2009) IL-33 amplifies the polarization of alternatively activated macrophages that contribute to airway inflammation. J Immunol 183:6469–6477

Le HT, Tran VG, Kim W, Kim J, Cho HR, Kwon B (2012) IL-33 priming regulates multiple steps of the neutrophil-mediated anti-*Candida albicans* response by modulating TLR and dectin-1 signals. J Immunol 189:287–295

Lefrancais E, Roga S, Gautier V, Gonzalez-de-Peredo A, Monsarrat B, Girard JP, Cayrol C (2012) IL-33 is processed into mature bioactive forms by neutrophil elastase and cathepsin G. Proc Natl Acad Sci USA 109:1673–1678

Li H, Tago K, Io K, Kuroiwa K, Arai T, Iwahana H, Tominaga S, Yanagisawa K (2000) The cloning and nucleotide sequence of human ST2L cDNA. Genomics 67:284–290

Li H, Wetten S, Li L, St Jean PL, Upmanyu R, Surh L, Hosford D, Barnes MR, Briley JD, Borrie M, Coletta N, Delisle R, Dhalla D, Ehm MG, Feldman HH, Fornazzari L, Gauthier S, Goodgame N, Guzman D, Hammond S, Hollingworth P, Hsiung GY, Johnson J, Kelly DD, Keren R, Kertesz A, King KS, Lovestone S, Loy-English I, Matthews PM, Owen MJ, Plumpton M, Pryse-Phillips W, Prinjha RK, Richardson JC, Saunders A, Slater AJ, St. George-Hyslop PH, Stinnett SW, Swartz JE, Taylor RL, Wherrett J, Williams J, Yarnall DP, Gibson RA, Irizarry MC, Middleton LT, Roses AD (2008) Candidate single-nucleotide polymorphisms from a genomewide association study of Alzheimer disease. Arch Neurol 65:45–53

Liew FY, Pitman NI, McInnes IB (2010) Disease-associated functions of IL-33: the new kid in the IL-1 family. Nat Rev Immunol 10:103–110

Lingel A, Weiss TM, Niebuhr M, Pan B, Appleton BA, Wiesmann C, Bazan JF, Fairbrother WJ (2009) Structure of IL-33 and its interaction with the ST2 and IL-1RAcP receptors: insight into heterotrimeric IL-1 signaling complexes. Structure 17:1398–1410

Lipsky PE, van der Heijde DM, St Clair EW, Furst DE, Breedveld FC, Kalden JR, Smolen JS, Weisman M, Emery P, Feldmann M, Harriman GR, Maini RN (2000) Infliximab and methotrexate in the treatment of rheumatoid arthritis. Anti-tumor necrosis factor trial in rheumatoid arthritis with concomitant therapy study group. N Engl J Med 343:1594–1602

Lohning M, Stroehmann A, Coyle AJ, Grogan JL, Lin S, Gutierrez-Ramos JC, Levinson D, Radbruch A, Kamradt T (1998) T1/ST2 is preferentially expressed on murine Th2 cells, independent of interleukin 4, interleukin 5, and interleukin 10, and important for Th2 effector function. Proc Natl Acad Sci USA 95:6930–6935

Louten J, Rankin AL, Li Y, Murphy EE, Beaumont M, Moon C, Bourne P, McClanahan TK, Pflanz S, de Waal MR (2011) Endogenous IL-33 enhances Th2 cytokine production and T-cell responses during allergic airway inflammation. Int Immunol 23:307–315

Luthi AU, Cullen SP, McNeela EA, Duriez PJ, Afonina IS, Sheridan C, Brumatti G, Taylor RC, Kersse K, Vandenabeele P, Lavelle EC, Martin SJ (2009) Suppression of interleukin-33 bioactivity through proteolysis by apoptotic caspases. Immunity 31:84–98

Martinez-Moczygemba M, Huston DP, Lei JT (2007) JAK kinases control IL-5 receptor ubiquitination, degradation, and internalization. J Leukoc Biol 81:1137–1148

Matsuba-Kitamura S, Yoshimoto T, Yasuda K, Futatsugi-Yumikura S, Taki Y, Muto T, Ikeda T, Mimura O, Nakanishi K (2010) Contribution of IL-33 to induction and augmentation of experimental allergic conjunctivitis. Int Immunol 22:479–489

Matsuda A, Okayama Y, Terai N, Yokoi N, Ebihara N, Tanioka H, Kawasaki S, Inatomi T, Katoh N, Ueda E, Hamuro J, Murakami A, Kinoshita S (2009) The role of interleukin-33 in chronic allergic conjunctivitis. Invest Ophthalmol Vis Sci 50:4646–4652

Matsuyama Y, Okazaki H, Tamemoto H, Kimura H, Kamata Y, Nagatani K, Nagashima T, Hayakawa M, Iwamoto M, Yoshio T, Tominaga S, Minota S (2010) Increased levels of interleukin 33 in sera and synovial fluid from patients with active rheumatoid arthritis. J Rheumatol 37:18–25

Matsuyama Y, Okazaki H, Hoshino M, Onishi S, Kamata Y, Nagatani K, Nagashima T, Iwamoto M, Yoshio T, Ohto-Ozaki H, Tamemoto H, Komine M, Sekiya H, Tominaga S, Minota S (2012) Sustained elevation of interleukin-33 in sera and synovial fluids from patients with rheumatoid arthritis non-responsive to anti-tumor necrosis factor: possible association with persistent IL-1beta signaling and a poor clinical response. Rheumatol Int 32:1397–1401

McLaren JE, Ramji DP (2009) Interferon gamma: a master regulator of atherosclerosis. Cytokine Growth Factor Rev 20:125–135

McLaren JE, Michael DR, Salter RC, Ashlin TG, Calder CJ, Miller AM, Liew FY, Ramji DP (2010) IL-33 reduces macrophage foam cell formation. J Immunol 185:1222–1229

Miller AM, Xu D, Asquith DL, Denby L, Li Y, Sattar N, Baker AH, McInnes IB, Liew FY (2008) IL-33 reduces the development of atherosclerosis. J Exp Med 205:339–346
Monticelli LA, Sonnenberg GF, Abt MC, Alenghat T, Ziegler CG, Doering TA, Angelosanto JM, Laidlaw BJ, Yang CY, Sathaliyawala T, Kubota M, Turner D, Diamond JM, Goldrath AW, Farber DL, Collman RG, Wherry EJ, Artis D (2011) Innate lymphoid cells promote lung-tissue homeostasis after infection with influenza virus. Nat Immunol 12:1045–1054
Moro K, Yamada T, Tanabe M, Takeuchi T, Ikawa T, Kawamoto H, Furusawa J, Ohtani M, Fujii H, Koyasu S (2010) Innate production of T(H)2 cytokines by adipose tissue-associated c-Kit(+) Sca-1(+) lymphoid cells. Nature (Lond) 463:540–544
Moussion C, Ortega N, Girard JP (2008) The IL-1-like cytokine IL-33 is constitutively expressed in the nucleus of endothelial cells and epithelial cells in vivo: a novel "alarmin"? PLoS One 3:e3331
Neill DR, Wong SH, Bellosi A, Flynn RJ, Daly M, Langford TK, Bucks C, Kane CM, Fallon PG, Pannell R, Jolin HE, McKenzie AN (2010) Nuocytes represent a new innate effector leukocyte that mediates type-2 immunity. Nature (Lond) 464:1367–1370
Nile CJ, Barksby E, Jitprasertwong P, Preshaw PM, Taylor JJ (2010) Expression and regulation of interleukin-33 in human monocytes. Immunology 130:172–180
Oboki K, Ohno T, Kajiwara N, Arae K, Morita H, Ishii A, Nambu A, Abe T, Kiyonari H, Matsumoto K, Sudo K, Okumura K, Saito H, Nakae S (2010) IL-33 is a crucial amplifier of innate rather than acquired immunity. Proc Natl Acad Sci USA 107:18581–18586
Ohno T, Oboki K, Kajiwara N, Morii E, Aozasa K, Flavell RA, Okumura K, Saito H, Nakae S (2009) Caspase-1, caspase-8, and calpain are dispensable for IL-33 release by macrophages. J Immunol 183:7890–7897
Ohno T, Morita H, Arae K, Matsumoto K, Nakae S (2012) Interleukin-33 in allergy. Allergy 67:1203–1214
Olson TS, Ley K (2002) Chemokines and chemokine receptors in leukocyte trafficking. Am J Physiol Regul Integr Comp Physiol 283:R7–R28
Onda H, Kasuya H, Takakura K, Hori T, Imaizumi T, Takeuchi T, Inoue I, Takeda J (1999) Identification of genes differentially expressed in canine vasospastic cerebral arteries after subarachnoid hemorrhage. J Cereb Blood Flow Metab 19:1279–1288
Palmer G, Gabay C (2011) Interleukin-33 biology with potential insights into human diseases. Nat Rev Rheumatol 7:321–329
Palmer G, Lipsky BP, Smithgall MD, Meininger D, Siu S, Talabot-Ayer D, Gabay C, Smith DE (2008) The IL-1 receptor accessory protein (AcP) is required for IL-33 signaling and soluble AcP enhances the ability of soluble ST2 to inhibit IL-33. Cytokine 42:358–364
Palmer G, Talabot-Ayer D, Lamacchia C, Toy D, Seemayer CA, Viatte S, Finckh A, Smith DE, Gabay C (2009) Inhibition of interleukin-33 signaling attenuates the severity of experimental arthritis. Arthritis Rheum 60:738–749
Pastorelli L, Garg RR, Hoang SB, Spina L, Mattioli B, Scarpa M, Fiocchi C, Vecchi M, Pizarro TT (2010) Epithelial-derived IL-33 and its receptor ST2 are dysregulated in ulcerative colitis and in experimental Th1/Th2 driven enteritis. Proc Natl Acad Sci USA 107:8017–8022
Pierson E, Simmons SB, Castelli L, Goverman JM (2012) Mechanisms regulating regional localization of inflammation during CNS autoimmunity. Immunol Rev 248:205–215
Polumuri SK, Jayakar GG, Shirey KA, Roberts ZJ, Perkins DJ, Pitha PM, Vogel SN (2012) Transcriptional regulation of murine IL-33 by TLR and non-TLR agonists. J Immunol 189:50–60
Prefontaine D, Lajoie-Kadoch S, Foley S, Audusseau S, Olivenstein R, Halayko AJ, Lemiere C, Martin JG, Hamid Q (2009) Increased expression of IL-33 in severe asthma: evidence of expression by airway smooth muscle cells. J Immunol 183:5094–5103
Pushparaj PN, Tay HK, H'Ng SC, Pitman N, Xu D, McKenzie A, Liew FY, Melendez AJ (2009) The cytokine interleukin-33 mediates anaphylactic shock. Proc Natl Acad Sci USA 106:9773–9778
Rocca A, Lamaze C, Subtil A, Dautry-Varsat A (2001) Involvement of the ubiquitin/proteasome system in sorting of the interleukin 2 receptor beta chain to late endocytic compartments. Mol Biol Cell 12:1293–1301

Roussel L, Erard M, Cayrol C, Girard JP (2008) Molecular mimicry between IL-33 and KSHV for attachment to chromatin through the H2A–H2B acidic pocket. EMBO Rep 9:1006–1012

Sakashita M, Yoshimoto T, Hirota T, Harada M, Okubo K, Osawa Y, Fujieda S, Nakamura Y, Yasuda K, Nakanishi K, Tamari M (2008) Association of serum interleukin-33 level and the interleukin-33 genetic variant with Japanese cedar pollinosis. Clin Exp Allergy 38:1875–1881

Salmond RJ, Mirchandani AS, Besnard AG, Bain CC, Thomson NC, Liew FY (2012) IL-33 induces innate lymphoid cell-mediated airway inflammation by activating mammalian target of rapamycin. J Allergy Clin Immunol 130:1159–1166.e6

Sanada S, Hakuno D, Higgins LJ, Schreiter ER, McKenzie AN, Lee RT (2007) IL-33 and ST2 comprise a critical biomechanically induced and cardioprotective signaling system. J Clin Invest 117:1538–1549

Savinko T, Matikainen S, Saarialho-Kere U, Lehto M, Wang G, Lehtimaki S, Karisola P, Reunala T, Wolff H, Lauerma A, Alenius H (2012) IL-33 and ST2 in atopic dermatitis: expression profiles and modulation by triggering factors. J Invest Dermatol 132:1392–1400

Schmitz J, Owyang A, Oldham E, Song Y, Murphy E, McClanahan TK, Zurawski G, Moshrefi M, Qin J, Li X, Gorman DM, Bazan JF, Kastelein RA (2005) IL-33, an interleukin-1-like cytokine that signals via the IL-1 receptor-related protein ST2 and induces T helper type 2-associated cytokines. Immunity 23:479–490

Seidelin JB, Bjerrum JT, Coskun M, Widjaya B, Vainer B, Nielsen OH (2010) IL-33 is upregulated in colonocytes of ulcerative colitis. Immunol Lett 128:80–85

Seki K, Sanada S, Kudinova AY, Steinhauser ML, Handa V, Gannon J, Lee RT (2009) Interleukin-33 prevents apoptosis and improves survival after experimental myocardial infarction through ST2 signaling. Circ Heart Fail 2:684–691

Shimizu M, Matsuda A, Yanagisawa K, Hirota T, Akahoshi M, Inomata N, Ebe K, Tanaka K, Sugiura H, Nakashima K, Tamari M, Takahashi N, Obara K, Enomoto T, Okayama Y, Gao PS, Huang SK, Tominaga S, Ikezawa Z, Shirakawa T (2005) Functional SNPs in the distal promoter of the ST2 gene are associated with atopic dermatitis. Hum Mol Genet 14:2919–2927

Shimpo M, Morrow DA, Weinberg EO, Sabatine MS, Murphy SA, Antman EM, Lee RT (2004) Serum levels of the interleukin-1 receptor family member ST2 predict mortality and clinical outcome in acute myocardial infarction. Circulation 109:2186–2190

Smithgall MD, Comeau MR, Yoon BR, Kaufman D, Armitage R, Smith DE (2008) IL-33 amplifies both Th1- and Th2-type responses through its activity on human basophils, allergen-reactive Th2 cells, iNKT and NK cells. Int Immunol 20:1019–1030

Spits H, Artis D, Colonna M, Diefenbach A, Di Santo JP, Eberl G, Koyasu S, Locksley RM, McKenzie ANJ, Mebius RE, Powrie F, Vivier E (2013) Innate lymphoid cells: a proposal for uniform nomenclature. Nat Rev Immunol 13:145–149

Sponheim J, Pollheimer J, Olsen T, Balogh J, Hammarstrom C, Loos T, Kasprzycka M, Sorensen DR, Nilsen HR, Kuchler AM, Vatn MH, Haraldsen G (2010) Inflammatory bowel disease-associated interleukin-33 is preferentially expressed in ulceration-associated myofibroblasts. Am J Pathol 177:2804–2815

Stolarski B, Kurowska-Stolarska M, Kewin P, Xu D, Liew FY (2010) IL-33 exacerbates eosinophil-mediated airway inflammation. J Immunol 185:3472–3480

Suzukawa M, Iikura M, Koketsu R, Nagase H, Tamura C, Komiya A, Nakae S, Matsushima K, Ohta K, Yamamoto K, Yamaguchi M (2008) An IL-1 cytokine member, IL-33, induces human basophil activation via its ST2 receptor. J Immunol 181:5981–5989

Talabot-Ayer D, Lamacchia C, Gabay C, Palmer G (2009) Interleukin-33 is biologically active independently of caspase-1 cleavage. J Biol Chem 284:19420–19426

Talabot-Ayer D, Calo N, Vigne S, Lamacchia C, Gabay C, Palmer G (2012) The mouse interleukin (Il)33 gene is expressed in a cell type- and stimulus-dependent manner from two alternative promoters. J Leukoc Biol 91:119–125

Taupin D, Podolsky DK (2003) Trefoil factors: initiators of mucosal healing. Nat Rev Mol Cell Biol 4:721–732

Tierney JB, Kharkrang M, La Flamme AC (2009) Type II-activated macrophages suppress the development of experimental autoimmune encephalomyelitis. Immunol Cell Biol 87:235–240

Tominaga S (1989) A putative protein of a growth specific cDNA from BALB/c-3T3 cells is highly similar to the extracellular portion of mouse interleukin 1 receptor. FEBS Lett 258:301–304
Tsuda H, Komine M, Karakawa M, Etoh T, Tominaga SI, Ohtsuki M (2012) Novel splice variants of IL-33: differential expression in normal and transformed cells. J Invest Dermatol 132: 2661–2664
Verri WA Jr, Souto FO, Vieira SM, Almeida SC, Fukada SY, Xu D, Alves-Filho JC, Cunha TM, Guerrero AT, Mattos-Guimaraes RB, Oliveira FR, Teixeira MM, Silva JS, McInnes IB, Ferreira SH, Louzada-Junior P, Liew FY, Cunha FQ (2010) IL-33 induces neutrophil migration in rheumatoid arthritis and is a target of anti-TNF therapy. Ann Rheum Dis 69:1697–1703
Weinberg EO, Shimpo M, De Keulenaer GW, MacGillivray C, Tominaga S, Solomon SD, Rouleau JL, Lee RT (2002) Expression and regulation of ST2, an interleukin-1 receptor family member, in cardiomyocytes and myocardial infarction. Circulation 106:2961–2966
Weinblatt ME, Kremer JM, Bankhurst AD, Bulpitt KJ, Fleischmann RM, Fox RI, Jackson CG, Lange M, Burge DJ (1999) A trial of etanercept, a recombinant tumor necrosis factor receptor:Fc fusion protein, in patients with rheumatoid arthritis receiving methotrexate. N Engl J Med 340:253–259
Wilhelm C, Hirota K, Stieglitz B, Van Snick J, Tolaini M, Lahl K, Sparwasser T, Helmby H, Stockinger B (2011) An IL-9 fate reporter demonstrates the induction of an innate IL-9 response in lung inflammation. Nat Immunol 12:1071–1077
Wills-Karp M, Rani R, Dienger K, Lewkowich I, Fox JG, Perkins C, Lewis L, Finkelman FD, Smith DE, Bryce PJ, Kurt-Jones EA, Wang TC, Sivaprasad U, Hershey GK, Herbert DR (2012) Trefoil factor 2 rapidly induces interleukin 33 to promote type 2 immunity during allergic asthma and hookworm infection. J Exp Med 209:607–622
Wong CK, Leung KM, Qiu HN, Chow JY, Choi AO, Lam CW (2012) Activation of eosinophils interacting with dermal fibroblasts by pruritogenic cytokine IL-31 and alarmin IL-33: implications in atopic dermatitis. PLoS One 7:e29815
Xu D, Jiang HR, Kewin P, Li Y, Mu R, Fraser AR, Pitman N, Kurowska-Stolarska M, McKenzie AN, McInnes IB, Liew FY (2008) IL-33 exacerbates antigen-induced arthritis by activating mast cells. Proc Natl Acad Sci USA 105:10913–10918
Xu D, Jiang HR, Li Y, Pushparaj PN, Kurowska-Stolarska M, Leung BP, Mu R, Tay HK, McKenzie AN, McInnes IB, Melendez AJ, Liew FY (2010) IL-33 exacerbates autoantibody-induced arthritis. J Immunol 184:2620–2626
Yamamoto K, Korenaga R, Kamiya A, Qi Z, Sokabe M, Ando J (2000) P2X(4) receptors mediate ATP-induced calcium influx in human vascular endothelial cells. Am J Physiol Heart Circ Physiol 279:H285–H292
Yanagisawa K, Tsukamoto T, Takagi T, Tominaga S (1992) Murine ST2 gene is a member of the primary response gene family induced by growth factors. FEBS Lett 302:51–53
Yanagisawa K, Takagi T, Tsukamoto T, Tetsuka T, Tominaga S (1993) Presence of a novel primary response gene ST2L, encoding a product highly similar to the interleukin 1 receptor type 1. FEBS Lett 318:83–87
Yasuda K, Muto T, Kawagoe T, Matsumoto M, Sasaki Y, Matsushita K, Taki Y, Futatsugi-Yumikura S, Tsutsui H, Ishii KJ, Yoshimoto T, Akira S, Nakanishi K (2012) Contribution of IL-33-activated type II innate lymphoid cells to pulmonary eosinophilia in intestinal nematode-infected mice. Proc Natl Acad Sci USA 109:3451–3456
Yasuoka S, Kawanokuchi J, Parajuli B, Jin S, Doi Y, Noda M, Sonobe Y, Takeuchi H, Mizuno T, Suzumura A (2011) Production and functions of IL-33 in the central nervous system. Brain Res 1385:8–17
Yoshimoto T (2010) Basophils as Th2-inducing antigen-presenting cells. Int Immunol 22:543–550
Zhang HF, Xie SL, Chen YX, Mai JT, Wang JF, Zhu WL, Zhu LG (2012) Altered serum levels of IL-33 in patients with advanced systolic chronic heart failure: correlation with oxidative stress. J Transl Med 10:120
Zhao J, Wei J, Mialki RK, Mallampalli DF, Chen BB, Coon T, Zou C, Mallampalli RK, Zhao Y (2012) F-box protein FBXL19-mediated ubiquitination and degradation of the receptor for IL-33 limits pulmonary inflammation. Nat Immunol 13:651–658

Chapter 12
Thymic Stromal Lymphopoietin (TSLP)

TSLP in Allergy, Autoimmunity, and Cancer

Steven F. Ziegler, Florence Roan, Bryan D. Bell, Thomas A. Stoklasek, Masayuki Kitajima, and Hongwei Han

Abstract Originally shown to promote the growth and activation of B cells, thymic stromal lymphopoietin (TSLP) is now known to have wide-ranging impacts on both hematopoietic and nonhematopoietic cell lineages, including dendritic cells (DCs), basophils, eosinophils, mast cells, $CD4^+$, $CD8^+$, and natural killer (NK) T cells, B cells, and epithelial cells. Although the role of TSLP in the promotion of TH2 responses has been extensively studied in the context of lung- and skin-specific allergic disorders, it is becoming increasingly clear that TSLP may impact multiple disease states within multiple organ systems, including the blockade of TH1/TH17 responses and the promotion of cancer and autoimmunity. This review will highlight recent advances in the understanding of TSLP signal transduction, as well as the role of TSLP in allergy, autoimmunity, and cancer. Importantly, these insights into TSLP's multifaceted roles could potentially allow for novel therapeutic manipulations of these disorders.

Keywords Allergy • Atopy • Cancer • Cytokines • TSLP

S.F. Ziegler (✉)
Immunology Program, Benaroya Research Institute, Seattle, WA 98191, USA

Department of Immunology, University of Washington School of Medicine, Seattle, WA, USA

Benaroya Research Institute, 1201 9th Avenue, Seattle, WA 98101, USA
e-mail: sziegler@benaroyaresearch.org

F. Roan
Immunology Program, Benaroya Research Institute, Seattle, WA 98191, USA

Department of Immunology, University of Washington School of Medicine, Seattle, WA, USA

Division of Allergy and Infectious Diseases, University of Washington School of Medicine, Seattle, WA, USA

B.D. Bell • T.A. Stoklasek • M. Kitajima • H. Han
Immunology Program, Benaroya Research Institute, Seattle, WA 98191, USA

T. Yoshimoto and T. Yoshimoto (eds.), *Cytokine Frontiers: Regulation of Immune Responses in Health and Disease*, DOI 10.1007/978-4-431-54442-5_12,

Abbreviations

AD	Atopic dermatitis
AR	Allergic rhinitis
BAL	Bronchoalveolar lavage
COPD	Chronic obstructive pulmonary disease
DC	Dendritic cell
FITC	Fluorescein isothiocyante
JAK	Janus kinase
LC	Langerhans cell
Nod2	Nucleotide-binding oligomerization domain-containing protein 2
SPINK5	Serine protease inhibitior Kazal-type 5
STAT	Signal transducers and activators of transcription
TSLP	Thymic stromal lymphopoietin

12.1 Introduction

Thymic stromal lymphopoietin (TSLP) is a member of the interleukin (IL)-2 cytokine family and a distant paralogue of IL-7 (Leonard 2002). Murine TSLP was discovered in thymic stromal cell line supernatants that supported B-cell development (Friend et al. 1994). Similar to IL-7, TSLP can stimulate thymocytes and promote B-cell lymphopoiesis. Accordingly, TSLP was initially studied as a B-cell growth factor (Levin et al. 1999). A human homologue was subsequently identified, and further characterization of the cytokine revealed a four-helix bundle structure containing six conserved cysteine residues and multiple potential sites for N-linked carbohydrate addition. As discussed later, in spite of only 43 % amino acid identity, human and murine TSLP share a significant degree of functional homology (Reche et al. 2001; Sims et al. 2000). During allergic inflammation, the primary producers of TSLP are epithelial cells, keratinocytes, and stromal cells, although recent data have demonstrated that both dendritic cells (DCs) and mast cells are capable of TSLP production (Soumelis et al. 2002; Watanabe et al. 2004; Ying et al. 2005; Kashyap et al. 2011; Moon et al. 2011).

Several groups identified a receptor capable of binding TSLP with low affinity (TSLPR subunit), which shares 24 % identity to the common γ-receptor chain (γ_c) (Pandey et al. 2000; Park et al. 2000). Upon further analyses, the functional receptor (TSLPR) was shown to include both the TSLPR subunit and the IL-7Rα chain in humans and mice (Quentmeier et al. 2001; Park et al. 2000). The functional TSLPR is expressed by a variety of hematopoietic cell populations, such as T cells, B cells, NK cells, monocytes, basophils, eosinophils, and DCs, as well as some nonhematopoietic cell lineages such as epithelial cells (Ziegler 2010; Reardon et al. 2011). Although classified as a hematopoietin receptor on the basis of structural homology, the TSLPR subunit contains notable differences from canonical hematopoietin receptors.

The TSLPR subunit contains the conserved box1 sequence, which regulates Janus protein tyrosine kinase (JAK) binding in other cytokine receptors, but lacks the conserved box2, and contains only one tyrosine (Y) residue four amino acids from its carboxy terminus (Park et al. 2000). Additionally, it contains a modified WSXWS motif and multiple potential N-linked glycosylation sites (Tonozuka et al. 2001).

12.2 TSLP Signaling

As a member of the hematopoietin receptor family, it was originally hypothesized that the TSLPR would utilize JAKs to activate STAT proteins downstream of the TSLPR. Indeed, TSLP stimulation of multiple cell lines leads to STAT5 phosphorylation. However, initial experiments in these cell lines showed that TSLPR signaling occurred in the absence of JAK activation, and dominant-negative forms of JAK-1 and -2 did not affect TSLP-mediated STAT5 activation (Isaksen et al. 1999; Levin et al. 1999). Several alternatives were implicated in TSLPR signaling, such as Src kinases and phosphinositol 3 kinase (Isaksen et al. 2002). However, two recent papers have demonstrated robust and sustained activation of JAK-1 and -2 following TSLP signaling in primary human dendritic cells and primary human and mouse $CD4^+$ T cells (Arima et al. 2010; Rochman et al. 2010). Surprisingly, in contrast to IL-7Rα and -γ_c in IL-7 signaling, which utilize JAK-1 and -3, the TSLPR subunit bound and utilized JAK-2 in concert with IL-7Rα-associated JAK-1. These latest findings resolve a long-standing question about the mode of TSLP signaling and show that TSLP-induced JAK activation precedes the activation of STAT proteins. In human peripheral blood-derived CD11c+ DCs, TSLP stimulation activated STAT 1, 3, 4, 5, and 6, as well as JAKs 1 and 2 (Arima et al. 2010). Similar results have been seen using mouse DCs, with the exception that no phosphorylation of Stat6 was seen (Bell et al. 2013). These data suggest that TSLP is capable of activating multiple STAT proteins. Whether TSLP utilizes similar signaling pathways in other cell lineages and how each STAT molecule contributes has yet to be elucidated.

12.3 TSLP-Responsive Cells

A plethora of cell types have been shown to be capable of responding to TSLP in vivo and in vitro: these include DCs, CD4 and CD8 T cells, B cells, mast cells, basophils, eosinophils, and natural killer T (NKT) cells. This long list of responding cell types suggests the important role of this cytokine in orchestrating the initial response to an epithelial insult. Although the normal function of TSLP is likely the maintenance of Th2-type homeostasis at barrier surfaces (Ziegler and Artis 2010), as is discussed here, dysregulated TSLP expression can result in the development of type 2 inflammatory responses, leading to allergic disease.

12.3.1 Dendritic Cells

It has now become apparent that a major TSLP-responsive cellular subset in both humans and mice are myeloid-derived dendritic cells (mDCs) (Reche et al. 2001; Zhou et al. 2005). Co-culture of TSLP-activated DCs with naïve syngeneic $CD4^+$ T cells led to T-cell proliferation but no differentiation, suggesting a role for TSLP in $CD4^+$ T-cell homeostasis (Watanabe et al. 2004). However, when TSLP-stimulated DCs primed $CD4^+$ T cells in an antigen-specific manner (e.g., in an allogeneic culture), the resulting T cells display characteristic features of Th2 differentiated cells [production of IL-4, IL-5, IL-13, and tumor necrosis factor (TNF)-α], with the exception that IL-10 production was not evident (Soumelis et al. 2002). These data suggest that TSLP-activated DCs primed for inflammatory Th2 cell differentiation. Interestingly, TSLP, in the absence of IL-12, induced OX40L expression on DCs, and OX40–OX40L interactions were critical for the ability of the DCs to drive Th2 cell differentiation (Ito et al. 2005). Consistent with a role in regulating Th2 cytokine responses, TSLP-activated DCs were also capable of supporting the maintenance and further polarization of $CRTH2^+$ Th2 effector memory cells (Wang et al. 2006). In contrast, autologous TSLP-activated DCs supported the expansion and functions of $CRTH2^+$ $CD4^+$ TH2 memory cells (Wang et al. 2006), but led to T-cell proliferation and elaboration of high levels of IL-2, but not IL-4, IL-5, or IL-13, when co-cultured with naïve T cells (Watanabe et al. 2004).

TSLP-conditioned DCs also augmented intestinal epithelial cell-mediated IgA2 class switching through the induction of APRIL (He et al. 2007). Finally, some in vitro studies have suggested a role for TSLP in the generation of tolerogenic DCs that can drive the differentiation of regulatory T cells (Tregs) (Watanabe et al. 2005; Besin et al. 2008; Iliev et al. 2009a, b), although other studies have indicated that TSLP may hinder the production and/or maintenance of FOXP3+ Tregs in vivo in certain disease processes (Lei et al. 2011; Duan et al. 2010).

12.3.2 T Lymphocytes

Early work from the Leonard laboratory showed that TSLPR-deficient mice had normal lymphocyte numbers, but that γ_c/TSLPR double-deficient mice had a more pronounced defect than γ_c-deficient mice alone (Al Shami et al. 2004). These investigators also showed that TSLP could drive the expansion of T and B cells when injected into γ_c-deficient mice, showing that TSLP can effect lymphoid homeostasis. Subsequent studies showed that TSLP can also act directly on CD4+ T cells, and in the presence of TCR stimulation, promoted proliferation and TH2 differentiation of naïve $CD4^+$ T cells through induction of IL-4 gene transcription (Omori and Ziegler 2007; Rochman et al. 2007). IL-4 further upregulated TSLPR on $CD4^+$ T cells, resulting in a positive feedback loop. Although IL-4 maintained TSLPR expression on both in vitro differentiated TH2 and TH17 cells, higher TSLPR levels were present on TH2 than on TH1 and TH17 cells, which correlated with the ability of TSLP to drive the proliferation and survival of activated TH2 cells (Kitajima et al. 2011). Naïve mouse $CD8^+$ T cells also express TSLPR, although TSLPR expression is low to absent on naïve human $CD8^+$ T cells; however, following activation, TSLPR expression is upregulated on both mouse and human $CD8^+$ T cells (Rochman and Leonard 2008; Akamatsu et al. 2008). In both $CD4^+$ and $CD8^+$ T cells, TSLP stimulation upregulated the survival protein Bcl-2 in a STAT5-dependent manner (Rochman and Leonard 2008; Rochman et al. 2010; Kitajima et al. 2011).

12.3.3 B Lymphocytes

The initial studies describing TSLP demonstrated that TSLP can support B-cell lymphopoiesis (Friend et al. 1994; Levin et al. 1999). In in vitro studies, pro-B cells derived from fetal liver, but not bone marrow, responded to TSLP, although pre-B cells from both origins could proliferate in response to TSLP (Vosshenrich et al. 2003). The role of TSLP in normal B-cell development or during inflammatory responses remains undefined. However, it is clear that aberrant TSLP signaling can have a significant impact on B cells, as has been demonstrated by the association of TSLPR mutations with a subtype of B-cell leukemia (Chapiro et al. 2010; Roll and Reuther 2010; Tasian and Loh 2011). In addition, elevated systemic TSLP has been shown to lead to aberrant B-cell development and function, with both direct effects on early B-cell development and indirect effects leading to autoimmune hemolytic anemia (Astrakhan et al. 2007; Iseki et al. 2012).

12.3.4 Innate Immune Cells

Multiple innate immune cells express the TSLPR and respond to TSLP. For example, TSLP can enhance cytokine production from mast cells, NKT cells, and eosinophils (Nagata et al. 2007; Allakhverdi et al. 2007; Wong et al. 2009). In addition, TSLP has very recently been shown to induce eosinophil extracellular traps (EETs), extrusions of mitochondrial DNA toxic granule molecules released in response to infection (Morshed et al. 2012). Finally, TSLP has also been shown to be important for the development and function of a subset of basophils (Siracusa et al. 2011). This subset is IL-3 independent and is recruited to sites of type 2 inflammation where is it speculated that they play a role in promoting Th2-type responses (Siracusa et al. 2011, 2012; Sokol et al. 2009). Thus, TSLP not only can directly promote type 2 responses through CD4 T-cell differentiation, it can also influence responses through the recruitment and activation of innate immune cells capable of producing cytokines involved in type 2 inflammation.

12.4 TSLP-Associated Diseases

The variety of TSLP-responsive cell types demonstrates that TSLP can impact type 2 inflammation through a myriad of different pathways. In addition, numerous studies in both humans and mice now implicate TSLP in a growing number of different disorders beyond allergic inflammation, including infection, cancer, and autoimmunity. The following sections describe the disorders associated with TSLP and what is known about the mechanisms through which TSLP may act.

12.4.1 Skin Disorders

Atopic dermatitis (AD) is a chronic inflammatory skin disease that affects an estimated 10–20 % of infants and young children in the United States (Leung et al. 2004; Boguniewicz and Leung 2011). Interestingly, there is suggestive evidence of linkage between single nucleotide polymorphisms (SNPs) in the TSLP gene and AD (Gao et al. 2010). In addition, although TSLP protein was undetectable in non-lesional skin in AD patients, TSLP was highly expressed in acute and chronic AD lesions (Soumelis et al. 2002). TSLP was also overexpressed in the skin of individuals with Netherton syndrome (NS), a severe skin disease characterized by atopic dermatitis-like lesions as well as other allergic manifestations that result from mutations in the serine peptidase inhibitor Kazal-type 5 (SPINK5) gene, which encodes the serine protease inhibitor lymphoepithelial Kazal-type-related inhibitor (LEKTI) (Briot et al. 2009).

In mice, overexpression of TSLP specifically in the skin was sufficient to induce a disease phenotype characterized by all the hallmark features of AD (Yoo et al. 2005). In the steady state, TSLP expression in the skin appears to be negatively regulated by retinoid X receptors (RXR), because keratinocyte-specific ablation of the retinoid X receptor isotypes RXRα and RXRβ resulted in upregulation of TSLP and development of AD-like skin inflammation (Li et al. 2005). RXRs heterodimerize with many nuclear receptor partners, including the vitamin D receptor and peroxisome proliferator-activated receptors. Administration of vitamin D or its analogues upregulated TSLP and resulted in the development of dermatitis (Li et al. 2006, 2009), suggesting that vitamin D administration may result in RXR derepression and recruitment of co-activators to promote transcription. Keratinocyte-specific deletion of Notch signaling, which causes severe epidermal differentiation defects, also resulted in high systemic levels of TSLP. However, TSLP expression in this model may be caused by responses to the resulting skin barrier defect rather than directly from the loss of keratinocyte-specific Notch signaling itself, because wild-type and mutant keratinocytes produced similar amounts of TSLP in in vitro cultures (Demehri et al. 2008). In SPINK5 knockout (SPINK5 −/−) mice, which reproduce many of the key features of NS, the absence of LEKTI resulted in unrestrained activity of the serine protease kallikrein 5, which directly activated proteinase-activated receptor 2 (PAR-2) and induced nuclear factor κB (NF-κB)-mediated overexpression of TSLP without contribution of the adaptive immune system (Briot et al. 2009; Kouzaki et al. 2009). Interestingly, in SPINK5/PAR-2 double knockout mice, TSLP expression was greatly diminished, although inflammation still occurred (Briot et al. 2010). Whether the cytokine milieu differs in the absence of TSLP remains to be determined.

TSLP may influence both the initiation and progression of allergic skin inflammation, but the relative contribution to these stages and the cellular requirements may differ depending on the context. Langerhans cell (LC) migration and activation was seen in human AD lesions in situ (Soumelis et al. 2002). Furthermore, TSLP has been shown to increase the number and maturation status of migratory LCs in human skin explants cultures and to condition LCs to prime co-cultured naïve $CD4^+$ T cells to adopt an inflammatory TH2 phenotype (Ebner et al. 2007). However, mouse models of AD implicate additional cell types in the initiation and promotion of AD by TSLP. A recent study by Oh et al. (2011) implicated TSLP in mediating skin fibrosis downstream of IL-13, in part through the stimulation of fibrocyte collagen production. In a model of allergic skin inflammation using epicutaneous (EC) sensitization to ovalbumin (OVA) on tape-stripped skin, TSLP acted directly on T cells during the challenge phase to potentiate TH2 cytokine production (He et al. 2008). T cells and eosinophils were also required for TSLP-mediated dermal inflammation induced through intradermal delivery of recombinant TSLP protein (Jessup et al. 2008). In contrast, TSLP was involved in both sensitization and challenge phases of FITC-mediated contact hypersensitivity, as ear swelling was minimal if blockade of TSLP occurred before sensitization, but was only modestly reduced when TSLP blockade occurred after sensitization but before challenge (Larson et al. 2010; Boehme et al. 2009). Although DC migration was intact in the absence of TSLP in EC sensitization, loss of TSLP signaling in the FITC CHS model was associated with reduced migration and activation of skin-derived antigen-bearing DCs. In addition, TSLP-responsive $CD4^+$ T cells were not required to induce a TH2 response in the CHS model (R.P. Larson and S.F. Ziegler, unpublished observations). In the setting of chronic high TSLP expression, skin inflammation also occurred in the absence of T cells (Yoo et al. 2005), possibly because of the ongoing stimulation of innate immune cells by TSLP.

TSLP has also been implicated in the phenomenon referred to as the atopic march, which describes the increased likelihood of individuals with AD of developing allergic rhinitis (AR) and asthma later in life (Bieber 2008). Several models of induced TSLP expression in mouse keratinocytes result in subsequent allergic airway inflammation following intranasal challenge, suggesting that TSLP may be an important factor contributing to this progression from AD to AR and asthma (Zhang et al. 2009; Demehri et al. 2009; Leyva-Castillo et al. 2012; Jiang et al. 2012). Although many of these methods used to induce TSLP expression result in artificially high systemic levels of TSLP that are not seen in AD patients, we have found that intradermal administration of TSLP triggers progression from atopic dermatitis to asthma in the absence of systemic TSLP (Han et al. 2012). In this study, TSLP was the airway response to antigen challenge was shown to be TSLP independent.

These models, as well as approaches that allow for more specific expression or deletion of TSLP, will be helpful in identifying the cellular targets of TSLP and the mechanisms involved in the progression from AD to AR and asthma.

12.4.2 Respiratory Diseases

The initial report demonstrating high TSLP expression in AD and potentiation of inflammatory TH2 responses by TSLP also suggested a potential role for TSLP in allergic airway disease (Soumelis et al. 2002). This hypothesis was supported by the demonstration that TSLP mRNA was present in human lung fibroblasts and bronchial epithelial and smooth muscle cells (Soumelis et al. 2002), and that aberrant levels of TSLP were associated with certain human respiratory disorders (Ying et al. 2005, 2008; Zhang et al. 2007; Kamekura et al. 2009; Semlali et al. 2010; Kimura et al. 2011; Shikotra et al. 2011; Xu et al. 2010). Lung epithelium and submucosa samples from asthmatics and chronic obstructive pulmonary disease (COPD) patients contained a greater number of TSLP mRNA-positive cells, and bronchoalveolar lavage (BAL) samples from these patients had higher concentration of TSLP protein compared to healthy controls (Ying et al. 2005, 2008; Semlali et al. 2010; Shikotra et al. 2011). Although the level of TSLP expression can be variable in asthmatic patients, it has been shown to correlate directly with TH2 cytokine and chemokine expression and inversely with lung function (Shikotra et al. 2011; Ying et al. 2008). Increased expression of TSLP in the nasal epithelium has also been found in biopsies from allergic rhinitis patients and was associated with TH2 cytokine production and eosinophilic infiltration in epithelial-associated tissue (Mou et al. 2009; Kamekura et al. 2009; Kimura et al. 2011; Xu et al. 2010). Genetic studies also support a critical role for TSLP in allergic airway disease. Several SNPs at the TSLP genomic locus found across multiple ethnic backgrounds were associated with increased asthma susceptibility or protection (Harada et al. 2009, 2011; Hunninghake et al. 2010; Bunyavanich et al. 2011; Torgerson et al. 2011; Shamim et al. 2007). One such SNP present in the genomic TSLP locus creates a novel AP-1 transcription factor-binding site that could potentially lead to increased TSLP transcription (Harada et al. 2009).

A role for TSLP in human asthma has been well supported by a variety of mouse models, such as the surfactant protein c (SPC)-TSLP mouse, in which TSLP is constitutively expressed by the lung epithelium under control of the SPC promoter (Zhou et al. 2005). With increasing age, these mice developed a progressive asthma-like disease characterized by lung infiltration of eosinophils and TH2 $CD4^+$ T cells, airway remodeling, and airway hyperreactivity. Disease in these mice was largely dependent on IL-4, IL-13, $CD4^+$ T cells, and antigen (Headley et al. 2009; Zhou et al. 2008). $CD4^+$ T cells and antigen were also required in an acute asthma model using intranasal administration of TSLP in conjunction with antigen (Seshasayee et al. 2007; Headley et al. 2009). In addition to driving allergic inflammation in the lung following direct TSLP administration, TSLP played a crucial role in the well-established ovalbumin (OVA)/alum allergic airway inflammation model. In this model, TSLP protein was found in the BAL and lung after intranasal OVA challenge, and disease symptoms were curtailed in the absence of TSLPR or when TSLP activity was blocked by antibody or recombinant TSLPR protein (Zhou et al. 2005; Al Shami et al. 2005; Shi et al. 2008; Li et al. 2010; Zhang et al. 2011). In an OVA-driven mouse model of allergic rhinitis, blocking TSLP also inhibited disease development (Miyata et al. 2008).

Most data currently point to a primary role for TSLP in the sensitization/priming stage of allergic airway disease. TSLP produced by activated human-derived lung cells stimulated human DCs to prime $CD4^+$ TH2 cell development and mast cell production of TH2-associated cytokines (Soumelis et al. 2002; Allakhverdi et al. 2007; Bleck et al. 2010). Furthermore, multiple studies have shown that TSLP-mediated DC activation was responsible for the disease phenotype observed in mouse models of asthma (Zhou et al. 2005; Seshasayee et al. 2007; Shi et al. 2008; Li et al. 2010; Zhang et al. 2011). TSLP-induced DC expression of costimulatory molecules, in particular OX40L, and DC production of TH2 chemokines, such as CCL17 and CCL21, are likely the predominant mechanisms of action (Zhou et al. 2005; Seshasayee et al. 2007) . However, TSLP may also influence the challenge stage of allergic airway disease by supporting TH2 $CD4^+$ T-cell cytokine production (Shi et al. 2008; Miyata et al. 2008; Li et al. 2010; Zhang et al. 2011; Al Shami et al. 2005; He et al. 2008). As already mentioned, TSLP may also influence the regulatory T-cell compartment. Several reports have shown the ability of TSLP to promote the development of thymic regulatory T cells (Tregs) in vitro (Mazzucchelli et al. 2008; Hanabuchi et al. 2010); however, in vivo, its role is less clear. In allergic airway disease, TSLP inhibited IL-10-mediated Treg function and the formation of inducible Tregs to exogenous antigen (Nguyen et al. 2010). Importantly, the BAL fluid from asthmatics inhibited pulmonary Treg function in a TSLP-dependent manner (Nguyen et al. 2010). In the OVA allergen model, TSLP was shown to interfere with tolerance by inhibiting the generation of allergen-specific Tregs (Lei et al. 2011). In the same model, nucleotide-binding oligomerization domain-containing protein 2 (Nod2), and to a lesser extent Nod1, stimulation blocked tolerance to OVA intranasal challenge in a TSLP- and OX40L-dependent manner (Duan et al. 2010). In this model, loss of TSLP signaling correlated with increased antigen-specific $FOXP3^+$ T cells following Nod2 stimulation.

A variety of stimuli, such as IL-4, IL-13, TNF-α, IL-1, bacterial peptidoglycan, lipoteichoic acid, double-stranded RNA (dsRNA), respiratory viruses, air pollutants, and allergens have been shown to induce TSLP expression by lung-derived parenchymal cells and immune cells (Soumelis et al. 2002; Allakhverdi et al. 2007; Lee and Ziegler 2007; Zhang et al. 2007; Bleck et al. 2010; Kouzaki et al. 2009; Smelter et al. 2010; Kashyap et al. 2011; Kato and Schleimer 2007). In particular, stimulation of Nod1 and Nod2 in nonhematopoietic cells were potent inducers of TH2 immunity via TSLP (Magalhaes et al. 2011). These stimuli likely all drive NF-κB-dependent expression of TSLP, as was shown to occur in human lung epithelial cells (Lee and Ziegler 2007). Furthermore, TSLP transcription was negatively regulated by 9-*cis*-retinoic acid via retinoid X receptors in lung cells

(Lee et al. 2008). Exposure to certain infectious agents or repeated environmental irritants may prime production of TSLP, leading to TH2-mediated human disease. For example, even in the absence of known lung disease, lung samples from smokers contained increased TSLP levels as compared to nonsmokers (Ying et al. 2008). In addition, lung epithelial cells from asthmatics produced more TSLP in response to dsRNA (viral analogue) stimulation in culture (Uller et al. 2010; Brandelius et al. 2011), which may explain, at least in part, why patients with asthma tend to suffer more airway dysfunction after respiratory infections compared to healthy individuals (Jackson and Johnston 2010). This aberrant TSLP production in response to lung insults may thus influence both the susceptibility of certain individuals to develop allergic respiratory diseases such as asthma and the clinical complications that arise after environmental insults to the lungs of these individuals.

Collectively, these data illustrate that aberrant lung expression of TSLP is associated with human allergic airway disease and can mimic asthma-like disease in mice. According to genetic studies and in vitro analyses, lung samples from individuals with asthma or COPD produce more TSLP in response to lung insult as compared to samples from healthy individuals. Clinical trials targeting TSLP in these conditions are currently underway. According to mouse asthma models, TSLP appears to influence the sensitization stage of allergic airway responses, but a more in depth examination of the influence of TSLP on the allergic effector response is required. Where and when TSLP acts during allergic airway disease will likely explain any trial results and dictate future therapeutic design.

12.4.3 Intestinal Inflammation

TSLP is constitutively expressed in both the mouse and human gastrointestinal tract, but can be further induced by a variety of cytokines, microbes, and microbial products (Rimoldi et al. 2005; Zaph et al. 2007; Taylor et al. 2009; He et al. 2007; Tanaka et al. 2010; Zeuthen et al. 2008; Humphreys et al. 2008). Mice carrying gene deletions specifically affecting the gut mucosa provide additional clues into the regulation of TSLP expression within the gut. TSLP mRNA levels were significantly decreased in mice with intestinal epithelial-specific deletion of Dicer (Biton et al. 2011), an enzyme involved in microRNA biosynthesis, or IκB kinase-β (Zaph et al. 2007). Both these knockout mice showed increased susceptibility to infection with the mouse whipworm *Trichuris muris*. TSLP expression was also decreased in mice carrying a missense mutation in the *Muc2* mucin gene that resulted in an epithelial defect and spontaneous colitis (Eri et al. 2011). In in vitro analyses of TSLP intestinal function, human colonic or gastric epithelial-derived TSLP has been implicated in conditioning DCs to drive development of inflammatory TH2 cells (Kido et al. 2010), regulatory T cells (Iliev et al. 2009a, b), or T-cell-independent IgA(2) class switching (He et al. 2007). Although supernatants from both human and mouse intestinal epithelial cells (IECs) can condition DCs to drive Treg differentiation, the requirements for TSLP may differ in humans and mice, because the presence of TSLP was required in mouse but not human IEC supernatants to drive a tolerigenic DC phenotype (Iliev et al. 2009a, b). Additional studies are just beginning to define whether and under what conditions TSLP may function in these pathways in vivo.

As is seen in atopic diseases of the skin and lung, aberrant expression of TSLP was also seen in allergic diseases of the gut. Polymorphisms in TSLP and the TSLPR were associated with the food allergy-related disorder eosinophilic esophagitis (EoE), and this association persisted when comparing EoE patients with allergic individuals without EoE (Rothenberg et al. 2010; Sherrill et al. 2010). Additionally, TSLP mRNA expression was higher in the esophagus of pediatric patients with EoE compared to controls and was decreased in homozygotes of the protective GG minor allele for the rs3806932 SNP. Some studies suggest, however, that TSLP not only plays an important role in the promotion of TH2 responses but is also a key player in maintaining intestinal homeostasis and modulation of TH1/TH17 inflammation. In contrast to the increased TSLP expression seen in EoE, decreased TSLP expression was seen in noninflamed colonic tissue in Crohn's disease (CD) and ulcerative colitis (UC), the two types of inflammatory bowel disease (IBD) (Noble et al. 2008, 2010; Rimoldi et al. 2005; Iliev et al. 2009a, b). However, studies of UC have indicated that in inflamed tissue, TSLP expression is upregulated compared with noninflamed tissue from either UC patients or controls (Noble et al. 2008; Tanaka et al. 2010).

Mouse models of TH2- and TH1-type inflammation also suggest important roles for TSLP in TH2-mediated immunity, maintenance of homeostasis, and modulation of TH1/TH17 responses within the gut. TSLP was required to induce diarrheal disease in a mouse model of food allergy (Blazquez et al. 2010) and protective TH2 responses to infection with *T. muris* (Zaph et al. 2007). However, TSLP was not required for oral tolerance to OVA, or for anaphylaxis and IL-4, IL-13, and IgE production following intragastric OVA/cholera toxin sensitization and challenge (Blazquez et al. 2010). Additionally, other helminths such as *Heligmosomoides polygyrus*, *Nippostrongylus brasiliensis*, and *Schistosoma mansoni* still induced TH2 responses in TSLPR knockout mice, although in some cases, these responses were modified or slightly attenuated (Massacand et al. 2009; Ramalingam et al. 2009). Thus, although TSLP may promote TH2 responses in the gut, it is not absolutely required for TH2-type inflammation. In contrast to *T. muris*, both *H. polygyrus* and *N. brasiliensis* produce excretory/secretory (ES) products that acted on DCs to attenuate IL-12/23p40 production. Of note, protective TH2 responses can be induced in *T. muris* infections in the absence of TSLP following the blockade of either interferon (IFN)-γ or IL-12/23p40 (Taylor et al. 2009; Massacand et al. 2009), suggesting that TSLP may play a prominent role in attenuating TH1 and TH17 responses.

Studies using mouse models of colitis have demonstrated important effects of TSLP in modulating the disease phenotype in intestinal inflammation, although there have been some conflicting results. In a chemical colitis model using dextran sulfate sodium (DSS), Taylor et al. (2009) showed that mice lacking TSLPR developed more acute weight loss and increased colonic inflammation that correlated with higher levels of IFN-γ and IL-17A within the mesenteric lymph nodes. In contrast, Reardon et al. reported comparable disease onset and severity in the DSS colitis model between mice that lack TSLP signaling versus controls. However, although wild-type mice recovered after DSS withdrawal, mice lacking either TSLP or its receptor demonstrated progressive disease and weight loss (Reardon et al. 2011). Reardon et al. showed that secretory leukocyte peptidase inhibitor (SLPI) was induced in DSS colitis in wild-type mice and that this induction was lost in TSLP knockout (TSLP KO) mice. Neutrophil elastase (NE) is a target of SLPI, and functions to degrade a number of substrates, including progranulin, a protein important in wound healing. Consistent with a role for TSLP in the inhibition of NE, TSLP KO mice displayed increased NE activity after treatment with DSS, and inhibition of NE reduced mortality in TSLP KO mice in this colitis model. While methodological differences may account for some of the discrepancies between these studies, a growing body of evidence demonstrates that differences in microbiota among various facilities can have profound effects on the development and function of the intestinal as well as systemic immune system (Gill and Finlay 2011). Thus, further exploration of how the gut microbiota affects TSLP expression and function may be warranted.

These studies support a role for TSLP in the promotion of TH2 responses in the gastrointestinal system, but also provide important evidence that TSLP plays a key role in the maintenance of immune homeostasis within the gut. Not only does TSLP function to attenuate TH1/TH17 responses, but also acts directly on the intestinal epithelium to support wound healing in colitis. Whether TSLP also contributes to wound healing and blockade of TH1/TH17 responses at other sites remains to be determined.

12.4.4 Cancer

A series of recent studies have implicated TSLP in the growth and metastasis of breast and pancreatic cancer, especially those that display an increased infiltration of TH2 cells (De Monte et al. 2011; Olkhanud et al. 2011; Pedroza-Gonzalez et al. 2011). Breast and pancreatic cancer cells and cancer-associated fibroblasts have been shown to produce TSLP in response to tumor-derived inflammatory cytokines and possibly other unidentified stimuli (De Monte et al. 2011; Olkhanud et al. 2011; Pedroza-Gonzalez et al. 2011). Furthermore, treatment of DCs with supernatants from these cells induced theTH2-attracting chemokines CCL17 and CCL22, as well as upregulation of DC costimulatory molecules CD80, CD86, OX40L and TSLPR, in a TSLP-dependent manner. Additionally, these primed DCs were able to promote TH2 polarizition of $CD4^+$ T cells in vitro. In support of these in vitro data, activated DCs and CCL17 and CCL22 were detected in the tumor and draining lymph nodes, but not nondraining lymph nodes, of human patients (De Monte et al. 2011). Importantly, a decreased ratio of TH1/TH2 cells in human pancreatic cancer cases was associated with disease progression and was an independent prognostic marker of reduced survival (De Monte et al. 2011). Although breast cancer cells with intact

TSLP expression were able to induce tumor growth and metastasis in mice, shRNA knockdown of TSLP in these cells resulted in clones with minimal growth or metastasis (Olkhanud et al. 2011). Tumor progression and metastasis of an injected breast cancer or melanoma cell line was also decreased in TSLPR-deficient mice compared to wild-type mice (Olkhanud et al. 2011).

Previous work has shown that TH2 cytokines promote disease progression through increased survival of cancer cells, M2 macrophage differentiation, and fibrosis (collagen degradation and synthesis) (Wynn 2004; Aspord et al. 2007; Mantovani et al. 2008; Joyce and Pollard 2009). TSLP may be linked to these phenomena in some human cancers, possibly based on its ability to drive TH2 differentiation and M2 macrophage differentiation (Ziegler 2010; Han and Ziegler, in manuscript). Alternatively, TSLP may promote tumor progression by controlling Treg migration. CCL22 production in human breast cancer is involved in the influx of tumor Tregs that may then alter the immunoregulatory environment (Gobert et al. 2009; Ménétrier-Caux et al. 2009). Further investigation is needed to identify the important sources and targets of TSLP within the tumor environment.

In addition to the association of TSLP with certain solid tumors, the TSLPR has been shown to be overexpressed in 5–10 % of childhood B-cell progenitor acute lymphoblastic leukemia (ALL) cases and approximately 60 % of acute lymphoblastic leukemia cases in children with Down's syndrome (Roll and Reuther 2010; Tasian and Loh 2011; Mullighan et al. 2009; Russell et al. 2009; Ensor et al. 2011). Approximately 15 % of adult and high-risk pediatric B-ALL that lack characteristic rearrangements demonstrated TSLPR overexpression (Yoda et al. 2010). In addition, some cases of activating TSLPR mutations were found (Chapiro et al. 2010). In almost all cases, TSLPR overexpression was associated with intrachromosomal deletion or rearrangement of the TSLPR/CRLF2 locus with the immunoglobulin heavy chain (IGH) locus, placing TSLPR/CRLF2 under alternate transcriptional control downstream of the P2YR8 promoter (Russell et al. 2009; Mullighan et al. 2009; Yoda et al. 2010). These rearrangements were highly correlated with the presence of JAK2 mutations and were associated with a poor prognosis (Roll and Reuther 2010; Mullighan et al. 2009; Russell et al. 2009; Cario et al. 2010; Harvey et al. 2010; Yoda et al. 2010; Ensor et al. 2011). In murine Ba/F3 cells, expression of TSLPR and JAK2 mutant alleles promoted growth factor-independent growth (Mullighan et al. 2009; Yoda et al. 2010). Mice with systemic overexpression of TSLP may provide a model for understanding the signaling mechanisms involved. In particular, loss of keratinocyte-specific Notch signaling resulted in high systemic levels of TSLP that correlated with a rapid expansion of pre-B cells in the early postnatal period that contributed to early mortality in these animals (Demehri et al. 2008). Interestingly, overexpression of TSLP early in the postnatal period was sufficient to drive a B-cell lymphoproliferative disorder, but administration or induction of TSLP after postnatal day 14 was not, although other studies have shown expansion of B-cell compartments following TSLP expression in adult mice (Astrakhan et al. 2007).

The association of TSLP and TSLP signaling pathways with hematological malignancies as well as solid tumors implicates TSLP and TSLPR in numerous regulatory pathways that support cell growth and survival in cancer. In B-ALL, activation of signaling pathways downstream of TSLP directly promotes the growth and survival of malignant cells, whereas in breast and pancreatic cancer, TSLP likely contributes to multiple components of the tumor environment that affect growth and metastasis as well as immune evasion. Several reports suggest that TSLP/TSLPR may be useful as a prognostic marker and may present a novel target for therapeutic intervention in cancer.

12.4.5 Other Autoimmune Diseases and Issues of Tolerance

Mouse models with constitutive or inducible overexpression of TSLP have demonstrated that TSLP can be associated with autoimmune phenomena. TSLP overexpression in these mice was associated with the development of cryoglobulinemic glomerulonephritis caused by increased production and kidney deposition of systemic polyclonal IgM and IgG via a monocyte/macrophage-dependent mechanism (Taneda et al. 2001; Astrakhan et al. 2007). In addition, these mice developed red blood cell-specific auto-antibodies and autoimmune hemolytic anemia in a $CD4^+$ T-cell- and IL-4-dependent manner (Iseki et al. 2012). Whether TSLP is involved in human mixed cryoglobulinemia or autoimmune hemolytic anemia is unknown.

As discussed earlier, TSLP expression was decreased in IBD, a disorder that is thought to arise from inappropriate immune activation against normally harmless microflora. Additionally, loss of TSLP signaling in a mouse model of autoimmune gastritis resulted in more severe disease (Nishiura et al. 2012). Although the impact of TSLP on colitis in mice appears more complex (Taylor et al. 2009; Reardon et al. 2011), this supports a model in which loss of TLSP, which can block TH1/TH17 responses, leads to increased inflammation. However, data from humans and mouse models suggest that TSLP may actively promote inflammation in TH1/TH17-associated autoimmune diseases such as rheumatoid arthritis (RA) and multiple sclerosis (MS). In a proteoglycan-induced arthritis mouse model of RA, TSLPR-deficient mice had reduced immunopathology associated with decreased levels of production of IL-17, IL-1β, and IL-6, but increased IFN-γ and IL-10 (Hartgring et al. 2011). Furthermore, blocking TSLP in a collagen-induced arthritis model ameliorated disease, whereas administering recombinant TSLP protein exacerbated disease (Koyama et al. 2007; Hartgring et al. 2011). Increased synovial concentrations of TSLP, as well as TNF-α, have also been seen in synovial fluid from RA patients compared to samples from patients with osteoarthritis. In in vitro studies, TSLP-primed human myeloid DCs induced proliferation of self-reactive $CD4^+$ T cells capable of TH1 or TH2 differentiation, and TSLP priming of DCs, in conjunction with TLR3 ligand, supported TH17 differentiation (Watanabe et al. 2004; Tanaka et al. 2009; Koyama et al. 2007). Thus, although the role of TSLP in RA is largely undefined, these data provide intriguing evidence of its possible involvement.

Single nucleotide polymorphisms (SNPs) in the IL-7Rα gene locus have been associated with multiple sclerosis (MS) and altered Treg numbers or function (Gregory et al. 2007; Lundmark et al. 2007). Although TSLPR pairs with IL-7Rα

and TSLP can affect Treg development, neither disease has yet been directly linked to TSLP. However, administration of TSLP or TSLP-treated bone marrow-derived DCs into nonobese diabetic mice prevented the development of diabetes in these mice (Besin et al. 2008), suggesting a possible role for TSLP in disease therapy. Although the mechanisms involved in protection from diabetes have not been determined, protection was associated with an increased number of Tregs.

One final link that has been made between TSLP and immune tolerance is in maternal–fetal tolerance during pregnancy (Li and Guo 2009). TSLP was produced and secreted by first-semester trophoblasts, and tissue from normal pregnancies demonstrated a TH2 bias and higher levels of TSLP expression than in samples from miscarriages (Guo et al. 2010; Pu et al. 2012; Wu et al. 2010). Thus, although TSLP expression and a TH2 bias may lead to disease progression in cancer, TSLP may contribute to tolerance at the maternal–fetal interface.

12.5 Conclusion

Much progress has been made in the understanding of TSLP biology and its role during TH2-type inflammation. Multiple cell lineages express the functional TSLPR that helps drive the immune response. More recent data have illustrated that TSLP is also involved in numerous disorders beyond allergy alone, and may play a role in maintaining homeostasis in diseases such as IBD or in disease progression in cancer and autoimmunity. To utilize the knowledge gained about the biological effects of TSLP, a better understanding of cell-specific signaling pathways must be delineated. Of utmost importance is deciphering whether TSLP invokes similar signaling pathways within different cells. Knowledge of the key targets and sources of TSLP in different disease states will also be important in furthering our comprehension of the pathophysiology of TSLP-associated disorders. Tools that can address these questions, such as approaches which use conditional deletion of the TSLPR and cytokine, will be important in the continued investigation of the role of TSLP during both atopic and nonatopic conditions.

Acknowledgments Support was provided by the Crohn's and Colitis Foundation of America (F.R.) and NIH grants 5T32AI007411-19 (B.D.B. and T.A.S.) and AI068731, AR056113, 5AR055695, HL098067, HL102708, AR059058 (S.F.Z.).

Conflict of Interest Statement The authors have no conflicts of interest concerning the work in this review.

References

Akamatsu T, Watanabe N, Kido M, Saga K, Tanaka J, Kuzushima K et al (2008) Human TSLP directly enhances expansion of CD8+ T cells. Clin Exp Immunol 154:98–106

Al Shami A, Spolski R, Kelly J, Fry T, Schwartzberg PL, Pandey A et al (2004) A role for thymic stromal lymphopoietin in CD4(+) T cell development. J Exp Med 200:159–168

Al Shami A, Spolski R, Kelly J, Keane-Myers A, Leonard WJ (2005) A role for TSLP in the development of inflammation in an asthma model. J Exp Med 202:829–839
Allakhverdi Z, Comeau MR, Jessup HK, Yoon BR, Brewer A, Chartier S et al (2007) Thymic stromal lymphopoietin is released by human epithelial cells in response to microbes, trauma, or inflammation and potently activates mast cells. J Exp Med 204:253–258
Arima K, Watanabe N, Hanabuchi S, Chang M, Sun SC, Liu YJ (2010) Distinct signal codes generate dendritic cell functional plasticity. Sci Signal 3:ra4
Aspord C, Pedroza-Gonzalez A, Gallegos M, Tindle S, Burton EC, Su D et al (2007) Breast cancer instructs dendritic cells to prime interleukin 13-secreting CD4+ T cells that facilitate tumor development. J Exp Med 204:1037–1047
Astrakhan A, Omori M, Nguyen T, Becker-Herman S, Iseki M, Aye T et al (2007) Local increase in thymic stromal lymphopoietin induces systemic alterations in B cell development. Nat Immunol 8:522–531
Bell BD, Kitajima M, Larson RP, Stoklasek TA, Dang K, Sakamoto K, Wagner KU, Reizis B, Hennighausen L, Ziegler SF (2013) The transcription factor STAT5 is critical in dendritic cells for the development of TH2 but not TH1 responses. Nat Immunol 14:364–371
Besin G, Gaudreau S, Ménard M, Guindi C, Dupuis G, Amrani A (2008) Thymic stromal lymphopoietin and thymic stromal lymphopoietin-conditioned dendritic cells induce regulatory T-cell differentiation and protection of NOD mice against diabetes. Diabetes 57:2107–2117
Bieber T (2008) Atopic dermatitis. N Engl J Med 358:1483–1494
Biton M, Levin A, Slyper M, Alkalay I, Horwitz E, Mor H et al (2011) Epithelial microRNAs regulate gut mucosal immunity via epithelium–T cell crosstalk. Nat Immunol 12:239–246
Blazquez AB, Mayer L, Berin MC (2010) Thymic stromal lymphopoietin is required for gastrointestinal allergy but not oral tolerance. Gastroenterology 139:1301–1309
Bleck B, Tse DB, Gordon T, Ahsan MR, Reibman J (2010) Diesel exhaust particle-treated human bronchial epithelial cells upregulate Jagged-1 and OX40 ligand in myeloid dendritic cells via thymic stromal lymphopoietin. J Immunol 185:6636–6645
Boehme SA, Franz-Bacon K, Chen EP, Sasik R, Sprague LJ, Ly TW et al (2009) A small molecule CRTH2 antagonist inhibits FITC-induced allergic cutaneous inflammation. Int Immunol 21:81–93
Boguniewicz M, Leung DY (2011) Atopic dermatitis: a disease of altered skin barrier and immune dysregulation. Immunol Rev 242:233–246
Brandelius A, Yudina Y, Calvén J, Bjermer L, Andersson M, Persson C et al (2011) dsRNA-induced expression of thymic stromal lymphopoietin (TSLP) in asthmatic epithelial cells is inhibited by a small airway relaxant. Pulm Pharmacol Ther 24:59–66
Briot A, Deraison C, Lacroix M, Bonnart C, Robin A, Besson C et al (2009) Kallikrein 5 induces atopic dermatitis-like lesions through PAR2-mediated thymic stromal lymphopoietin expression in Netherton syndrome. J Exp Med 206:1135–1147
Briot A, Lacroix M, Robin A, Steinhoff M, Deraison C, Hovnanian A (2010) Par2 inactivation inhibits early production of TSLP, but not cutaneous inflammation, in Netherton syndrome adult mouse model. J Invest Dermatol 130:2736–2742
Bunyavanich S, Melen E, Wilk JB, Granada M, Soto-Quiros ME, Avila L et al (2011) Thymic stromal lymphopoietin (TSLP) is associated with allergic rhinitis in children with asthma. Clin Mol Allergy 9:1
Cario G, Zimmermann M, Roney R, Gesk S, Vater I, Harbott J et al (2010) Presence of the P2RY8-CRLF2 rearrangement is associated with a poor prognosis in non-high-risk precursor B-cell acute lymphoblastic leukemia in children treated according to the ALL-BFM 2000 protocol. Blood 115:5393–5397
Chapiro E, Russell L, Lainey E, Kaltenbach S, Ragu C, Della-Valle V et al (2010) Activating mutation in the TSLPR gene in B-cell precursor lymphoblastic leukemia. Leukemia 24:642–645
De Monte L, Reni M, Tassi EICD, Papa I, Recalde H, Braga M et al (2011) Intratumor T helper type 2 cell infiltrate correlates with cancer-associated fibroblast thymic stromal lymphpoietin production and reduced survival in pancreatic cancer. J Exp Med 208:469–478

Demehri S, Liu Z, Lee J, Lin MH, Crosby SD, Roberts CJ et al (2008) Notch-deficient skin induces a lethal systemic B-lymphoproliferative disorder by secreting TSLP, a sentinel for epidermal integrity. PLoS Biol 6:e123

Demehri S, Morimoto M, Holtzman MJ, Kopan R (2009) Skin-derived TSLP triggers progression from epidermal-barrier defects to asthma. PLoS Biol 7:e1000067

Duan W, Mehta AK, Magalhaes JG, Ziegler SF, Dong C, Philpott DJ et al (2010) Innate signals from Nod2 block respiratory tolerance and program T(H)2-driven allergic inflammation. J Allergy Clin Immunol 126:1284–1293

Ebner S, Nguyen VA, Forstner M, Wang YH, Wolfram D, Liu YJ et al (2007) Thymic stromal lymphopoietin converts human epidermal Langerhans cells into antigen-presenting cells that induce proallergic T cells. J Allergy Clin Immunol 119:982–990

Ensor HM, Schwab C, Russell LJ, Richards SM, Morrison H, Masic D et al (2011) Demographic, clinical, and outcome features of children with acute lymphoblastic leukemia and CRLF2 deregulation: results from the MRC ALL97 clinical trial. Blood 117:2129–2136

Eri RD, Adams RJ, Tran TV, Tong H, Das I, Roche DK et al (2011) An intestinal epithelial defect conferring ER stress results in inflammation involving both innate and adaptive immunity. Mucosal Immunol 4:354–364

Friend SL, Hosier S, Nelson A, Foxworthe D, Williams DE, Farr A (1994) A thymic stromal cell line supports in vitro development of surface IgM+ B cells and produces a novel growth factor affecting B and T lineage cells. Exp Hematol 22:321–328

Gao PS, Rafaels NM, Mu D, Hand T, Murray T, Boguniewicz M et al (2010) Genetic variants in thymic stromal lymphopoietin are associated with atopic dermatitis and eczema herpeticum. J Allergy Clin Immunol 125:1403–1407

Gill N, Finlay BB (2011) The gut microbiota: challenging immunology. Nat Rev Immunol 11:636–637

Gobert M, Treilleux I, Bendriss-Vermare N, Bachelot T, Goddard-Leon S, Arfi V et al (2009) Regulatory T cells recruited through CCL22/CCR4 are selectively activated in lymphoid infiltrates surrounding primary breast tumors and lead to an adverse clinical outcome. Cancer Res 69:2000–2009

Gregory SG, Schmidt S, Seth P, Oksenberg JR, Hart J, Prokop A et al (2007) Interleukin 7 receptor alpha chain (IL7R) shows allelic and functional association with multiple sclerosis. Nat Genet 39:1083–1091

Guo PF, Du MR, Wu HX, Lin Y, Jin LP, Li DJ (2010) Thymic stromal lymphopoietin from trophoblasts induces dendritic cell-mediated regulatory TH2 bias in the decidua during early gestation in humans. Blood 116:2061–2069

Han H, Xu W, Headley MB, Jessup HK, Lee KS, Omori M et al (2012) Thymic stromal lymphopoietin (TSLP)-mediated dermal inflammation aggravates experimental asthma. Mucosal Immunol 5(3):342–351

Hanabuchi S, Ito T, Park WR, Watanabe N, Shaw JL, Roman E et al (2010) Thymic stromal lymphopoietin-activated plasmacytoid dendritic cells induce the generation of FOXP3+ regulatory T cells in human thymus. J Immunol 184:2999–3007

Harada M, Hirota T, Jodo AI, Doi S, Kameda M, Fujita K et al (2009) Functional analysis of the thymic stromal lymphopoietin variants in human bronchial epithelial cells. Am J Respir Cell Mol Biol 40:368–374

Harada M, Hirota T, Jodo AI, Hitomi Y, Sakashita M, Tsunoda T et al (2011) TSLP promoter polymorphisms are associated with susceptibility to bronchial asthma. Am J Respir Cell Mol Biol 44(6):787–793

Hartgring SA, Willis CR, Dean CE, Broere F, van Eden W, Bijlsma JW et al (2011) Critical proinflammatory role of thymic stromal lymphopoietin and its receptor in experimental autoimmune arthritis. Arthritis Rheum 63:1878–1887

Harvey RC, Mullighan CG, Chen IM, Wharton W, Mikhail FM, Carroll AJ et al (2010) Rearrangement of CRLF2 is associated with mutation of JAK kinases, alteration of IKZF1, Hispanic/Latino ethnicity, and a poor outcome in pediatric B-progenitor acute lymphoblastic leukemia. Blood 115:5312–5321

He B, Xu W, Santini PA, Polydorides AD, Chiu A, Estrella J et al (2007) Intestinal bacteria trigger T cell-independent immunoglobulin A(2) class switching by inducing epithelial-cell secretion of the cytokine APRIL. Immunity 26:812–826

He R, Oyoshi MK, Garibyan L, Kumar L, Ziegler SF, Geha RS (2008) TSLP acts on infiltrating effector T cells to drive allergic skin inflammation. Proc Natl Acad Sci USA 105: 11875–11880

Headley MB, Zhou B, Shih WX, Aye T, Comeau MR, Ziegler SF (2009) TSLP conditions the lung immune environment for the generation of pathogenic innate and antigen-specific adaptive immune responses. J Immunol 182:1641–1647

Humphreys NE, Xu D, Hepworth MR, Liew FY, Grencis RK (2008) IL-33, a potent inducer of adaptive immunity to intestinal nematodes. J Immunol 180:2443–2449

Hunninghake GM, Soto-Quiros ME, Avila L, Kim HP, Lasky-Su J, Rafaels N et al (2010) TSLP polymorphisms are associated with asthma in a sex-specific fashion. Allergy 65(12): 1566–1575

Iliev ID, Mileti E, Matteoli G, Chieppa M, Rescigno M (2009a) Intestinal epithelial cells promote colitis-protective regulatory T cell differentiation through dendritic cell conditioning. Mucosal Immunol 2:340–350

Iliev ID, Spandoni I, Mileti E, Matteoli G, Sonzogni A, Sampietro GM et al (2009b) Human intestinal epithelial cells promote the differentiation of tolerogenic dendritic cells. Gut 58:1481–1489

Isaksen DE, Baumann H, Trobridge PA, Farr AG, Levin SD, Ziegler SF (1999) Requirement for stat5 in thymic stromal lymphopoietin-mediated signal transduction. J Immunol 163: 5971–5977

Isaksen DE, Baumann H, Zhou B, Nivollet S, Farr AG, Levin SD et al (2002) Uncoupling of proliferation and Stat5 activation in thymic stromal lymphopoietin-mediated signal transduction. J Immunol 168:3288–3294

Iseki M, Omori-Miyake M, Xu W, Sun X, Takaki S, Rawlings DJ et al (2012) Thymic stromal lymphopoietin (TSLP)-induced polyclonal B-cell activation and autoimmunity are mediated by CD4+ T cells and IL-4. Int Immunol 24:183–195

Ito T, Wang YH, Duramad O, Hori T, Delespesse GJ, Watanabe N et al (2005) TSLP-activated dendritic cells induce an inflammatory T helper type 2 cell response through OX40 ligand. J Exp Med 202:1213–1223

Jackson DJ, Johnston SL (2010) The role of viruses in acute exacerbations of asthma. J Allergy Clin Immunol 125:1178–1187

Jessup HK, Brewer AW, Omori M, Rickel EA, Budelsky AL, Yoon BR et al (2008) Intradermal administration of thymic stromal lymphopoietin induces a T cell- and eosinophil-dependent systemic Th2 inflammatory response. J Immunol 181:4311–4319

Jiang H, Hener P, Li J, Li M (2012) Skin thymic stromal lymphopoietin promotes airway sensitization to inhalant house dust mites leading to allergic asthma in mice. Allergy 67:1078–1082

Joyce JA, Pollard JW (2009) Microenvironmental regulation of metastasis. Nat Rev Cancer 9:239–252

Kamekura R, Kojima T, Koizumi J, Ogasawara N, Kurose M, Go M et al (2009) Thymic stromal lymphopoietin enhances tight-junction barrier function of human nasal epithelial cells. Cell Tissue Res 338:282–293

Kashyap M, Rochman Y, Spolski R, Samsel L, Leonard WJ (2011) Thymic stromal lymphopoietin is produced by dendritic cells. J Immunol 187:1207–1211

Kato A, Schleimer RP (2007) Beyond inflammation: airway epithelial cells are at the interface of innate and adaptive immunity. Curr Opin Immunol 19:711–720

Kido M, Tanaka J, Aoki N, Iwamoto S, Nishiura H, CHiba T et al (2010) *Helicobacter pylori* promotes the production of thymic stromal lymphopoietin by gastric epithelial cells and induces dendritic cell-mediated inflammatory Th2 responses. Infect Immun 78:108–114

Kimura S, Pawankar R, Mori S, Nonaka M, Masuno S, Yagi T et al (2011) Increased expression and role of thymic stromal lymphopoietin in nasal polyposis. Allergy Asthma Immunol Res 3:186–193

Kitajima M, Lee HC, Nakayama T, Ziegler SF (2011) TSLP enhances the function of helper type 2 cells. Eur J Immunol 41:1862–1871

Kouzaki H, O'Grady SM, Lawrence CB, Kita H (2009) Proteases induce production of thymic stromal lymphopoietin by airway epithelial cells through protease-activated receptor-2. J Immunol 183:1427–1434

Koyama K, Ozawa T, Hatsushika K, Ando T, Takano S, Wako M et al (2007) A possible role for TSLP in inflammatory arthritis. Biochem Biophys Res Commun 357:99–104

Larson RP, Zimmerli SC, Comeau MR, Itano A, Iseki M, Hauser C et al (2010) Dibutyl phthalate-induced thymic stromal lymphopoietin is required for Th2 contact hypersensitivity responses. J Immunol 184:2974–2984

Lee HC, Ziegler SF (2007) Inducible expression of the proallergic cytokine thymic stromal lymphopoietin in airway epithelial cells is controlled by NFkappaB. Proc Natl Acad Sci USA 104:914–919

Lee HC, Headley MB, Iseki M, Ikuta K, Ziegler SF (2008) Cutting edge: Inhibition of NF-kappaB-mediated TSLP expression by retinoid X receptor. Cutt Edge J Immunol 181:5189–5193

Lei L, Zhang Y, Yao W, Kaplan MH, Zhou B (2011) Thymic stromal lymphopoietin interferes with airway tolerance by suppressing the generation of antigen-specific regulatory T cells. J Immunol 186:2254–2261

Leonard WJ (2002) TSLP: finally in the limelight. Nat Immunol 3:605–607

Leung DY, Boguniewicz M, Howell MD, Nomura I, Hamid QA (2004) New insights into atopic dermatitis. J Clin Invest 113:651–657

Levin SD, Koelling RM, Friend SL, Isaksen DE, Ziegler SF, Perlmutter RM et al (1999) Thymic stromal lymphopoietin: a cytokine that promotes the development of IgM+ B cells in vitro and signals via a novel mechanism. J Immunol 162:677–683

Leyva-Castillo JM, Hener P, Jiang H, Li M (2012) TSLP produced by keratinocytes promotes allergen sensitization through skin and thereby triggers atopic march in mice. J Invest Dermatol 133(1):154–163

Li DJ, Guo PF (2009) The regulatory role of thymic stromal lymphopoietin (TSLP) in maternal-fetal immune tolerance during early human pregnancy. J Reprod Immunol 83:106–108

Li M, Messaddeq N, Teletin M, Pasquali JL, Metzger D, Chambon P (2005) Retinoid X receptor ablation in adult mouse keratinocytes generates an atopic dermatitis triggered by thymic stromal lymphopoietin. Proc Natl Acad Sci USA 102:14795–14800

Li M, Hener P, Zhang Z, Kato S, Metzger D, Chambon P (2006) Topical vitamin D3 and low-calcemic analogs induce thymic stromal lymphopoietin in mouse keratinocytes and trigger an atopic dermatitis. Proc Natl Acad Sci USA 103:11736–11741

Li M, Hener P, Zhang Z, Ganti KP, Metzger D, Chambon P (2009) Induction of thymic stromal lymphopoietin expression in keratinocytes is necessary for generating an atopic dermatitis upon application of the active vitamin D3 analogue MC903 on mouse skin. J Invest Dermatol 129:498–502

Li Y, Li HJ, Ji F, Zhang X, Wang R, Hao JQ et al (2010) Thymic stromal lymphopoietin promotes lung inflammation through activation dendritic cells. J Asthma 47:117–123

Lundmark F, Duvefelt K, Iacobaeus E, Kockum I, Wallström E, Khademi M et al (2007) Variation in interleukin 7 receptor alpha chain (IL7R) influences risk of multiple sclerosis. Nat Genet 39:1108–1113

Magalhaes JG, Rubino SJ, Travassos LH, Le Bourhis L, Duan W, Sellge G et al (2011) Nucleotide oligomerization domain-containing proteins instruct T cell helper type 2 immunity through stromal activation. Proc Natl Acad Sci USA 108:14896–14901

Mantovani A, Romero P, Palucka AK, Marincola FM (2008) Tumour immunity: effector response to tomour and role of the microenvironment. Lancet 371:771–783

Massacand JC, Stettler RC, Meier R, Humphreys NE, Grencis RK, Marsland BJ et al (2009) Helminth products bypass the need for TSLP in Th2 immune responses by directly modulating dendritic cell function. Proc Natl Acad Sci USA 106:13968–13973

Mazzucchelli R, Hixon JA, Spolski R, Chen X, Li WQ, Hall VL et al (2008) Development of regulatory T cells requires IL-7Ralpha stimulation by IL-7 or TSLP. Blood 112:3283–3292

Ménétrier-Caux C, Gobert M, Caux C (2009) Differences in tumor regulatory T-cell localization and activation status impact patient outcome. Cancer Res 69:7895–7898

Miyata M, Hatsushika K, Ando T, Shimokawa N, Ohnuma Y, Katoh R et al (2008) Mast cell regulation of epithelial TSLP expression plays an important role in the development of allergic rhinitis. Eur J Immunol 38:1487–1492

Moon PD, Choi IH, Kim HM (2011) Naringenin suppresses the production of thymic stromal lymphopoietin through the blockade of RIP2 and caspase-1 signal cascade in mast cells. Eur J Pharmacol 671:128–132

Morshed M, Yousefi S, Stöckle C, Simon HU, Simon D (2012) Thymic stromal lymphopoietin stimulates the formation of eosinophil extracellular traps. Allergy 67:1127–1137

Mou Z, Xia J, Tan Y, Wang X, Zhang Y, Zhou B et al (2009) Overexpression of thymic stromal lymphopoietin in allergic rhinitis. Acta Otolaryngol 129:297–301

Mullighan CG, Collins-Underwood JR, Phillips LA, Loudin MG, Liu W, Zhang J et al (2009) earrangement of CRLF2 in B-progenitor- and Down syndrome-associated acute lymphoblastic leukemia. Nat Genet 41:1243–1246

Nagata Y, Kamijuku H, Taniguchi M, Ziegler S, Seino K (2007) Differential role of thymic stromal lymphopoietin in the induction of airway hyperreactivity and Th2 immune response in antigen-induced asthma with respect to natural killer T cell function. Int Arch Allergy Immunol 144:305–314

Nguyen KD, Vanichsarn C, Nadeau KC (2010) TSLP directly impairs pulmonary Treg function: association with aberrant tolerogenic immunity in asthmatic airway. Allergy Asthma Clin Immunol 6:4

Nishiura H, Kido M, Aoki N, Iwamoto S, Maruoka R, Ikeda A et al (2012) Increased susceptibility to autoimmune gastritis in thymic stromal lymphopoietin receptor-deficient mice. J Immunol 188:190–197

Noble CL, Abbas AR, Cornelius J, Lees CW, Ho GT, Toy K et al (2008) Regional variation in gene expression in the healthy colon is dysregulated in ulcerative colitis. Gut 57:1398–1405

Noble CL, Abbas AR, Lees CW, Cornelius J, Toy K, Modrusan Z et al (2010) Characterization of intestinal gene expression profiles in Crohn's disease by genome-wide microarray analysis. Inflamm Bowel Dis 16:1717–1728

Oh MH, Oh SY, Yu J, Myers AC, Leonard WJ, Liu YJ et al (2011) IL-13 induces skin fibrosis in atopic dermatitis by thymic stromal lymphopoietin. J Immunol 186:7232–7242

Olkhanud PB, Rochman Y, Bodogai M, Malchinkhuu E, Wejksza K, Xu M et al (2011) Thymic stromal lymphopoietin is a key mediator of breast cancer progression. J Immunol 186:5656–5662

Omori M, Ziegler S (2007) Induction of IL-4 expression in CD4(+) T cells by thymic stromal lymphopoietin. J Immunol 178:1396–1404

Pandey A, Ozaki K, Baumann H, Levin SD, Puel A, Farr AG et al (2000) Cloning of a receptor subunit required for signaling by thymic stromal lymphopoietin. Nat Immunol 1:59–64

Park LS, Martin U, Garka K, Gliniak B, Di Santo JP, Muller W et al (2000) Cloning of the murine thymic stromal lymphopoietin (TSLP) receptor: formation of a functional heteromeric complex requires interleukin 7 receptor. J Exp Med 192:659–670

Pedroza-Gonzalez A, Xu K, Wu TC, Aspord C, Tindle S, Marches F et al (2011) Thymic stromal lymphopoietin fosters human breast tumor growth by promoting type 2 inflammation. J Exp Med 208:479–490

Pu HH, Duan J, Wang Y, Fan DX, Li DJ, Jin LP (2012) Thymic stromal lymphopoietin promotes the proliferation of human trophoblasts via phosphorylated STAT3-mediated c-Myc upregulation. Placenta 33:387–391

Quentmeier H, Drexler HG, Fleckenstein D, Zaborski M, Armstrong A, Sims JE et al (2001) Cloning of human thymic stromal lymphopoietin (TSLP) and signaling mechanisms leading to proliferation. Leukemia 15:1286–1292

Ramalingam TR, Pesce JT, Mentink-Kane MM, Madala S, Cheever AW, Comeau MR et al (2009) Regulation of helminth-induced Th2 responses by thymic stromal lymphopoietin. J Immunol 182:6452–6459

Reardon C, Lechmann M, Brüstle A, Gareau MG, Shuman N, Philpott D et al (2011) Thymic stromal lymphopoetin-induced expression of the endogenous inhibitory enzyme SLPI mediates recovery from colonic inflammation. Immunity 35:223–235

Reche PA, Soumelis V, Gorman DM, Clifford T, Liu M, Travis M et al (2001) Human thymic stromal lymphopoietin preferentially stimulates myeloid cells. J Immunol 167:336–343

Rimoldi M, Chieppa M, Salucci V, Avogadri F, Sonzogni A, Sampietro GM et al (2005) Intestinal immune homeostasis is regulated by the corsstalk between epithelial cells and dendritic cells. Nat Immunol 6:507–514

Rochman Y, Leonard WJ (2008) The role of thymic stromal lymphpoietin in CD8+ T cell homeostasis. J Immunol 181:7699–7705

Rochman I, Watanabe N, Arima K, Liu YJ, Leonard WJ (2007) Cutting edge: direct action of thymic stromal lymphopoietin on activated human CD4+ T cells. J Immunol 178:6720–6724

Rochman Y, Kashyap M, Robinson GW, Sakamoto K, Gomez-Rodriguez J, Wagner KU et al (2010) Thymic stromal lymphopoietin-mediated STAT5 phosphorylation via kinases JAK1 and JAK2 reveals a key difference from IL-7-induced signaling. Proc Natl Acad Sci USA 107:19455–19460

Roll JD, Reuther GW (2010) CRLF2 and JAK2 in B-progenitor acute lymphoblastic leukemia: a novel association in oncogenesis. Cancer Res 70:7347–7352

Rothenberg ME, Spergel JM, Sherrill JD, Annaiah K, Martin LJ, Cianferoni A et al (2010) Common variants at 5q22 associate with pediatric eosinophilic esophagitis. Nat Genet 42:289–291

Russell LJ, Capasso M, Vater I, Akasaka T, Bernard OA, Calasanz MJ et al (2009) Deregulated expression of cytokine receptor gene, CRLF2, is involved in lymphoid transformation in B-cell precursor acute lymphoblastic leukemia. Blood 114:2688–2698

Semlali A, Jacques E, Koussih L, Gounni AS, Chakir J (2010) Thymic stromal lymphopoietin-induced human asthmatic airway epithelial cell proliferation through an IL-13-dependent pathway. J Allergy Clin Immunol 125:844–850

Seshasayee D, Lee WP, Zhou M, Shu J, Suto E, Zhang J et al (2007) In vivo blockade of OX40 ligand inhibits thymic stromal lymphopoietin driven atopic inflammation. J Clin Invest 117:3868–3878

Shamim Z, Muller K, Svejgaard A, Poulsen LK, Bodtger U, Ryder LP (2007) Association between genetic polymorphisms in the human interleukin-7 receptor alpha-chain and inhalation allergy. Int J Immunogenet 34:149–151

Sherrill JD, Gao PS, Stucke EM, Blanchard C, Collins MH, Putnam PE et al (2010) Variants of thymic stromal lymphopoietin and its receptor associate with eosinophilic esophagitis. J Allergy Clin Immunol 126:160–165

Shi L, Leu SW, Xu F, Zhou X, Yin H, Cai L et al (2008) Local blockade of TSLP receptor alleviated allergic disease by regulating airway dendritic cells. Clin Immunol 129:202–210

Shikotra A, Choy DF, Ohri CM, Doran E, Butler C, Hargadon B et al (2011) Increased expression of immunoreactive thymic stromal lymphopoietin in patients with severe asthma. J Allergy Clin Immunol 129(1):104-11.e1–104-11.e9

Sims JE, Williams DE, Morrissey PJ, Garka K, Foxworthe D, Price V et al (2000) Molecular cloning and biological characterization of a novel murine lymphoid growth factor. J Exp Med 192:671–680

Siracusa MC, Saenz SA, Hill DA, Kim BS, Headley MB, Doering TA et al (2011) TSLP promotes interleukin-3-independent basophil haematopoiesis and type 2 inflammation. Nature (Lond) 477:229–233

Siracusa MC, Wojno ED, Artis D (2012) Functional heterogeneity in the basophil cell lineage. Adv Immunol 115:141–159

Smelter DF, Sathish V, Thompson MA, Pabelick CM, Vassallo R, Prakash YS (2010) Thymic stromal lymphopoietin in cigarette smoke-exposed human airway smooth muscle. J Immunol 185:3035–3040
Sokol CL, Chu NQ, Yu S, Nish SA, Laufer TM, Medzhitov R (2009) Basophils function as antigen-presenting cells for an allergen-induced T helper type 2 response. Nat Immunol 10:713–720
Soumelis V, Reche PA, Kanzler H, Yuan W, Edward G, Homey B et al (2002) Human epithelial cells trigger dendritic cell mediated allergic inflammation by producing TSLP. Nat Immunol 3:673–680
Tanaka J, Watanabe N, Kido M, Saga K, Akamatsu T, Nishio A et al (2009) Human TSLP and TLR3 ligands promote differentiation of Th17 cells with a central memory phenotype under Th2-polarizing conditions. Clin Exp Allergy 39:89–100
Tanaka J, Saga K, Kido M, Nishiura H, Akamatsu T, CHiba T et al (2010) Proinflammatory Th2 cytokines induce production of thymic stromal lymphopoietin in human colonic epithelial cells. Dig Dis Sci 55:1896–1904
Taneda S, Segerer S, Hudkins KL, Cui Y, Wen M, Segerer M et al (2001) Cryoglobulinemic glomerulonephritis in thymic stromal lymphopoietin transgenic mice. Am J Pathol 159: 2355–2369
Tasian SK, Loh ML (2011) Understanding the biology of CRLF2-overexpressing acute lymphoblastic leukemia. Crit Rev Oncog 16:13–24
Taylor BC, Zaph C, Troy AE, Du Y, Guild KJ, Comeau MR et al (2009) TSLP regulates intestinal immunity and inflammation in mouse models of helminth infection and colitis. J Exp Med 206:655–667
Tonozuka Y, Fujio K, Sugiyama T, Nosaka T, Hirai M, Kitamura T (2001) Molecular cloning of a human novel type I cytokine receptor related to delta1/TSLPR. Cytogenet Cell Genet 93:23–25
Torgerson DG, Ampleford EJ, Chiu GY, Gauderman WJ, Gignoux CR, Graves PE et al (2011) Meta-analysis of genome-wide association studies of asthma in ethnically diverse North American populations. Nat Genet 43:887–892
Uller L, Leino M, Bedke N, Sammut D, Green B, Lau L et al (2010) Double-stranded RNA induces disproportionate expression of thymic stromal lymphopoietin versus interferon-beta in bronchial epithelial cells from donors with asthma. Thorax 65:626–632
Vosshenrich CA, Cumano A, Muller W, Di Santo JP, Vieira P (2003) Thymic stromal-derived lymphopoietin distinguishes fetal from adult B cell development. Nat Immunol 4:773–779
Wang YH, Ito T, Wang YH, Homey B, Watanabe N, Martin R et al (2006) Maintenance and polarization of human TH2 central memory T cells by thymic stromal lymphopoietin-activated dendritic cells. Immunity 24:827–838
Watanabe N, Hanabuchi S, Soumelis V, Yuan W, Ho S, Waal-Malefyt R et al (2004) Human thymic stromal lymphopoietin promotes dendritic cell-mediated CD4+ T cell homeostatic expansion. Nat Immunol 5:426–434
Watanabe N, Wang YH, Lee HK, Ito T, Wang YH, Cao W et al (2005) Hassall's corpuscles instruct dendritic cells to induce CD4+CD25+ regulatory T cells in human thymus. Nature (Lond) 436:1181–1185
Wong CK, Hu S, Cheung PF, Lam CW (2009) TSLP induces chemotactic and pro-survival effects in eosinophils: implications in allergic inflammation. Am J Respir Cell Mol Biol 43(3): 305–315
Wu HX, Guo PF, Jin LP, Liang SS, Li DJ (2010) Functional regulation of thymic stromal lymphopoietin on proliferation and invasion of trophoblasts in human first-trimester pregnancy. Hum Reprod 25:1146–1152
Wynn TA (2004) Fibrotic disease: the T(H)1/T(H)2 paradign. Nat Rev Immunol 4:583–594
Xu G, Zhang L, Wang DY, Xu R, Liu Z, Han DM et al (2010) Opposing roles of IL-17A and IL-25 in the regulation of TSLP production in human nasal epithelial cells. Allergy 65:581–589
Ying S, O'Connor B, Ratoff J, Meng Q, Mallett K, Cousins D et al (2005) Thymic stromal lymphopoietin expression is increased in asthmatic airways and correlates with expression of Th2-attracting chemokines and disease severity. J Immunol 174:8183–8190

Ying S, O'Connor B, Ratoff J, Meng Q, Cousins D, Zhang G et al (2008) Expression and cellular provenance of thymic stromal lymphopoietin and chemokines in patients with severe asthma and chronic obstructive pulmonary disease. J Immunol 181:2790–2798

Yoda A, Yoda Y, Chiaretti S, Bar-Natan M, Mani K, Rodig SJ et al (2010) Functional screening identifies CRLF2 in precursor B-cell acute lymphoblastic leukemia. Proc Natl Acad Sci USA 107:252–257

Yoo J, Omori M, Gyarmati D, Zhou B, Aye T, Brewer A et al (2005) Spontaneous atopic dermatitis in mice expressing an inducible thymic stromal lymphopoietin transgene specifically in the skin. J Exp Med 202:541–549

Zaph C, Troy AE, Taylor BC, Berman-Booty LD, Guild KJ, Du Y et al (2007) Epithelial-cell-intrinsic IKK-á expression regulates intestinal immune homeostasis. Nature (Lond) 446:552–556

Zeuthen LH, Fink LN, Frokiaer H (2008) Epithelial cells prime the immune response to an array of gut-derived commensals towards a tolerogenic phenotype through distinct actions of thymic stromal lymphopoietin and transforming growth factor-beta. Immunology 123:197–208

Zhang K, Shan L, Rahman MS, Unruh H, Halayko AJ, Gounni AS (2007) Constitutive and inducible thymic stromal lymphopoietin expression in human airway smooth muscle cells: role in chronic obstructive pulmonary disease. Am J Physiol Lung Cell Mol Physiol 293:L375–L382

Zhang Z, Hener P, Frossard N, Kato S, Metzger D, Li M et al (2009) Thymic stromal lymphopoietin overproduced by keratinocytes in mouse skin aggravates experimental asthma. Proc Natl Acad Sci USA 106:1536–1541

Zhang F, Huang G, Hu B, Song Y, Shi Y (2011) A soluble thymic stromal lymphopoietin (TSLP) antagonist, TSLPR-immunoglobulin, reduces the severity of allergic disease by regulating pulmonary dendritic cells. Clin Exp Immunol 164:256–264

Zhou B, Comeau MR, De Smedt T, Liggitt HD, Dahl ME, Lewis DB et al (2005) Thymic stromal lymphopoietin as a key initiator of allergic airway inflammation in mice. Nat Immunol 6:1047–1053

Zhou B, Headley MB, Aye T, Tocker J, Comeau MR, Ziegler SF (2008) Reversal of thymic stromal lymphopoietin-induced airway inflammation through inhibition of Th2 responses. J Immunol 181:6557–6562

Ziegler SF (2010) The role of thymic stromal lymphopoietin (TSLP) in allergic disorders. Curr Opin Immunol 22:795–799

Ziegler SF, Artis D (2010) Sensing the outside world: TSLP regulates barrier immunity. Nat Immunol 11:289–293

Part III
Cytokines in Immunological Tolerance and Anti-inflammation

Chapter 13
Interleukin-10: Cytokines in Anti-inflammation and Tolerance

Ashleigh Howes, Philippa Stimpson, Paul Redford, Leona Gabrysova, and Anne O'Garra

Abstract Interleukin (IL)-10 is an immunosuppressive cytokine produced by many cells of the innate and adaptive immune response, including macrophages, dendritic cells, B cells, and T cells. A major role of IL-10 is to act as a feedback regulator of the immune response by inhibiting the production of inflammatory cytokines by innate cells such as macrophages and dendritic cells. Thus, the production of IL-10 in response to microbial flora is required to inhibit the development of colitis and, during infection, to inhibit an over-exuberant immune response that may potentially result in morbidity and mortality to the host. Thus, appropriate levels of IL-10 should be induced to inhibit host damage, but not to the level that they inhibit a protective immune response to a pathogen which could result in chronic infection. Regulation of the immune response by IL-10 has been shown to play a role in control of allergic, inflammatory, and autoimmune diseases, and further knowledge may advance our ability to use antagonists or agonists of IL-10 for immune therapies.

Keywords Anti-inflammatory • Cytokine • Dendritic cells • Macrophages • T cells

13.1 Introduction to Interleukin-10 and the Need for Immune Regulation

The immune system is composed of a complex network of specialized cell types. These cells act in a coordinated fashion to protect the host from infection with a range of pathogens including bacteria, viruses, parasites, and fungi.

The AOG lab is funded by MRC grant U117565642. Ashleigh Howes and Philippa Stimpson contributed equally.

A. Howes • P. Stimpson • P. Redford • L. Gabrysova • A. O'Garra (✉)
Division of Immunoregulation, The MRC National Institute for Medical Research, The Ridgeway, Mill Hill, London, NW7 1AA, UK
e-mail: aogarra@nimr.mrc.ac.uk

T. Yoshimoto and T. Yoshimoto (eds.), *Cytokine Frontiers: Regulation of Immune Responses in Health and Disease*, DOI 10.1007/978-4-431-54442-5_13, © Springer Japan 2014

An immune response consists of the innate and adaptive phases with major overlap of these two responses. The immune response is initiated by the detection of microbes in the tissues by resident macrophages, dendritic cells (DCs), and other cells of the innate immune system. These cells express pattern recognition receptors (PRR) on their surfaces that recognize pathogens and initiate a response. Macrophages and DCs are phagocytic cells, and so, if activated and other microbes, can attempt to contain the infection by engulfing and destroying microbes. Within hours of infection, macrophages, and DCs produce pro-inflammatory cytokines [e.g., tumor necrosis factor (TNF)-α, IL-1, IL-12, and IL-6] and chemokines (e.g., CXCL8, CCL3, and CCL4) to attract other immune cells to the region such as monocytes, neutrophils, natural killer (NK) cells, eosinophils, and mast cells. These infiltrating cells use various effector mechanisms to attempt to eliminate the pathogen. This process causes inflammation at the site of infection.

The innate immune response is critical for the early containment of infection. However, complete elimination of the pathogen and long-lasting immunity to reinfection requires activation of an adaptive immune response: this involves the recruitment and activation of lymphocytes, including $CD4^+$ T helper cells, $CD8^+$ T killer cells, and B cells. In an uninfected individual these cells are largely located in the lymphoid organs. DCs are largely responsible for initiating the adaptive response by taking up antigens and then migrating from the site of infection to nearby lymph nodes and activating specific naive T cells that reside there. DCs activate T cells by (1) presenting antigens derived from the pathogen in the context of major histocompatibility (MHC) molecules to the T-cell receptor (TCR), (2) producing cytokines that guide T-cell differentiation, and (3) providing co-stimulatory signals through the binding of co-stimulatory molecules upregulated on the surface of the DC to their receptors on the T cell.

Within the lymph node, activated $CD4^+$ T cells expand and differentiate into specific T helper (Th) cell subsets including Th1, Th2, and Th17 cells. Each subset is characterized by the production of hallmark cytokines: interferon (IFN)-γ for Th1 (Murphy et al. 2000); IL-4, IL-5, and IL-13 for Th2 (Zheng and Flavell 1997); and IL-17 for Th17 (Ivanov et al. 2006). Each subset is important in the defence against specific types of pathogen. For example, Th1 cells are important to activate macrophages for the killing of intracellular pathogens, Th2 cells combat parasitic helminths, and Th17 cells are major components of the immune response against extracellular bacterial or fungal infections. Activated $CD8^+$ T cells also expand within the lymph node. These cells have cytotoxic abilities mediated by factors including perforins and granzymes that enable them to kill infected cells. Activated $CD4^+$ and $CD8^+$ T cells then migrate back to the site of infection where they carry out their effector functions, the major extent of which are mediated by the action of their hallmark cytokines. This action often involves enhancing the recruitment and activation of innate cells by the specific cytokine. For example, Th1 cells produce IFN-γ, which promotes macrophage cytokine production and antimicrobial activity.

B cells can be activated by a combination of direct antigen recognition, DC help, and T-cell help. Activated B cells differentiate to plasma cells and produce antibodies or immunoglobulins (Ig). There are several isotypes of antibody including IgM, IgG,

IgE, IgD, and IgA. Each isotype is associated with a particular function, for example, the targetting of pathogens for killing by other immune cells such as macrophages (IgG), or providing protection against pathogens at mucosal surfaces (IgA).

Together, these cellular expansions and influxes mediate inflammation, which promotes the clearance of infection. However, for the immune system to work safely and effectively it is essential that:

1. Inflammation is regulated so that it is allowed to eliminate the pathogen, but does not become excessive and cause immune-mediated damage (immunopathology) to the host.
2. The immune system is restrained from generating inappropriate responses to nonharmful foreign particles such as pollen or commensal bacteria that colonize the gut without causing disease.
3. The immune system is prevented from generating inappropriate immune responses to the body itself (autoimmunity).

For these criteria to be upheld, regulatory mechanisms exist to dampen immune responses when necessary. One of these mechanisms is the production of the anti-inflammatory cytokine interleukin-10 (IL-10). IL-10 was first characterized as a protein that inhibited the production of cytokines from Th1 cells via an indirect mechanism (Fiorentino et al. 1989). Because of this suppressive function, IL-10 was initially termed cytokine synthesis inhibitory factor (CSIF). After more than 20 years of research, IL-10 is now considered one of the most important anti-inflammatory cytokines, with roles in numerous infections, allergies, and autoimmune conditions to limit the immune responses and prevent host damage (Moore et al. 2001). IL-10 is actually the founding member of the IL-10 family cytokines, which also include IL-19, IL-20, IL-22, IL-24, IL-26, IL-28A and –B, and IL-29. The functions of these cytokines range from immune defense to tissue remodeling and wound healing, but IL-10 is the only cytokine in this family that is currently well established to have direct immunosuppressive activity (Ouyang et al. 2011). In this chapter, current knowledge of the sources, regulation, and functions of IL-10 in different immunological settings is discussed.

13.2 Production, Regulation, and Molecular Function of IL-10

13.2.1 Cellular Sources of IL-10

Th2 cells were the first cells described to produce IL-10 (Fiorentino et al. 1989). However, since then, in vitro and in vivo studies have collectively shown that almost all immune cell types have the capacity to produce IL-10 (Saraiva and O'Garra 2010).

In the innate immune system, macrophages and DCs produce IL-10 in response to microbial stimuli. Of note, plasmacytoid DCs (pDC), which are specialized in antiviral activity, are an exception as they have not yet been shown to make IL-10 (Boonstra et al. 2006; Kaiser et al. 2009). Mast cells, eosinophils, NK cells, and neutrophils are also sources of IL-10 (Saraiva and O'Garra 2010).

In the adaptive immune system, all subsets of effector $CD4^+$ T cells including Th1, Th2, and Th17 cells are able to produce IL-10 (Saraiva and O'Garra 2010). $Foxp3^+$ $CD4^+$ regulatory T cells (Tregs) comprise another T-cell subset, important in maintaining immunological tolerance and suppressing immune responses to infection. Tregs expressing the forkhead box protein 3 (Foxp3) transcription factor can also produce IL-10, which at least in part mediates their suppressive activity (Bacchetta et al. 2007; Maynard and Weaver 2008; Lloyd and Hawrylowicz 2009). $CD8^+$ T cells are also able to produce IL-10 (Tanchot et al. 1998; Gilliet and Liu 2002). Although B cells have mainly been studied in the context of antibody production, they can produce IL-10 (O'Garra et al. 1990; O'Garra et al. 1992), and it is recently becoming appreciated that B-cell-derived IL-10 may have an important role in infection and autoimmune diseases (Fillatreau 2011; Mauri and Bosma 2012).

Thus, during the course of an immune response, IL-10 can potentially be produced by almost any cell type in the innate or adaptive phase.

13.2.2 Induction and Molecular Regulation of IL-10 Production

Immune cells must be stimulated to produce IL-10. However, the exact stimulus required and how much IL-10 is produced is dependent on the cell type. Several points of regulation exist along the pathways that lead to IL-10 production, making it a highly controlled process. This control is important as inappropriate production of IL-10 can be detrimental to the host because of the powerful suppressive activity of this cytokine.

13.2.2.1 The Regulation of IL-10 in Macrophages and DCs

The production of IL-10 and indeed other cytokines by macrophages and DCs is largely dependent on stimulation by PRRs. These receptors recognize conserved molecular motifs on or within pathogens. There are several families of PRR, the most studied of which is the Toll-like receptor (TLR) family (Akira and Takeda 2004). Activation of signalling through several TLRs has been described to induce IL-10 in macrophages and DCs (Saraiva and O'Garra 2010): these include TLR4, which binds lipopolysaccharide (LPS) present in the cell wall of gram-negative bacteria; TLR2, which binds lipoproteins present in the cell wall of gram-positive bacteria; and TLR9, which binds specific motifs present in bacterial DNA (Saraiva and O'Garra 2010). Although both macrophages and DCs express these receptors (although at different levels), macrophages often produce notably more IL-10 than DCs, highlighting the concept that distinct cell types have different capacities to produce IL-10 (Boonstra et al. 2006; Kaiser et al. 2009).

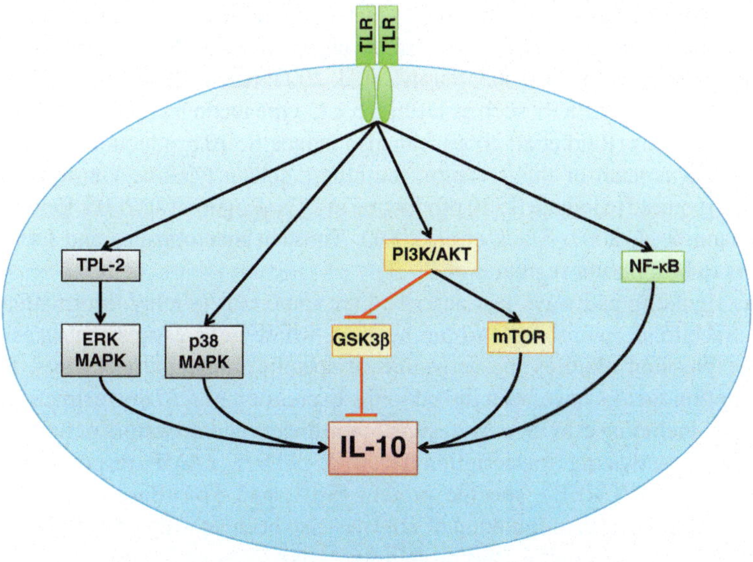

Fig. 13.1 Interleukin-10 (IL-10) is regulated by several pathways downstream of Toll-like receptor (*TLR*) stimulation in macrophages and dendritic cells (DCs). Ligand binding of TLRs activates the MAP kinases ERK or p38, which positively regulate IL-10. TPL-2 is important for the upstream activation of ERK. The PI3K/AKT pathway activates mTOR, which promotes IL-10 production. The PI3K/AKT pathway additionally inhibits GSK3β, an inhibitor of IL-10. The NF-κB pathway also modulates IL-10 production. The combination of signals from these various pathways determines the overall level of IL-10 production. Note that several of these pathways also mediate the production of pro-inflammatory cytokines

Downstream of TLRs, several signaling pathways are activated and many of these are involved in the regulation of IL-10 (Fig. 13.1). Mitogen-activated protein-kinase (MAPK) pathways, which lead to the activation of signaling proteins known as MAP kinases, are important for the positive regulation of IL-10. In particular, extracellular-regulated kinase (ERK) and p38 MAP kinases have been shown to be critical in the production of IL-10 in murine macrophages and DCs (Yi et al. 2002; Dillon et al. 2004; Banerjee et al. 2006; Jarnicki et al. 2008; Kim et al. 2008; Kaiser et al. 2009) (Fig. 13.1). In fact, differential ERK activation has been linked to the differing level of IL-10 produced by macrophages and DCs (Kaiser et al. 2009). ERK is activated by the upstream kinase tumor progression locus 2 (TPL-2), and several studies have also shown the importance of TPL-2 in the regulation of IL-10 (Banerjee et al. 2006; Kaiser et al. 2009). It has also been shown in human monocytes and DCs that p38 positively regulates IL-10 (Foey et al. 1998; Jarnicki et al. 2008). The phosphatidylinositide 3-kinase (PI3K)/AKT pathway is also activated by the TLR signaling pathway and promotes IL-10 production in two ways: first, by activating the kinase mammalian target of rapamycin (mTOR), and second, by inhibiting glycogen synthase kinase beta (GSK3β), a negative regulator of IL-10 (Ohtani et al. 2008; Weichhart et al. 2008) (Fig. 13.1). The NF-κB pathway, which is well established to mediate pro-inflammatory cytokine production downstream of TLRs (Kawai and Akira 2007), has also been implicated in the modulation of IL-10 production (Kanters et al. 2003; Saraiva et al. 2005; Cao et al. 2006) (Fig. 13.1).

Non-TLR family PRRs such as Dectin-1, a C-type lectin receptor that recognizes fungal β-glucans (Reid et al. 2009), can also induce IL-10 production (Rogers et al. 2005). Downstream of this receptor, the kinase spleen tyrosine kinase (Syk) and ERK are required to induce IL-10 production in DCs (Rogers et al. 2005; LeibundGut-Landmann et al. 2007; Slack et al. 2007). Thus, in macrophages and DCs, ERK appears to be a master regulator of IL-10.

The signaling pathways just described are important to relay information from the extracellular environment to the nucleus, where gene expression takes place. Gene expression requires the activation of specific transcription factors. Several transcription factors have been linked to the expression of *Il10* downstream of TLR activation including c-FOS, c-Maf, signal transducers and activators of transcription 3 (STAT3), activating transcription factor 1 (ATF1), cAMP response element-binding protein (CREB), specific protein (Sp)1, and Sp3 (Saraiva and O'Garra 2010). Importantly, the regulation of *Il10* has also been shown to be affected by the structure of the chromatin (the protein and DNA complexes that form chromosomes) at the *Il10* locus. For example, in macrophages, the activation of ERK induces chromatin remodeling, which exposes the *Il10* promoter to transcription factors (Zhang et al. 2006).

IL-10 production is further regulated by posttranscriptional mechanisms: these affect the stability of *Il10* mRNA and therefore the amount of *Il10* mRNA available to translate into IL-10 protein. Recently it has been shown that certain micro RNAs (small RNAs that can bind larger mRNAs) are able to directly or indirectly affect the stability of *Il10* mRNA. For example, the micro RNA hsa-miR-106a binds to *Il10* mRNA and promotes its degradation (Sharma et al. 2009). *Il10* mRNA is also a target of the mRNA-binding protein tristetraproline (TTP) which promotes *Il10* mRNA degradation (Stoecklin et al. 2008). In fact, the inhibition of TTP is a mechanism employed by p38 to promote IL-10 production (Tudor et al. 2009). Posttranscriptional regulation is therefore important in determining the level of IL-10 production. Although most studies investigating this aspect of IL-10 regulation have been conducted in macrophages, these mechanisms may also operate in other innate and adaptive cell types.

In addition to the recognition of microbial products by PRRs, other cytokines present in the immediate cellular environment can modulate the level of IL-10 production. For example, if macrophages are stimulated through TLR ligation in the presence of IFN-γ, produced by various cell types such as Th1 cells, IL-10 production is reduced compared to macrophages receiving signals through TLR alone (Hu et al. 2006). In contrast, if macrophages are stimulated through TLR in the presence of type I IFN, which can be produced by the macrophage itself or other surrounding cells, the production of IL-10 is enhanced (Chang et al. 2007a; Howes et al., unpublished data; Ewbank et al., unpublished data).

Thus, several signals can be integrated to fine tune innate IL-10 production according to the challenge to the immune response encountered.

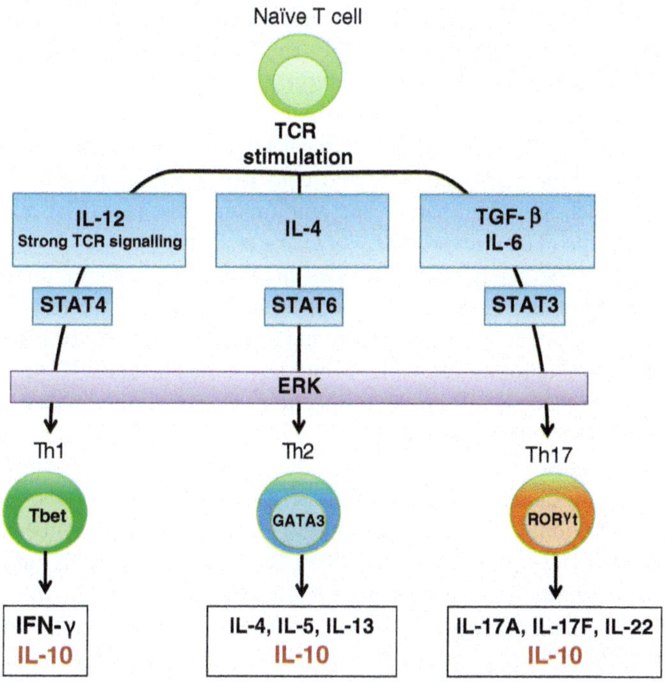

Fig. 13.2 Distinct and common factors mediate the regulation of IL-10 in Th1, Th2, and Th17 cells. Upon activation, naïve CD4+ T cells differentiate into Th1, Th2, and Th17 cells. The factors required for differentiation have been shown to also be required for the induction of IL-10 by each subset (*blue boxes*). ERK is thought to be required by all T-cell subsets to produce IL-10 (*purple box*). *TCR*, T-cell receptor

13.2.2.2 Pathways of IL-10 Regulation in T Helper Cells

Upon activation under specific conditions, naïve CD4+ T cells differentiate along distinct molecular pathways into different T helper cell subsets (Th1, Th2, and Th17). Each subset expresses a unique combination of cytokines that promotes their specific effector functions. The distinct differentiation pathways are determined by the cytokines present in the extracellular environment, the downstream signaling of which is dependent upon STAT molecules (Leonard and O'Shea 1998). The expression of a subset determining transcription factor, or master regulator, is also required for commitment to a particular T-cell lineage. Th1 cell development is driven predominantly by IL-12, which signals through STAT4 together with the Th1 master regulator Tbet (Murphy et al. 2000; Szabo et al. 2000). Th2 cell differentiation requires IL-4, which signals through STAT6, and expression of the master regulator GATA-binding protein 3 (GATA3) (Zheng and Flavell 1997). Th17 cell development is dependent on transforming growth factor (TGF)-β and IL-6, the latter of which signals through STAT3, and induction of the master regulator RORγt (Ivanov et al. 2006; Veldhoen et al. 2006) (Fig. 13.2). Although IL-10 was first described to be produced by mouse Th2 cells, it has recently been shown to be produced by all Th cell subsets and is therefore not a subset-specific cytokine (Saraiva and O'Garra 2010). Of note, the recently described Th9 subset that produces IL-9 as its signature cytokine also produces IL-10 (Veldhoen et al. 2008); this suggests that IL-10 production is a general mechanism used by Th cells to keep immune responses in check as inappropriate or uncontrolled effector T cells have been implicated in pathological responses such as inflammatory and autoimmune diseases (Th1 and Th17) or allergy and asthma (Th2) (reviewed by O'Garra et al. 2008).

However, the specific signaling cascades and transcription factor complexes that determine IL-10 production by the different Th subsets are still unclear (Saraiva and O'Garra 2010). Historically, Th2 cells are the most studied as IL-10 is co-regulated with their hallmark cytokines IL-4, IL-5, and IL-13. In Th2 cells, IL-4, STAT6, and GATA3 are critical for IL-10 production (Zhu et al. 2004; Shoemaker et al. 2006; Chang et al. 2007b). The master regulator GATA3 promotes IL-10 expression in Th2 cells by altering the chromatin structure of the *Il10* gene; however, GATA3 alone does not transactivate the *Il10* promoter (Shoemaker et al. 2006), illustrating the need for additional factors. An example is Jun, which has also been shown to regulate IL-10 in Th2 cells (Wang et al. 2005). Importantly, GATA3 is only expressed in Th2 cells, implying other mechanisms regulate IL-10 production in the different Th cell subsets that do not express GATA3. For example, in Th1 cells, IL-12, STAT4, and strong TCR signaling are essential for IL-10 production (Saraiva et al. 2009). Additionally, in Th1 cells, a role for the Notch signaling pathway in the induction of IL-10 has been identified (Rutz et al. 2008), and recently it has been shown that the Notch ligand Jagged1 mediates CD46-driven induction of IL-10-secreting Th1 cells in humans (Le Friec et al. 2012). Conversely, Ets-1 has been described to suppress IL-10 production in Th1 cells (Grenningloh et al. 2005; Lee et al. 2012), adding further complexity to the regulation of IL-10. In Th17 cells, TGF-β, IL-6, and STAT3 are required for IL-10 production (Stumhofer et al. 2007). In addition, c-Maf has also been shown to promote IL-10 production when introduced into Th cells by retroviral transduction (Xu et al. 2009). Furthermore, other cytokines other than those required for T-cell differentiation such as IL-21 or IL-27 can enhance IL-10 in the various Th subsets (Spolski et al. 2009; Hall et al. 2012).

Thus, in Th1, Th2, and Th17 subsets, it seems that following stimulation via the TCR, factors required for the differentiation of the actual subsets themselves are also required for the induction of IL-10 in addition to other modulatory factors. On the other hand, a common requirement for the MAP kinase ERK has been shown in all Th cell subsets for the production of IL-10 (Saraiva et al. 2009) (Fig. 13.2). Therefore, similar to macrophages and DCs, ERK seems to be a dominant regulator of IL-10 in Th cells.

A picture of IL-10 regulation is therefore emerging in which some factors, for example, ERK, may be required by all cell types to make IL-10, whereas other IL-10-promoting factors may be cell type specific and may also determine the levels of IL-10 produced. Further studies are required to decipher the complex IL-10 regulatory networks in various cell types.

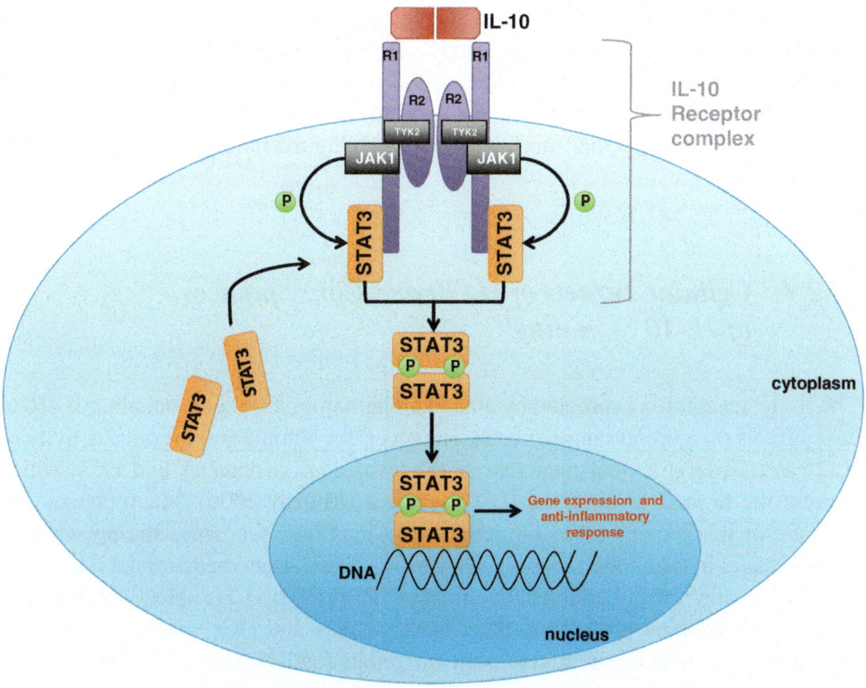

Fig. 13.3 The IL-10 receptor activates STAT3, which induces the anti-inflammatory response. When IL-10 binds to the IL-10 receptor, STAT3 molecules are recruited to the intracellular receptor domains. Here receptor-associated kinases phosphorylate (*P*) and activate the docked STAT3 molecules. Activated STAT3 molecules then dimerize and translocate to the nucleus where they bind DNA and activate the transcription of genes. These gene products mediate the anti-inflammatory response of IL-10

13.2.3 Signaling Pathways Induced by the IL-10 Receptor

The IL-10 receptor is composed of IL-10R1 and IL-10R2 chains. The IL-10R1 chain mediates receptor binding of the IL-10 molecule whereas the IL-10R2 chain is important for signal transduction (Moore et al. 2001). Some of the other IL-10 family cytokines (IL-22, IL-26, IL-28A, IL-28B, and IL-29) also use the IL-10R2 chain as part of their receptor, but only IL-10 additionally uses IL-10R1. Thus, IL-10R1 imparts specificity for IL-10 to the receptor complex (Ouyang et al. 2011).

IL-10 predominantly signals through the JAK/STAT pathway. Specifically, Janus kinase 1 (Jak1) and tyrosine kinase 2 (Tyk2) are the tyrosine kinases that activate STAT3 on binding of IL-10 to its receptor (Moore et al. 2001) (Fig. 13.3). The expression of STAT3-dependent genes mediates the anti-inflammatory response to IL-10 (Murray 2006), primarily by inhibiting the transcription of pro-inflammatory genes (Murray 2005). For example, it has been shown that IL-10 signaling can prevent the production of TNF by blocking the elongation of *Tnf* mRNA in human macrophages (Smallie et al. 2010). Despite much investigation, however, the full identification of these IL-10/STAT3-induced genes and how they function is still ongoing. Genome-wide studies that assess the genes regulated by STAT3 in response to IL-10, such as conducted recently (Hutchins et al. 2012), will help further our knowledge of the molecular mechanisms underlying the immunosuppressive activity of IL-10.

13.2.4 *Cellular Targets of IL-10 and Consequences of IL-10 Signaling*

The IL-10 receptor is expressed by almost all hematopoietic cells, enabling IL-10 to have effects on a broad range of cells. However, the immunosuppressive activity of IL-10 is to a large extent mediated at the level of macrophages and DCs, which express the highest level of the IL-10 receptor (Murray 2006). Macrophages and DCs are at the center of the immune response as they recognise the presence of microbes and initiate inflammation. They are also important mediators of the activation, differentiation and effector functions of Th cells. IL-10 is able to inhibit cytokine and chemokine production from macrophages and DCs, and their ability to migrate to lymphoid organs and present antigen to T cells (Moore et al. 2001). Thus, IL-10 can directly suppress innate immune cells and indirectly inhibit the activation and effector functions of T cells. This suppression of macrophage and DC activity leads to a broad dampening of the immune response. A recent study has suggested that IL-10 can directly inhibit Th17 cell activity (Huber et al. 2011) or promote Treg cell activity (Murai et al. 2009; Chaudhry et al. 2011), however the concept of IL-10 acting directly on T cells is less well established.

Although IL-10 is considered a predominantly immunosuppressive cytokine, IL-10 can enhance the cytotoxicity of $CD8^+$ T cells and promote survival and antibody production from B cells (Moore et al. 2001). These roles of IL-10 may have effects during infection, anti-cancer immune responses, or autoimmune disorders, for example, Systemic Lupus Erythematosus (SLE) (O'Garra et al. 2008).

13.3 Interleukin 10 and the Regulation of Immune Responses

Immune responses require regulation so that infections are adequately cleared without causing immune mediated damage to the host. An excessive inflammatory response can be induced by certain pathogens and result in downstream immunopathology. In addition, induction of an inflammatory response to an inappropriate antigen such as self-antigen or harmless commensals can result in immunopathology. IL-10 has a vital role in regulating the balance between a protective immune

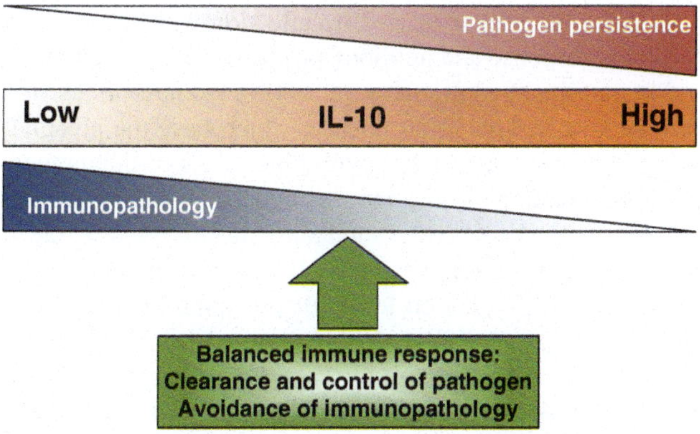

Fig. 13.4 IL-10 regulates the immune response to maintain a balance between immunopathology and chronic infection. Low levels of IL-10 induction in an immune response can lead to excessive pro-inflammatory cytokine production. In extreme situations, this can lead to immunopathology (*left*). In contrast, high levels of IL-10 induction can suppress the immune response such that there are low levels of inflammatory cytokines; this can facilitate chronic infection, but protects the host from immunopathology (*right*). An adequate balance in the immune response is achieved when IL-10 production controls the immune response to avoid immunopathology but does not contribute to chronic infection (*middle*)

response and immunopathology (Mege et al. 2006) (Fig. 13.4). In these situations, IL-10 is beneficial and dampens the immune response, preventing damage to the host. In contrast, an inappropriately high level of IL-10 production can be detrimental to the host, restraining the immune response, resulting in inadequate clearance of a pathogen, contributing to the establishment of chronic infection (Filippi and von Herrath 2008) (Fig. 13.4).

How IL-10 affects the immune response in different situations is discussed as well as the therapeutic potential of the manipulation of IL-10 levels.

13.3.1 Regulation of the Immune Response During Infection

13.3.1.1 IL-10 Can Protect Against Immunopathology Resulting from an Excessive Response to a Pathogen

The role of IL-10 in regulating immune responses to infectious diseases to protect against immunopathology has been demonstrated in studies using IL-10-deficient mice, which suffer severe immunopathology when infected with certain pathogens. For example, IL-10 is necessary to inhibit an excessive immune response in infection with *Plasmodium* spp., *Toxoplasma gondii*, *Trypanosoma cruzi*, and *Schistosoma mansonii* (Moore et al. 2001). Furthermore, a role for IL-10 in human disease is suggested through genetic associations and correlating levels of IL-10 with infection (Moore et al. 2001; Cyktor and Turner 2011).

Malaria is a parasitic disease caused by infection with species of *Plasmodium*. An IFN-γ-mediated response and antibody production are important in the immune response to this parasite (Langhorne et al. 2008). Complications of the infection include severe anemia and cerebral malaria, which have the potential to be fatal. These complications have been linked with high levels of inflammatory cytokines and excessive inflammation. In the mouse model of malaria, in which mice are infected with *Plasmodium chabaudi chabaudi*, IL-10 plays a role in protection against an excessive Th1 immune response. IL-10-deficient mice exhibit increased mortality associated with increased levels of pro-inflammatory cytokines: IFN-γ, TNF-α, and IL-12 (Li et al. 1999). Interestingly it has been found that IFN-γ-producing $Foxp3^-$ $CD4^+$ Th1 cells are the essential source of IL-10 in *P. chabaudi chabaudi* infection, illustrating that Th1 effector cells can feedback to negatively regulate themselves (do Rosario et al. 2012). In humans, several studies have implicated IL-10 in controlling immunopathology. Lower serum IL-10 levels correlate with worse outcome in *Plasmodium falciparum* infection and higher IL-10 production correlates with protection against severe anemia, although other studies associate IL-10 with a more severe disease phenotype in humans (do Rosario and Langhorne 2012). Therefore, the role of IL-10 in human malaria is currently not clear, possibly because local as opposed to systemic levels of IL-10 may be more related to the outcome of infection and thus more detailed studies are required for a better understanding of the role of IL-10 in human malarial infection (do Rosario and Langhorne 2012).

Another example of IL-10 restraining excessive immune responses is in the mouse model of *Toxoplasma gondii* (T. gondii) infection. This parasite also induces a strong Th1-mediated response, and IL-10-deficient mice exhibit enhanced susceptibility and mortality with increased levels of IFN-γ and IL-12 in the serum, as well as increased IFN-γ, IL-12, IL-1β, and TNF-α mRNA levels in the lung (Gazzinelli et al. 1996). Similar to *P. chabaudi chabaudi* infection, it was also found that the main source of IL-10 in *T. gondii* infection was $Foxp3^-$ $CD4^+$ Th1 cells (Jankovic et al. 2007). T-cell-derived IL-10 is therefore critical in protecting against immunopathology during toxoplasmosis.

These examples illustrate an important role for IL-10 in the control of an excessive Th1 response. IL-10 can also contribute to the control of inappropriate Th2 responses, demonstrated in the parasitic helminth infection with *Schistosoma mansoni* (Hoffmann et al. 2000). A Th2 response against *S. mansoni* contributes to the establishment of a granulomatous response capable of walling off the schistosome eggs. In addition to the protective role of this granulomatous response in sequestering eggs, it can lead to severe fibrosis (Wilson et al. 2007). Mice deficient in IL-10 showed increased granuloma size in acute infection but no phenotype in chronic infection (Wynn et al. 1998). However, mice deficient in both IL-10 and IL-12 (an important cytokine for the induction of Th1 cells) mount an enhanced Th2 response and show an increase in granuloma size and level of hepatic fibrosis (Hoffmann et al. 2000). In addition, in human infection with *S. mansoni*, the most severe fibrosis around the portal vein was associated with low levels of IL-10 and IFN-γ (Wynn 2004). This finding indicates that both IL-10 and the Th1 response may co-operate in controlling hepatic fibrosis induced by the Th2 response (Hoffmann et al. 2000).

13.3.1.2 IL-10 Can Suppress Protective Immune Responses Against Pathogens

Because of its powerful suppressive activity, IL-10 production can be unsuitable in the response to certain pathogens, since it can results in suboptimal control and clearance of the pathogen. Examples of such pathogens include *Mycobacteria* spp., *Leishmania* spp., *Listeria monocytogenes*, and lymphocytic choriomeningitis virus (LCMV) (Moore et al. 2001; Filippi and von Herrath 2008).

Mycobacterium tuberculosis (Mtb) causes the disease tuberculosis (TB). Upon infection an individual can develop an active disease or establish an asymptomatic latent disease. This asymptomatic latent disease can persist for a lifetime or alternatively can reactivate to active disease. An individual has an approximately 10 % risk of reactivation in a lifetime, and this risk is increased if they become immunocompromised (North and Jung 2004). IL-10 is implicated in inappropriately suppressing the immune response to Mtb (Redford et al. 2011). IL-10-deficient mice display an enhanced Th1 immune response to aerosol challenge with Mtb (Redford et al. 2010) and lower bacterial loads compared to wild-type mice. In addition, more susceptible mouse strains express higher levels of IL-10 in chronic infection, such as the CBA/J mouse strain, and blockade of IL-10 signaling leads to a decreased bacterial load and increased survival (Beamer et al. 2008; Redford et al. 2010). Furthermore, in a resistant mouse strain, artificially enhancing IL-10 expression increased susceptibility (Turner et al. 2002). These studies link susceptibility to Mtb infection with the level of IL-10 production and suggest that IL-10 is inappropriately suppressing an otherwise protective response.

In active TB patients, elevated levels of IL-10 are seen in the lungs, serum, sputum, and bronchoalveolar lavage fluid (BAL) (Reviewed in Redford et al. 2011). $CD4^{+}$ cells isolated from the BAL were seen to produce both IFN-γ and IL-10 more frequently in active TB patients compared to healthy controls or inactive subjects (Gerosa et al. 1999), therefore associating IL-10 production with active disease. In addition, IL-10 has been associated with bacterial loads as the level of CFP32, an Mtb antigen, correlated with level of IL-10 in active TB patients (Huard et al. 2003). Different strains of Mtb have been demonstrated to have different capacities to induce IL-10 upon infection of human monocyte-derived macrophages and thus further their own survival through immune deviation (Newton et al. 2006). However, these studies are mainly correlative, and more work is required to fully elucidate the role of IL-10 in human TB (Redford et al. 2011).

Infection with parasites of *Leishmania* spp. is another situation where IL-10 inappropriately suppresses a protective response. These parasites can cause acute disease but also a chronic latent infection that can reactivate and cause serious clinical disease. Protective immunity is mediated through a Th1 response. IL-10 has been shown to suppress this response in leishmaniasis, contributing to development of a persistent infection. IL-10-deficient mice, but not wild-type mice, were able to achieve a sterile immunity (Belkaid et al. 2001). IL-10 derived from Treg subsets and $Foxp3^{-}$ $CD4^{+}$ Th1 cells have been demonstrated to mediate the suppression, resulting in persistent infection, and it is possible that these different sources may result from infection with different leishmanial substrains (Belkaid et al. 2002; Anderson et al. 2007). In humans, high levels of IL-10 are associated with visceral leishmaniasis, a severe form of disease, and these levels decreased dramatically with successful treatment (Nylen and Sacks 2007). These data from human and mouse studies support a role for IL-10 in the pathogenesis of this disease and the establishment of chronic infection.

IL-10 can also suppress the immune response to certain viruses and promote their persistence. Indeed, in humans an increase in IL-10 levels is seen in response to several persistent viruses, including hepatitis C virus (HCV) and Epstein–Barr virus (EBV) (Filippi and von Herrath 2008). IL-10-deficient mice infected with LCMV were able to maintain an effector response and clear the infection (Brooks et al. 2006). IL-10-receptor blockade during LCMV infection increased the number of virus-specific T cells and maintained their responsiveness, accompanied by a sharp decrease in viral titer (Brooks et al. 2006; Ejrnaes et al. 2006). These data indicate that IL-10 suppresses the immune response such that persistence is promoted.

Further support for IL-10 in facilitating the establishment of chronic infection is that some viruses encode homologues of IL-10, most likely acquired from the host. Several herpesviruses and pox viruses encode viral IL-10 (vIL-10), displaying varying amounts of sequence identity to the human IL-10 (Slobedman et al. 2009). EBV encodes a vIL-10 (Moore et al. 1990). The EBV vIL-10 has retained much of the immunosuppressive activity and is able to suppress the production of inflammatory cytokines, reduce macrophage activation, and suppress antigen presentation but has lost many of the immunostimulatory properties of mammalian IL-10 (Hsu et al. 1990), suggesting a mechanism of immune evasion that contributes to the ability to persist (Mege et al. 2006). Another virus that encodes an IL-10 homologue is human cytomegalovirus (HCMV) (Kotenko et al. 2000; Lockridge et al. 2000). In contrast to EBV vIL-10, HCMV vIL-10 shows a much lower sequence identity to human IL-10; however, HCMV vIL-10 has retained most of the immunomodulatory functions of mammalian IL-10 (Slobedman et al. 2009).

In summary, IL-10 can suppress a potentially harmful immune response directed against a pathogen such as *Plasmodium* or *Toxoplasma*, preventing immunopathology. In contrast, during infection with pathogens such as Mtb, *Leishmania* spp., and LCMV, IL-10 suppresses the immune response such that the host is not able to effectively clear the infection.

13.3.2 Role of IL-10 in Colitis and Inflammatory Bowel Disease

The gut is naturally colonized by organisms that comprise the gut microbiome, and the immune system must be regulated to avoid an inappropriate response against these bacteria. IL-10 plays an essential role in mediating tolerance to these innocuous and beneficial microorganisms (Maloy and Powrie 2011).

IL-10-deficient mice display extensive intestinal inflammation, spanning the duodenum, jejunum, and colon, when their intestinal tracts are colonized by particular bacteria, and this inflammation is less marked, encompassing only the colon, when these mice are kept in specific pathogen-free (SPF) conditions (Kuhn et al. 1993; Berg et al. 1996). When IL-10-deficient mice were kept in germfree conditions, they did not develop colitis, demonstrating that colonization by bacteria is required for colitis to develop in the absence of IL-10 (Sellon et al. 1998). This finding therefore indicates a role for IL-10 in mediating tolerance to the normal intestinal microflora. T-cell-derived IL-10 has been found to be necessary for the regulation of intestinal inflammation through the study of mice deficient in IL-10, specifically in the T-cell compartment (Roers et al. 2004). Subsequently, it has been found that IL-10 derived from both Foxp3^{+} CD4^{+} T cells and Foxp3^{-} CD4^{+} T cells plays a role in intestinal homeostasis and contributes to tolerance to commensal bacteria and food antigens (Izcue et al. 2009).

In humans, inflammatory bowel disease (IBD) encompasses Crohn's disease (CD) and Ulcerative Colitis (UC). Mutations in the *Il10ra* or *Il10rb* genes are associated with early-onset severe IBD (Glocker et al. 2009). In addition, Genome-Wide Association Studies (GWAS) have identified polymorphisms in the *Il10* gene to be associated with UC and CD (Franke et al. 2008; Franke et al. 2010; Maloy and Powrie 2011). Thus, mouse and human data support the necessity of IL-10 in mediating intestinal homeostasis and tolerance to commensal bacteria.

13.3.3 Role of IL-10 in Allergy

Allergy occurs when an individual is sensitized to an allergen such as pollen or house dust mites, resulting in an adaptive, allergen-specific response. This specific response is typically Th2 mediated, characterized by IL-4, IL-5, IL-9, IL-13, and high levels of IgE. When the individual subsequently encounters the allergen, IgE cross links on mast cells, leading to degranulation and release of inflammatory mediators, contributing to pathology. Responses to aeroallergens such as pollen are associated with allergic asthma (Hawrylowicz and O'Garra 2005).

IL-10 is suggested to be important in the control of allergic responses. In presensitized mice, intranasal administration of IL-10 simultaneously with antigen challenge can prevent the accumulation of eosinophils and neutrophils and TNF-α production in the lung (Zuanyamorim et al. 1995). Similarly, adoptive transfer of antigen-specific, IL-10-transfected T cells into sensitized mice can inhibit airway inflammation (Oh et al. 2002). In humans, lower levels of IL-10 were seen in the BAL of asthma patients compared to healthy individuals (Borish et al. 1996). Furthermore, treatment for asthma includes use of glucocorticoids, which induce IL-10 in human T cells (Hawrylowicz and O'Garra 2005). These studies suggest that IL-10 is capable of controlling the allergic Th2-mediated immune response and that insufficient levels of IL-10 may contribute to allergic disease.

13.3.4 Role of IL-10 in Autoimmunity

Autoimmunity occurs when an immune response is mounted against a self-antigen, causing immune-mediated damage to healthy tissue. Multiple sclerosis (MS) is an autoimmune disease affecting the central nervous system (CNS). MS is caused by immune-mediated destruction of myelin, which normally insulates nerves, resulting in impairment of neural transmission. The disease can follow several courses, the most common being a relapsing–remitting course, with periods free from symptoms (remission) and periods in which symptoms are most severe (relapse). The mouse model for MS is experimental autoimmune encephalomyelitis (EAE). A role for IL-10 in suppressing the pathogenic immune response in EAE is suggested by observations that in wild-type mice an adjuvant must be given with the myelin peptide to induce EAE, whereas in IL-10-deficient mice EAE can be induced without an adjuvant (Bettelli et al. 1998). In addition, IL-10-deficient mice display enhanced susceptibility to the induction of EAE, with higher clinical scores and enhanced proliferation of, and cytokine production by, myelin peptide-specific T cells (Bettelli et al. 1998). Furthermore, IL-10 is expressed during recovery from inflammation, and IL-10-deficient mice are unable to enter remission (Moore et al. 2001; O'Garra et al. 2008).

IL-10 may also have a role in suppressing the immune response in rheumatoid arthritis (RA). RA is an autoimmune condition mainly affecting the joints, resulting in painful, swollen, and deformed joints. In animal models of RA, IL-10 had a protective role through decreasing expression of inflammatory cytokines, reducing swelling and destruction of the joint. In addition, in human patients with RA, IL-10 is expressed in inflamed joints (Moore et al. 2001).

Systemic lupus erythematosis (SLE) is an autoimmune disease affecting multiple organ systems, characterized by high levels of autoantibodies against nuclear antigens such as double-stranded DNA, which result in immune complex deposition (Beebe et al. 2002). In contrast to MS and RA where IL-10 suppresses the pathological response, IL-10 is thought to be involved in the pathogenesis of SLE. As discussed earlier, IL-10 can promote survival of B cells and antibody production by these cells (Moore et al. 2001). The mouse strain NZB/WF1 is prone to the development of SLE. Treatment of these mice with anti-IL-10 antibodies substantially delayed the onset of autoimmunity in these mice, whereas administration of IL-10 accelerated onset (Ishida et al. 1994). In humans the level of serum IL-10 has been correlated to disease status. Serum IL-10 levels were seen to be elevated in SLE patients with active disease compared to controls, and following treatment the level of IL-10 in the serum correlated with change in the disease activity, as measured by a clinical scoring system (Beebe et al. 2002).

To summarize, IL-10 is vital in suppressing an inappropriate immune response to self in certain contexts, or to a harmless environmental antigen such as pollen or commensal bacteria. However, in SLE IL-10 may be involved in the pathogenesis of disease.

13.3.5 Role of IL-10 in Cancer

The role of IL-10 in cancer is controversial, with studies supporting both the idea that IL-10 contributes to immunosuppression and the idea that the immunostimulatory properties of IL-10 on B cells and $CD8^+$ T cells promote antitumor immune responses (Mocellin et al. 2005).

In humans, IL-10 expression has been observed in several malignant tumor types including breast carcinoma, renal cell carcinoma, colorectal adenocarcinoma, B-cell lymphoma, and melanoma (O'Garra et al. 2008). IL-10 expression early in the tumor microenvironment could suppress antigen presentation by DCs and therefore the development of an antitumor response. IL-10-producing $CD4^+$ T cells have been identified in some human cancers, where they act to suppress the antitumor immune response (O'Garra et al. 2008). In addition, in a model using a colon carcinoma cell line, treatment with anti-IL-10R antibody and the TLR9 ligand CpG led to an enhanced tumor-specific immune response (Vicari et al. 2004). However, administration of IL-10 in in vivo models of melanoma (Berman et al. 1996) or breast cancer (Kundu et al. 1996) can also lead to tumor regression (Mocellin et al. 2003; Mocellin et al. 2005), and transfection of certain tumor cell lines with IL-10 leads to a decrease in their tumor-forming capability (Mocellin et al. 2003). Furthermore, IL-10 can also suppress angiogenesis, a process required for development of a blood supply to the tumor, and can therefore suppress the establishment of a tumor (Mocellin et al. 2005).

Thus, it seems that IL-10 may have both immunosuppressive and immunostimulatory effects in cancer. The location, source, timing, and level of IL-10 expression may partly explain the contradictory results observed (Mocellin et al. 2005; O'Garra et al. 2008)

13.3.6 Therapeutic Potential of IL-10

Manipulation of the levels of IL-10 or the ability of IL-10 to signal has therapeutic potential. Antagonism of IL-10 could be used to alleviate excessive immunosuppression. Alternatively, administration or induction of IL-10 could be used to promote tolerance and avoid excessive immune activation.

13.3.6.1 Administration or Induction of IL-10 in Treatment of Autoimmunity, Allergy, and Asthma

Preliminary studies in humans where IL-10 was administered as treatment for various inflammatory conditions including Crohn's disease and psoriasis showed trends toward efficacy. Subsequent studies indicated that the therapeutic potential was only moderate and that treatment incurred side effects, potentially caused by the immunostimulatory effects of IL-10 (O'Garra et al. 2008). In EAE studies, different models of IL-10 treatment gave contradictory results. If a virus engineered to express IL-10 was delivered intracranially just before the expected onset of disease, animals were resistant to the development of EAE. However, if the IL-10-expressing virus was delivered systemically, no such resistance was seen (Cua et al. 2001). The location of IL-10 expression or administration may therefore be important for its efficacy as a therapy (O'Garra et al. 2008). Furthermore, induction of IL-10-secreting Th1 cells through intranasal administration of self-antigen in mice modified to express a TCR specific for this antigen can render mice resistant to the induction of EAE (Gabrysova et al. 2009).

Therapies used in the treatment of allergy and asthma have been seen to induce antigen-specific or nonspecific T-cell-derived IL-10, and this IL-10 induction may contribute to effective treatments (Hawrylowicz and O'Garra 2005).

There is some evidence to suggest that successful immunotherapy, which involves the administration of increasing doses of the allergen, leads to the induction of IL-10-producing T cells and a decrease in the antigen-specific Th2 response (Akdis et al. 1998; Hawrylowicz and O'Garra 2005). Furthermore, glucocorticoid steroids, commonly used in the treatment of inflammatory diseases, can induce IL-10 production in vitro as well as in human patients (Hawrylowicz and O'Garra 2005). However, $CD4^+$ T cells from some patients have been shown to have a limited ability to produce IL-10 in response to steroids, and these patients fail to show improvement with steroid treatment. The effect of vitamin D combined with glucocorticoid treatment can lead to enhanced IL-10 secretion in patients with steroid-resistant asthma (Xystrakis et al. 2006; O'Garra et al. 2008).

13.3.6.2 Antagonizing IL-10 to Enhance Vaccination

Immunomodulation by IL-10 is seen in the response to various infectious pathogens, as discussed earlier. IL-10 receptor (IL-10R) blockade could be used to increase the efficacy of vaccines against infectious disease. In vaccination studies, IL-10R blockade has increased the response seen to several different vaccines. For example, the Th1 response and protection against *Leishmania* infection induced by a viral vector expressing leishmanial proteins was enhanced by IL-10R blockade (Darrah et al. 2010). Therapeutic vaccination in LCMV infection was only effective at increasing the antiviral immune response and enhancing clearance of the persistent pathogen in the context of IL-10R blockade (Brooks et al. 2008). Furthermore, IL-10R blockade during bacillus Calmette–Guérin (BCG) vaccination lowered bacterial load following Mtb infection and enhanced the IFN-γ response (Pitt et al. 2012). BCG is also used therapeutically to induce a Th1 immune response promoting immunity against bladder cancer (Luo 2012). IL-10R blockade with BCG in a mouse model of bladder cancer increased the level of IFN-γ mRNA in the bladder and enhanced clearance of the cancer (Bockholt et al. 2012).

Thus, manipulation of IL-10 during vaccination regimes could represent a strategy to enhance vaccines where the protection induced by vaccination is inadequate. However, such therapies would need to be closely monitored to minimize immunopathology.

13.4 Summary and Concluding Remarks

IL-10 is a widely studied predominantly immunosuppressive cytokine. IL-10 is produced by most immune cell types and can modulate the function of many cell types directly or indirectly. Depending on the context, IL-10 can be beneficial or detrimental to the host. When beneficial to the host, IL-10 suppresses inflammation, avoiding immunopathology. When detrimental to the host, IL-10 suppresses a protective immune response, leading to chronic infection. Because IL-10 can regulate the immune response, IL-10 has therapeutic potential. Blockade of IL-10 signaling could be beneficial in enhancing the immune response to several pathogens and could enhance vaccination efficacy. Administration or induction of IL-10, on the other hand, could promote tolerance to allergens or to self in allergic disease or autoimmunity.

There are still further questions that need addressing regarding the role of IL-10 in anti-inflammation and tolerance. For the specific disease contexts discussed here, we need a more detailed understanding of the sources, mechanisms of action, downstream functions, and pathways regulating IL-10 production and function. In conclusion, by furthering our understanding of the biology of IL-10 in anti-inflammation and immune tolerance, we will be more capable of exploiting or manipulating IL-10 as a potential therapy in the clinic.

References

Akdis CA, Blesken T, Akdis M, Wuthrich B, Blaser K (1998) Role of interleukin 10 in specific immunotherapy. J Clin Invest 102(1):98–106

Akira S, Takeda K (2004) Toll-like receptor signalling. Nat Rev Immunol 4(7):499–511

Anderson CF, Oukka M, Kuchroo VJ, Sacks D (2007) CD4(+)CD25(−)Foxp3(−) Th1 cells are the source of IL-10-mediated immune suppression in chronic cutaneous leishmaniasis. J Exp Med 204(2):285–297

Bacchetta R, Gambineri E, Roncarolo MG (2007) Role of regulatory T cells and FOXP3 in human diseases. J Allergy Clin Immunol 120(2):227–235, quiz 236–237

Banerjee A, Gugasyan R, McMahon M, Gerondakis S (2006) Diverse Toll-like receptors utilize Tpl2 to activate extracellular signal-regulated kinase (ERK) in hemopoietic cells. Proc Natl Acad Sci USA 103(9):3274–3279

Beamer GL, Flaherty DK, Assogba BD, Stromberg P, Gonzalez-Juarrero M, de Waal Malefyt R, Vesosky B, Turner J (2008) Interleukin-10 promotes *Mycobacterium tuberculosis* disease progression in CBA/J mice. J Immunol 181(8):5545–5550

Beebe AM, Cua DJ, de Waal Malefyt R (2002) The role of interleukin-10 in autoimmune disease: systemic lupus erythematosus (SLE) and multiple sclerosis (MS). Cytokine Growth Factor Rev 13(4–5):403–412

Belkaid Y, Hoffmann KF, Mendez S, Kamhawi S, Udey MC, Wynn TA, Sacks DL (2001) The role of interleukin (IL)-10 in the persistence of Leishmania major in the skin after healing and the therapeutic potential of anti-IL-10 receptor antibody for sterile cure. J Exp Med 194(10):1497–1506

Belkaid Y, Piccirillo CA, Mendez S, Shevach EM, Sacks DL (2002) CD4+CD25+ regulatory T cells control Leishmania major persistence and immunity. Nature (Lond) 420(6915):502–507

Berg DJ, Davidson N, Kuhn R, Muller W, Menon S, Holland G, Thompson-Snipes L, Leach MW, Rennick D (1996) Enterocolitis and colon cancer in interleukin-10-deficient mice are associated with aberrant cytokine production and CD4(+) TH1-like responses. J Clin Invest 98(4): 1010–1020

Berman RM, Suzuki T, Tahara H, Robbins PD, Narula SK, Lotze MT (1996) Systemic administration of cellular IL-10 induces an effective, specific, and long-lived immune response against established tumors in mice. J Immunol 157(1):231–238

Bettelli E, Das MP, Howard ED, Weiner HL, Sobel RA, Kuchroo VK (1998) IL-10 is critical in the regulation of autoimmune encephalomyelitis as demonstrated by studies of IL-10- and IL-4-deficient and transgenic mice. J Immunol 161(7):3299–3306

Bockholt NA, Knudson MJ, Henning JR, Maymi JL, Weady P, Smith GJ 3rd, Eisenbraun MD, Fraser JD, O'Donnell MA, Luo Y (2012) Anti-interleukin-10R1 monoclonal antibody enhances bacillus Calmette–Guerin induced T-helper type 1 immune responses and antitumor immunity in a mouse orthotopic model of bladder cancer. J Urol 187(6):2228–2235

Boonstra A, Rajsbaum R, Holman M, Marques R, Asselin-Paturel C, Pereira JP, Bates EE, Akira S, Vieira P, Liu YJ, Trinchieri G, O'Garra A (2006) Macrophages and myeloid dendritic cells, but not plasmacytoid dendritic cells, produce IL-10 in response to MyD88- and TRIF-dependent TLR signals, and TLR-independent signals. J Immunol 177(11):7551–7558

Borish L, Aarons A, Rumbyrt J, Cvietusa P, Negri J, Wenzel S (1996) Interleukin-10 regulation in normal subjects and patients with asthma. J Allergy Clin Immunol 97(6):1288–1296

Brooks DG, Trifilo MJ, Edelmann KH, Teyton L, McGavern DB, Oldstone MBA (2006) Interleukin-10 determines viral clearance or persistence in vivo. Nat Med 12(11):1301–1309

Brooks DG, Lee AM, Elsaesser H, McGavern DB, Oldstone MBA (2008) IL-10 blockade facilitates DNA vaccine-induced T cell responses and enhances clearance of persistent virus infection. J Exp Med 205(3):533–541

Cao S, Zhang X, Edwards JP, Mosser DM (2006) NF-kappaB1 (p50) homodimers differentially regulate pro- and anti-inflammatory cytokines in macrophages. J Biol Chem 281(36): 26041–26050

Chang EY, Guo B, Doyle SE, Cheng G (2007a) Cutting edge: involvement of the type I IFN production and signaling pathway in lipopolysaccharide-induced IL-10 production. J Immunol 178(11):6705–6709

Chang HD, Helbig C, Tykocinski L, Kreher S, Koeck J, Niesner U, Radbruch A (2007b) Expression of IL-10 in Th memory lymphocytes is conditional on IL-12 or IL-4, unless the IL-10 gene is imprinted by GATA-3. Eur J Immunol 37(3):807–817

Chaudhry A, Samstein RM, Treuting P, Liang Y, Pils MC, Heinrich JM, Jack RS, Wunderlich FT, Bruning JC, Muller W, Rudensky AY (2011) Interleukin-10 signaling in regulatory T cells is required for suppression of Th17 cell-mediated inflammation. Immunity 34(4):566–578

Cua DJ, Hutchins B, LaFace DM, Stohlman SA, Coffman RL (2001) Central nervous system expression of IL-10 inhibits autoimmune encephalomyelitis. J Immunol 166(1):602–608

Cyktor JC, Turner J (2011) Interleukin-10 and immunity against prokaryotic and eukaryotic intracellular pathogens. Infect Immun 79(8):2964–2973

Darrah PA, Hegde ST, Patel DT, Lindsay RWB, Chen L, Roederer M, Seder RA (2010) IL-10 production differentially influences the magnitude, quality, and protective capacity of Th1 responses depending on the vaccine platform. J Exp Med 207(7):1421–1433

Dillon S, Agrawal A, Van Dyke T, Landreth G, McCauley L, Koh A, Maliszewski C, Akira S, Pulendran B (2004) A Toll-like receptor 2 ligand stimulates Th2 responses in vivo, via induction of extracellular signal-regulated kinase mitogen-activated protein kinase and c-Fos in dendritic cells. J Immunol 172(8):4733–4743

do Rosario APF, Langhorne J (2012) T cell-derived IL-10 and its impact on the regulation of host responses during malaria. Int J Parasitol 42(6):549–555

do Rosario APF, Lamb T, Spence P, Stephens R, Lang A, Roers A, Muller W, O'Garra A, Langhorne J (2012) IL-27 promotes IL-10 production by effector Th1 CD4(+) T cells: a critical mechanism for protection from severe immunopathology during malaria infection. J Immunol 188(3):1178–1190

Ejrnaes M, Filippi CM, Martinic MM, Ling EM, Togher LM, Crotty S, von Herrath MG (2006) Resolution of a chronic viral infection after interleukin-10 receptor blockade. J Exp Med 203(11):2461–2472

Filippi CM, von Herrath MG (2008) IL-10 and the resolution of infections. J Pathol 214(2): 224–230

Fillatreau S (2011) Novel regulatory functions for Toll-like receptor-activated B cells during intracellular bacterial infection. Immunol Rev 240(1):52–71

Fiorentino DF, Bond MW, Mosmann TR (1989) Two types of mouse T helper cell. IV. Th2 clones secrete a factor that inhibits cytokine production by Th1 clones. J Exp Med 170(6): 2081–2095

Foey AD, Parry SL, Williams LM, Feldmann M, Foxwell BM, Brennan FM (1998) Regulation of monocyte IL-10 synthesis by endogenous IL-1 and TNF-alpha: role of the p38 and p42/44 mitogen-activated protein kinases. J Immunol 160(2):920–928

Franke A, Balschun T, Karlsen TH, Sventoraityte J, Nikolaus S, Mayr G, Domingues FS, Albrecht M, Nothnagel M, Ellinghaus D, Sina C, Onnie CM, Weersma RK, Stokkers PC, Wijmenga C, Gazouli M, Strachan D, McArdle WL, Vermeire S, Rutgeerts P, Rosenstiel P, Krawczak M, Vatn MH, Mathew CG, Schreiber S (2008) Sequence variants in IL10, ARPC2 and multiple other loci contribute to ulcerative colitis susceptibility. Nat Genet 40(11):1319–1323

Franke A, McGovern DP, Barrett JC, Wang K, Radford-Smith GL, Ahmad T, Lees CW, Balschun T, Lee J, Roberts R, Anderson CA, Bis JC, Bumpstead S, Ellinghaus D, Festen EM, Georges M, Green T, Haritunians T, Jostins L, Latiano A, Mathew CG, Montgomery GW, Prescott NJ, Raychaudhuri S, Rotter JI, Schumm P, Sharma Y, Simms LA, Taylor KD, Whiteman D, Wijmenga C, Baldassano RN, Barclay M, Bayless TM, Brand S, Buning C, Cohen A, Colombel JF, Cottone M, Stronati L, Denson T, De Vos M, D'Inca R, Dubinsky M, Edwards C, Florin T, Franchimont D, Gearry R, Glas J, Van Gossum A, Guthery SL, Halfvarson J, Verspaget HW, Hugot JP, Karban A, Laukens D, Lawrance I, Lemann M, Levine A, Libioulle C, Louis E, Mowat C, Newman W, Panes J, Phillips A, Proctor DD, Regueiro M, Russell R, Rutgeerts P, Sanderson J, Sans M, Seibold F, Steinhart AH, Stokkers PC, Torkvist L, Kullak-Ublick G, Wilson D, Walters T, Targan SR, Brant SR, Rioux JD, D'Amato M, Weersma RK, Kugathasan S, Griffiths AM, Mansfield JC, Vermeire S, Duerr RH, Silverberg MS, Satsangi J, Schreiber S, Cho JH, Annese V, Hakonarson H, Daly MJ, Parkes M (2010) Genome-wide meta-analysis increases to 71 the number of confirmed Crohn's disease susceptibility loci. Nat Genet 42(12):1118–1125

Gabrysova L, Nicolson KS, Streeter HB, Verhagen J, Sabatos-Peyton CA, Morgan DJ, Wraith DC (2009) Negative feedback control of the autoimmune response through antigen-induced differentiation of IL-10-secreting Th1 cells. J Exp Med 206(8):1755–1767

Gazzinelli RT, Wysocka M, Hieny S, Scharton-Kersten T, Cheever A, Kuhn R, Muller W, Trinchieri G, Sher A (1996) In the absence of endogenous IL-1O, mice acutely infected with *Toxoplasma gondii* succumb to a lethal immune response dependent on CD4(+) T cells and accompanied by overproduction of IL-12, IFN-gamma, and TNF-alpha. J Immunol 157(2):798–805

Gerosa F, Nisii C, Righetti S, Micciolo R, Marchesini M, Cazzadori A, Trinchieri G (1999) CD4(+) T cell clones producing both interferon-gamma and interleukin-10 predominate in bronchoalveolar lavages of active pulmonary tuberculosis patients. Clin Immunol 92(3):224–234

Gilliet M, Liu YJ (2002) Generation of human CD8 T regulatory cells by CD40 ligand-activated plasmacytoid dendritic cells. J Exp Med 195(6):695–704

Glocker EO, Kotlarz D, Boztug K, Gertz EM, Schaffer AA, Noyan F, Perro M, Diestelhorst J, Allroth A, Murugan D, Hatscher N, Pfeifer D, Sykora KW, Sauer M, Kreipe H, Lacher M, Nustede R, Woellner C, Baumann U, Salzer U, Koletzko S, Shah N, Segal AW, Sauerbrey A, Buderus S, Snapper SB, Grimbacher B, Klein C (2009) Inflammatory bowel disease and mutations affecting the interleukin-10 receptor. N Engl J Med 361(21):2033–2045

Grenningloh R, Kang BY, Ho IC (2005) Ets-1, a functional cofactor of T-bet, is essential for Th1 inflammatory responses. J Exp Med 201(4):615–626

Hall AO, Silver JS, Hunter CA (2012) The immunobiology of IL-27. Adv Immunol 115:1–44

Hawrylowicz CM, O'Garra A (2005) Potential role of interleukin-10-secreting regulatory T cells in allergy and asthma. Nat Rev Immunol 5(4):271–283

Hoffmann KF, Cheever AW, Wynn TA (2000) IL-10 and the dangers of immune polarization: excessive type 1 and type 2 cytokine responses induce distinct forms of lethal immunopathology in murine schistosomiasis. J Immunol 164(12):6406–6416

Hsu DH, de Waal Malefyt R, Fiorentino DF, Dang MN, Vieira P, de Vries J, Spits H, Mosmann TR, Moore KW (1990) Expression of interleukin-10 activity by Epstein–Barr virus protein BCRF1. Science 250(4982):830–832

Hu X, Paik PK, Chen J, Yarilina A, Kockeritz L, Lu TT, Woodgett JR, Ivashkiv LB (2006) IFN-gamma suppresses IL-10 production and synergizes with TLR2 by regulating GSK3 and CREB/AP-1 proteins. Immunity 24(5):563–574

Huard RC, Chitale S, Leung M, Lazzarini LC, Zhu H, Shashkina E, Laal S, Conde MB, Kritski AL, Belisle JT, Kreiswirth BN, Lapa e Silva JR, Ho JL (2003) The *Mycobacterium tuberculosis* complex-restricted gene cfp32 encodes an expressed protein that is detectable in tuberculosis patients and is positively correlated with pulmonary interleukin-10. Infect Immun 71(12): 6871–6883

Huber S, Gagliani N, Esplugues E, O'Connor W Jr, Huber FJ, Chaudhry A, Kamanaka M, Kobayashi Y, Booth CJ, Rudensky AY, Roncarolo MG, Battaglia M, Flavell RA (2011) Th17 cells express interleukin-10 receptor and are controlled by Foxp3(−) and Foxp3+ regulatory CD4+ T cells in an interleukin-10-dependent manner. Immunity 34(4):554–565

Hutchins AP, Poulain S, Miranda-Saavedra D (2012) Genome-wide analysis of STAT3 binding in vivo predicts effectors of the anti-inflammatory response in macrophages. Blood 119(13): e110–e119

Ishida H, Muchamuel T, Sakaguchi S, Andrade S, Menon S, Howard M (1994) Continuous administration of anti-interleukin-10 antibodies delays onset of autoimmunity in NZB/W F1-mice. J Exp Med 179(1):305–310

Ivanov II, McKenzie BS, Zhou L, Tadokoro CE, Lepelley A, Lafaille JJ, Cua DJ, Littman DR (2006) The orphan nuclear receptor RORgammat directs the differentiation program of proinflammatory IL-17+ T helper cells. Cell 126(6):1121–1133

Izcue A, Coombes JL, Powrie F (2009) Regulatory lymphocytes and intestinal inflammation. Annu Rev Immunol 27:313–338

Jankovic D, Kullberg MC, Feng CG, Goldszmid RS, Collazo CM, Wilson M, Wynn TA, Kamanaka M, Flavell RA, Sheri A (2007) Conventional T-bet(+)Foxp3(−) Th1 cells are the major source of host-protective regulatory IL-10 during intracellular protozoan infection. J Exp Med 204(2):273–283

Jarnicki AG, Conroy H, Brereton C, Donnelly G, Toomey D, Walsh K, Sweeney C, Leavy O, Fletcher J, Lavelle EC, Dunne P, Mills KH (2008) Attenuating regulatory T cell induction by TLR agonists through inhibition of p38 MAPK signaling in dendritic cells enhances their efficacy as vaccine adjuvants and cancer immunotherapeutics. J Immunol 180(6):3797–3806

Kaiser F, Cook D, Papoutsopoulou S, Rajsbaum R, Wu X, Yang HT, Grant S, Ricciardi-Castagnoli P, Tsichlis PN, Ley SC, O'Garra A (2009) TPL-2 negatively regulates interferon-beta production in macrophages and myeloid dendritic cells. J Exp Med 206(9):1863–1871

Kanters E, Pasparakis M, Gijbels MJ, Vergouwe MN, Partouns-Hendriks I, Fijneman RJ, Clausen BE, Forster I, Kockx MM, Rajewsky K, Kraal G, Hofker MH, de Winther MP (2003) Inhibition of NF-kappaB activation in macrophages increases atherosclerosis in LDL receptor-deficient mice. J Clin Invest 112(8):1176–1185

Kawai T, Akira S (2007) TLR signaling. Semin Immunol 19(1):24–32

Kim C, Sano Y, Todorova K, Carlson BA, Arpa L, Celada A, Lawrence T, Otsu K, Brissette JL, Arthur JS, Park JM (2008) The kinase p38 alpha serves cell type-specific inflammatory functions in skin injury and coordinates pro- and anti-inflammatory gene expression. Nat Immunol 9(9):1019–1027

Kotenko SV, Saccani S, Izotova LS, Mirochnitchenko OV, Pestka S (2000) Human cytomegalovirus harbors its own unique IL-10 homolog (cmvIL-10). Proc Natl Acad Sci USA 97(4): 1695–1700

Kuhn R, Lohler J, Rennick D, Rajewsky K, Muller W (1993) Interleukin-10-deficient mice develop chronic enterocolitis. Cell 75(2):263–274

Kundu N, Beaty TL, Jackson MJ, Fulton AM (1996) Antimetastatic and antitumor activities of interleukin 10 in a murine model of breast cancer. J Natl Cancer Inst 88(8):536–541

Langhorne J, Ndungu FM, Sponaas AM, Marsh K (2008) Immunity to malaria: more questions than answers. Nat Immunol 9(7):725–732

Le Friec G, Sheppard D, Whiteman P, Karsten CM, Shamoun SA, Laing A, Bugeon L, Dallman MJ, Melchionna T, Chillakuri C, Smith RA, Drouet C, Couzi L, Fremeaux-Bacchi V, Kohl J, Waddington SN, McDonnell JM, Baker A, Handford PA, Lea SM, Kemper C (2012) The CD46-Jagged1 interaction is critical for human TH1 immunity. Nat Immunol 13(12): 1213–1221

Lee CG, Kwon HK, Sahoo A, Hwang W, So JS, Hwang JS, Chae CS, Kim GC, Kim JE, So HS, Hwang ES, Grenningloh R, Ho IC, Im SH (2012) Interaction of Ets-1 with HDAC1 represses IL-10 expression in Th1 cells. J Immunol 188(5):2244–2253

LeibundGut-Landmann S, Gross O, Robinson MJ, Osorio F, Slack EC, Tsoni SV, Schweighoffer E, Tybulewicz V, Brown GD, Ruland J, Reis e Sousa C (2007) Syk- and CARD9-dependent coupling of innate immunity to the induction of T helper cells that produce interleukin 17. Nat Immunol 8(6):630–638

Leonard WJ, O'Shea JJ (1998) Jaks and STATs: biological implications. Annu Rev Immunol 16:293–322

Li C, Corraliza I, Langhorne J (1999) A defect in interleukin-10 leads to enhanced malarial disease in *Plasmodium chabaudi chabaudi* infection in mice. Infect Immun 67(9):4435–4442

Lloyd CM, Hawrylowicz CM (2009) Regulatory T cells in asthma. Immunity 31(3):438–449

Lockridge KM, Zhou SS, Kravitz RH, Johnson JL, Sawai ET, Blewett EL, Barry PA (2000) Primate cytomegaloviruses encode and express an IL-10-like protein. Virology 268(2): 272–280

Luo Y (2012) Blocking IL-10 enhances bacillus Calmette–Guerin induced T helper type 1 immune responses and anti-bladder cancer immunity. Oncoimmunology 1(7):1183–1185

Maloy KJ, Powrie F (2011) Intestinal homeostasis and its breakdown in inflammatory bowel disease. Nature (Lond) 474(7351):298–306

Mauri C, Bosma A (2012) Immune regulatory function of B cells. Annu Rev Immunol 30: 221–241

Maynard CL, Weaver CT (2008) Diversity in the contribution of interleukin-10 to T-cell-mediated immune regulation. Immunol Rev 226:219–233

Mege JL, Meghari S, Honstettre A, Capo C, Raoult D (2006) The two faces of interleukin 10 in human infectious diseases. Lancet Infect Dis 6(9):557–569

Mocellin S, Panelli MC, Wang E, Nagorsen D, Marincola FM (2003) The dual role of IL-10. Trends Immunol 24(1):36–43

Mocellin S, Marincola FM, Young HA (2005) Interleukin-10 and the immune response against cancer: a counterpoint. J Leukoc Biol 78(5):1043–1051

Moore KW, Vieira P, Fiorentino DF, Trounstine ML, Khan TA, Mosmann TR (1990) Homology of cytokine synthesis inhibitory factor (IL-10) to the Epstein–Barr virus gene BCRFI. Science 248(4960):1230–1234

Moore KW, de Waal Malefyt R, Coffman RL, O'Garra A (2001) Interleukin-10 and the interleukin-10 receptor. Annu Rev Immunol 19:683–765

Murai M, Turovskaya O, Kim G, Madan R, Karp CL, Cheroutre H, Kronenberg M (2009) Interleukin 10 acts on regulatory T cells to maintain expression of the transcription factor Foxp3 and suppressive function in mice with colitis. Nat Immunol 10(11):1178–1184

Murphy KM, Ouyang W, Farrar JD, Yang J, Ranganath S, Asnagli H, Afkarian M, Murphy TL (2000) Signaling and transcription in T helper development. Annu Rev Immunol 18:451–494

Murray PJ (2005) The primary mechanism of the IL-10-regulated antiinflammatory response is to selectively inhibit transcription. Proc Natl Acad Sci USA 102(24):8686–8691

Murray PJ (2006) Understanding and exploiting the endogenous interleukin-10/STAT3-mediated anti-inflammatory response. Curr Opin Pharmacol 6(4):379–386

Newton SM, Smith RJ, Wilkinson KA, Nicol MP, Garton NJ, Staples KJ, Stewart GR, Wain JR, Martineau AR, Fandrich S, Smallie T, Foxwell B, Al-Obaidi A, Shafi J, Rajakumar K,

Kampmann B, Andrew PW, Ziegler-Heitbrock L, Barer MR, Wilkinson RJ (2006) A deletion defining a common Asian lineage of *Mycobacterium tuberculosis* associates with immune subversion. Proc Natl Acad Sci USA 103(42):15594–15598

North RJ, Jung YJ (2004) Immunity to tuberculosis. Annu Rev Immunol 22:599–623

Nylen S, Sacks D (2007) Interleukin-10 and the pathogenesis of human visceral leishmaniasis. Trends Immunol 28(9):378–384

O'Garra A, Stapleton G, Dhar V, Pearce M, Schumacher J, Rugo H, Barbis D, Stall A, Cupp J, Moore K et al (1990) Production of cytokines by mouse B cells: B lymphomas and normal B cells produce interleukin 10. Int Immunol 2(9):821–832

O'Garra A, Chang R, Go N, Hastings R, Haughton G, Howard M (1992) Ly-1 B (B-1) cells are the main source of B cell-derived interleukin 10. Eur J Immunol 22(3):711–717

O'Garra A, Barrat FJ, Castro AG, Vicari A, Hawrylowicz C (2008) Strategies for use of IL-10 or its antagonists in human disease. Immunol Rev 223:114–131

Oh JW, Seroogy CM, Meyer EH, Akbari O, Berry G, Fathman CG, Dekruyff RH, Umetsu DT (2002) CD4 T-helper cells engineered to produce IL-10 prevent allergen-induced airway hyperreactivity and inflammation. J Allergy Clin Immunol 110(3):460–468

Ohtani M, Nagai S, Kondo S, Mizuno S, Nakamura K, Tanabe M, Takeuchi T, Matsuda S, Koyasu S (2008) Mammalian target of rapamycin and glycogen synthase kinase 3 differentially regulate lipopolysaccharide-induced interleukin-12 production in dendritic cells. Blood 112(3): 635–643

Ouyang W, Rutz S, Crellin NK, Valdez PA, Hymowitz SG (2011) Regulation and functions of the IL-10 family of cytokines in inflammation and disease. Annu Rev Immunol 29:71–109

Pitt JM, Stavropoulos E, Redford PS, Beebe AM, Bancroft GJ, Young DB, O'Garra A (2012) Blockade of IL-10 signaling during bacillus Calmette–Guerin vaccination enhances and sustains Th1, Th17, and innate lymphoid IFN-gamma and IL-17 responses and increases protection to *Mycobacterium tuberculosis* infection. J Immunol 189(8):4079–4087

Redford PS, Boonstra A, Read S, Pitt J, Graham C, Stavropoulos E, Bancroft GJ, O'Garra A (2010) Enhanced protection to *Mycobacterium tuberculosis* infection in IL-10-deficient mice is accompanied by early and enhanced Th1 responses in the lung. Eur J Immunol 40(8):2200–2210

Redford PS, Murray PJ, O'Garra A (2011) The role of IL-10 in immune regulation during *M. tuberculosis* infection. Mucosal Immunol 4(3):261–270

Reid DM, Gow NA, Brown GD (2009) Pattern recognition: recent insights from Dectin-1. Curr Opin Immunol 21(1):30–37

Roers A, Siewe L, Strittmatter E, Deckert M, Schluter D, Stenzel W, Gruber AD, Krieg T, Rajewsky K, Muller W (2004) T cell-specific inactivation of the interleukin 10 gene in mice results in enhanced T cell responses but normal innate responses to lipopolysaccharide or skin irritation. J Exp Med 200(10):1289–1297

Rogers NC, Slack EC, Edwards AD, Nolte MA, Schulz O, Schweighoffer E, Williams DL, Gordon S, Tybulewicz VL, Brown GD, Reis e Sousa C (2005) Syk-dependent cytokine induction by Dectin-1 reveals a novel pattern recognition pathway for C type lectins. Immunity 22(4): 507–517

Rutz S, Janke M, Kassner N, Hohnstein T, Krueger M, Scheffold A (2008) Notch regulates IL-10 production by T helper 1 cells. Proc Natl Acad Sci USA 105(9):3497–3502

Saraiva M, O'Garra A (2010) The regulation of IL-10 production by immune cells. Nat Rev Immunol 10(3):170–181

Saraiva M, Christensen JR, Tsytsykova AV, Goldfeld AE, Ley SC, Kioussis D, O'Garra A (2005) Identification of a macrophage-specific chromatin signature in the IL-10 locus. J Immunol 175(2): 1041–1046

Saraiva M, Christensen JR, Veldhoen M, Murphy TL, Murphy KM, O'Garra A (2009) Interleukin-10 production by Th1 cells requires interleukin-12-induced STAT4 transcription factor and ERK MAP kinase activation by high antigen dose. Immunity 31(2):209–219

Sellon RK, Tonkonogy S, Schultz M, Dieleman LA, Grenther W, Balish E, Rennick DM, Sartor RB (1998) Resident enteric bacteria are necessary for development of spontaneous colitis and immune system activation in interleukin-10-deficient mice. Infect Immun 66(11):5224–5231

Sharma A, Kumar M, Aich J, Hariharan M, Brahmachari SK, Agrawal A, Ghosh B (2009) Posttranscriptional regulation of interleukin-10 expression by hsa-miR-106a. Proc Natl Acad Sci USA 106(14):5761–5766

Shoemaker J, Saraiva M, O'Garra A (2006) GATA-3 directly remodels the IL-10 locus independently of IL-4 in CD4+ T cells. J Immunol 176(6):3470–3479

Slack EC, Robinson MJ, Hernanz-Falcon P, Brown GD, Williams DL, Schweighoffer E, Tybulewicz VL, Reis e Sousa C (2007) Syk-dependent ERK activation regulates IL-2 and IL-10 production by DC stimulated with zymosan. Eur J Immunol 37(6):1600–1612

Slobedman B, Barry PA, Spencer JV, Avdic S, Abendroth A (2009) Virus-encoded homologs of cellular interleukin-10 and their control of host immune function. J Virol 83(19):9618–9629

Smallie T, Ricchetti G, Horwood NJ, Feldmann M, Clark AR, Williams LM (2010) IL-10 inhibits transcription elongation of the human TNF gene in primary macrophages. J Exp Med 207(10):2081–2088

Spolski R, Kim HP, Zhu W, Levy DE, Leonard WJ (2009) IL-21 mediates suppressive effects via its induction of IL-10. J Immunol 182(5):2859–2867

Stoecklin G, Tenenbaum SA, Mayo T, Chittur SV, George AD, Baroni TE, Blackshear PJ, Anderson P (2008) Genome-wide analysis identifies interleukin-10 mRNA as target of tristetraprolin. J Biol Chem 283(17):11689–11699

Stumhofer JS, Silver JS, Laurence A, Porrett PM, Harris TH, Turka LA, Ernst M, Saris CJ, O'Shea JJ, Hunter CA (2007) Interleukins 27 and 6 induce STAT3-mediated T cell production of interleukin 10. Nat Immunol 8(12):1363–1371

Szabo SJ, Kim ST, Costa GL, Zhang X, Fathman CG, Glimcher LH (2000) A novel transcription factor, T-bet, directs Th1 lineage commitment. Cell 100(6):655–669

Tanchot C, Guillaume S, Delon J, Bourgeois C, Franzke A, Sarukhan A, Trautmann A, Rocha B (1998) Modifications of CD8+ T cell function during in vivo memory or tolerance induction. Immunity 8(5):581–590

Tudor C, Marchese FP, Hitti E, Aubareda A, Rawlinson L, Gaestel M, Blackshear PJ, Clark AR, Saklatvala J, Dean JL (2009) The p38 MAPK pathway inhibits tristetraprolin-directed decay of interleukin-10 and pro-inflammatory mediator mRNAs in murine macrophages. FEBS Lett 583(12):1933–1938

Turner J, Gonzalez-Juarrero M, Ellis DL, Basaraba RJ, Kipnis A, Orme IM, Cooper AM (2002) In vivo IL-10 production reactivates chronic pulmonary tuberculosis in C57BL/6 mice. J Immunol 169(11):6343–6351

Veldhoen M, Hocking RJ, Atkins CJ, Locksley RM, Stockinger B (2006) TGFbeta in the context of an inflammatory cytokine milieu supports de novo differentiation of IL-17-producing T cells. Immunity 24(2):179–189

Veldhoen M, Uyttenhove C, van Snick J, Helmby H, Westendorf A, Buer J, Martin B, Wilhelm C, Stockinger B (2008) Transforming growth factor-beta 'reprograms' the differentiation of T helper 2 cells and promotes an interleukin 9-producing subset. Nat Immunol 9(12):1341–1346

Vicari AP, Vanbervliet B, Massacrier C, Chiodoni C, Vaure C, Ait-Yahia S, Dercamp C, Matsos F, Reynard O, Taverne C, Merle P, Colombo MP, O'Garra A, Trinchieri G, Caux C (2004) In vivo manipulation of dendritic cell migration and activation to elicit antitumour immunity. Novartis Found Symp 256:241–254, discussion 254–269

Wang ZY, Sato H, Kusam S, Sehra S, Toney LM, Dent AL (2005) Regulation of IL-10 gene expression in Th2 cells by Jun proteins. J Immunol 174(4):2098–2105

Weichhart T, Costantino G, Poglitsch M, Rosner M, Zeyda M, Stuhlmeier KM, Kolbe T, Stulnig TM, Horl WH, Hengstschlager M, Muller M, Saemann MD (2008) The TSC-mTOR signaling pathway regulates the innate inflammatory response. Immunity 29(4):565–577

Wilson MS, Mentink-Kane MM, Pesce JT, Ramalingam TR, Thompson R, Wynn TA (2007) Immunopathology of schistosomiasis. Immunol Cell Biol 85(2):148–154

Wynn TA (2004) Fibrotic disease and the T(H)1/T(H)2 paradigm. Nat Rev Immunol 4(8): 583–594

Wynn TA, Cheever AW, Williams ME, Hieny S, Caspar P, Kuhn R, Muller W, Sher A (1998) IL-10 regulates liver pathology in acute murine *Schistosomiasis mansoni* but is not required for immune down-modulation of chronic disease. J Immunol 160(9):4473–4480

Xu J, Yang Y, Qiu G, Lal G, Wu Z, Levy DE, Ochando JC, Bromberg JS, Ding Y (2009) c-Maf regulates IL-10 expression during Th17 polarization. J Immunol 182(10):6226–6236

Xystrakis E, Kusumakar S, Boswell S, Peek E, Urry Z, Richards DF, Adikibi T, Pridgeon C, Dallman M, Loke TK, Robinson DS, Barrat FJ, O'Garra A, Lavender P, Lee TH, Corrigan C, Hawrylowicz CM (2006) Reversing the defective induction of IL-10-secreting regulatory T cells in glucocorticoid-resistant asthma patients. J Clin Invest 116(1):146–155

Yi AK, Yoon JG, Yeo SJ, Hong SC, English BK, Krieg AM (2002) Role of mitogen-activated protein kinases in CpG DNA-mediated IL-10 and IL-12 production: central role of extracellular signal-regulated kinase in the negative feedback loop of the CpG DNA-mediated Th1 response. J Immunol 168(9):4711–4720

Zhang X, Edwards JP, Mosser DM (2006) Dynamic and transient remodeling of the macrophage IL-10 promoter during transcription. J Immunol 177(2):1282–1288

Zheng W, Flavell RA (1997) The transcription factor GATA-3 is necessary and sufficient for Th2 cytokine gene expression in CD4 T cells. Cell 89(4):587–596

Zhu J, Min B, Hu-Li J, Watson CJ, Grinberg A, Wang Q, Killeen N, Urban JF Jr, Guo L, Paul WE (2004) Conditional deletion of Gata3 shows its essential function in T(H)1-T(H)2 responses. Nat Immunol 5(11):1157–1165

Zuanyamorim C, Haile S, Leduc D, Dumarey C, Huerre M, Vargaftig BB, Pretolani M (1995) Interleukin-10 inhibits antigen-induced cellular recruitment into the airways of sensitized mice. J Clin Invest 95(6):2644–2651

Chapter 14
Interleukin-27: Regulation of Immune Responses and Disease Development by a Pleiotropic Cytokine with Pro- and Anti-inflammatory Properties

Izuru Mizoguchi, Kaname Higuchi, Kana Mitobe, Ren Tsunoda, Junichiro Mizuguchi, and Takayuki Yoshimoto

Abstract Interleukin (IL)-27 is a member of the IL-6/IL-12 heterodimeric cytokine family and has pleiotropic properties, including both pro-inflammatory and anti-inflammatory functions depending on the context. IL-27 promotes early helper T (Th)1 differentiation and generation of cytotoxic T cells and IL-10-producing regulatory T cells. In addition, IL-27 inhibits differentiation of $CD4^+$ T cells into Th1, Th2, Th9, and Th17 cells; suppresses their responses; and limits production of pro-inflammatory cytokines. Blocking the interaction between IL-27 and its receptor or genetic knockout of their subunits exacerbates inflammatory responses in infectious and autoimmune diseases, whereas IL-27 injection or forced expression induces potent antitumor activity against a variety of mouse and human tumors and suppresses the development of inflammatory, allergic, and autoimmune diseases. In this chapter, we review the molecular characterization of IL-27 and its receptor, its pro- and anti-inflammatory properties, and the therapeutic implications.

Keywords EBI3 • gp130 • IL-27 • p28 • WSX-1

I. Mizoguchi • K. Mitobe • R. Tsunoda • T. Yoshimoto (✉)
Department of Immunoregulation, Institute of Medical Science,
Tokyo Medical University, 6-1-1 Shinjuku, Shinjuku-ku, Tokyo, 160-8402, Japan
e-mail: yoshimot@tokys-med.ac.jp

K. Higuchi
Department of Immunoregulation, Institute of Medical Science, Tokyo Medical University,
6-1-1 Shinjuku, Shinjuku-ku, Tokyo, 160-8402, Japan

Department of Immunology, Tokyo Medical University, 6-1-1 Shinjuku, Shinjuku-ku,
Tokyo, 160-8402, Japan

J. Mizuguchi
Department of Immunology, Tokyo Medical University, 6-1-1 Shinjuku, Shinjuku-ku,
Tokyo, 160-8402, Japan

T. Yoshimoto and T. Yoshimoto (eds.), *Cytokine Frontiers: Regulation of Immune Responses in Health and Disease*, DOI 10.1007/978-4-431-54442-5_14,

14.1 Introduction

The interleukin (IL)-6/IL-12 cytokine family is unique in that it is a heterodimeric cytokine composed of two different subunits (Trinchieri 2003; Kastelein et al. 2007) (Fig. 14.1). IL-12 is composed of p35 and p40 subunits, its receptor (R) consists of two subunits IL-12Rβ1 and β2, and IL-12 activates signal transducer and activator of transcription (STAT) 4, which binds to the cytoplasmic region of IL-12Rβ2 (Trinchieri 2003). The p40 subunit is also covalently bound with an IL-12 p35-related protein, p19, to form IL-23 (Oppmann et al. 2000). The receptor for IL-23 is composed of one of the IL-12R subunits, IL-12Rβ1 and an IL-12Rβ2-like receptor subunit designated IL-23R (Parham et al. 2002). IL-23 activates STAT3 and STAT4, and STAT3 activation is required for IL-17 production by T cells with IL-23 (Yang et al. 2007). IL-27 consists of an IL-12 p35-related protein, p28, which is also called IL-30, and an IL-12 p40-related protein, Epstein–Barr virus (EBV)-induced gene 3 (EBI3), which has been previously identified as one of the molecules induced by EBV infection (Sprecher et al. 1998; Chen et al. 2000; Pflanz et al. 2002). IL-27R is composed of IL-27Rα (WSX-1/T-cell cytokine receptor, TCCR), which has a WSXWS sequence and is homologous to the IL-12Rβ2 subunit, and glycoprotein (gp)130, a common receptor subunit for IL-6 family cytokines (Pflanz et al. 2004). EBI3 was previously reported to associate with p35 as well to form the

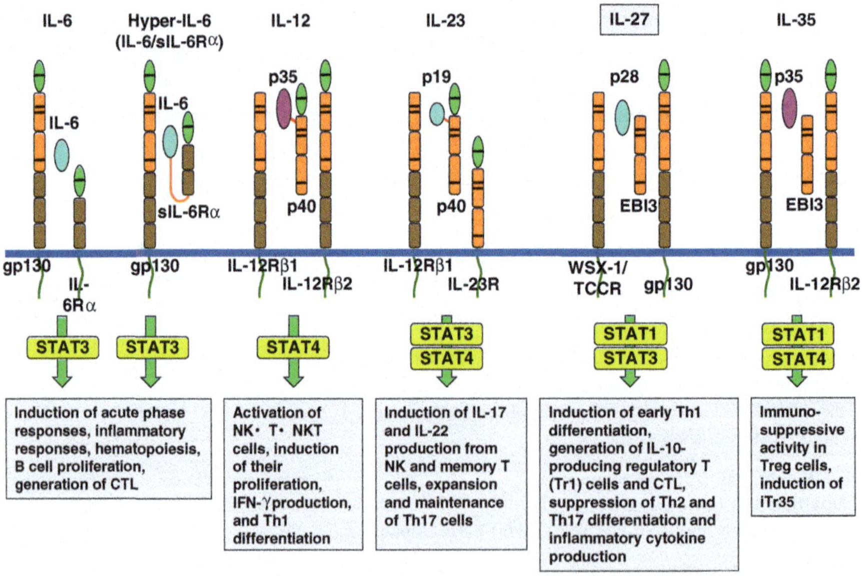

Fig. 14.1 Molecular characteristics of the interleukin (IL)-6/IL-12 family cytokines: IL-6, IL-12, IL-27, and IL-35. This cytokine family is unique in that it is a heterodimeric cytokine composed of two different subunits, which may be substituted for other subunits of different cytokines belonging to this family

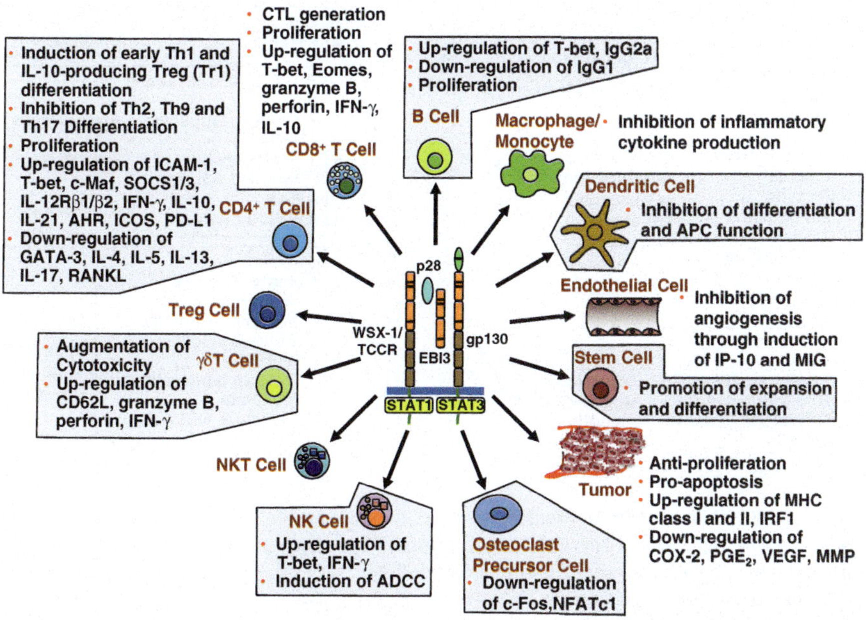

Fig. 14.2 IL-27 is a pleiotropic cytokine acting on a variety of cells by upregulating or downregulating various signal transduction molecules including transcriptional factors and downstream cytokines. IL-27 activates STAT1 and STAT3, which directly bind to the cytoplasmic regions of WSX-1 and gp130, respectively, leading to pro- and anti-inflammatory responses

heterodimeric molecule EBI3/p35, but its function remained unknown until recently (Devergne et al. 1996, 1997). EBI3/p35 has since been demonstrated to be produced by regulatory T (Treg) cells and to play a suppressive role in the Treg cells; it was named IL-35 (Collison et al. 2007). IL-35 converts naive T cells into inducible IL-35-producing regulatory T (iTr35) cells (Collison et al. 2010). IL-35 signals through a unique heterodimer of receptor chains IL-12Rβ2 and gp130 or homodimers of each chain (Collison et al. 2012). Signaling through IL-35R requires the transcription factors STAT1 and STAT4, which form a unique heterodimer that binds to distinct sites in the promoters of the genes encoding IL-12 subunits p35 and EBI3.

IL-27 is one of the pleiotropic cytokines, and its role in regulating immune response is complex. The stimulatory and inhibitory effects of IL-27 act differentially on various kinds of cells such as T cells, B cells, macrophages, dendritic cells (DCs) (Villarino et al. 2004), hematopoietic stem cells (Seita et al. 2008), and so on, depending on the context (Fig. 14.2). As a signaling molecule, IL-27 efficiently activates both STAT1 and STAT3, which bind to distinct IL-27R subunits, IL-27Rα and gp130, respectively (Lucas et al. 2003; Takeda et al. 2003; Kamiya et al. 2004; Yoshimura et al. 2006; Owaki et al. 2008). IL-27 promotes the early induction of helper T (Th)1 differentiation (Yoshida et al. 2001; Takeda et al. 2003; Owaki et al.

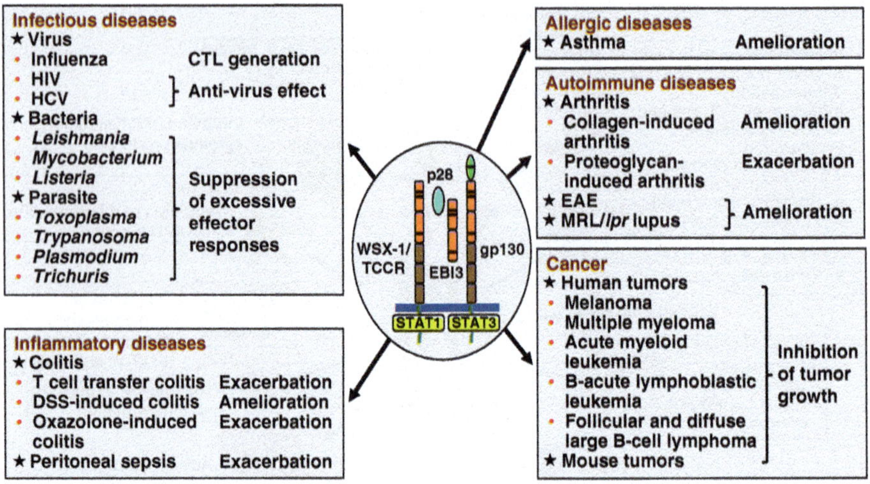

Fig. 14.3 IL-27 ameliorates or exacerbates a variety of diseases by regulating immune responses including Th1, Th2, Th9, Th17, Tr1, Treg, cytotoxic T lymphocytes (CTL), antibody production, macrophage and DC activation, proliferation, and cell survival. *DSS* dextran sodium sulfate, *MRL/lpr* MRL/*lymphoproliferation*

2005) and generation of cytotoxic T lymphocytes (CTLs) (Morishima et al. 2005, 2010), resulting in augmentation of antitumor effects and inhibition of infection with influenza virus (Fig. 14.3). In addition, IL-27 inhibits the differentiation of naive CD4^{+} T cells into Th2 cells (Miyazaki et al. 2005; Yoshimoto et al. 2007; Fujita et al. 2009; Dokmeci et al. 2011), Th9 cells (Murugaiyan et al. 2012), and Th17 cells (Batten et al. 2006; Stumhofer et al. 2006; Fitzgerald et al. 2007a). IL-27 also suppresses their responses, leading to amelioration of allergic diseases including asthma and autoimmune diseases such as encephalomyelitis and arthritis. Moreover, IL-27 enhances IL-10 production from these Th cells and generation of IL-10-producing regulatory (Tr1) T cells (Awasthi et al. 2007; Fitzgerald et al. 2007b; Stumhofer et al. 2007) and inhibits the production of pro-inflammatory cytokines (Hamano et al. 2003; Villarino et al. 2003).

14.2 Molecular Characterization of IL-27

14.2.1 IL-27 Production

EBI3 was initially identified as one of the molecules induced in responses to EBV (Devergne et al. 1996), and subsequent studies revealed that EBI3 is expressed in activated DCs, monocytes, and macrophages by lipopolysaccharide (LPS) (Hashimoto et al. 2000). Other cell types including plasma cells, endothelial cells, microglia, and placental trophoblasts express EBI3 and p28. A variety of Toll-like receptor (TLR) agonists [LPS, poly(I:C), CpG] can induce mRNA expression of EBI3 and p28 in human and mouse antigen-presenting cells (Hashimoto et al. 2000; Smits et al. 2004; Wirtz et al. 2005; Liu et al. 2007). LPS markedly induces EBI3 and p28 expression in synergy with interferon (IFN)-γ through distinct mechanisms (Wirtz et al. 2005; Liu et al. 2007). In response to LPS, nuclear factor (NF)-κB binds to the p28 and EBI3 promoters through activation of myeloid differentiation factor 88. Alternatively, LPS activates IFN regulatory factor (IRF)3 through the adaptor Toll/IL-1R-related domain-containing adaptor-inducing IFN (TRIF) (Molle et al. 2007). EBI3 expression is also induced by the signal via TRIF but not mediated by IRF3. In addition, IFN-γ or IFN-α/β can enhance p28 expression by STAT1/IRF1 or STAT1/STAT2/IRF9 (ISGF3), respectively (Pirhonen et al. 2007; Molle et al. 2010; Zhang et al. 2010).

Recently, simultaneous activation of two types of cells, CD4$^+$ T cells and macrophages, by anti-CD3/CD28 and CpG in the same microenvironment was reported to be crucial for inducing the robust expression of p28 via the CD40/CD154 signaling pathway independent of IFN-γ (Dibra et al. 2012). Moreover, EBI3 expression in tumors was demonstrated to be a novel diagnosis marker in lung cancer (Nishino et al. 2011), Burkitt lymphoma, and diffuse large B-cell lymphoma (Gonin et al. 2011). In the lung cancer cells, reduction in EBI3 expression by small interfering RNA suppresses their cell proliferation, while induction of exogenous EBI3 confers growth-promoting activity, indicating that EBI3 expression in tumors plays a pivotal role in tumorigenesis. The physiological significance of the abundant p28 production or the molecular mechanism whereby EBI3 promotes tumorigenesis remains to be elucidated.

14.2.2 IL-27R Expression and Signaling

One IL-27R, WSX-1, was first identified as a receptor with homology to gp130 (Sprecher et al. 1998). The same receptor was also cloned and termed TCCR (Chen et al. 2000). Subsequently, another IL-27R partner was identified to be gp130, a signal-transducing receptor that is common in IL-6 superfamily cytokines (Pflanz et al. 2004). gp130 is widely expressed in a variety of cell types, and its expression level varies depending on the cell activation status (Taga and Kishimoto 1997). WSX-1 is also widely expressed by T cells, B cells, natural killer (NK) cells, macrophages, DCs (Villarino et al. 2005), and hematopoietic stem cells (Seita et al. 2008).

Stimulation of naive CD4$^+$ T cells with IL-27 induces the activation of Janus kinase (JAK)1, JAK2, and TYK2, and subsequently the activation of various STATs including STAT1, STAT3, and STAT5 (Takeda et al. 2003; Kamiya et al. 2004; Owaki et al. 2008). As a negative feedback mechanism that is important for the balance of immune responses, IL-27 also upregulates suppressor of cytokine signaling (SOCS)1 and 3 (Owaki et al. 2006b; Brender et al. 2007; Liu and Rohowsky-Kochan 2011). Although STAT1 and STAT3 are similar proteins often activated by the same stimuli, they have very different effects on cell growth and survival. It is currently considered that STAT1 is a tumor suppressor while STAT3 is an oncogene. Because IL-27 efficiently activates both STAT1 and STAT3 through its distinct receptor subunits, IL-27α and gp130, physiological consequences in response to IL-27 may be more complicated (Lucas et al. 2003; Takeda et al. 2003; Kamiya et al. 2004; Owaki et al. 2008). Their relative abundance, which may vary substantially in different types of cells under different conditions, seems to affect how cells behave in response to IL-27 thereafter.

14.3 Pro-Inflammatory Properties of IL-27

14.3.1 IL-27 Promotes Th1 Responses

IL-27 upregulates the expression of T-cell-specific T-box transcription factor (T-bet), intercellular adhesion molecule (ICAM)-1, and subsequent IL-12Rβ2 through STAT1 activation and synergies with IL-12 in primary IFN-γ production (Pflanz et al. 2002; Hibbert et al. 2003; Takeda et al. 2003; Kamiya et al. 2004; Owaki et al. 2005, 2006a). Subsequent studies revealed that IL-27 augments Th1 differentiation via p38 mitogen-activated protein kinase (MAPK)/T-bet-dependent pathway and also via ICAM-1/lymphocyte function-associated antigen 1 (LFA-1)/extracellular signal-regulated kinase (ERK)1/2-dependent pathway (Owaki et al. 2006a). This action is in contrast to IL-12, which induces it only via the p38 MAPK/T-bet-dependent pathway. In vivo studies using TCCR- or WSX-1-deficient mice demonstrated that these deficient mice are more susceptible to *Listeria monocytogenes* and *Leishmania major*, presumably because of defects in Th1 responses (Chen et al. 2000; Yoshida et al. 2001). Similar but clearer results were demonstrated in proteoglycan (PG)-induced arthritis in which TCCR-deficient mice showed reduced production of IFN-γ and PG-specific IgG2a, indicating that IL-27 plays a critical role in the development of a pathological Th1 response in certain situations (Cao et al. 2008).

14.3.2 IL-27 Enhances CD8⁺ T-Cell Responses

In a manner similar to the action of IL-27 on $CD4^+$ T cells, IL-27 activates various STATs, including STAT1 and STAT3, and augments the expression not only of T-bet and IL-12Rβ2 but also effector molecules such as granzyme B and perforin in mouse and human naive $CD8^+$ T cells stimulated with anti-CD3 and anti-CD28 (Morishima et al. 2005, 2010; Schneider et al. 2011). IL-27 also induces synergistic IFN-γ production with IL-12 and proliferation of naive $CD8^+$ T cells. Moreover, IL-27 enhances the generation of allogenic CTLs and augments their expression of effector molecules such as perforin and granzyme B in a STAT1-dependent manner (Morishima et al. 2005). As downstream signaling molecules of STAT1, both T-bet and Eomesdermin (Eomes) are necessary for the IL-27-mediated generation of CTLs and exertion of their effector functions.

There is ample evidence showing that CTLs play a central role in the clearance of pathogenic viruses. In hepatitis C virus (HCV) infection, vigorous HCV-specific CTL responses exist in patients resolving acute HCV infection, and enhancement of HCV-specific CTL generation is considered to be one of the strategies to clear the virus. Adjuvant effects of IL-27 were demonstrated in the prime-boost immunization, consisting of priming and the first boosting with the HCV-core expression plasmid, which was followed by a second boosting with recombinant adenovirus expressing the HCV core for induction of HCV core-specific CTLs in HLA-A*0201 transgenic BALB/c mice (Matsui et al. 2004). Moreover, IL-27Rα signals but not IFN-γR signals in $CD8^+$ T cells were reported to be indispensable for the T-bet-dependent IFN-γ production during infection with influenza, indicating that IL-27 plays a critical role in driving effector $CD8^+$ T-cell responses (Mayer et al. 2008). A number of in vivo studies further showed that tumor cells engineered to express IL-27 induced tumor-specific CTL responses and tumor regression with memory responses to subsequent challenge in some cases (Xu et al. 2010), more clearly indicating the efficacy of IL-27 in promoting $CD8^+$ T-cell effector function in vivo.

14.3.3 IL-27 Augments B-Cell Responses

IL-27 was initially demonstrated to promote IgG2a secretion, but inhibit IgG1 secretion, by mouse B cells through STAT1, as does IFN-γ (Yoshimoto et al. 2004). In human B cells, IL-27 was reported to promote IgG1 production (Boumendjel et al. 2006). Consistent with these results, WSX-1-deficient mice were demonstrated to have reduced IgG2a production in infection with *L. monocytogenes* (Chen et al. 2000) and in the PG-induced model of arthritis (Cao et al. 2008). Moreover, WSX-1-deficient mice have exacerbated IgE production after airway ovalbumin (OVA) challenge (Miyazaki et al. 2005). In addition to these possible direct effects of IL-27 on B cells, IL-27 can also induce $CD4^+$ T cells to produce IL-21. IL-21 then promotes the B-cell expression of B-lymphocyte-induced maturation protein-1 (Blimp-1) and B-cell lymphoma 6 protein (Bcl-6), critical for plasma cell differentiation, and enhances follicular Th (Tfh) cell survival as well; however, IL-27 is not necessary for the differentiation of Tfh cells (Ozaki et al. 2004; Pot et al. 2009; Batten et al. 2010). TCCR-deficient mice were shown to have reduced germinal center B-cell responses, diminished production of class-switched antibodies and hapten-specific antibody, and less severe pathology in a pristine-induced lupus model (Batten et al. 2010). Collectively, IL-27 plays a critical role in B-cell responses by its ability to act directly on B cells as well as through its effect on $CD4^+$ T cells.

14.3.4 IL-27 Limits Regulatory T Cells

Treg cells are suppressive CD4⁺ T cells that express the transcription factor Foxp3, maintain self-tolerance during homeostasis, and limit immunopathology following infection. IL-27 was previously demonstrated to reduce the percentage of inducible Foxp3⁺ Treg cells differentiated from naive CD4⁺ T cells by transforming growth factor (TGF)-β and IL-2 (Neufert et al. 2007; Huber et al. 2008). In a colitis model, transfer of TCCR-deficient $CD4^+CD45RB^{hi}$ cells showed reduced colitis with an increased percentage of $Foxp3^+$ Treg cells in the gut compared with transfer of wild-type cells (Cox et al. 2011). Moreover, the role of IL-27 in the regulation of Treg cells was investigated in vivo using a transgenic mouse model in which IL-27 p28 and EBI3 subunits were overexpressed (IL-27 Tg) (Wojno et al. 2011). IL-27 Tg mice almost completely diminished Treg cells in lymphoid organs and succumbed to systemic inflammation. Although IL-27 does not downregulate Foxp3 expression in Treg cells, IL-27 inhibits reconstitution of the Treg cell population when IL-27 is present during generation of Treg cells. IL-27 Tg mice have a marked defect in their capacity to produce IL-2, indicating that IL-27 may influence differentiating Treg cells indirectly through IL-2 modulation (Wojno et al. 2011). Paradoxically, however, it has been recently reported that IL-27 promotes the expansion of Treg cells and enhances their expression of T-bet and CXCR3 (Hall et al. 2012a) as described later (Sect. 14.4.4).

14.4 Anti-inflammatory Properties of IL-27

14.4.1 IL-27 Inhibits Th1 Responses

In contrast to the pro-inflammatory properties already described, including promoting Th1 activity, CTL generation, and antibody production, IL-27 also possesses anti-inflammatory properties to suppress excessive Th1 immune responses during infection (Villarino et al. 2005). Although WSX-1-deficient mice infected with *Toxoplasma gondii* succumbed within 2 weeks of challenge, they efficiently controlled parasite replication via increased IFN-γ production with enhanced $CD8^+$ T and $CD4^+$ T-cell responses (Villarino et al. 2003). Similar exacerbated Th1 responses with disease pathology in WSX-1/TCCR-deficient mice were demonstrated in a variety of infections with *Trypanosoma cruzi*, *Plasmodium bergheri*, *Mycobacterium tuberculosis*, *L. major*, and *L. donovanii* (Hamano et al. 2003; Rosas et al. 2006; Findlay et al. 2010). WSX-1-deficient mice infected with *T. cruzi* developed increased Th1 responses and exacerbated IFN-γ production accompanied by liver injury and lethality (Hamano et al. 2003). These infected mice also showed enhanced Th2 responses and therefore increased parasite burden by the elevated IL-4 production. In the studies with *T. gondii*, depletion of $CD4^+$ T cells was demonstrated to rescue acute lethality in these mice, clearly indicating that IL-27 directly limits Th1 responses (Villarino et al. 2003). Moreover, WSX-1-deficient macrophages infected with *Mycobacterium tuberculosis* produced more IL-12p40, IL-6, and tumor necrosis factor (TNF)-α because of the exacerbated Th1 responses (Holscher et al. 2005). Thus, IL-27 plays an important role in limiting excess Th1 immune responses.

14.4.2 IL-27 Suppresses Th2/Th9/T17 Responses

In an experimental asthma model, WSX-1-deficient mice showed exacerbated Th2 response-mediated asthma (Miyazaki et al. 2005). Consistent with these results, administration of IL-27 ameliorated asthma, indicating a novel therapeutic approach for Th2-mediated allergic disorders (Yoshimoto et al. 2007). The molecular mechanism whereby IL-27 inhibits Th2 responses is considered to mediate inhibition of GATA-3, a master transcriptional factor for development of Th2 responses. Moreover, IL-27 stimulation induced marked reduction in the capacity of the host mouse to mount a Th2 response against *Strongyloides venezuelensis* infection, resulting in impaired intestinal mastocytosis and delayed parasite expulsion (Yoshimoto et al. 2007).

In addition, a number of reports demonstrated that IL-27 suppresses Th17 responses in addition to Th1 and Th2 responses (Batten et al. 2006; Stumhofer et al. 2006; Fitzgerald et al. 2007a). Th17 cells, a subset of Th cells that produce proinflammatory cytokine IL-17 and IL-22, are associated with chronic inflammation of autoimmune diseases such as rheumatoid arthritis, experimental autoimmune encephalomyelitis (EAE), and inflammatory bowel disease. WSX-1-, EBI3-, or p28-deficient mice showed exacerbated EAE with an increased population of Th17 cells (Stumhofer et al. 2006; Diveu et al. 2009). IL-27 inhibits the production of IL-17 from $CD4^+$ T cells in a STAT1-dependent and partially STAT3-dependent manner, but independent of T-bet and SOCS3 (Stumhofer et al. 2006). IL-27 also inhibits production of IL-22 and granulocyte macrophage colony-stimulating factor, which are important for effector functions of Th17 cells (Yang et al. 2008; Young et al. 2012). Moreover, IL-27 directly suppresses the expression of RORα and RORγt, which are necessary for Th17 development (Yang et al. 2008; Diveu et al. 2009; El-behi et al. 2009). One of the molecular mechanisms whereby IL-27 inhibits the production of IL-17 is considered to be mediated by IL-10 (Batten et al. 2008), as described next (Sect. 14.4.3). In addition, the IL-27 priming of naive T cells was recently reported to upregulate expression of programmed death ligand 1 (PD-L1) in a STAT1-dependent manner (Hirahara et al. 2012). When cocultured with naive $CD4^+$ T cells, IL-27-primed T cells inhibited their differentiation into Th17 cells in *trans* through a PD-1–PD-L1 interaction and also suppressed development of EAE with the coadministration of IL-27-primed T cells expressing PD-L1. Th17 cells were also demonstrated to be important for osteoclastogenesis, which is therefore a therapeutic target for IL-27 (Pickens et al. 2011). IL-27 inhibits the expression of receptor activator of NFκ-B ligand (RANKL) in $CD4^+$ T cells in part through STAT3 (Kamiya et al. 2011), and also suppresses osteoclastogenesis via

STAT1-dependent downregulation of the transcription factor c-Fos and resulting nuclear factor of activated T cells, cytoplasmic 1 (NFATc1), which are induced by RANKL, in osteoclasts (Furukawa et al. 2009; Kalliolias et al. 2010).

Recently, Th9 cells have been established as a separate lineage of Th cells distinct from conventional Th1, Th2, and Th17 cells. Th9 cells are characterized by expression of IL-9 and are considered to be mediators of the inflammation associated with autoimmune diseases such as EAE. IL-27 was demonstrated to suppress the development of Th9 cells, which is partially dependent on the transcriptional factors STAT1 and T-bet (Murugaiyan et al. 2012). Furthermore, IL-27 treatment completely abrogates the encephalitogenicity of Th9 cells in the EAE model. Of note, IFN-γ also directly suppresses Th9 differentiation but also inhibits it through the induction of IL-27 from DCs (Murugaiyan et al. 2012).

14.4.3 IL-27 Promotes IL-10 Production

IL-10 is a potent anti-inflammatory cytokine, which plays a critical role in suppressing excessive immune responses. During infection, IL-10 is a key factor in limiting infection-induced pathology caused by pathogens such as *Toxoplasma gondii* (Gazzinelli et al. 1996). WSX-1-deficient mice were shown to have enhanced pathology with reduced levels of IL-10 in toxoplasmic encephalitis, malaria infection, and EAE (Awasthi et al. 2007; Fitzgerald et al. 2007b; Stumhofer et al. 2007; Freitas do Rosario et al. 2012). Consistent with these in vivo results, in vitro studies revealed that IL-27 promotes IL-10 production by various Th cells including Th1, Th2, Th17, and Treg cells (Stumhofer et al. 2007). In an adoptive transfer model of EAE, encepalitogenic T cells failed to induce disease when treated with IL-27 in an IL-10-dependent manner (Fitzgerald et al. 2007b). Moreover, IL-27 was demonstrated to augment IL-10 production by $CD4^+$ T cells in collaboration with TGF-β and cause them to differentiate into IL-10-producing regulatory T cells, the so-called Tr1 cells (Awasthi et al. 2007). Tr1 cells do not express the master transcription factor Foxp3, which is distinct from Treg cells (Pot et al. 2009, 2011). DCs primed in vivo by induced $Foxp3^+$ Tregs cells were reported to be likely tolerogenic DCs and to secrete IL-27, resulting in suppression of immune responses through generation of the Tr1 cells (Awasthi et al. 2007). The differentiation into Tr1 cells by IL-27 is highly dependent on STAT1 (Xu et al. 2009), and IL-27 was demonstrated to induce expression of aryl hydrocarbon receptor (AHR) and the transcription factor c-Maf, both of which can induce the production of IL-10 and IL-21 by $CD4^+$ T cells (Apetoh et al. 2010). Of note, IL-27-induced c-Maf expression activates IL-21 production, which acts as an autocrine growth factor for the expansion and/or maintenance of Tr1 cells, and IL-27 also enhances the costimulatory receptor inducible costimulatory (ICOS) expression, further promoting IL-27-mediated generation of Tr1 cells (Pot et al. 2009).

It was recently demonstrated that $CD4^+$ T-cell assistance in the form of IL-2 is required for IL-10 production by CTLs, but not for the induction of CTL effector cytokines (Sun et al. 2011), because IL-2 derived from $CD4^+$ helper T cells cooperates with innate immune cell-derived IL-27 to amplify IL-10 production by CTLs through a Blimp-1-dependent mechanism. Moreover, during primary respiratory viral infection, $CD8^+$ T effector T cells release the immunosuppressive cytokine IL-10, which limits inflammation and immunopathology and is essential for host survival. In contrast, it was recently demonstrated that $CD8^+$ T-cell-derived IL-10 is absent in a recall response because of a persistent loss of IL-27 responsiveness in $CD8^+$ memory T cells caused by downregulation of the common cytokine receptor, gp130 (Perona-Wright et al. 2012). Thus, IL-27 plays a critical role in suppressing excessive immune responses and pathology through IL-10 production by $CD4^+$ and $CD8^+$ T cells.

14.4.4 IL-27 Increases Treg Cell Expansion

In contrast to the antagonistic effect of IL-27 on Treg cells as previously mentioned, a very recent report has paradoxically demonstrated that IL-27 does not limit the generation of Treg cell population but rather promotes Treg cell expansion and expression of T-bet and CXCR3 (Hall et al. 2012a). In wild-type mice challenged with *T. gondii* or other pathogens, a population of Treg cells that expressed T-bet, CXCR3, and IL-10 emerged and limited T effector responses. In contrast, in WSX-1-deficient mice, the Treg population was reduced at primary sites of infection but not at peripheral sites, such as in the spleen, where IFN-γ has a similar and more prominent role. Transcriptional profiling was demonstrated to highlight that IL-27 appears to have a more dominant impact than IFN-γ on Treg cell expression of immunosuppressive genes such as IL-10 (Hall et al. 2012a). Further studies are necessary to clarify the precise role IL-27 in generation of Treg cells.

14.5 Therapeutic Implications of IL-27

14.5.1 Antitumor Effects of IL-27

Since the antitumor efficacy of IL-27 was first evaluated in 2004 (Hisada et al. 2004), accumulating evidence has revealed that IL-27 has potent antitumor activity, which is mediated by multiple mechanisms including $CD8^+$ T cells (Chiyo et al. 2004, 2005; Hisada et al. 2004; Salcedo et al. 2004, 2009), NK cells (Oniki et al. 2006; Liu et al. 2008; Hu et al. 2009), antibody-dependent cell-mediated cytotoxicity (ADCC) (Matsui et al. 2009), anti-angiogenesis (Shimizu et al. 2006; Airoldi and Ribatti 2011), direct suppression of tumor growth (Yoshimoto et al. 2008), and inhibition of cyclooxygenase-2 (COX-2) and resultant prostaglandin E_2 (PGE_2) expression (Ho et al. 2009), depending on the characteristics of individual tumors (Fig. 14.4). Because each subunit of IL-27 is not chemically bound as are IL-12 and IL-23, a single chain expression vector was constructed using a flexible linker such as

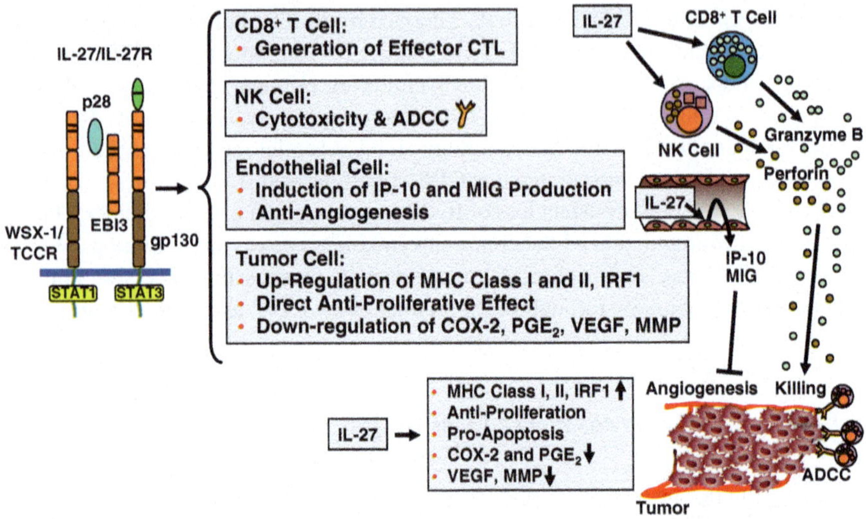

Fig. 14.4 Potent antitumor activity of IL-27 via multiple mechanisms including $CD8^+$ T cells, NK cells, antibody-dependent cell-mediated cytotoxicity (ADCC), anti-angiogenesis, and direct antiproliferation depending on individual properties of tumors

Gly_4Ser_3 (Pflanz et al. 2002; Hisada et al. 2004). For instance, highly immunogenic murine colon carcinoma colon 26 cells transfected with the IL-27 expression vector showed greatly reduced tumor growth, which was mainly mediated by $CD8^+$ T cells, IFN-γ, and T-bet (Hisada et al. 2004). Moreover, the overall effect of endogenous IL-27 was revealed to contribute to the antitumor effect, which was demonstrated using WSX-1-deficient mice (Shinozaki et al. 2009). Although endogenous IL-27 promotes generation of tumor-specific CTL in $CD8^+$ T cells, it inhibits APC function in DCs. Thus, IL-27 plays a particular important role in the generation of CTL (Morishima et al. 2010). On the other hand, the antitumor effects of IL-27 against poorly immunogenic tumors such as B16F10 melanoma are mediated by various mechanisms through NK cells (Oniki et al. 2006; Liu et al. 2008; Hu et al. 2009), angiogenesis (Shimizu et al. 2006), and its direct effects on tumors (Yoshimoto et al. 2008; Ho et al. 2009). IL-27 not only activates NK cells but also induces tumor-specific immunoglobulin, which cooperatively elicits ADCC activity (Matsui et al. 2009). IL-27 also possesses potent anti-angiogenic activity on melanomas as does IFN-γ, which contributes to its antitumor and antimetastatic activities (Shimizu et al. 2006). IL-27 was demonstrated to directly act on human umbilical vein endothelial cells and induce production of the anti-angiogenic chemokines such as IFN-γ-inducible protein 10 (IP-10, CXCR10) and monokine induced by IFN-γ (MIG, CXCR9). IL-27 also possesses potent direct antiproliferative activity on melanomas through WSX-1/STAT1 and in part through IRF1, but not IRF8, which is analogous to IFN-γ (Yoshimoto et al. 2008).

Furthermore, IL-27 was demonstrated to inhibit the growth of human tumors including melanoma, multiple myeloma, B-acute lymphoblastic leukemia (B-ALL)

cells, follicular lymphoma, diffuse large B-cell lymphoma, and acute myeloid leukemia (Yoshimoto et al. 2008; Cocco et al. 2010, 2012a, b; Airoldi and Ribatti 2011; Canale et al. 2011; Giuliani and Airoldi 2011; Zorzoli et al. 2012). IL-27 strongly inhibited tumor growth and in vivo tumorigenicity of multiple myeloma cells through suppression of angiogenesis (Cocco et al. 2010). IL-27 also reduced in vitro B-ALL cell proliferation and angiogenesis and augments apoptosis. Moreover, IL-27 severely hampered the leukemic spreading induced in nonobese diabetic/severe combined immunodeficient (NOD/SCID)/IL-2R$\gamma^{-/-}$ mice by injection with B-ALL cells from pediatric patients (Canale et al. 2011). Similarly, acute myeloid leukemia cells injected into (NOD/SCID)/IL-2R$\gamma^{-/-}$ mice gave rise to leukemia dissemination that was also severely inhibited by IL-27 (Zorzoli et al. 2012). These antitumor effects were caused by significant reduction of angiogenic and spreading-related genes, including vascular endothelial growth factors (VEGFs), angiopoietins, and matrix metalloproteinases (MMPs), and also by upregulation of angiostatic molecules, such as tissue inhibitor of MMP (Zorzoli et al. 2012).

14.5.2 Therapeutic Effects of IL-27 on Various Diseases

Although IL-27 possesses both pro-inflammatory and anti-inflammatory properties, therapeutic application against various diseases such as cancer, allergy, and autoimmune diseases are anticipated. Because systemic injection of IL-12 induces severe toxicity, clinical trials of IL-12 were halted (Cohen 1995; Leonard et al. 1997). In contrast, IL-27 treatment seems to have far fewer adverse effects such as splenomegaly and liver injury (Hisada et al. 2004; Oniki et al. 2006), although further studies with systemic administration of IL-27 and detailed analyses in other organs are necessary. For therapeutic application, injection of IL-27 expression vector or minicircle was demonstrated to reduce tumor growth of head and neck squamous cell carcinoma SCCVII and neuroblastoma TBJ, indicating that IL-27 could be utilized in the clinical setting (Matsui et al. 2009; Salcedo et al. 2009). IL-27 also provides us with a potential therapeutic means for treating Th2-associated diseases such as bronchial asthma (Yoshimoto et al. 2007). Intranasal administration of IL-27 inhibited OVA-induced airway hyperresponsiveness and inflammation in OVA-sensitized animals (Yoshimoto et al. 2007). Moreover, in vivo daily IL-27 treatment for a week after *L. major* infection protected BALB/c mice from footpad swelling by diminishing parasite burden via reciprocal regulation of Th1 and Th2 responses (Yoshimoto et al. 2007). On the other hand, IL-27 was demonstrated to inhibit the replication of human immunodeficiency virus type 1 (HIV-1) in CD4^{+} T cells and monocyte-derived macrophages as well as HCV replication by inducing antiviral genes including RNA-dependent protein kinase, oligoadenylate synthetase, and myxovirus protein in the same manner as IFN-α (Imamichi et al. 2008; Greenwell-Wild et al. 2009; Frank et al. 2010; Guzzo et al. 2010). Conversely, IL-27-induced gene expression is also impaired in human immunodeficiency virus (HIV) infection; this suggests that HIV-mediated dysregulation of IL-27 functions occurs during HIV infection, which is a new viral pathogenic mechanism contributing to the widespread impairment of immune responses observed in HIV pathogenesis (Guzzo et al. 2012).

In addition, IL-27 treatment attenuates collagen-induced arthritis (CIA) (Niedbala et al. 2008), and local injection of an adenovirus expressing IL-27 into the ankles of mice with CIA reduced clinical scores (Pickens et al. 2011). In murine models of EAE, IL-27 delivery sustained through osmotic pumps delayed the onset of EAE and ameliorated established disease (Fitzgerald et al. 2007b; El-behi et al. 2009). Injection of DNA minicircles of IL-27 suppressed the development of EAE (Stumhofer et al. 2010). There are also reports showing that blocking IL-27 activity by neutralizing IL-27R-Fc fusion protein limits disease progression in peritoneal sepsis (Wirtz et al. 2006). Similar blocking of IL-27 activity with reduced Th1 cell-mediated liver injury was demonstrated by using a mutant form of IL-27 in which the p28 subunit was mutated and unable to interact with gp130 (Rousseau et al. 2010). Thus, injection of IL-27 protein or its expression vector, or conversely blocking the interaction between IL-27 and its receptor, could be therapeutically applied for the potential treatment of various diseases.

14.6 Future Prospects and Concluding Remarks

There are conflicting reports as to antitumor effects of WSX-1 overexpression on tumors (Fig. 14.5). In certain tumors overexpression of WSX-1 induces the expression of MHC class I-related chain A, which is a ligand for NK group 2, member D, and promotes NK cell-mediated cytotoxicity (Dibra et al. 2009). In contrast, overexpression of WSX-1 in different tumor cells inhibits antitumor effector responses and promotes tumor growth (Dibra et al. 2011); this may be caused by homodimerization of WSX-1, which was also reported to possess hematopoietic cell-transforming activity (Pradhan et al. 2007). Furthermore, p28 independent of EBI3 was recently demonstrated to possess antagonistic effects of gp130-mediated signaling against the activity of IL-6, IL-11, and IL-27, analogous to p40 against IL-12 activity (Shimozato et al. 2009; Stumhofer et al. 2010). Moreover, it was reported that p28 also associates with p40, and the composite molecule functions as an antagonist against IL-12 and IL-27 (Wang et al. 2012). In marked contrast, p28 was demonstrated to induce agonistic activity associating with cytokine-like factor (CLF) through a tripartite receptor composed of gp130, WSX-1, and IL-6Rα, leading to activation of NK and T cells (Crabe et al. 2009). Moreover, p28 forms a heterodimer with soluble IL-6Rα and induces trans-signaling activity through the gp130 homodimer (Garbers et al. 2012). Thus, formation of monomers, homodimers, or heterodimers with different subunits, including those belonging to the other IL-6/IL-12 family cytokines and their receptors, further makes the interpretation of IL-27 biology more complex and remaining to be clarified (Fig. 14.5).

In conclusion, IL-27 is a pleiotropic cytokine and promotes both pro- and anti-inflammatory effects depending on context (Kastelein et al. 2007; Yoshida and

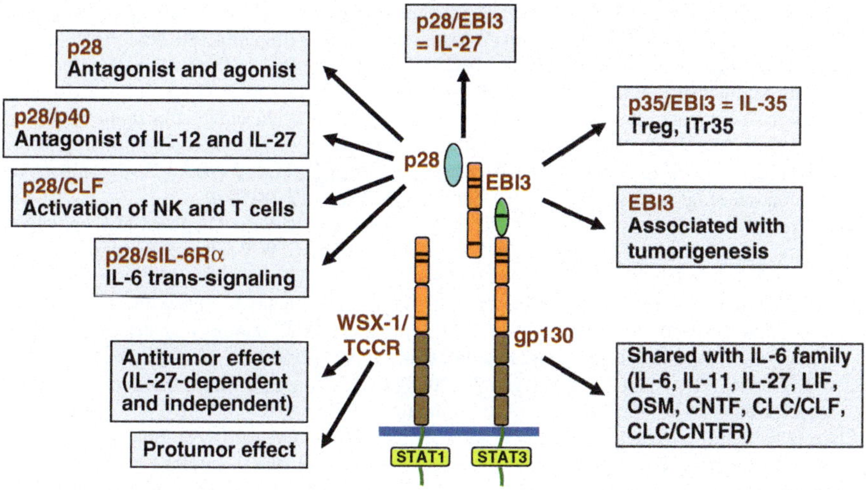

Fig. 14.5 Complexity of the interpretation of IL-27 biology resulting from possible formation of monomers, homodimers, or heterodimers with different subunits, including those belonging to the other IL-6/IL-12 family cytokines and their receptors. *CLC* cytokine-like factor, *CNTF* ciliary neurotrophic factor, *CNTFR* CNTF receptor, *LIF* leukemia inhibitory factor, *OSM* oncostatin M, *sIL-6Rα* soluble IL-6R**α**

Miyazaki 2008; Hall et al. 2012b; Hunter and Kastelein 2012). Although IL-27 was shown to have lower toxicity in mouse models, likely the result of the low induction of IFN-γ in vivo (Hisada et al. 2004; Oniki et al. 2006), clinical trials to evaluate toxicity and efficacy of IL-27 are necessary. Understanding the biology of IL-27 in more detail and the context that determines the outcome, promotion or suppression, of immune responses to IL-27 is very important and remains to be investigated. Taken together, a number of studies reporting IL-27, including preclinical studies as described here, support the possibility that IL-27 is a novel, promising therapeutic agent against various diseases such as cancer, allergies, autoimmune disorders, and infection.

References

Airoldi I, Ribatti D (2011) Regulation of angiostatic chemokines driven by IL-12 and IL-27 in human tumors. J Leukoc Biol 90:875–882

Apetoh L, Quintana FJ, Pot C, Joller N, Xiao S, Kumar D, Burns EJ, Sherr DH, Weiner HL, Kuchroo VK (2010) The aryl hydrocarbon receptor interacts with c-Maf to promote the differentiation of type 1 regulatory T cells induced by IL-27. Nat Immunol 11:854–861

Awasthi A, Carrier Y, Peron JP, Bettelli E, Kamanaka M, Flavell RA, Kuchroo VK, Oukka M, Weiner HL (2007) A dominant function for interleukin 27 in generating interleukin 10-producing anti-inflammatory T cells. Nat Immunol 8:1380–1389

Batten M, Li J, Yi S, Kljavin NM, Danilenko DM, Lucas S, Lee J, de Sauvage FJ, Ghilardi N (2006) Interleukin 27 limits autoimmune encephalomyelitis by suppressing the development of interleukin 17-producing T cells. Nat Immunol 7:929–936

Batten M, Kljavin NM, Li J, Walter MJ, de Sauvage FJ, Ghilardi N (2008) Cutting edge: IL-27 is a potent inducer of IL-10 but not FoxP3 in murine T cells. J Immunol 180:2752–2756
Batten M, Ramamoorthi N, Kljavin NM, Ma CS, Cox JH, Dengler HS, Danilenko DM, Caplazi P, Wong M, Fulcher DA, Cook MC, King C, Tangye SG, de Sauvage FJ, Ghilardi N (2010) IL-27 supports germinal center function by enhancing IL-21 production and the function of T follicular helper cells. J Exp Med 207:2895–2906
Boumendjel A, Tawk L, Malefijt Rde W, Boulay V, Yssel H, Pene J (2006) IL-27 induces the production of IgG1 by human B cells. Eur Cytokine Netw 17:281–289
Brender C, Tannahill GM, Jenkins BJ, Fletcher J, Columbus R, Saris CJ, Ernst M, Nicola NA, Hilton DJ, Alexander WS, Starr R (2007) SOCS3 regulates CD8 T cell proliferation by inhibition of IL-6 and IL-27. Blood 110:2528–2536
Canale S, Cocco C, Frasson C, Seganfreddo E, Di Carlo E, Ognio E, Sorrentino C, Ribatti D, Zorzoli A, Basso G, Dufour C, Airoldi I (2011) Interleukin-27 inhibits pediatric B-acute lymphoblastic leukemia cell spreading in a preclinical model. Leukemia 25:1815–1824
Cao Y, Doodes PD, Glant TT, Finnegan A (2008) IL-27 induces a Th1 immune response and susceptibility to experimental arthritis. J Immunol 180:922–930
Chen Q, Ghilardi N, Wang H, Baker T, Xie MH, Gurney A, Grewal IS, de Sauvage FJ (2000) Development of Th1-type immune responses requires the type I cytokine receptor TCCR. Nature (Lond) 407:916–920
Chiyo M, Shimozato O, Iizasa T, Fujisawa T, Tagawa M (2004) Antitumor effects produced by transduction of dendritic cells-derived heterodimeric cytokine genes in murine colon carcinoma cells. Anticancer Res 24:3763–3767
Chiyo M, Shimozato O, Yu L, Kawamura K, Iizasa T, Fujisawa T, Tagawa M (2005) Expression of IL-27 in murine carcinoma cells produces antitumor effects and induces protective immunity in inoculated host animals. Int J Cancer 115:437–442
Cocco C, Giuliani N, Di Carlo E, Ognio E, Storti P, Abeltino M, Sorrentino C, Ponzoni M, Ribatti D, Airoldi I (2010) Interleukin-27 acts as multifunctional antitumor agent in multiple myeloma. Clin Cancer Res 16:4188–4197
Cocco C, Di Carlo E, Zupo S, Canale S, Zorzoli A, Ribatti D, Morandi F, Ognio E, Airoldi I (2012a) Complementary IL-23 and IL-27 anti-tumor activities cause strong inhibition of human follicular and diffuse large B-cell lymphoma growth in vivo. Leukemia 26:1365–1374
Cocco C, Pistoia V, Airoldi I (2012b) Anti-leukemic properties of IL-12, IL-23 and IL-27: differences and similarities in the control of pediatric B acute lymphoblastic leukemia. Crit Rev Oncol Hematol 83:310–318
Cohen J (1995) IL-12 deaths: explanation and a puzzle. Science 270:908
Collison LW, Workman CJ, Kuo TT, Boyd K, Wang Y, Vignali KM, Cross R, Sehy D, Blumberg RS, Vignali DA (2007) The inhibitory cytokine IL-35 contributes to regulatory T-cell function. Nature (Lond) 450:566–569
Collison LW, Chaturvedi V, Henderson AL, Giacomin PR, Guy C, Bankoti J, Finkelstein D, Forbes K, Workman CJ, Brown SA, Rehg JE, Jones ML, Ni HT, Artis D, Turk MJ, Vignali DA (2010) IL-35-mediated induction of a potent regulatory T cell population. Nat Immunol 11:1093–1101
Collison LW, Delgoffe GM, Guy CS, Vignali KM, Chaturvedi V, Fairweather D, Satoskar AR, Garcia KC, Hunter CA, Drake CG, Murray PJ, Vignali DA (2012) The composition and signaling of the IL-35 receptor are unconventional. Nat Immunol 13:290–299
Cox JH, Kljavin NM, Ramamoorthi N, Diehl L, Batten M, Ghilardi N (2011) IL-27 promotes T cell-dependent colitis through multiple mechanisms. J Exp Med 208:115–123
Crabe S, Guay-Giroux A, Tormo AJ, Duluc D, Lissilaa R, Guilhot F, Mavoungou-Bigouagou U, Lefouili F, Cognet I, Ferlin W, Elson G, Jeannin P, Gauchat JF (2009) The IL-27 p28 subunit binds cytokine-like factor 1 to form a cytokine regulating NK and T cell activities requiring IL-6R for signaling. J Immunol 183:7692–7702
Devergne O, Hummel M, Koeppen H, Le Beau MM, Nathanson EC, Kieff E, Birkenbach M (1996) A novel interleukin-12 p40-related protein induced by latent Epstein–Barr virus infection in B lymphocytes. J Virol 70:1143–1153

Devergne O, Birkenbach M, Kieff E (1997) Epstein-Barr virus-induced gene 3 and the p35 subunit of interleukin 12 form a novel heterodimeric hematopoietin. Proc Natl Acad Sci USA 94: 12041–12046

Dibra D, Cutrera JJ, Xia X, Birkenbach MP, Li S (2009) Expression of WSX1 in tumors sensitizes IL-27 signaling-independent natural killer cell surveillance. Cancer Res 69:5505–5513

Dibra D, Cutrera J, Xia X, Li S (2011) WSX1 expression in tumors induces immune tolerance via suppression of effector immune cells. PLoS One 6:e19072

Dibra D, Cutrera JJ, Li S (2012) Coordination between TLR9 signaling in macrophages and CD3 signaling in T cells induces robust expression of IL-30. J Immunol 188:3709–3715

Diveu C, McGeachy MJ, Boniface K, Stumhofer JS, Sathe M, Joyce-Shaikh B, Chen Y, Tato CM, McClanahan TK, de Waal MR, Hunter CA, Cua DJ, Kastelein RA (2009) IL-27 blocks RORc expression to inhibit lineage commitment of Th17 cells. J Immunol 182:5748–5756

Dokmeci E, Xu L, Robinson E, Golubets K, Bottomly K, Herrick CA (2011) EBI3 deficiency leads to diminished T helper type 1 and increased T helper type 2 mediated airway inflammation. Immunology 132:559–566

El-Behi M, Ciric B, Yu S, Zhang GX, Fitzgerald DC, Rostami A (2009) Differential effect of IL-27 on developing versus committed Th17 cells. J Immunol 183:4957–4967

Findlay EG, Greig R, Stumhofer JS, Hafalla JC, de Souza JB, Saris CJ, Hunter CA, Riley EM, Couper KN (2010) Essential role for IL-27 receptor signaling in prevention of Th1-mediated immunopathology during malaria infection. J Immunol 185:2482–2492

Fitzgerald DC, Ciric B, Touil T, Harle H, Grammatikopolou J, Sarma JD, Gran B, Zhang GX, Rostami A (2007a) Suppressive effect of IL-27 on encephalitogenic Th17 cells and the effector phase of experimental autoimmune encephalomyelitis. J Immunol 179:3268–3275

Fitzgerald DC, Zhang GX, El-Behi M, Fonseca-Kelly Z, Li H, Yu S, Saris CJ, Gran B, Ciric B, Rostami A (2007b) Suppression of autoimmune inflammation of the central nervous system by interleukin 10 secreted by interleukin 27-stimulated T cells. Nat Immunol 8:1372–1379

Frank AC, Zhang X, Katsounas A, Bharucha JP, Kottilil S, Imamichi T (2010) Interleukin-27, an anti-HIV-1 cytokine, inhibits replication of hepatitis C virus. J Interferon Cytokine Res 30: 427–431

Freitas do Rosario AP, Lamb T, Spence P, Stephens R, Lang A, Roers A, Muller W, O'Garra A, Langhorne J (2012) IL-27 promotes IL-10 production by effector Th1 CD4+ T cells: a critical mechanism for protection from severe immunopathology during malaria infection. J Immunol 188:1178–1190

Fujita H, Teng A, Nozawa R, Takamoto-Matsui Y, Katagiri-Matsumura H, Ikezawa Z, Ishii Y (2009) Production of both IL-27 and IFN-γ after the treatment with a ligand for invariant NK T cells is responsible for the suppression of Th2 response and allergic inflammation in a mouse experimental asthma model. J Immunol 183:254–260

Furukawa M, Takaishi H, Takito J, Yoda M, Sakai S, Hikata T, Hakozaki A, Uchikawa S, Matsumoto M, Chiba K, Kimura T, Okada Y, Matsuo K, Yoshida H, Toyama Y (2009) IL-27 abrogates receptor activator of NF-κ B ligand-mediated osteoclastogenesis of human granulocyte-macrophage colony-forming unit cells through STAT1-dependent inhibition of c-Fos. J Immunol 183:2397–2406

Garbers C, Spudy B, Aparicio-Siegmund S, Waetzig GH, Sommer J, Holscher C, Rose-John S, Grotzinger J, Lorenzen I, Scheller J (2012) An interleukin-6 receptor-dependent molecular switch mediates signal transduction of the IL-27 cytokine subunit p28 (IL-30) via a gp130 receptor homodimer. J Biol Chem 288(6):4346–4354

Gazzinelli RT, Wysocka M, Hieny S, Scharton-Kersten T, Cheever A, Kuhn R, Muller W, Trinchieri G, Sher A (1996) In the absence of endogenous IL-10, mice acutely infected with *Toxoplasma gondii* succumb to a lethal immune response dependent on CD4+ T cells and accompanied by overproduction of IL-12, IFN-γ and TNF-α. J Immunol 157:798–805

Giuliani N, Airoldi I (2011) Novel insights into the role of interleukin-27 and interleukin-23 in human malignant and normal plasma cells. Clin Cancer Res 17:6963–6970

Gonin J, Larousserie F, Bastard C, Picquenot JM, Couturier J, Radford-Weiss I, Dietrich C, Brousse N, Vacher-Lavenu MC, Devergne O (2011) Epstein–Barr virus-induced gene 3

(EBI3): a novel diagnosis marker in Burkitt lymphoma and diffuse large B-cell lymphoma. PLoS One 6:e24617
Greenwell-Wild T, Vazquez N, Jin W, Rangel Z, Munson PJ, Wahl SM (2009) Interleukin-27 inhibition of HIV-1 involves an intermediate induction of type I interferon. Blood 114:1864–1874
Guzzo C, Hopman WM, Che Mat NF, Wobeser W, Gee K (2010) Impact of HIV infection, highly active antiretroviral therapy, and hepatitis C coinfection on serum interleukin-27. AIDS 24:1371–1374
Guzzo C, Hopman WM, Che Mat NF, Wobeser W, Gee K (2012) IL-27-induced gene expression is downregulated in HIV-infected subjects. PLoS One 7:e45706
Hall AO, Beiting DP, Tato C, John B, Oldenhove G, Lombana CG, Pritchard GH, Silver JS, Bouladoux N, Stumhofer JS, Harris TH, Grainger J, Wojno ED, Wagage S, Roos DS, Scott P, Turka LA, Cherry S, Reiner SL, Cua D, Belkaid Y, Elloso MM, Hunter CA (2012a) The cytokines interleukin 27 and interferon-γ promote distinct Treg cell populations required to limit infection-induced pathology. Immunity 37:511–523
Hall AO, Silver JS, Hunter CA (2012b) The immunobiology of IL-27. Adv Immunol 115:1–44
Hamano S, Himeno K, Miyazaki Y, Ishii K, Yamanaka A, Takeda A, Zhang M, Hisaeda H, Mak TW, Yoshimura A, Yoshida H (2003) WSX-1 is required for resistance to *Trypanosoma cruzi* infection by regulation of proinflammatory cytokine production. Immunity 19:657–667
Hashimoto SI, Suzuki T, Nagai S, Yamashita T, Toyoda N, Matsushima K (2000) Identification of genes specifically expressed in human activated and mature dendritic cells through serial analysis of gene expression. Blood 96:2206–2214
Hibbert L, Pflanz S, De Waal MR, Kastelein RA (2003) IL-27 and IFN-alpha signal via Stat1 and Stat3 and induce T-Bet and IL-12Rβ2 in naive T cells. J Interferon Cytokine Res 23:513–522
Hirahara K, Ghoreschi K, Yang XP, Takahashi H, Laurence A, Vahedi G, Sciume G, Hall AO, Dupont CD, Francisco LM, Chen Q, Tanaka M, Kanno Y, Sun HW, Sharpe AH, Hunter CA, O'Shea JJ (2012) Interleukin-27 priming of T cells controls IL-17 production in trans via induction of the ligand PD-L1. Immunity 36:1017–1030
Hisada M, Kamiya S, Fujita K, Belladonna ML, Aoki T, Koyanagi Y, Mizuguchi J, Yoshimoto T (2004) Potent antitumor activity of interleukin-27. Cancer Res 64:1152–1156
Ho MY, Leu SJ, Sun GH, Tao MH, Tang SJ, Sun KH (2009) IL-27 directly restrains lung tumorigenicity by suppressing cyclooxygenase-2-mediated activities. J Immunol 183:6217–6226
Holscher C, Holscher A, Ruckerl D, Yoshimoto T, Yoshida H, Mak T, Saris C, Ehlers S (2005) The IL-27 receptor chain WSX-1 differentially regulates antibacterial immunity and survival during experimental tuberculosis. J Immunol 174:3534–3544
Hu P, Hu HD, Chen M, Peng ML, Tang L, Tang KF, Matsui M, Belladonna ML, Yoshimoto T, Zhang DZ, Xiang R, Ren H (2009) Expression of interleukins-23 and 27 leads to successful gene therapy of hepatocellular carcinoma. Mol Immunol 46:1654–1662
Huber M, Steinwald V, Guralnik A, Brustle A, Kleemann P, Rosenplanter C, Decker T, Lohoff M (2008) IL-27 inhibits the development of regulatory T cells via STAT3. Int Immunol 20:223–234
Hunter CA, Kastelein R (2012) Interleukin-27: balancing protective and pathological immunity. Immunity 37:960–969
Imamichi T, Yang J, Huang DW, Brann TW, Fullmer BA, Adelsberger JW, Lempicki RA, Baseler MW, Lane HC (2008) IL-27, a novel anti-HIV cytokine, activates multiple interferon-inducible genes in macrophages. AIDS 22:39–45
Kalliolias GD, Zhao B, Triantafyllopoulou A, Park-Min KH, Ivashkiv LB (2010) Interleukin-27 inhibits human osteoclastogenesis by abrogating RANKL-mediated induction of nuclear factor of activated T cells c1 and suppressing proximal RANK signaling. Arthritis Rheum 62:402–413
Kamiya S, Owaki T, Morishima N, Fukai F, Mizuguchi J, Yoshimoto T (2004) An indispensable role for STAT1 in IL-27-induced T-bet expression but not proliferation of naive $CD4^+$ T cells. J Immunol 173:3871–3877
Kamiya S, Okumura M, Chiba Y, Fukawa T, Nakamura C, Nimura N, Mizuguchi J, Wada S, Yoshimoto T (2011) IL-27 suppresses RANKL expression in $CD4^+$ T cells in part through STAT3. Immunol Lett 138:47–53

Kastelein RA, Hunter CA, Cua DJ (2007) Discovery and biology of IL-23 and IL-27: related but functionally distinct regulators of inflammation. Annu Rev Immunol 25:221–242

Leonard JP, Sherman ML, Fisher GL, Buchanan LJ, Larsen G, Atkins MB, Sosman JA, Dutcher JP, Vogelzang NJ, Ryan JL (1997) Effects of single-dose interleukin-12 exposure on interleukin-12-associated toxicity and interferon-γ production. Blood 90:2541–2548

Liu H, Rohowsky-Kochan C (2011) Interleukin-27-mediated suppression of human Th17 cells is associated with activation of STAT1 and suppressor of cytokine signaling protein 1. J Interferon Cytokine Res 31:459–469

Liu J, Guan X, Ma X (2007) Regulation of IL-27 p28 gene expression in macrophages through MyD88- and interferon-γ-mediated pathways. J Exp Med 204:141–152

Liu L, Wang S, Shan B, Shao L, Sato A, Kawamura K, Li Q, Ma G, Tagawa M (2008) IL-27-mediated activation of natural killer cells and inflammation produced antitumour effects for human oesophageal carcinoma cells. Scand J Immunol 68:22–29

Lucas S, Ghilardi N, Li J, de Sauvage FJ (2003) IL-27 regulates IL-12 responsiveness of naive $CD4^+$ T cells through Stat1-dependent and -independent mechanisms. Proc Natl Acad Sci USA 100:15047–15052

Matsui M, Moriya O, Belladonna ML, Kamiya S, Lemonnier FA, Yoshimoto T, Akatsuka T (2004) Adjuvant activities of novel cytokines, interleukin-23 (IL-23) and IL-27, for induction of hepatitis C virus-specific cytotoxic T lymphocytes in HLA-A*0201 transgenic mice. J Virol 78:9093–9104

Matsui M, Kishida T, Nakano H, Yoshimoto K, Shin-Ya M, Shimada T, Nakai S, Imanishi J, Yoshimoto T, Hisa Y, Mazda O (2009) Interleukin-27 activates natural killer cells and suppresses NK-resistant head and neck squamous cell carcinoma through inducing antibody-dependent cellular cytotoxicity. Cancer Res 69:2523–2530

Mayer KD, Mohrs K, Reiley W, Wittmer S, Kohlmeier JE, Pearl JE, Cooper AM, Johnson LL, Woodland DL, Mohrs M (2008) Cutting edge: T-bet and IL-27R are critical for in vivo IFN-γ production by CD8 T cells during infection. J Immunol 180:693–697

Miyazaki Y, Inoue H, Matsumura M, Matsumoto K, Nakano T, Tsuda M, Hamano S, Yoshimura A, Yoshida H (2005) Exacerbation of experimental allergic asthma by augmented Th2 responses in WSX-1-deficient mice. J Immunol 175:2401–2407

Molle C, Nguyen M, Flamand V, Renneson J, Trottein F, De Wit D, Willems F, Goldman M, Goriely S (2007) IL-27 synthesis induced by TLR ligation critically depends on IFN regulatory factor 3. J Immunol 178:7607–7615

Molle C, Goldman M, Goriely S (2010) Critical role of the IFN-stimulated gene factor 3 complex in TLR-mediated IL-27p28 gene expression revealing a two-step activation process. J Immunol 184:1784–1792

Morishima N, Owaki T, Asakawa M, Kamiya S, Mizuguchi J, Yoshimoto T (2005) Augmentation of effector $CD8^+$ T cell generation with enhanced granzyme B expression by IL-27. J Immunol 175:1686–1693

Morishima N, Mizoguchi I, Okumura M, Chiba Y, Xu M, Shimizu M, Matsui M, Mizuguchi J, Yoshimoto T (2010) A pivotal role for interleukin-27 in $CD8^+$ T cell functions and generation of cytotoxic T lymphocytes. J Biomed Biotechnol 2010:605483

Murugaiyan G, Beynon V, Pires Da Cunha A, Joller N, Weiner HL (2012) IFN-γ limits Th9-mediated autoimmune inflammation through dendritic cell modulation of IL-27. J Immunol 189:5277–5283

Neufert C, Becker C, Wirtz S, Fantini MC, Weigmann B, Galle PR, Neurath MF (2007) IL-27 controls the development of inducible regulatory T cells and Th17 cells via differential effects on STAT1. Eur J Immunol 37:1809–1816

Niedbala W, Cai B, Wei X, Patakas A, Leung BP, McInnes IB, Liew FY (2008) Interleukin 27 attenuates collagen-induced arthritis. Ann Rheum Dis 67:1474–1479

Nishino R, Takano A, Oshita H, Ishikawa N, Akiyama H, Ito H, Nakayama H, Miyagi Y, Tsuchiya E, Kohno N, Nakamura Y, Daigo Y (2011) Identification of Epstein–Barr virus-induced gene 3 as a novel serum and tissue biomarker and a therapeutic target for lung cancer. Clin Cancer Res 17:6272–6286

Oniki S, Nagai H, Horikawa T, Furukawa J, Belladonna ML, Yoshimoto T, Hara I, Nishigori C (2006) Interleukin-23 and interleukin-27 exert quite different antitumor and vaccine effects on poorly immunogenic melanoma. Cancer Res 66:6395–6404

Oppmann B, Lesley R, Blom B, Timans JC, Xu Y, Hunte B, Vega F, Yu N, Wang J, Singh K, Zonin F, Vaisberg E, Churakova T, Liu M, Gorman D, Wagner J, Zurawski S, Liu Y, Abrams JS, Moore KW, Rennick D, de Waal-Malefyt R, Hannum C, Bazan JF, Kastelein RA (2000) Novel p19 protein engages IL-12p40 to form a cytokine, IL-23, with biological activities similar as well as distinct from IL-12. Immunity 13:715–725

Owaki T, Asakawa M, Morishima N, Hata K, Fukai F, Matusi M, Mizuguchi J, Yoshimoto T (2005) A role for IL-27 in early regulation of Th1 differentiation. J Immunol 175:2191–2200

Owaki T, Asakawa M, Fukai F, Mizuguchi J, Yoshimoto T (2006a) IL-27 induces Th1 differentiation via p38 MAPK/T-bet- and intercellular adhesion molecule-1/LFA-1/ERK1/2-dependent pathways. J Immunol 177:7579–7587

Owaki T, Asakawa M, Kamiya S, Takeda K, Fukai F, Mizuguchi J, Yoshimoto T (2006b) IL-27 suppresses CD28-mediated IL-2 production through suppressor of cytokine signaling 3. J Immunol 176:2773–2780

Owaki T, Asakawa M, Morishima N, Takeda K, Fukai F, Mizuguchi J, Yoshimoto T (2008) STAT3 is indispensable to IL-27-mediated cell proliferation but not to IL-27-induced Th1 differentiation and suppression of proinflammatory cytokine production. J Immunol 180:2903–2911

Ozaki K, Spolski R, Ettinger R, Kim HP, Wang G, Qi CF, Hwu P, Shaffer DJ, Akilesh S, Roopenian DC, Morse HC 3rd, Lipsky PE, Leonard WJ (2004) Regulation of B cell differentiation and plasma cell generation by IL-21, a novel inducer of Blimp-1 and Bcl-6. J Immunol 173:5361–5371

Parham C, Chirica M, Timans J, Vaisberg E, Travis M, Cheung J, Pflanz S, Zhang R, Singh KP, Vega F, To W, Wagner J, O'Farrell AM, McClanahan T, Zurawski S, Hannum C, Gorman D, Rennick DM, Kastelein RA, de Waal MR, Moore KW (2002) A receptor for the heterodimeric cytokine IL-23 is composed of IL-12Rβ1 and a novel cytokine receptor subunit, IL-23R. J Immunol 168:5699–5708

Perona-Wright G, Kohlmeier JE, Bassity E, Freitas TC, Mohrs K, Cookenham T, Situ H, Pearce EJ, Woodland DL, Mohrs M (2012) Persistent loss of IL-27 responsiveness in $CD8^+$ memory T cells abrogates IL-10 expression in a recall response. Proc Natl Acad Sci USA 109:18535–18540

Pflanz S, Timans JC, Cheung J, Rosales R, Kanzler H, Gilbert J, Hibbert L, Churakova T, Travis M, Vaisberg E, Blumenschein WM, Mattson JD, Wagner JL, To W, Zurawski S, McClanahan TK, Gorman DM, Bazan JF, de Waal Malefyt R, Rennick D, Kastelein RA (2002) IL-27, a heterodimeric cytokine composed of EBI3 and p28 protein, induces proliferation of naive $CD4^+$ T cells. Immunity 16:779–790

Pflanz S, Hibbert L, Mattson J, Rosales R, Vaisberg E, Bazan JF, Phillips JH, McClanahan TK, de Waal Malefyt R, Kastelein RA (2004) WSX-1 and glycoprotein 130 constitute a signal-transducing receptor for IL-27. J Immunol 172:2225–2231

Pickens SR, Chamberlain ND, Volin MV, Mandelin AM 2nd, Agrawal H, Matsui M, Yoshimoto T, Shahrara S (2011) Local expression of interleukin-27 ameliorates collagen-induced arthritis. Arthritis Rheum 63:2289–2298

Pirhonen J, Siren J, Julkunen I, Matikainen S (2007) IFN-alpha regulates Toll-like receptor-mediated IL-27 gene expression in human macrophages. J Leukoc Biol 82:1185–1192

Pot C, Jin H, Awasthi A, Liu SM, Lai CY, Madan R, Sharpe AH, Karp CL, Miaw SC, Ho IC, Kuchroo VK (2009) Cutting edge: IL-27 induces the transcription factor c-Maf, cytokine IL-21, and the costimulatory receptor ICOS that coordinately act together to promote differentiation of IL-10-producing Tr1 cells. J Immunol 183:797–801

Pot C, Apetoh L, Awasthi A, Kuchroo VK (2011) Induction of regulatory Tr1 cells and inhibition of T(H)17 cells by IL-27. Semin Immunol 23:438–445

Pradhan A, Lambert QT, Reuther GW (2007) Transformation of hematopoietic cells and activation of JAK2-V617F by IL-27R, a component of a heterodimeric type I cytokine receptor. Proc Natl Acad Sci USA 104:18502–18507

Rosas LE, Satoskar AA, Roth KM, Keiser TL, Barbi J, Hunter C, de Sauvage FJ, Satoskar AR (2006) Interleukin-27R (WSX-1/T-cell cytokine receptor) gene-deficient mice display enhanced resistance to leishmania donovani infection but develop severe liver immunopathology. Am J Pathol 168:158–169

Rousseau F, Basset L, Froger J, Dinguirard N, Chevalier S, Gascan H (2010) IL-27 structural analysis demonstrates similarities with ciliary neurotrophic factor (CNTF) and leads to the identification of antagonistic variants. Proc Natl Acad Sci USA 107:19420–19425

Salcedo R, Stauffer JK, Lincoln E, Back TC, Hixon JA, Hahn C, Shafer-Weaver K, Malyguine A, Kastelein R, Wigginton JM (2004) IL-27 mediates complete regression of orthotopic primary and metastatic murine neuroblastoma tumors: role for $CD8^+$ T cells. J Immunol 173:7170–7182

Salcedo R, Hixon JA, Stauffer JK, Jalah R, Brooks AD, Khan T, Dai RM, Scheetz L, Lincoln E, Back TC, Powell D, Hurwitz AA, Sayers TJ, Kastelein R, Pavlakis GN, Felber BK, Trinchieri G, Wigginton JM (2009) Immunologic and therapeutic synergy of IL-27 and IL-2: enhancement of T cell sensitization, tumor-specific CTL reactivity and complete regression of disseminated neuroblastoma metastases in the liver and bone marrow. J Immunol 182:4328–4338

Schneider R, Yaneva T, Beauseigle D, El-Khoury L, Arbour N (2011) IL-27 increases the proliferation and effector functions of human naive $CD8^+$ T lymphocytes and promotes their development into Tc1 cells. Eur J Immunol 41:47–59

Seita J, Asakawa M, Ooehara J, Takayanagi S, Morita Y, Watanabe N, Fujita K, Kudo M, Mizuguchi J, Ema H, Nakauchi H, Yoshimoto T (2008) Interleukin-27 directly induces differentiation in hematopoietic stem cells. Blood 111:1903–1912

Shimizu M, Shimamura M, Owaki T, Asakawa M, Fujita K, Kudo M, Iwakura Y, Takeda Y, Luster AD, Mizuguchi J, Yoshimoto T (2006) Antiangiogenic and antitumor activities of IL-27. J Immunol 176:7317–7324

Shimozato O, Sato A, Kawamura K, Chiyo M, Ma G, Li Q, Tagawa M (2009) The secreted form of p28 subunit of interleukin (IL)-27 inhibits biological functions of IL-27 and suppresses anti-allogeneic immune responses. Immunology 128:e816–e825

Shinozaki Y, Wang S, Miyazaki Y, Miyazaki K, Yamada H, Yoshikai Y, Hara H, Yoshida H (2009) Tumor-specific cytotoxic T cell generation and dendritic cell function are differentially regulated by interleukin 27 during development of anti-tumor immunity. Int J Cancer 124:1372–1378

Smits HH, van Beelen AJ, Hessle C, Westland R, de Jong E, Soeteman E, Wold A, Wierenga EA, Kapsenberg ML (2004) Commensal gram-negative bacteria prime human dendritic cells for enhanced IL-23 and IL-27 expression and enhanced Th1 development. Eur J Immunol 34:1371–1380

Sprecher CA, Grant FJ, Baumgartner JW, Presnell SR, Schrader SK, Yamagiwa T, Whitmore TE, O'Hara PJ, Foster DF (1998) Cloning and characterization of a novel class I cytokine receptor. Biochem Biophys Res Commun 246:82–90

Stumhofer JS, Laurence A, Wilson EH, Huang E, Tato CM, Johnson LM, Villarino AV, Huang Q, Yoshimura A, Sehy D, Saris CJ, O'Shea JJ, Hennighausen L, Ernst M, Hunter CA (2006) Interleukin 27 negatively regulates the development of interleukin 17-producing T helper cells during chronic inflammation of the central nervous system. Nat Immunol 7:937–945

Stumhofer JS, Silver JS, Laurence A, Porrett PM, Harris TH, Turka LA, Ernst M, Saris CJ, O'Shea JJ, Hunter CA (2007) Interleukins 27 and 6 induce STAT3-mediated T cell production of interleukin 10. Nat Immunol 8:1363–1371

Stumhofer JS, Tait ED, Quinn WJ 3rd, Hosken N, Spudy B, Goenka R, Fielding CA, O'Hara AC, Chen Y, Jones ML, Saris CJ, Rose-John S, Cua DJ, Jones SA, Elloso MM, Grotzinger J, Cancro MP, Levin SD, Hunter CA (2010) A role for IL-27p28 as an antagonist of gp130-mediated signaling. Nat Immunol 11:1119–1126

Sun J, Dodd H, Moser EK, Sharma R, Braciale TJ (2011) $CD4^+$ T cell help and innate-derived IL-27 induce Blimp-1-dependent IL-10 production by antiviral CTLs. Nat Immunol 12:327–334

Taga T, Kishimoto T (1997) Gp130 and the interleukin-6 family of cytokines. Annu Rev Immunol 15:797–819

Takeda A, Hamano S, Yamanaka A, Hanada T, Ishibashi T, Mak TW, Yoshimura A, Yoshida H (2003) Cutting edge: role of IL-27/WSX-1 signaling for induction of T-bet through activation of STAT1 during initial Th1 commitment. J Immunol 170:4886–4890
Trinchieri G (2003) Interleukin-12 and the regulation of innate resistance and adaptive immunity. Nat Rev Immunol 3:133–146
Villarino A, Hibbert L, Lieberman L, Wilson E, Mak T, Yoshida H, Kastelein RA, Saris C, Hunter CA (2003) The IL-27R (WSX-1) is required to suppress T cell hyperactivity during infection. Immunity 19:645–655
Villarino AV, Huang E, Hunter CA (2004) Understanding the pro- and anti-inflammatory properties of IL-27. J Immunol 173:715–720
Villarino AV, Larkin J 3rd, Saris CJ, Caton AJ, Lucas S, Wong T, de Sauvage FJ, Hunter CA (2005) Positive and negative regulation of the IL-27 receptor during lymphoid cell activation. J Immunol 174:7684–7691
Wang RX, Yu CR, Mahdi RM, Egwuagu CE (2012) Novel IL27p28/IL12p40 cytokine suppressed experimental autoimmune uveitis by inhibiting autoreactive Th1/Th17 cells and promoting expansion of regulatory T cells. J Biol Chem 287:36012–36021
Wirtz S, Becker C, Fantini MC, Nieuwenhuis EE, Tubbe I, Galle PR, Schild HJ, Birkenbach M, Blumberg RS, Neurath MF (2005) EBV-induced gene 3 transcription is induced by TLR signaling in primary dendritic cells via NF-κ B activation. J Immunol 174:2814–2824
Wirtz S, Tubbe I, Galle PR, Schild HJ, Birkenbach M, Blumberg RS, Neurath MF (2006) Protection from lethal septic peritonitis by neutralizing the biological function of interleukin 27. J Exp Med 203:1875–1881
Wojno ED, Hosken N, Stumhofer JS, O'Hara AC, Mauldin E, Fang Q, Turka LA, Levin SD, Hunter CA (2011) A role for IL-27 in limiting T regulatory cell populations. J Immunol 187:266–273
Xu J, Yang Y, Qiu G, Lal G, Wu Z, Levy DE, Ochando JC, Bromberg JS, Ding Y (2009) c-Maf regulates IL-10 expression during Th17 polarization. J Immunol 182:6226–6236
Xu M, Mizoguchi I, Morishima N, Chiba Y, Mizuguchi J, Yoshimoto T (2010) Regulation of antitumor immune responses by the IL-12 family cytokines, IL-12, IL-23, and IL-27. Clin Dev Immunol 2010:pii: 832454
Yang J, Yang M, Htut TM, Ouyang X, Hanidu A, Li X, Sellati R, Jiang H, Zhang S, Li H, Zhao J, Ting AT, Mayer L, Unkeless JC, Labadia ME, Hodge M, Li J, Xiong H (2008) Epstein–Barr virus-induced gene 3 negatively regulates IL-17, IL-22 and RORγ t. Eur J Immunol 38:1204–1214
Yang XO, Panopoulos AD, Nurieva R, Chang SH, Wang D, Watowich SS, Dong C (2007) STAT3 regulates cytokine-mediated generation of inflammatory helper T cells. J Biol Chem 282:9358–9363
Yoshida H, Hamano S, Senaldi G, Covey T, Faggioni R, Mu S, Xia M, Wakeham AC, Nishina H, Potter J, Saris CJ, Mak TW (2001) WSX-1 is required for the initiation of Th1 responses and resistance to *L. major* infection. Immunity 15:569–578
Yoshida H, Miyazaki Y (2008) Regulation of immune responses by interleukin-27. Immunol Rev 226:234–247
Yoshimoto T, Okada K, Morishima N, Kamiya S, Owaki T, Asakawa M, Iwakura Y, Fukai F, Mizuguchi J (2004) Induction of IgG2a class switching in B cells by IL-27. J Immunol 173:2479–2485
Yoshimoto T, Yoshimoto T, Yasuda K, Mizuguchi J, Nakanishi K (2007) IL-27 suppresses Th2 cell development and Th2 cytokines production from polarized Th2 cells: a novel therapeutic way for Th2-mediated allergic inflammation. J Immunol 179:4415–4423
Yoshimoto T, Morishima N, Mizoguchi I, Shimizu M, Nagai H, Oniki S, Oka M, Nishigori C, Mizuguchi J (2008) Antiproliferative activity of IL-27 on melanoma. J Immunol 180:6527–6535
Yoshimura T, Takeda A, Hamano S, Miyazaki Y, Kinjyo I, Ishibashi T, Yoshimura A, Yoshida H (2006) Two-sided roles of IL-27: induction of Th1 differentiation on naive $CD4^+$ T cells

versus suppression of proinflammatory cytokine production including IL-23-induced IL-17 on activated CD4+ T cells partially through STAT3-dependent mechanism. J Immunol 177: 5377–5385

Young A, Linehan E, Hams E, O'Hara Hall AC, McClurg A, Johnston JA, Hunter CA, Fallon PG, Fitzgerald DC (2012) Cutting edge: suppression of GM-CSF expression in murine and human T cells by IL-27. J Immunol 189:2079–2083

Zhang J, Qian X, Ning H, Yang J, Xiong H, Liu J (2010) Activation of IL-27 p28 gene transcription by interferon regulatory factor 8 in cooperation with interferon regulatory factor 1. J Biol Chem 285:21269–21281

Zorzoli A, Di Carlo E, Cocco C, Ognio E, Ribatti D, Ferretti E, Dufour C, Locatelli F, Montagna D, Airoldi I (2012) Interleukin-27 inhibits the growth of pediatric acute myeloid leukemia in NOD/SCID/Il2rγ$^{-/-}$ mice. Clin Cancer Res 18:1630–1640

Chapter 15
Interleukin-35: A Novel Mediator of Peripheral Tolerance

Greg M. Delgoffe and Dario A.A. Vignali

Abstract Interleukin-35 is a potent suppressive cytokine of the IL-12 family. Although other members of the IL-12 family are produced mainly by antigen-presenting cells (APCs), IL-35 is produced by regulatory T (Treg) cells and suppresses cell proliferation. It has been shown to play an important role in many disease models and has been recently shown to have additional functions aside from inhibition of proliferation, including inducing its own expression in non-Treg cells. In this chapter, we discuss the history and current status of IL-35 biology, as well as suggest where the field might move in the future.

Keywords IL-35 • STAT heterodimer • Suppression • Treg

15.1 The IL-12 Family Members of Heterodimeric Cytokines

Cytokines comprise a critical mediator of cell-to-cell communication. Although T-cell–APC contacts in the form of pMHC:TCR and costimulation are important for initiating the immune response, cytokines are essential for shaping it, inducing differentiation into particular programmed effector lineages and aiding in the generation of memory. In addition, cytokines can have direct effector function by inducing cell activation or apoptosis, modulation of proliferation, and alteration of metabolism.

Although some cytokine families are predominantly pro-inflammatory or anti-inflammatory, the IL-12 cytokines family is unusual in having particularly broad immunomodulatory activities. The IL-12 family belongs to the IL-6 superfamily of type 1 cytokines, which is characterized by cytokines with multisubunit receptors

G.M. Delgoffe • D.A.A. Vignali (✉)
Department of Immunology, St. Jude Children's Research Hospital, 262 Danny Thomas Place, Mail Stop 351, Memphis, TN, 38105, USA
e-mail: vignali.lab@stjude.org

T. Yoshimoto and T. Yoshimoto (eds.), *Cytokine Frontiers: Regulation of Immune Responses in Health and Disease*, DOI 10.1007/978-4-431-54442-5_15, © Springer Japan 2014

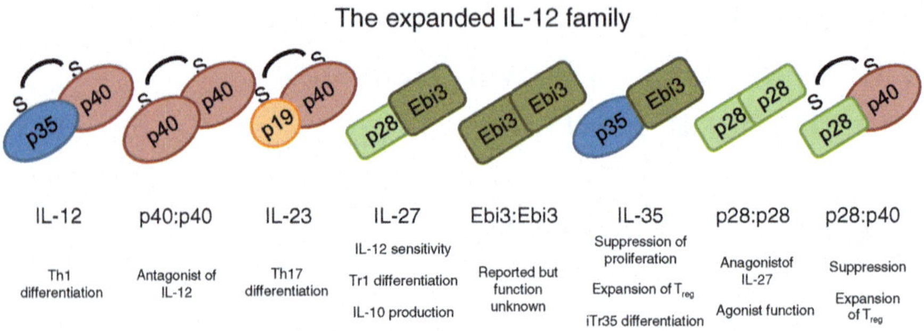

Fig. 15.1 The growing interleukin (IL)-12 family. The IL-12 family of cytokines is characterized by subunit sharing and pairing. Some, such as IL-12 and IL-23, are linked via disulfide linkage, whereas others (those utilizing Ebi3) are held together by other structural forces. The functions of the IL-12 family are many and varied, often related to differentiation. The presence of homodimers of subunits has been reported, and these often have been implicated in acting as natural antagonists of bioactive cytokine signaling

(Jones and Vignali 2011; Vignali and Kuchroo 2012).These cytokines have a wide variety of functions in the hematopoietic lineages.

The IL-12 family was formed with the discovery of IL-12, a heterodimer of p35 (encoded by *Il12a*) and p40 (encoded by *Il12b*), linked by disulfide bonds (Fig. 15.1). This family is characterized by heterodimers of alpha-subunits consistent with the structure of type 1 cytokines and beta-subunits that are structurally similar to soluble cytokine receptors (Jones and Vignali 2011; Vignali and Kuchroo 2012). This structure thought to have evolved from the IL-6 superfamily of cytokines. Many IL-6 superfamily members utilize gp130 as a signal-transducing subunit (IL-6, IL-11, CNTF, LIF, OSM, CLF, etc.). Beta-subunits structurally resemble signal-transducing subunits as well as cytokine-binding proteins (Jones and Vignali 2011). IL-12 was shown to be critical for induction of T helper 1 (Th1) cell differentiation and interferon (IFN)-γ production. The generation of knockout mice for *Il12a* and *Il12b* confirms the physiological importance of IL-12 in a variety of in vivo models, including experimental autoimmune encephalomyelitis (EAE) (Bettelli et al. 2007). As the Th1/Th2 paradigm suggested that Th1 cells would be important for EAE disease pathogenesis, it was anticipated that $Il12b^{-/-}$ mice would be resistant to EAE. However, the observation that $Il12a^{-/-}$ were susceptible to EAE was surprising and suggested a more complex biology. This idea prompted researchers to suggest that perhaps p40 was playing a role as a subunit for another cytokine, leading to the discovery of the p19 alpha-chain and the demonstration that it pairs with the p40 beta-chain via a disulfide linkage to form IL-23. This new cytokine, while originally thought to activate memory cells, was shown to induce the differentiation of Th17 cells, a predominantly inflammatory T cell subset. In the coming years, new cytokines have be added to the IL-12 family based on structural homology. IL-27, a heterodimer of p28 (encoded by *Il27*) and Epstein–Barr virus-induced gene 3 (Ebi3, encoded by *Ebi3*), which lacks disulfide linkage, was shown to be produced by recently activated antigen-presenting cells (APCs) (Fig. 15.1). It was first thought to be a proinflammatory cytokine, because stimulation of naïve T cells with IL-27 induces *Il12rb2* expression, greatly enhancing T-cell sensitivity to IL-12 (Devergne et al. 1996; Hall et al. 2012). However, recent data have highlighted a more immunomodulatory role for IL-27. For instance, IL-27 can induce the generation of IL-10-producing Tr1 cells while inhibiting Th17 and peripherally induced Treg (pTreg) induction. It is also notable that several IL-12 family member subunits have been suggested to have functional roles as homodimers. IL12p40 homodimers, p28 homodimers (IL-30), and Ebi3 homodimers have been reported with various functions resulting from their antagonistic activity via receptor blockade (Hall et al. 2012; Kempe et al. 2009; Vignali and Kuchroo 2012). Recently, p28 has been shown to have an agonistic function in T cells, although it remains to be determined whether these homodimers have physiological roles in vivo (Garbers et al. 2013).

It is clear that the IL-12 cytokine family mediates a broad and varied portfolio of immunological events (Vignali and Kuchroo 2012). Thus, it seems likely that new insights are likely to emerge and new members of this family may also remain to be discovered. However, in this chapter we focus exclusively on interleukin-35, the newest and perhaps most elusive member of this family.

15.2 Discovery of IL-35 and Identification of Its Cellular Source

When a second IL-12 family cytokine beta-chain, Ebi3, was discovered in 1996 (Devergne et al. 1996), there was considerable interest in identifying its "natural" binding partner. This quest continued for some time as p28 was identified 6 years after the discovery of Ebi3 (Pflanz et al. 2002). In 1997, it was shown that Ebi3 could pair with p35 in transfection experiments (Devergne et al. 1997). Although it was suggested that placental trophoblasts might coexpress Ebi3 and p35 (Devergne et al. 1997, 2001), it was not clear whether this was a bona fide heterodimeric cytokine, what its predominant cellular source might be, and its function in vivo.

A decade later, detailed analysis of the Treg cell gene expression "signature" revealed *Ebi3* as an upregulated gene and potential Foxp3 target (Collison et al. 2007; Hill et al. 2007; Knoechel et al. 2006; Zheng et al. 2007). Given that Ebi3 is a p40-related member of the IL-12 family, a cytokine family that was thought to be exclusively produced by APCs, increased expression in Treg cells was surprising. In 2007, the Vignali lab identified IL-35 as a p35:Ebi3 heterodimer important for Treg cell function (Collison et al. 2007). Importantly, *Ebi3*- and *Il12a*-deficient Treg cells, stimulated in isolation or co-cultured with wild-type conventional cells, displayed limited suppressive capabilities compared with their wild-type controls. Furthermore, recombinant 2A-linked "native" or a single-chain "hyperkine" IL-35 suppressed T-cell proliferation. Around the same time, another group used a p35:Ebi3 single-chain hyperkine-Fc fusion protein to abrogate collagen-induced arthritis in mice (Niedbala et al. 2007). However, this study seemed to indicate a role for IL-35 in the expansion of Tregs rather than the suppression of proliferation.

These contrasting data in vitro confirmed some earlier observations, but a physiological role and source for IL-35 in vivo seemed to be lacking (Devergne et al. 1997; Niedbala et al. 2007). In all, these early observations suggested that IL-35 may be an important member of the suppressive cytokine "community," joining transforming growth factor (TGF)-β and IL-10 as soluble mediators of suppression.

15.3 The Function(s) of IL-35

The discovery of IL-35 as a Treg-derived suppressive cytokine has sparked considerable interest. Although there is clearly much to learn about this intriguing cytokine, several recent studies have begun to provide important insight into its function (Fig. 15.2). The primary function of IL-35 appears to be the induction of cell-cycle arrest in many cell types (Bettini et al. 2012; Collison et al. 2007; Wirtz et al. 2011). The ability of IL-35 to suppress cellular proliferation may be a critical component of the Treg cell armament, especially in inflammatory environments (Bettini et al. 2012; Collison et al. 2007, 2009; Wirtz et al. 2011). It has been suggested that IL-35 can expand Treg cells, although the mechanism of this expansion is unknown and the observation has yet to be confirmed in vivo (Niedbala et al. 2007). It is possible that that Treg cells are merely resistant to IL-35 and thus are selected for in this context, or it may be that IL-35 blocks factors that would hamper Treg cell function.

Many studies have implicated IL-35 in various disease models. Initial studies suggested that IL-35 could ameliorate inflammatory environments, such as in models of inflammatory bowel disease (IBD) (Collison et al. 2007). Early in vivo

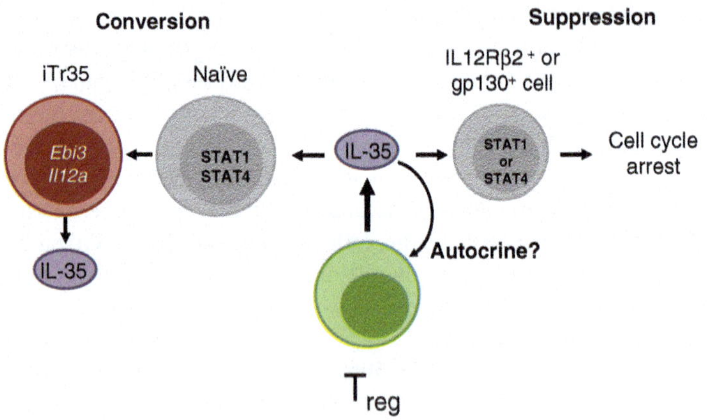

Fig. 15.2 Multiple functions of IL-35. IL-35-treated T cells expressing gp130 or IL12Rβ2 signal via STAT1 or STAT4, respectively, and induce cell-cycle arrest (suppression). Treg cell-derived IL-35 can also suppress activated T cells via a STAT1:STAT4 heterodimer, and induce *Ebi3* and *Il12a* transcription and IL-35 production (iTr35 conversion). Finally, IL-35 may be able to act on Treg cells in an autocrine fashion to program stability, but these findings have yet to be fully elucidated

functions of IL-35 were attributed to the expansion of regulatory cells or the suppression of Th17 cells, specifically in the progression of experimental arthritis (Kochetkova et al. 2010; Niedbala et al. 2007; Yang et al. 2008). IL-35 was also shown to be important in inhibiting the clearance of *Listeria* infection (Yang et al. 2008). *Ebi3*$^{-/-}$ animals have exacerbated central nervous system (CNS) pathology in response to coronavirus infection (Tirotta et al. 2013). Use of an IL-35 expression construct in mice led to attenuated autoimmune-induced CNS demyelination (Zandian et al. 2011), decreased experimental colitis (Wirtz et al. 2011), and reduced severity of airway inflammation and antibody production in a model of dust mite allergy (Huang et al. 2011). Transgenic NOD mice expressing IL-35 in pancreatic beta cells in the islets of Langerhans have substantially decreased diabetes and immune pathology (Bettini et al. 2012). Further analysis of IL-35 function in vivo is clearly required, but the general consensus is that IL-35 exhibits an inhibitory phenotype in a broad range of disease models.

The suppressive cytokines TGF-β and IL-10 have the capacity to convert CD4$^+$ T cells in to in vitro-generated induced Treg (iTreg) that suppress via TGF-β or IL-10, respectively (Vignali et al. 2008). It seemed logical that IL-35 might have a similar function as a mediator of "infectious tolerance," the notion that cells which are suppressed can themselves become suppressive, passing the tolerant state from one cell to another (Gravano and Vignali 2012). Indeed, naive CD4$^+$ T cells stimulated in the presence of IL-35 are not only suppressed but are converted to an IL-35-producing iTreg population termed iTr35 (Collison et al. 2010). Interestingly, these cells do not express Foxp3 or suppress via TGF-β or IL-10 and are implicated in several disease models, especially cancer, as tumors are enriched for Foxp3$^-$ IL-35-producing cells.

Although there is mounting evidence supporting an inhibitory role for murine IL-35, evidence supporting a role for human IL35 remains limited. It has been suggested that human IL-35 may be suppressive (Collison et al. 2010; Jones et al. 2012), but initial studies could not detect IL-35 production by human Treg cells ex vivo or following short-term stimulation (Bardel et al. 2008). Furthermore, retroviral-mediated overexpression of FOXP3 in human T cells could not induce *EBI3* or *IL12A* expression (Allan et al. 2008). However, most Treg cells isolated from human samples are not particularly suppressive ex vivo and do not always express high levels of FOXP3. Indeed, IL-35 has been detectable in human samples from a variety of disease states (Langhans et al. 2010; Liu et al. 2011; Seyerl et al. 2010). Furthermore, hepatitis C virus-specific human Tregs have been shown to suppress via IL-35 in response to HCV epitopes (Langhans et al. 2010). IL-35 has also been shown to be produced by human Tregs in response to rhinovirus infection, implicating PD-L1 and CD169 in its induction (Seyerl et al. 2010). As in the mouse, it also appears that human CD4$^+$ T cells exposed to IL-35 can express and mediate suppression via IL-35 (Collison et al. 2010). Although most of the focus has been on production of IL-35 by CD4$^+$ regulatory T-cell populations, an intriguing recent study has suggested that a human CD8$^+$ T-cell subpopulation can regulate antitumor immunity in prostate cancer via IL-35 production and CTLA-4 (Olson et al. 2012). This finding is particularly intriguing because CD8$^+$ cells express a functional receptor for IL-35 that could induce IL-35 production (Collison et al. 2012). Also recently, a study has suggested that IL-35 may be produced by tumor cells, inducing cell-cycle arrest and promoting apoptosis among themselves as well as tumor-infiltrating cells (Long et al. 2013). Some cancer lines were also shown to express functional receptors for IL-35 as well as IL-35 itself. Furthermore, IL-35-negative tumor lines in which IL-35 was overexpressed gained the ability to mediate cell-cycle arrest and apoptosis (Long et al. 2013). This ability may serve as a mechanism for tumors to evade the immune system and establish long-term tolerance, promoting disease progression. Such observations are encouraging, but it will be critical to fully evaluate the functional capacity and scope of human IL-35, moving forward to determine the potential of its therapeutic utilization or blockade.

15.4 Cellular Targets of IL-35: Lessons from Its Receptor

Based on structural homology between the subunits of IL-35 and other members of the IL-12 family, it was likely that IL-35 would utilize one or more of the receptor subunits associated with this family. Utilizing cells deficient in these receptor subunits, we demonstrated a role for gp130 and IL-12Rβ2 in IL-35-mediated suppression (Collison et al. 2012). However, although gp130- and IL-12Rβ2 double-deficient cells were completely resistant to IL-35-mediated suppression, those lacking only a single chain of the receptor could still signal and suppress, especially in vivo. This observation was surprising as other members of the IL-12 family lose their signaling capacity in the absence of a single receptor chain (Garbers et al. 2012; Vignali and Kuchroo 2012). Intriguingly, signaling via gp130 or IL-12Rβ2 alone was not sufficient to mediate IL-35 production and iTr35 conversion. Subsequent biochemical analysis and microscopy approaches revealed that IL-35 can utilize three receptors: gp130 homodimers and IL-12Rβ2 homodimers, which were sufficient to mediate suppression, and gp130:IL12Rβ2 heterodimers, which mediate suppression and iTr35 conversion (Fig. 15.3). This conversion occurs through STAT1:STAT4 heterodimers, which can bind to regions within the *Ebi3* and *Il12a* promoters and initiate transcription. What remains unclear is how gp130 or IL-12Rβ2 homodimers can stimulate STAT1 or STAT4 activation and mediate suppression in the absence of the other chain.

IL-12Rβ2 is primarily restricted to the hematopoietic lineage and is really only expressed on a small proportion of cells [activated conventional $CD4^+$ and $CD8^+$ T cells, natural killer (NK) cells, dendritic cells (DCs), and some B cells). Interestingly, $CD8^+$ regulatory cells generated in prostate tumors have been shown to suppress via IL-35 production, suggesting that perhaps the presence of IL-12Rβ2 and gp130 on activated $CD8^+$ T cells may sensitize them for conversion to IL-35 production (Olson et al. 2012). However, gp130 is ubiquitously expressed throughout the body. As IL-35 can mediate suppression through gp130 homodimers, this implies that most cells could be susceptible to IL-35-mediated suppression. This is an exciting thought but it highlights questions regarding the potency of IL-35 in vivo and

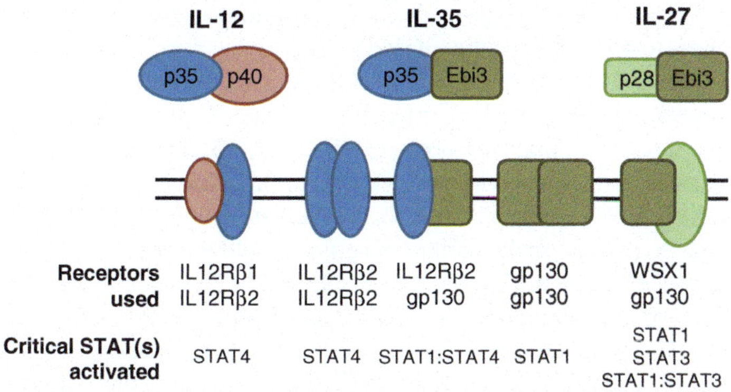

Fig. 15.3 IL-35 utilizes three receptors. Although IL-12 and IL-27 utilize a single receptor consisting of two chains, IL-35 is unique in that it can also signal, at least in part, through homodimers of its receptor chains. However, in the absence of one of the receptor chains, its function is limited merely to suppression, whereas the full receptor of gp130:IL-12Rβ2 can generate STAT1:STAT4 heterodimers, which program IL-35 production and iTr35 conversion

implications for clinical use. Does IL-35-mediated Treg suppression require closer contact and direct delivery to its target cells? Is IL-35 removed from the microenvironment by the myriad of receptor-positive cells in vivo? Is there a method for therapeutic intervention by which one could specifically deliver IL-35 to its intended targets, sparing $gp130^{+}$ bystander cells?

15.5 Conundrums Associated with IL-35

IL-35 has proven to be an extraordinarily hard cytokine to study. A key issue is that purified biologically active recombinant IL-35 is still not commercially available and has not been reported. This restriction is in part because IL-35 is poorly secreted and may be somewhat unstable (Jones et al. 2012). Generally Fc fusion proteins or hyperkines are used to study heterodimeric cytokines (such as IL-27), but attempts to generate functional IL-35 in this manner have had mixed results. Although 293T cells coexpressing p35 and Ebi3 secrete bioactive IL-35, the amount is low and it has not been successful purified (Collison et al. 2007, 2009; Jones et al. 2012). Important studies are being undertaken currently to determine the fundamental biology of IL-35, which may help shed light on problems with its production and purification as a research reagent. Interestingly, p35 utilizes two potential start codons, the first encoding a nonclassical, extraordinarily long signal peptide (Kollet et al. 2001; Vaidyanathan et al. 2001, 2003), raising the possibility that IL-35 may be processed differently than are other cytokines. These hypotheses parallel the initial observations that Ebi3 may also be membrane associated and have alternative processing (Devergne et al. 1996). Solving these inherent problems in IL-35 expression, production, and purification may help elucidate the biology of this mysterious cytokine. Such studies may pave the way for future use of IL-35 as a research reagent and a tool for therapeutic intervention.

Given that IL-35 is a multisubunit cytokine, genetic models of simple gene deletion complicate any interpretation of published data (Collison and Vignali 2008). For instance, *Il12a*$^{-/-}$ mice lack both IL-35 and IL-12, thus clouding whether the phenotype observed is caused by one cytokine or the other. An attractive idea would be to find sites in Ebi3 and p35 that would mediate binding with IL-12, IL-27, or IL-35, respectively, and make mutant mice that would thus be deficient in only one cytokine, rather than both (Vignali and Kuchroo 2012). Given that IL-35 is a heterodimeric cytokine, one might anticipate that mutations could be made that would abrogate binding to one binding partner versus the other. However, extensive mutagenesis of Ebi3 and p35 has suggested that this may not be the case. Although IL-12 and IL-27 have conserved regions of homology crucial to subunit pairing, p35 and Ebi3 appear to pair in an unconventional manner (Jones et al. 2012). However, using point mutants discovered by these mutagenesis studies could pave the way for generation of mice that are specifically deficient in IL-12 or IL-27, leaving IL-35 pairing and function intact.

15.6 Future Directions in IL-35 Biology

Many important questions still remain regarding IL-35 beyond the recombinant production issues detailed here.

First, a more detailed, temporally- and cell type-restricted assessment of IL-35 function in a variety of disease settings is essential, using a combination of IL-35-specific neutralizing monoclonal antibodies (mAbs) and floxed *Ebi3* and *Il12a* mice.

Second, a detailed analysis of which cell types produce IL-35, and when, is needed, using flow cytometric cytokine-staining approaches and reporter mice. Similar studies also need to be performed for IL-35 receptor expression and function.

Third, understanding the mechanism by which IL-35 promotes cell-cycle arrest is of critical importance. It is especially important to delineate whether other functions of IL-35 (such as iTr35 conversion) are dependent on the capacity to be "suppressed" or are independent of these signaling mechanisms.

Fourth, although IL-35 has been shown to signal via a STAT1:STAT4 heterodimer, it is still unclear how a heterodimer is formed (as opposed to STAT homodimers). Evaluating potential mechanisms of heterodimer formation in the context of IL-35 can, hopefully, shed light on the general role of STAT heterodimers in immunity. Furthermore, elucidating how STAT heterodimers function distinctly from homodimers has implications for all of cytokine biology. Given that many cytokines with distinct functions utilize the same STATs, determining the mechanism by which STAT heterodimers promote alternative transcription and cellular function will provide fundamental insight into how these critical pathways are modulated.

Fifth, if IL-35 is shown to expand Treg populations (Niedbala et al. 2007), it is extremely important to interrogate any "autocrine" activity of IL-35 on Tregs. Tregs do express gp130, but they do not express high levels of IL-12Rβ2, suggesting that they would be insensitive to the "converting" functions of IL-35. It may stand to reason that IL-35 acts to block gp130-utilizing cytokines that would otherwise hamper suppressive function, such as IL-6.

Finally, clearly it is extremely important to determine to the fullest extent the role that IL-35 plays in human Treg function, given the opposing findings in this matter. Only then might the full therapeutic utility and potential of IL-35 be realized.

Acknowledgments This work was supported by the National Institutes of Health (R01 AI091977 and AI039480 to D.A.A.V.; F32 AI098383 to G.M.D.), NCI Comprehensive Cancer Center Support CORE grant (CA21765, to D.A.A.V.), and the American Lebanese Syrian Associated Charities (ALSAC, to D.A.A.V.).

References

Allan SE, Song-Zhao GX, Abraham T, McMurchy AN, Levings MK (2008) Inducible reprogramming of human T cells into Treg cells by a conditionally active form of FOXP3. Eur J Immunol 38:3282–3289

Bardel E, Larousserie F, Charlot-Rabiega P, Coulomb-L'Hermine A, Devergne O (2008) Human CD4+ CD25+ Foxp3+ regulatory T cells do not constitutively express IL-35. J Immunol 181:6898–6905

Bettelli E, Korn T, Kuchroo VK (2007) Th17: the third member of the effector T cell trilogy. Curr Opin Immunol 19:652–657

Bettini M, Castellaw AH, Lennon GP, Burton AR, Vignali DA (2012) Prevention of autoimmune diabetes by ectopic pancreatic beta-cell expression of interleukin-35. Diabetes 61:1519–1526

Collison LW, Vignali DA (2008) Interleukin-35: odd one out or part of the family? Immunol Rev 226:248–262

Collison LW, Workman CJ, Kuo TT, Boyd K, Wang Y, Vignali KM, Cross R, Sehy D, Blumberg RS, Vignali DA (2007) The inhibitory cytokine IL-35 contributes to regulatory T-cell function. Nature (Lond) 450:566–569

Collison LW, Pillai MR, Chaturvedi V, Vignali DA (2009) Regulatory T cell suppression is potentiated by target T cells in a cell contact, IL-35- and IL-10-dependent manner. J Immunol 182:6121–6128

Collison LW, Chaturvedi V, Henderson AL, Giacomin PR, Guy C, Bankoti J, Finkelstein D, Forbes K, Workman CJ, Brown SA et al (2010) IL-35-mediated induction of a potent regulatory T cell population. Nat Immunol 11:1093–1101

Collison LW, Delgoffe GM, Guy CS, Vignali KM, Chaturvedi V, Fairweather D, Satoskar AR, Garcia KC, Hunter CA, Drake CG et al (2012) The composition and signaling of the IL-35 receptor are unconventional. Nat Immunol 13:290–299

Devergne O, Hummel M, Koeppen H, Le Beau MM, Nathanson EC, Kieff E, Birkenbach M (1996) A novel interleukin-12 p40-related protein induced by latent Epstein–Barr virus infection in B lymphocytes. J Virol 70:1143–1153

Devergne O, Birkenbach M, Kieff E (1997) Epstein-Barr virus-induced gene 3 and the p35 subunit of interleukin 12 form a novel heterodimeric hematopoietin. Proc Natl Acad Sci USA 94:12041–12046

Devergne O, Coulomb-L'Hermine A, Capel F, Moussa M, Capron F (2001) Expression of Epstein–Barr virus-induced gene 3, an interleukin-12 p40-related molecule, throughout human pregnancy:

involvement of syncytiotrophoblasts and extravillous trophoblasts. Am J Pathol 159: 1763–1776
Garbers C, Hermanns HM, Schaper F, Muller-Newen G, Grotzinger J, Rose-John S, Scheller J (2012) Plasticity and cross-talk of interleukin 6-type cytokines. Cytokine Growth Factor Rev 23:85–97
Garbers C, Spudy B, Aparicio-Siegmund S, Waetzig GH, Sommer J, Holscher C, Rose-John S, Grotzinger J, Lorenzen I, Scheller J (2013) An interleukin-6 receptor-dependent molecular switch mediates signal transduction of the IL-27 cytokine subunit p28 (IL-30) via a gp130 receptor homodimer. J Biol Chem 288:4346–4354
Gravano DM, Vignali DA (2012) The battle against immunopathology: infectious tolerance mediated by regulatory T cells. Cell Mol Life Sci 69:1997–2008
Hall AO, Silver JS, Hunter CA (2012) The immunobiology of IL-27. Adv Immunol 115:1–44
Hill JA, Feuerer M, Tash K, Haxhinasto S, Perez J, Melamed R, Mathis D, Benoist C (2007) Foxp3 transcription-factor-dependent and -independent regulation of the regulatory T cell transcriptional signature. Immunity 27:786–800
Huang CH, Loo EX, Kuo IC, Soh GH, Goh DL, Lee BW, Chua KY (2011) Airway inflammation and IgE production induced by dust mite allergen-specific memory/effector Th2 cell line can be effectively attenuated by IL-35. J Immunol 187:462–471
Jones LL, Vignali DA (2011) Molecular interactions within the IL-6/IL-12 cytokine/receptor superfamily. Immunol Res 51:5–14
Jones LL, Chaturvedi V, Uyttenhove C, Van Snick J, Vignali DA (2012) Distinct subunit pairing criteria within the heterodimeric IL-12 cytokine family. Mol Immunol 51:234–244
Kempe S, Heinz P, Kokai E, Devergne O, Marx N, Wirth T (2009) Epstein–Barr virus-induced gene-3 is expressed in human atheroma plaques. Am J Pathol 175:440–447
Knoechel B, Lohr J, Zhu S, Wong L, Hu D, Ausubel L, Abbas AK (2006) Functional and molecular comparison of anergic and regulatory T lymphocytes. J Immunol 176:6473–6483
Kochetkova I, Golden S, Holderness K, Callis G, Pascual DW (2010) IL-35 stimulation of CD39+ regulatory T cells confers protection against collagen II-induced arthritis via the production of IL-10. J Immunol 184:7144–7153
Kollet J, Witek C, Gentry JD, Liu X, Schwartzbach SD, Petro TM (2001) Deletional analysis of the murine IL-12 p35 promoter comparing IFN-gamma and lipopolysaccharide stimulation. J Immunol 167:5653–5663
Langhans B, Braunschweiger I, Arndt S, Schulte W, Satoguina J, Layland LE, Vidovic N, Hoerauf A, Oldenburg J, Sauerbruch T, Spengler U (2010) Core-specific adaptive regulatory T-cells in different outcomes of hepatitis C. Clin Sci (Lond) 119:97–109
Liu F, Tong F, He Y, Liu H (2011) Detectable expression of IL-35 in CD4+ T cells from peripheral blood of chronic hepatitis B patients. Clin Immunol 139:1–5
Long J, Zhang X, Wen M, Kong Q, Lv Z, An Y, Wei XQ (2013) IL-35 over-expression increases apoptosis sensitivity and suppresses cell growth in human cancer cells. Biochem Biophys Res Commun 430(1):364–369
Niedbala W, Wei XQ, Cai B, Hueber AJ, Leung BP, McInnes IB, Liew FY (2007) IL-35 is a novel cytokine with therapeutic effects against collagen-induced arthritis through the expansion of regulatory T cells and suppression of Th17 cells. Eur J Immunol 37:3021–3029
Olson BM, Jankowska-Gan E, Becker JT, Vignali DA, Burlingham WJ, McNeel DG (2012) Human prostate tumor antigen-specific CD8+ regulatory T cells are inhibited by CTLA-4 or IL-35 blockade. J Immunol 189:5590–5601
Pflanz S, Timans JC, Cheung J, Rosales R, Kanzler H, Gilbert J, Hibbert L, Churakova T, Travis M, Vaisberg E et al (2002) IL-27, a heterodimeric cytokine composed of EBI3 and p28 protein, induces proliferation of naive CD4(+) T cells. Immunity 16:779–790
Seyerl M, Kirchberger S, Majdic O, Seipelt J, Jindra C, Schrauf C, Stockl J (2010) Human rhinoviruses induce IL-35-producing Treg via induction of B7-H1 (CD274) and sialoadhesin (CD169) on DC. Eur J Immunol 40:321–329
Tirotta E, Duncker P, Oak J, Klaus S, Tsukamoto MR, Gov L, Lane TE (2013) Epstein–Barr virus-induced gene 3 negatively regulates neuroinflammation and T cell activation following coronavirus-induced encephalomyelitis. J Neuroimmunol 254(1-2):110–116

Vaidyanathan H, Gentry JD, Weatherman A, Schwartzbach SD, Petro TM (2001) Differential response of the murine IL-12 p35 gene to lipopolysaccharide compared with interferon-gamma and CD40 ligation. Cytokine 16:1–9

Vaidyanathan H, Zhou Y, Petro TM, Schwartzbach SD (2003) Intracellular localization of the p35 subunit of murine IL-12. Cytokine 21:120–128

Vignali DA, Kuchroo VK (2012) IL-12 family cytokines: immunological playmakers. Nat Immunol 13:722–728

Vignali DA, Collison LW, Workman CJ (2008) How regulatory T cells work. Nat Rev Immunol 8:523–532

Wirtz S, Billmeier U, McHedlidze T, Blumberg RS, Neurath MF (2011) Interleukin-35 mediates mucosal immune responses that protect against T-cell-dependent colitis. Gastroenterology 141:1875–1886

Yang J, Yang M, Htut TM, Ouyang X, Hanidu A, Li X, Sellati R, Jiang H, Zhang S, Li H et al (2008) Epstein–Barr virus-induced gene 3 negatively regulates IL-17, IL-22 and ROR gamma t. Eur J Immunol 38:1204–1214

Zandian M, Mott KR, Allen SJ, Dumitrascu O, Kuo JZ, Ghiasi H (2011) Use of cytokine immunotherapy to block CNS demyelination induced by a recombinant HSV-1 expressing IL-2. Gene Ther 18:734–742

Zheng Y, Josefowicz SZ, Kas A, Chu TT, Gavin MA, Rudensky AY (2007) Genome-wide analysis of Foxp3 target genes in developing and mature regulatory T cells. Nature (Lond) 445: 936–940

Index

T. Yoshimoto and T. Yoshimoto (eds.), *Cytokine Frontiers: Regulation of Immune Responses in Health and Disease*, DOI 10.1007/978-4-431-54442-5, © Springer Japan 2014

The manufacturer's authorised representative in the EU is Springer
Nature Customer Service Centre GmbH, Europaplatz 3, 69115 Heidelberg,
Germany. If you have any concerns regarding our products, please
contact ProductSafety@springernature.com

Printed and bound by CPI Group (UK) Ltd, Croydon, CR0 4YY
22/09/2026
02221056-0003